POLYMER SCIENCE AND TECHNOLOGY
Volume 13

ULTRAFILTRATION MEMBRANES AND APPLICATIONS

POLYMER SCIENCE AND TECHNOLOGY

A Continuation Order Plan is available for this series. A continuation order will bring delivery of each new volume immediately upon publication. Volumes are billed only upon actual shipment. For further information please contact the publisher.

POLYMER SCIENCE AND TECHNOLOGY
Volume 13

Symposium on

ULTRAFILTRATION MEMBRANES AND APPLICATIONS

Edited by
Anthony R. Cooper
Dynapol
Palo Alto, California

PLENUM PRESS · NEW YORK AND LONDON

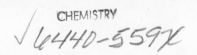

CHEMISTRY

6440-559X

Library of Congress Cataloging in Publication Data

Symposium on Ultrafiltration Membranes and Applications, Washington, D.C., 1979.
Ultrafiltration membranes and applications.

(Polymer science and technology; v. 13)
"Proceedings of the Symposium on Ultrafiltration Membranes and Applications,
sponsored by the American Chemical Society, and held in Washington, D.C., September 9—14, 1979."
Includes index.
1. Ultrafiltration—Congresses. 2. Membranes (Technology)—Congresses. I. Cooper,
Anthony R. II. American Chemical Society. III. Title. IV. Series. [DNLM: 1. Ultrafiltration—Congresses. QD63.F5 S989u 1979]
TP156.F5S95 1979   660.2'84245   80-18685
ISBN 0-306-40548-2

Proceedings of the Symposium on Ultrafiltration Membranes and Applications,
sponsored by the American Chemical Society, and held in Washington, D.C.,
September 9—14, 1979.

© 1980 Plenum Press, New York
A Division of Plenum Publishing Corporation
227 West 17th Street, New York, N.Y. 10011

Printed in the United States of America

To Richard and Julie
for the time I missed with them

PREFACE

This book is a record of a symposium, "Ultrafiltration Membranes and Applications," which was held at the 178th National Meeting of the American Chemical Society in Washington, D.C., September 11-13, 1979.

In organizing these sessions, I hoped to provide a comprehensive survey of the current state of ultrafiltration theory, the most recent advances in membrane technology, and a thorough treatment of existing applications and future directions for ultrafiltration.

For me, the symposium was an outstanding success. It was a truly international forum with stimulating presentations and an enthusiastic audience. I hope that some of this spirit has spilled over into this volume, which is intended to reach a much wider audience.

I am indebted to the Division of Colloid and Surface Chemistry of the American Chemical Society for their sponsorship.

ANTHONY R. COOPER
Palo Alto, California
March, 1980

CONTENTS

PART V.   BIOMEDICAL APPLICATIONS
OF ULTRAFILTRATION

PART VI.   ULTRAFILTRATION APPLICATIONS
IN ENVIRONMENTAL PROBLEMS

ULTRAFILTRATION MEMBRANES

AND APPLICATIONS

FIFTEEN YEARS OF ULTRAFILTRATION:

PROBLEMS AND FUTURE PROMISES OF AN ADOLESCENT TECHNOLOGY

Alan S. Michaels

Department of Chemical Engineering
Stanford University
Stanford, California   94305

I.   HISTORICAL PERSPECTIVES

Despite the fact that the term "ultrafiltration" first appeared in the colloid chemical literature toward the end of the last century, the emergence of ultrafiltration as a viable and practicable separation process had its origins with the development of the first synthetic, high-hydraulic-permeability, macrosolute-retentive ultrafiltration membrane in 1963. This development was an evolutionary consequence of the discovery of the asymmetric cellulose acetate reverse osmosis membrane of the late 1950's, and the discovery of polyelectrolyte complex hydrogels and their anomalous water-transport properties in research at MIT in the early 1960's. It is a curious fact that the UF membrane development was a direct consequence of a collaboration between Amicon Corporation and Dorr-Oliver, Inc. initiated in 1963 for the purpose of developing an economical process for the large-scale removal of colloidal and macromolecular impurities from secondary sewage effluent -- an application which yet remains to be widely exploited. These first-generation membranes, which were shown to be asymmetric structures resembling the Loeb/Sourirajan RO membranes, were relatively water-impermeable by present-day UF membrane standards, but were nonetheless several orders of magnitude more permeable than homogeneous gel-membrane structures with comparable solute retentivity. Despite the significant strides that have since been made in fabricating more permeable and durable membrane structures, these polyion complex membranes continue to fill an important place in laboratory and medical ultrafiltration practice due to their high retentivity for relatively small molecules, and their relative unreactivity with proteins, nucleic acids, and other biological polymers.

1

Even during the earliest stages of this development, it became evident that ultrafiltration could, with the advent of these new membrane structures, compete with such conventional processes as dialysis, vacuum-evaporation, lyophilization, and related processes for the concentration and demineralization of aqueous solutions of biological and other hydrophilic macromolecules. Thus, by 1965, the first laboratory-scale ultrafiltration membranes, and rudimentary ultrafiltration cells and auxiliary hardware for laboratory use, appeared on the market. Within five years of that introduction, laboratory-scale ultrafiltration became firmly entrenched around the world for a host of macromolecular processing needs, and today there is hardly an institution engaged in chemical, medical, or biological research that does not use the process.

The early findings that synthetic ultrafiltration membranes displayed the unique ability quantitatively to retain proteins of the molecular weight of serum albumin, while freely passing all biological microsolutes and intermediate-molecular-weight proteins, suggested the utility of ultrafiltration as an alternative to hemodialysis for the removal of blood-borne impurities from patients with chronic uremia. Thus, in 1966, the concept of "hemo-diafiltration" -- now recognized by the simpler term of "hemofiltration" -- became the focus of the first biomedical application of ultrafiltration. As will become evident later in this symposium, hemofiltration is today regarded as a feasible, and perhaps in some cases preferable, alternative to hemodialysis, and is currently the subject of worldwide clinical scrutiny.

The ten-year period between 1965 and 1975 was one of intensive research and development efforts to fabricate ultrafiltration membranes which were more water-permeable, more mechanically rugged and chemically durable, of more controllable solute-retentivity, and more easily, reproducibly, and cheaply manufactured. Successful accomplishment of these goals followed the discovery that a wide variety of chemically and thermally stable engineering thermoplasts and fiber-forming polymers (including but not limited to polyvinyl chloride, polyacrylonitrile, polycarbonate, polysulfone, and a variety of aliphatic and aromatic polyamides) could, by controlled solvent-casting and nonsolvent precipitation techniques, be converted into a broad spectrum of asymmetric, ultramicroporous membranes with high hydraulic permeability, precisely controllable permselectivity, and excellent mechanical, chemical, and thermal stability. As a consequence, it became possible to produce high-flux membranes with the capacity to retain molecules as small as 500 daltons, and others capable of passing macromolecules as large as 300,000 daltons; membranes capable of repeated exposure to autoclaving temperatures or strongly acidic or alkaline environments without deterioration; and membranes capable of complete dehydration without loss in transport properties. Even more importantly, these improved fabrication methods and membrane materials were adaptable to large-scale,

continuous membrane-fabrication processes, thereby making possible
for the first time the adaptation of membrane ultrafiltration to
high-volume industrial separation applications.

Until the late 1960's, the principal geometric configurations
of available ultrafiltration membranes were flat-sheet structures
(most commonly laminates with fibrous felts to provide superior
mechanical strength), or relatively large-diameter tubular structures. The rather special and stringent requirements of ultrafiltration devices for use in hemofiltration, where high area/volume ratio
devices of low priming volume with minimum blood-trauma were critically important needs, ultimately resulted in the development of
techniques for fabricating asymmetric membrane hollow-fibers whose
barrier-layers were confined to the inside walls of fibers. In view
of the low hydraulic pressure requirements for ultrafiltration, it
thus became possible to ultrafilter fluids through these fibers by
radial outward flow through the fiber walls, without risk of fiber-
rupture. This, therefore, allowed the fabrication of hollow-fiber-
bundle ultrafiltration membrane modules of extraordinarily high
throughput capacity per unit volume, capable of fabrication in large
quantity at quite low cost per unit area. Today, a wide variety of
ultrafiltration membranes of differing solute-cutoff characteristics
and polymer composition are available in hollow fiber modules of
widely variable fiber dimensions and active membrane area.

Through curious circumstances, the development of refractory,
high-water-permeability asymmetric membranes coincided with growing
recognition of the limitations of cellulose acetate as a water-
desalination membrane, and the concurrent development of a series of
new polymeric materials exhibiting superior salt rejection, albeit
with disappointingly low intrinsic water permeability. These events
led to the quite logical attempt to fabricate laminates of ultrathin
films of highly salt-rejective polymers deposited on high-permea-
bility, asymmetric ultrafiltration membrane substrates as a new
route to produce high-flux reverse-osmosis membranes with superior
desalination capacity. This has led to a fortuitous, and eminently
successful, marriage of reverse-osmosis and ultrafiltration membrane
technology in the form of an expanding family of high-performance,
"ultrathin barrier" RO membrane laminates, which are today finding
their way increasingly into large-scale, high-performance, reverse-
osmosis water desalination systems.

Inasmuch as the early commercial outlet for ultrafiltration was
in small-scale laboratory applications, early-generation membrane
devices and auxiliary hardware consisted of quite simple, gas-pres-
surized, stirred cells, and state-of-the-art fluid-control components capable of handling a few milliliters to a few liters of solution. While such equipment was seriously "under-engineered" from the
standpoint of maximizing membrane throughput, and minimizing energy
consumption and processing time, these limitations were of relative-

ly minor importance to the adaptation of ultrafiltration to laboratory practice and equipment of this type continues to occupy an important place in laboratory ultrafiltration to this day. Nonetheless, in the past eight to ten years, important innovations in laboratory ultrafiltration hardware have emerged which now make possible
highly reliable, unattended processing of microliter quantities of
solution (which is proving to be a boon to medical diagnostic procedures), as well as multiliter-capacity systems designed for pilot
scale processing of vaccines and other biologicals.

The problem of concentration polarization at the ultrafiltration membrane surface, which severely limits transmembrane throughput under normal operating conditions, was encountered and recognized very early in the development of ultrafiltration equipment,
and attempts to mitigate the problem by applying the knowledge and
experience gained in concentration-polarization control in reverse
osmosis demineralization of water were made in the design of the
early high-capacity ultrafiltration module designs produced by Dorr-
Oliver in the mid-1960's. These modules consisted of closely-spaced,
parallel-membrane-bearing plates between which feed-fluid was forced
to flow under turbulent conditions. Later, turbulent-flow tubular
modules were introduced (by Abcor, Inc. among others) and more recently, plate-and-frame high-membrane-area modules using replaceable
sheet-membrane stock have been introduced (by DDS and Millipore
Corp.) for use in intermediate to large scale ultrafiltration systems. The advantages of laminar flow through ultrathin channels
bounded by membranes were first tested and verified in reverse-
osmosis systems, and the concept was ultimately embodied into small
scale sheet-membrane modules (by Amicon Corp.) for laboratory and
pilot-scale production applications. The advent of hollow-fiber
ultrafiltration membrane structures in the late 1960's made possible
the fabrication of high-membrane-area, high-throughput, laminar-
feed-flow modules with superior polarization control. Another extremely important feature of hollow fiber membrane modules is their
capacity to be subjected to reverse-flow depolarization for cleaning purposes, a unique advantage not shared with conventional sheet-
membrane devices.

The impetus for the development of improved, broader-retention-
spectrum membranes during the late 1960's was provided by expanding
interest in uses of ultrafiltration as a laboratory tool in life-
science research; in contrast, the impetus for development of more
efficient, higher-capacity, and more economical ultrafiltration membrane-modules and devices during this same period was largely motivated by growing interest in the applications of ultrafiltration in
the pharmaceutical, food, and waste-treatment industries. As might
be expected, ultrafiltration membrane development significantly out-
paced large scale device- and system-development during this period,
since there was an immediate market-outlet for improved membranes in
the laboratory field, where high-performance ultrafiltration equip-

ment was not essential to exploiting the benefits of improved membranes in small-scale applications. To those of us involved in the
early stages of engineering development of high-output ultrafiltration systems, this state-of-affairs was particularly frustrating,
since an expanding array of membranes with astonishingly high water-
permeability, precisely controllable rejection spectra, and excellent thermal, chemical, and mechanical durability were in hand without any way to embody them into economically-priced membrane modules
with operating performance characteristics and service-lifetimes attractive for high-volume industrial applications.

Undoubtedly the most important and fortuitous event in the
evolution of industrial ultrafiltration was the development in the
early 1960s of the electrocoat process for the primer-coating of complex metal shapes, which became a focus of keen interest of the automotive and appliance manufacturers, and their principal suppliers of
coatings and finishes. This process involves the electrophoretic
deposition of colloidal, resinous particles in aqueous dispersion on
metal surfaces under the action of an impressed DC potential, whereupon an extremely uniform, coherent, and defect-free coating can be
deposited on very complex metal shapes quite rapidly. The potential
economic benefits of this means for primer-coating of automobile and
appliance frames and components over traditional manual methods for
spray-coating of such objects was readily appreciated by the industry. However, the successful commercialization of this process was
critically dependent upon the concurrent development of reliable and
economic means for recovering colloidally dispersed paint from the
rinsings from electrocoated parts, and for the removal of soluble
impurities and corrosion products from the electrocoating baths. In
the absence of such recovery or purification capabilities, the economic advantages of the electrocoat process would have been unat-
tainable. The potential of ultrafiltration for facilitating paint-
recovery and abating wastewater pollution on the one hand, and permitting efficient removal of bath impurities on the other, led to
a unique "push-pull" development undertaking, involving close collaboration between the electrocoat-system proponents and industrial
ultrafiltration systems-development organizations. This collaboration resulted in the first (and even today, commercially most significant) application of large scale ultrafiltration systems. Fortu-
nately, in this particular application, only the most primitive requirements of the ultrafiltration process -- namely, the separation
of dispersed colloidal particles from water and microsolutes -- are
required, and the problems of membrane-fouling by anodic (negatively
charged) paint dispersions are relatively minor. As a consequence,
relatively simple tubular or plate-and-frame membrane-modules have
been able to deal with these systems quite efficiently and economically, and have resulted in the electrocoat process' becoming the
first reliable, large-scale, profitable application of ultrafiltration. It is, however, interesting to note that recent developments
in electrocoat technology have led to a preference for cathodic in

lieu of anodic deposition methods, with a corresponding shift to
positively charged (as differentiated from negatively charged) paint
dispersions. Cathodic paint dispersions are far more troublesome
upon ultrafiltration (due to much more marked membrane-fouling) than
are anodic paints, with the result that considerable efforts have
had to be expended in developing modified membranes suitable for
long-term service with cathodic paint systems. These requirements
have reestablished a significant new competitive element into the
marketing of ultrafiltration systems for the electrocoat industry,
and is likely to change the distribution of sales among marketers to
this industry over the next few years.

Another important industrial application of ultrafiltration
which enjoyed early industrial acceptance, and has become well-established among pharmaceutical manufacturers is the purification and
concentration of biological macromolecules such as vaccines, peptide
hormones, and plasma proteins. In this application, the alternative
traditional process of membrane-dialysis, ion-exchange-resin demineralization, lyophilization, and selective precipitation are sufficiently costly, labor-intensive, and inefficient in product-recovery
to have rendered ultrafiltration competitively attractive, even with
relatively inefficient systems and hardware. In these applications,
the value of the product on a unit-activity basis is so high that
the capital and operating costs of separation and purification
facilities become of relatively minor importance compared with the
ultimate yield of active product. The ability of ultrafiltration
processing to minimize losses of labile biologicals due to denaturation or decomposition has proved to be the most important determi-
nant in the preferential selection of this process of biologicals
production.

Among the earliest potential applications of ultrafiltration
in the food-processing industries has been the recovery and concentration of protein values from cheese whey. Interest in this process
has arisen from two important benefits: The first is the elimination
of a troublesome water-pollution problem (since cheese whey has
traditionally been disposed of as a waste stream from dairy processing), and the recovery at relatively high concentration and high
purity of a high-nutritional-value milk protein, potentially market-
able as a high-dollar-value food-supplement or processed food component. Despite the keen interest and intensive developmental activity
in this field, the acceptance of ultrafiltration as a process element in commercial whey processing operations has been very slow to
develop. The principal reasons for this appear to be that (1) whey
proteins are notorious ultrafiltration-membrane-fouling substances,
so that it is difficult to maintain high membrane throughput capacities over reasonable operating schedules with existing ultrafiltration modules and membranes; and (2) the regulatory and sanitary requirements for processing plants require frequent cleaning and sanitizing operations to be performed on process equipment, which

processes have proven to be quite destructive of current-generation membranes. Hence, the necessarily high cost of frequent membrane replacement, and the relative inefficiency of membrane cleaning and regeneration procedures, have made ultrafiltration systems normally too costly for large-scale, general-purpose whey protein recovery operations. The recent development of "super-refractory" membrane structures which display superior resistance to breakdown by active-chlorine-containing sanitizing solutions, and the development of back-washable, hollow-fiber membranes modules and membrane modification techniques which minimize protein fouling, offer considerable promise for circumventing these limitations, and thus rendering whey protein recovery by ultrafiltration both practical and economic on a large scale.

Yet another industrially important application of ultrafiltration which has achieved quite important commercial status today is the production of ultrapure water for use in the electronics and pharmaceutical industries, in the preparation of sterile fluids for medical use, and in the pretreatment of water for colloid-removal prior to demineralization by ion exchange or reverse osmosis. In these applications, the sole requirement of the ultrafiltration process is to remove trace-concentrations of colloidal or macromolecular debris and impurities whose presence in subsequent processing would adversely affect purification procedures, or product reliability or safety. In these applications, the concentrations of impurities to be removed by the ultrafiltration membrane are so low that virtually the full water-throughput-capacity of the membranes can be realized at quite low operating pressures, and high water-throughput can be sustained for very long periods of continuous operation. This naturally renders ultrafiltration a highly reliable and quite inexpensive water-treatment method.

It has now also been conclusively demonstrated that ultrafiltration membranes with solute-cutoffs in the range of 10,000 to 30,000 daltons are reliably and quantitatively retentive toward both virus particles and pyrogenic mucopolysaccharides, with the result that water filtered in this fashion can be assured of being substantially sterile, and virus- and pyrogen-free. Fully integrated and automated ultrafiltration systems for the production of sterile, pyrogen-free water for use in the production of parenteral pharmaceutical products, or for hospital use, are now either commercially available, or in advanced development.

Finally, the use of ultrafiltration membranes in artificial kidneys is rapidly becoming an established commercial reality. Consequent to the recognition that ultrafiltration of blood of uremic patients appears to allow removal of toxic metabolic wastes which are not readily removed by conventional hemodialysis has led to the development of a number of hemodialyzers containing high-flux ultrafiltration membranes which can be concurrently operated as dialyzers

and hemoultrafilters. These devices are operated in much the same
manner as conventional hemodialyzers, but require quite precise control of transmembrane hydrostatic pressure differences in order accurately to control ultrafiltration rates. Also under intense clinical investigation is the procedure described as sequential ultrafil-
tration/dialysis, wherein the patient is first subjected to hemofiltration (with consequent depletion of his blood plasma volume), followed by a subsequent period of hemodialysis without alteration in
plasma volume. Several clinical groups are of the opinion that this
protocol is more effective in controlling chronic uremia than is conventional hemodialysis. Lastly, the process of continuous hemofiltration concomitant with infusion of sterile reconstituting fluid has
reached semicommercial status in Europe, where the procedure is
being evaluated on a large population of end-stage renal failure
patients. While the procedure is recognized as being more complex
and requiring closer control than conventional hemodialysis, there
is evidence of sufficient therapeutic benefit from this procedure to
render it a desirable alternative to hemodialysis for a significant
fraction of the artificial kidney patient population.

II. PRINCIPAL LIMITATIONS OF ULTRAFILTRATION AS A LARGE-SCALE SEPARATION PROCESS

Many of us involved in the early phases of ultrafiltration development during the early mid-1960's had projected significant penetration of this membrane separation process into the chemical
process industry within the ensuing decade. Those predictions --
which anticipated a market in the mid-70's for membranes and membrane process equipment of the order of $30-50 million annually --
have been largely unfulfilled. The explanation for this shortfall
appears to reside with some inherent, persisting limitations (both
technical and economic) in membranes and membrane module design
which must be overcome if ultrafiltration is to become firmly established in cost-sensitive, high-volume-output, industrial applications.

Probably the most serious technical limitation of the current
state-of-the-art ultrafiltration systems is the inescapable problem
of permeation flux-depression by solute polarization at the upstream
membrane surfaces. This phenomenon manifests itself by a tenfold or
greater reduction in membrane hydraulic permeability relative to the
measured pure-water permeation rate, along with a very marked negative dependence of permeability on the concentration of retained
solutes in the upstream fluid. The ultimate impact of these phenomena is significantly to increase membrane area-requirements for a
given plant throughput capacity, to require increased pressures (and
thus pumping energy) to maintain reasonable membrane flows, and to
limit the maximum obtainable retentate concentration which can be
produced under reasonable operating conditions. Superimposed upon

the polarization process (which occurs almost instantaneously on initiation of ultrafiltration) there is a further and more serious problem of "membrane fouling," which is manifested by slow, continuous decline in permeation flux that is substantially independent of feed-solute concentration and upstream hydrodynamic conditions. This fouling process, which has variously been ascribed to membrane pore-plugging or to the formation of a slowly-consolidating, gelatinous solute-layer on the upstream membrane surface, is very unpredictable, and appears to vary markedly in severity depending on the composition of the membrane, the nature of the membrane-retained solutes present in the feed solution, and such other variables as solution pH, ionic strength, electrolyte composition, solution temperature, and operating pressure. Under the worst of circumstances, fouling processes of this type can cause a 90% irreversible flux-decline (relative to the initial flux observed with a "clean" membrane) within a period measured in days or weeks. While in many cases these fouling contaminants can be removed from the membrane surface by appropriate cleaning and scouring procedures which do not necessitate membrane module-disassembly, and with recovery of a large fraction of the original membrane permeability, there are many circumstances where such cleaning procedures are relatively ineffective, or where the rigorousness of the scouring process required to decontaminate the membrane surface is so severe that it compromises membrane- integrity and lifetime. In all events, membrane cleaning and rejuvenation procedures cut into operating time, consume reagents, and in many cases, degrade membranes and other system components, thereby contributing to processing costs and system complexity.

Yet another undesirable consequence of membrane polarization and fouling processes is the inability to make effective use of the macromolecular fractionation capabilities of ultrafiltration membranes for the large-scale resolution of macromolecular mixtures such as blood plasma proteins. This loss in fractionation capability is still poorly understood, although it is probably attributable to partial membrane pore-obstruction, or secondary membrane formation via macrosolute deposition. Consequently, the potentially exciting utilization of membrane ultrafiltration for large-scale complex macromolecular mixture-separations which are currently performed by such techniques as gel permeation, adsorption, or ion-exchange chromatography, selective precipitation, or electrophoresis, remains tantalizingly beyond reach at present.

III. AN ANALYSIS OF THE POLARIZATION/FOULING PROBLEM

It is clear that a rational attack on the polarization/fouling problem depends upon a reasonably clear understanding of the molecular and microscopic features of the fouling process. Our present state of knowledge of the phenomena involved remains quite

rudimentary, but there is a consistent body of experimental information from which some logical mechanistic deductions can be drawn.

For some years, the argument has raged among various researchers as to whether the instantaneous flux-loss on macrosolute
ultrafiltration can be explained by classical fluid mechanical and
mass transfer theory, wherein the elevation of the solute concentration at the membrane surface is accompanied by an osmotic pressure
increase which effectively reduces the net hydraulic gradient available for water transport across the membrane; or whether solute-
accumulation on the upstream membrane surface creates a "gel-film"
of sufficiently high hydraulic resistance to reduce the true hydraulic pressure available at the membrane surface for transmembrane
flow. Under conditions of sufficiently low permeation flux, sufficiently high mass-transfer rates in the upstream fluid channel, and
adequately low macrosolute concentration in solution, there is no
question but that the classical mass-transfer model of the polarization process is eminently adequate to explain the observations. However, as permeation rates increase, upstream mass transfer conditions deteriorate, and macrosolute concentration rises, an operating
regime is encountered where the case for classical polarization vis-
a-vis gel-film polarization becomes moot. Finally, at high fluxes,
solute concentrations, and poor feed-control flow conditions, and
particularly for macrosolutes or colloids of extremely low free-
solution diffusivities, the Sherwood Numbers become so astronomical-
ly large that no rational extension of the classical polarization
model could even approximately represent the observed real ultrafiltration behavior. Moreover, the extreme sensitivity of permeation
flux to such parameters as membrane chemical composition, temperature, pH, and ionic strength, under conditions where these variables
appear to have little effect upon the diffusivity of the macrosolute
in bulk solution, are incompatible with the classical polarization
arguments. In a qualitative sense, these inconsistencies can be adequately reconciled by postulating that membrane-fouling is a hybrid
process of particle filtration and classical polarization, the
former mechanism dominating the events occurring in a zone very
close to the upstream membrane surface, while the latter dominates
events taking place at a greater distance from the membrane. According to this picture, specific interactions between the ultrafiltered
macromolecules or colloidal particles and the surface of the underlying membrane (which can result in membrane-pore obstruction or con-
striction, or adsorption with marked conformational change in the
macromolecule), changes in macromolecule configuration, or short-
range forces of interaction between macromolecules, can all have a
marked effect upon the morphology of the concentrated macromolecular
cake which accumulates on the membrane surface; clearly, these
solute/membrane and solute/solute interactions can have major influence upon the hydraulic resistance to solvent-flow through the membrane. In addition, these interactions can cause major changes in
the effective pore size and pore size distribution of the initial

membrane, with consequent major alteration in the solute rejection
properties of the membrane.

It is thus evident that, while fluid mechanical and mass-transfer processes dominate solute and solvent diffusion and convection
in solution, and are undoubtedly of importance in determining membrane transport kinetics during ultrafiltration, surface and colloid
chemical phenomena which markedly influence solute/membrane and
solute/solute interactions may prove to be far more consequential in
determining the dynamics of solute and solvent transport through
ultrafiltration membranes than has heretofore been suggested. Striking evidence of the importance of surface and colloidal phenomena in
influencing ultrafiltration is provided by observed astonishing differences in membrane-fouling tendency by cationic (as contrasted
with anionic) electrocoat paint dispersions (wherein it is found
that flux-decline is far more rapid with cationic dispersions ultrafiltered through conventional polysulfone or acrylic membranes), and
the common observation that serum plasma proteins have a far more
depressing effect upon the hydraulic permeability of the more hydrophobic polysulfone membranes than upon relatively hydrophilic poly-
ion-complex or cellulosic membrane structures. These observations
clearly point to the importance of coulombic forces and hydrophobic
molecular interactions in governing ultrafiltration dynamics, and
the essentiality of gaining a clearer understanding of the nature of
such interactions as a means for solving the remaining problems
facing successful exploitation of large-volume ultrafiltration
processes.

Beyond the aforementioned adverse affects of solute/membrane
interactions on membrane hydraulic permeability, these interactions
are frequently found to exert marked effects upon the macro- and
microsolute-rejection characteristics of these membranes. For example, recent studies of unsteady-state permeation of serum albumin
solutions through asymmetric ultrafiltration membranes which are
normally regarded as albumin-retentive indicate that, upon their
initial exposure to albumin solutions, the membranes are quite leaky
to the protein, but within periods of time measured in minutes, the
albumin transport drops to exceedingly low values. This is most
logically interpreted as indicating that the membranes contain large
pores which are permeable to albumin, but that these pores rapidly
become partially or completely obstructed to albumin passage by the
absorption of the protein at the pore-entrances or on the pore-
walls. In this case, solute-adsorption by the membrane can be regarded as a beneficial event, since it enhances the retentivity of
the membrane for that particular solute. On the other hand, efforts
to separate mixtures of two serum-proteins such as albumin and globulin by ultrafiltration through a membrane of sufficiently large pore
size to transmit freely the albumin while virtually quantitatively
retaining the globulin have been frustrated by the finding that upon
ultrafiltration of the protein mixture through such membrane, virtu-

ally quantitative retention of both proteins occurs. This effect was first observed with a Dynel asymmetric membrane, and was ascribed to polarization of the upstream membrane surface by a film of concentrated globulin whose pore-structure was too fine to permit the passage of the smaller albumin molecule. In the face, however, of evidence we have recently found that serum albumin is rather strongly adsorbed to Dynel membranes, with substantial attendant reduction in membrane water-permeability, one cannot exclude the possibility that the loss of albumin/globulin separative capacity may rather be due to preferential adsorption of globulin molecules onto or within the pores of the membrane structure, thereby blocking the passage of either the larger or smaller solute molecule. While the differences between these two hypothesized mechanisms of solute-interference with rejection may appear to be trivial, establishment of the actual mechanism may turn out to be crucial to the future of ultrafiltration for macromolecular fractionation: If the former hypothesis is correct, no obvious modifications in either membrane properties or process operating conditions are likely to mitigate the separation problem; whereas, if the latter hypothesis is the case, suitable surface-treatment or chemical modification of the membrane to reduce or eliminate macrosolute-adsorption should markedly increase the separation efficiency for macromolecular mixtures, and thereby open up a lucrative new market for large-scale ultrafiltration.

It is of note, however, that certain macromolecular mixtures are amenable to fractionation by ultrafiltration. These, generally, are polydisperse mixtures of linear macromolecules of quite high hydrophilicity, such as dextran, hydroxyethyl cellulose, and polyvinyl pyrrolidone. These are, of course, predominantly linear macromolecules which are well-solvated and chain-extended in aqueous solution; the fact that the hydraulic permeability of most conventional asymmetric ultrafiltration membranes to solutions of these polymers is very nearly equal to that measured for pure water verifies the low adsorptivity of these polymers to these membranes, and the highly hydrated nature of the polarization layers which may be formed from these macromolecules. Indeed, the absence of pore-obstruction via solute-adsorption, and the structural openness of the polarization layers, may account for the ability of ultrafiltration to separate mixtures of this type.

Yet another manifestation of the influence of macrosolute/ membrane interactions upon membrane characteristics is the frequently-observed effect of the presence in solution of membrane- retained polyelectrolytes upon the permeation of ionic microsolutes through the membrane under normal ultrafiltration conditions. Usually, the presence of such polyelectrolytes is accompanied by the development of significant microion-rejection by a membrane which would normally display no such retention capacity. The phenomenon has been quite properly ascribed to Donnan ion-exclusion by the polyelectrolyte polarization-layer formed on the upstream membrane surface; the same

result would, of course, be observed if the polyelectrolyte molecules were to adsorb on the upstream membrane surface, or the walls
of the membrane pores. It is of interesting historical note that the
phenomenon of Donnan ion-exclusion by adsorbed or pore-obstructing
polyelectrolytes applied to porous supports was first utilized in
the so-called "dynamic membrane" reverse-osmosis desalination concept, which anteceded by many years the development of high- flux
asymmetric ultrafiltration membranes. The combination of these two
developments now provides the basis for a new and improved, low-
energy-demand, high-capacity, water desalting process suitable for
use with relatively low-salinity (5,000 ppm or less) feed waters
which is now receiving increased attention.

IV.   CIRCUMVENTING ULTRAFILTRATION PROCESS LIMITATIONS --
      CURRENT STATUS AND FUTURE PROSPECTS

   The problems of reduced throughput capacity, increased power
consumption, compromised separation capability, and reduced membrane
service lifetime associated with macrosolute- and colloid-fouling of
ultrafiltration membranes have stubbornly resisted adequate solution
despite ten years of engineering experience in pilot- and full-scale
industrial situations. To be sure, the problems have been mitigated
by relatively simple operating procedures which take cognizance of
the nature of the basic problems, and make use of techniques which
have been well-established in conventional continuous particle-
filtration practice. For example, it is becoming increasingly well-
recognized that, by operating ultrafiltration membrane modules at
relatively high feed-side flow rates (with or without fluid recirculation) which maintain high hydraulic shear-stresses at the membrane
surface, by deliberately adding particulate solids to the feed-side
fluid to provide scouring action at the membrane surface, and by
operating under the lowest practicable transmembrane driving pressures, permeation rates can be maintained at respectable levels for
reasonably long time-periods without serious membrane deterioration.
Also, it has been demonstrated that by intermittent and periodic
flushing of the feed-channels and upstream membrane surfaces with
cleaning solutions of suitably adjusted temperature and pH, and frequently containing enzymes or surfactants to facilitate degradation
or disaggregation of membrane-adherent deposits, it is possible to
restore in large measure the losses in permeability attendant upon
membrane fouling, without materially shortening membrane lifetime.
One of the more appealing features of hollow-fiber membrane modules
for use with even heavily contaminated feed streams is their ability
to withstand negative pressure-differences across the fiber wall, a
property which permits these modules to be operated in a "backwash"
mode, which significantly aids unplugging of membrane-pores and de-
tachment of adhering deposits. An even more intriguing operating-
mode for such modules is the so-called "blocked-permeate/reverse-
feed- flow" mode, wherein permeate generated at the inlet-end of the

module is forced to flow backwards through the fiber-walls into the
fiber lumen at the discharge-end of the module. This process backflushes the distal half of the fiber-bundle; by subsequently reversing the feed-flow through the bundle, the proximal half of the
fibers can be similarly backwashed. In such manner, modules can be
speedily and conveniently rejuvenated without the necessity of adding extra pumps, piping systems, valves, and wash-solutions to the
system.

The growing recognition, however, of the importance of colloidal and surface chemical phenomena in the membrane fouling
process has focused research attention upon techniques of membrane-
modification and feed solution properties-control as holding the key
to major improvements in ultrafiltration process performance and economics. For example, there is growing evidence to suggest that techniques for altering the hydrophilicity and/or electrostatic charge
of ultrafiltration membranes may provide the solution to irreversible membrane-fouling. By preparing membranes from polymer-blends or
block-copolymers, at least one component of which is highly hydrophilic and/or ionogenic, or by posttreatment of ultrafiltration membranes by chemical means to introduce hydroxyl, amide, or ionic functionality into the polymer surface, it is possible to produce membrane structures with decidedly improved resistance to macrosolute-
and colloid-fouling. Yet another approach to the fouling problem,
which is under considerable scrutiny today, and may provide a very
convenient and versatile solution to the problem, is pretreatment of
membranes with dilute aqueous solutions of selected water-soluble
polymers which form highly hydrated, high-hydraulic-permeability surface films on the surfaces of the membrane by either filtration or
adsorption. These films offer promise of preventing the adsorption
of fouling macromolecules or colloidal particles onto the membrane
surface, and preventing the adhesion of such substances to the membrane. Such pretreatment has the important advantage of being able
to be performed easily by flowing dilute solutions of the selected
polymer through the ultrafiltration module prior to initiation of
feed-solution processing. Yet another advantage of this pretreatment-
-technique is the ability periodically to recondition and retreat
the membrane in-situ in the modules, without serious interference
with normal plant operations.

Finally, our growing awareness of the importance of solution-
composition variables such as pH, ionic strength, and the presence
of low concentrations of such water-miscible organic solvents as
alcohol, upon the state of aggregation, conformation, and charge of
macromolecules in solution can undoubtedly be put to much more effective use in reducing membrane-obstruction by macromolecular adsorption, and increasing the water- and solute-permeability of polarization layers, by rendering such layers more open-structured, larger-
pored, and less cohesive. Progress which is being made in these
directions is to be found in numerous contributions to this
symposium.

V.   NEW HORIZONS FOR ULTRAFILTRATION AS A SEPARATION PROCESS

As is probably obvious from the foregoing discussion, the development of operational methods or modified membranes which can further mitigate or eliminate the problem of membrane-fouling will expand the spectrum of large-volume applications of ultrafiltration into areas where process economics have prevented its introduction. These include such areas as feed-water pretreatment for water-demineralization processes, secondary and tertiary treatment of sewage plant effluents, and removal and/or recovery of macromolecular and colloidal contaminants from aqueous chemical- and food-processing-plant effluents. The current energyresource crisis, coupled with the increasing pressures to reduce environmental pollution, is also contributing significantly to reevaluation of ultrafiltration for high-volume, aqueous-stream concentrations and purifications. An interesting illustration of this trend is the resurgence of interest in the ultrafiltrative process for cheese-whey deproteinization, consequent to growing interest in the production of fuel-grade ethanol by direct fermentation of lactose. In the past, the principal justifications for whey ultrafiltration were pollution-abatement, and the resale- value of the recovered protein-concentrate as a food- and animal feed-supplement. The return to the dairy processor from these two sources has in the past been barely adequate to offset the costs of installation and operation of whey ultrafiltration facilities. With alcohol from lactose now surfacing as an additional revenue-source from whey treatment, the economics of ultrafiltrative processing for protein removal (even with the present-day costs of UF processing) are becoming increasingly more attractive.

Also emerging are completely new technologies which promise to open new vistas for ultrafiltration applications. This trend is exemplified by recent developments in immuno-chemistry, which are leading to exquisitely sensitive and selective methods for detecting and quantifying complex biochemicals (such as antigens, antibodies, peptide hormones, and the like) in blood and other biological fluids, and which are leading to revolutionary improvements in diagnostic medicine. The use of ultrafiltration of fluid samples as a preliminary concentration- and purification-step prior to immunochemical processing, or as a means of rapidly separating immunocomplexes from residual haptens, has already been proved to increase substantially the sensitivity of these assay procedures, and in many cases greatly to reduce the time required for the conduct of such tests. While, obviously, the quantity of fluid processed in these diagnostic procedures is measured only in fractions of a milliliter, the potential volume of tests to be performed, and the value added to the test procedure by the ultrafiltration step so great, that this particular area promises to be one of the most lucrative future markets for ultrafiltration membranes and equipment.

Another extremely important and potentially very-large-volume application for ultrafiltration lies in a process-area where basic operational principles behind ultrafiltration -- that is, the continuous separation of a solution or dispersion into a solute-rich and solute-poor fraction -- is applied to the removal and concentration of _particulate solids_ in aqueous suspension. This process is alternatively termed "crossflow filtration," and comprises flowing a particulate suspension over the surface of a filtration membrane under hydrostatic conditions favoring transport of suspending solvent through the membrane, while the concentrated suspension remaining behind is force-convected across the membrane surface, and out of the filtration device. This process is uniquely effective for solid/liquid separations wherein the dispersed solid phase is very difficult to remove by conventional filtration procedures due to the low hydraulic permeability of the resulting filter cake, or where small density- differences between suspended phase and the suspending medium make gravity-sedimentation or centrifugation inoperative or uneconomic. A particularly promising specific application of this process is the separation of cells and cellular debris from whole fermentation broths, where it has been found possible to achieve exceedingly high recovery of particle-free filtrate, and thus produce highly concentrated particulate suspensions as retentates. Crossflow filtration offers promise of replacing conventional filter-aid filtration of gelatinous suspensions of this type, with significant reduction in filtration costs, minimization of waste-disposal problems, and in some cases recovery of a byproduct cell-concentrate which can be economically spray-dried to yield a saleable, nutritious animal-feed supplement. Other potentially important advantages of crossflow filtration in fermentation processes is enhanced product recovery, greater product purity, and significantly reduced product-isolation costs.

A rather unique, and quite exciting special application of the crossflow filtration principle has recently generated strong interest on the part of the medical community and the pharmaceutical industry, and will be described in greater detail during this symposium: This is the process of _membrane-plasmapheresis_ -- the continuous filtrative separation of the cellular components of whole blood (human or animal) from plasma, yielding a cell-free filtrate containing a full complement of plasma proteins, and an undamaged cell-concentrate. This technique of plasmapheresis offers promise of supplanting alternative methods (such as refrigerated centrifugation) for safely and cheaply harvesting plasma from blood donors, and in combination with affinity-sorption techniques, may provide a simple and effective means for selective removal of immunocomplexes from the blood of patients with allergies, autoimmune diseases, and cancer. Such "subtractive therapies" constitute the frontier of a promising new approach to the effective treatment of these otherwise incurable diseases.

A third area of application where large-scale ultrafiltration

may soon prove to play an important role is in hydrometallurgical
processing, where the selective separation and concentration of
trace-metals (present in tailings-streams or leach-solutions) is desired. The gradual depletion of our principal mineral resources is
compelling us to develop recovery methods applicable to increasingly
leaner ore-bodies, and to the ocean and subsurface brines as potential sources of metal values. The combined use of highly selective
chemical-complexation techniques in combination with ultrafiltration
has been recognized as a potentially efficient and inexpensive way
to recover metal-values from such solutions. The principle involved
is quite straightforward: The desired metal-ion-containing solution
is treated with a water- soluble polymer which chelates or complexes
with the desired ion, thereby forming a metal-rich polymeric complex, which can be ultrafiltratively removed from the solution to
yield a polymer- and metal-rich concentrate. This concentrate is subsequently chemically treated to decouple the metal ion from the
polymer, and the resulting solution can then be subjected to a
second ultrafiltration, yielding a metal-free polymer concentrate,
and a filtrate rich in the desired metal. The polymer-concentrate
is recycled to the feed-end of the system for reuse in the complexation step, while the metal salt concentrate obtained as product can
be treated by conventional electrochemical, precipitative, or pyro-
metallurgical processes to effect the ultimate isolation of the elemental metal. The potential of this technique of "complexation ultrafiltration" for the production of high-dollar-value metals and other
elements from high dilute process streams seems quite bright.

VI.   NEW APPLICATIONS FOR ULTRAFILTRATION MEMBRANES

     In addition to the use of ultrafiltration membranes and
processes for novel separations, the rather special transport and
separative properties of these membranes have suggested several
exciting and promising applications for them which do not involve
ultrafiltration per se, yet perform useful process functions. One of
these applications is the use of asymmetric, hollow-fiber ultrafiltration membranes as "artificial capillary beds" for the continuous
culture of mammalian cells. By inoculating viable mammalian cells
into the shellside space of a hollow fiber bundle, and passing appro-
priately-constituted nutrient solution through the fiber lumens, it
has been found possible to properly nourish, and maintain a satisfactory living environment for, such cell populations within the fiber-
interstices; if such cells synthesize and secrete a biological product, it becomes possible to produce such biochemical continuously in
such a device as long as the membrane fiber wall is permeable to
that product. Devices of this type are now being used for the continuous culture of pancreatic islet-cells for the synthesis of insulin, and for the culture of cells of other secretory organs from
animals and man for the continuous production of other peptide hormones, enzymes, and the like. Success has also been realized in

culturing liver cells within such a hollow-fiber device, and using
such an "artificial liver" as an extracorporeal blood-detoxification
device for the treatment of acute liver dysfunction in animals.
Potential use of such hollow-fiber, mammalian-cell-containing devices as implantable replacement organs for the treatment of human
diseases, or for the large-scale manufacture of human biological
products, is receiving increasing attention today.

The successful immobilization of mammalian cells in membrane
hollow fiber devices has naturally led to the development of immobilized enzyme reactors based upon this same systems concept. In this
application, a specific enzyme or mixture of enzymes designed to perform a particular series of chemical conversions is physically immobilized within the porous sponge-layer of such hollow-fibers, and
substrate-containing solution forced to flow continuously through
the lumens of these fibers. By the combined processes of diffusion
and convection, substrate molecules are brought into contact with
the enzymes where the conversion reactions occur, and the products
of reaction are similarly diffused and convected out of the device
into the effluent lumen-fluid. The principal requirement for successful operation of this type of immobilized enzyme reactor, of course,
is that the membrane- wall be substantially impermeable to enzyme
molecules, while displaying unrestricted transport of either substrate or product molecules. The utility of hollow fiber reactors
for enzymatic hydrolysis and isomerization reactions has already
been demonstrated, and broader use of the procedure and technique
for industrially important food- and pharmaceutical-conversion reactions can be expected. There has also been growing interest in the
possible use of the asymmetric hollow-fiber system for the immobilization of either homogeneous or heterogeneous catalysts, with the
objective of developing a broad spectrum of "catalytic wall" reactors suitable for use in many large-scale, continuous chemical synthesis operations.

Finally, the success in utilizing hollow fiber ultrafiltration
membranes for mammalian-cell culture and enzyme immobilization has
suggested the possibility of employing these fiber structures as
matrices for the immobilization of viable microorganisms for the continuous production of chemicals by biochemical means. In this manner, it should be possible to supply nutrients and oxygen to such
cell cultures at far higher rates, and with greater efficiency, than
is possible in conventional tank-fermentations, and similarly, to
remove rapidly and efficiently the biochemical products and metabolic wastes elaborated by the microorganisms. The end result of
microorganism maintenance under these conditions should be substantially enhanced biochemical productivity by these organisms, compared to their performance under normal fermentation conditions. The
major impetus for the development of microbial reactors based on
this hollow fiber device concept has come from advances in genetic
manipulation of microorganism populations by the techniques of DNA

recombination, whereby it has been possible to manipulate the biosynthetic machinery of a given species of microorganism to produce
specific biochemical substances which have great value and utility,
but which would normally not be synthesized by that microorganism.
Geneticists and chemical engineers at Stanford University are now
actively engaged in just such a collaboration, and the possibilities
of developing an entirely new dimension in chemical synthesis by
microbiologic means seems to be quite bright indeed. Recent exploratory experiments with both coliform bacteria and bacilli in hollow
fibers confirm that both of these species of organism can be grown
in very high population-densities in the porous-wall structure of
the fibers, and display sustained and enhanced biosynthetic capability in this environment. While results are still preliminary, and
the evidence of feasibility fragmentary, this development has
already begun to capture the imagination and enthusiasm of the scientific, engineering, and industrial communities.

VII. WHITHER ULTRAFILTRATION?

This writer has not infrequently been criticized for his per-
sistent optimism about the future of ultrafiltration, beginning with
his first "crystal-ball-gazing" publication in Chemical Engineering
Progress in 1968, and continuing with several intervening publications prior to this review. In retrospect, it would appear that the
criticism has been justified, to the extent that he has anticipated
an earlier market-penetration of the technology than has actually
taken place. This review involves considerably less prognostication
about the future of ultrafiltration than its predecessors, although
his optimism about the future of the technology remains unabated.
Needless to say, the scope and diversity of the contributions to
this symposium, to which this particular presentation is merely a
prelude, will attest to the major strides which have been made in
advancing this technology over the past decade, and to the extent to
which the utility of ultrafiltration as a unit operation has become
recognized and appreciated by the Chemical Engineering Profession.

PRODUCTION, SPECIFICATION, AND SOME TRANSPORT CHARACTERISTICS OF
CELLULOSE ACETATE ULTRAFILTRATION MEMBRANES FOR AQUEOUS FEED
SOLUTIONS

S. Sourirajan, Takeshi Matsuura, Fu-Hung Hsieh
and Gary R. Gildert

Division of Chemistry
National Research Council of Canada
Ottawa, Canada K1A OR9

INTRODUCTION

A precise and detailed understanding of the scientific basis
of ultrafiltration separation in all its aspects is a necessary
condition for its fullest technological development and practical
utilization. There are several approaches to the subject [1-3].
This paper presents one other approach; that an ultrafiltration
membrane is also a reverse osmosis membrane is the essence of
this approach. This means that, just as in reverse osmosis, in
ultrafiltration also, both the chemical nature and the porous
structure of the membrane surface in contact with the feed
solution govern the separation and permeability characteristics
of the membrane under the operating conditions of the process.
At the membrane-solution interface, whatever solute-solvent-
membrane material interactions can prevail, they do prevail whether
the solute is small or big, the solvent is aqueous or nonaqueous,
the average pore size on the membrane surface is small or big,
the osmotic pressure of the feed solution is significant or
insignificant, or the operating pressure used for the process is
high or low. This means then that all the factors governing
solute and solvent transport through reverse osmosis membranes
also govern such transport through ultrafiltration membranes.
Consequently, the science of reverse osmosis [4] offers a natural
basis for the scientific and technical development of ultrafiltration in all its aspects.

In industrial practice, ultrafiltration membranes and
processes are commonly and arbitrarily distinguished from reverse
osmosis membranes and processes. For example, in ultrafiltration,

21

the average pore sizes on membrane surfaces are relatively bigger,
the operating pressures used are relatively lower (usually less
than 690 kPa gauge ($<\sim$100 psig)), and the molecular weights of
solutes in the feed solutions used are relatively higher;
further, ultrafiltration membranes are also often used simply as
filters for colloidal or particulate matter suspended in feed
liquids. The development of ultrafiltration membranes and
processes is hence particularly concerned with the above aspects
of the subject. Following the preferential sorption-capillary
flow mechanism for reverse osmosis, and reverse osmosis approach
to ultrafiltration, some significant progress has been made in
this laboratory during recent years in the areas of production,
specification and transport characteristics of cellulose acetate
ultrafiltration membranes. A brief report on this progress is
the subject of this paper.

PRODUCTION OF CELLULOSE ACETATE ULTRAFILTRATION MEMBRANES

Every step in the process of casting asymmetric porous
cellulose acetate reverse osmosis membranes has important effects
on the average size and distribution of pores on the surface
("skin") layer of the resulting membranes [5]. The composition
and temperature of the film casting solution, and the composition
and temperature of the gelation medium are important film casting
variables which can be made use of for producing useful cellulose
acetate ultrafiltration membranes.

It has been shown [5-7] that decrease in solvent/polymer
(S/P) ratio, increase in nonsolvent/solvent (N/S) ratio and
increase in nonsolvent/polymer (N/P) ratio in the casting solution
composition, and also decrease in temperature of the casting
solution tend to increase the average size of pores on the
surface of resulting membranes in the as cast condition; further,
increase in S/P ratio in the casting solution composition, and
increase in temperature of the casting solution tend to increase
the effective number of such pores on the membrane surface.
Often, during such variations, an increase in the size of pores
on the membrane surface results also in a simultaneous decrease
in the effective number of such pores. The foregoing criteria
offer a wide choice for the composition of the film casting
solution and other film casting conditions for making ultrafiltration membranes suitable for different applications. The
effects of variations of S/P, N/S, and N/P ratios indicated above
on the average size of pores on the surface of resulting membranes,
as illustrated in Fig. 1, are of particular interest in the
choice of composition of the film casting solution for making
ultrafiltration membranes. For example, the composition of the
film casting solution [8] given in Table 1 has S/P, N/S and N/P
weight ratios of 4.25, 0.352 and 1.5 respectively. From the
correlations of S/P, N/S and N/P with data on NaCl separation and

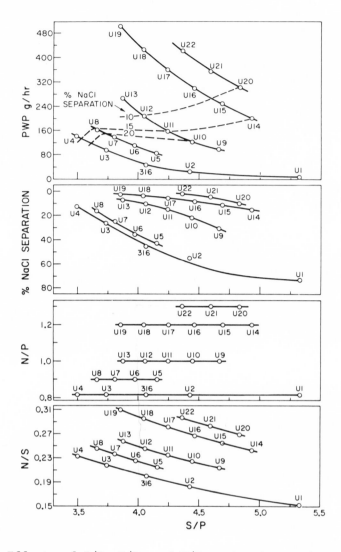

Fig. 1. Effects of S/P, N/S, and N/P ratios in casting solution
composition on performance of resulting membranes.
Polymer (P): cellulose acetate (E-398-3), solvent (S):
acetone, nonsolvent (N): aqueous magnesium perchlorate
($Mg(ClO_4)_2$: H_2O = 1:8.5); temperature of film casting
solution: $0°C$; temperature of film casting atmosphere,
$20°C$; solvent evaporation period, minimum (2 to 5 s);
solute concentration in feed, 200 ppm; operating pressure,
689.5 kPa gauge (100 psig); effective film area, 13.2 cm^2.
Data of Kutowy and Sourirajan [7]. Reprinted with
permission of copyright ©️ owner, John Wiley & Sons, Inc.

Table 1.   A Composition (Wt%) of Film Casting Solution for
           Making Cellulose Acetate Ultrafiltration Membranes[a]

Cellulose acetate (Eastman 400-25):   14.8
Acetone: 63.0
Water: 19.9
Magnesium perchlorate: 2.3

---

[a]Data of Kutowy, Thayer and Sourirajan [8].

pure water permeation rate (PWP) given in Fig. 1, it is evident
that the above composition for film casting solution is
particularly suitable for making ultrafiltration membranes.

    Gelation control has also been shown to be an effective
means of changing the average size of pores on the surface of
resulting membranes. By using alcohol-water mixtures as gelation
media, at different temperatures and in a wide range of alcohol
concentrations, cellulose acetate reverse osmosis and ultrafiltration membranes with different surface porosities have been
obtained [8-11]. Fig. 2 illustrates the effects of temperature
and ethyl alcohol concentration in the EtOH-H$_2$O gelation medium
on reverse osmosis performance of cellulose acetate membranes
obtained from a casting solution of composition given in Table 1.
The results show that by using EtOH-H$_2$O gelation media with an
alcohol concentration greater than that corresponding to the
initial minimum in water flux, the average pore size on the
membrane surface can be increased resulting in membranes which
are useful for many ultrafiltration applications. Such membranes
have been produced and their practical utility has been
demonstrated [8].

    The results shown in Fig. 2 have led to the differential
gelation technique for establishing asymmetric porosity in making
integrally supported tubular cellulose acetate ultrafiltration
membranes [12,13] obtained from a casting solution of composition
given in Table 1. In this technique [13], an ethyl alcohol-water
solution at laboratory temperature, and composition corresponding
to mole fraction of ethyl alcohol in the range 0.1 to 0.15 is the
gelation medium for the inside surface of the tubular cellulose
acetate membrane cast on a porous support tube; similarly, another
ethyl alcohol-water solution at laboratory or higher temperature,
and composition corresponding to mole fraction of ethyl alcohol
greater than 0.8, is the gelation medium for the casting solution
underneath the inside surface layer of the membrane, including
the casting solution that has penetrated into the voids of the
porous support tube during film casting. The latter gelation

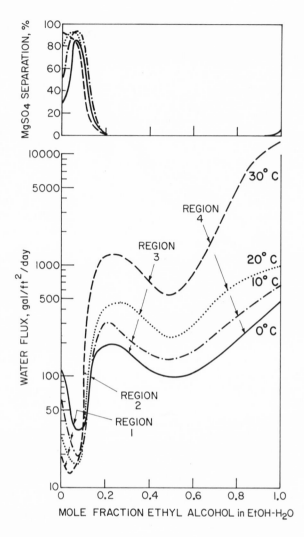

Fig. 2. Effects of temperature and ethyl alcohol concentration
        in the gelation medium on membrane performance. Casting
        solution composition, same as given in Table I; operating
        pressure, 689.5 kPa gauge (100 psig); feed solution, 300
        ppm $MgSO_4$-H_2O. 1 gal/ft^2 day = 0.0407 m^3/m^2 day. Data
        of Kutowy et al. [8].

medium is made to percolate continuously from the outside of the
porous support tube into its voids by capillary action during
film casting. Such supported tubular cellulose acetate ultrafiltration membranes have been produced and their utility
demonstrated [13] for the major recovery of clear reusable water

from coal mine wastewaters laden with coal tailings and fine clay
particles.

SPECIFICATION OF CELLULOSE ACETATE ULTRAFILTRATION MEMBRANES

The subject of membrane specification and predictability of
membrane performance in reverse osmosis is discussed in detail in
the literature [14]. At any given temperature and pressure, a
cellulose acetate reverse osmosis membrane can be specified in
terms of its pure water permeability constant A and solute
transport parameter $D_{AM}/K\delta$ for NaCl (chosen as a convenient
reference solute) from experimental data on pure water permeation
rate (PWP), product rate (PR) and fraction solute separation f.
The quantity $(D_{AM}/K\delta)_{NaCl}$ is related to the corresponding quantity
$\ln C^*_{NaCl}$ and the polar free energy parameter $-\Delta\Delta G/RT$ for Na^+
and Cl^- ions by the expression [4,15]:

$$\ln (D_{AM}/K\delta)_{NaCl} = \ln C^*_{NaCl} + \Sigma \left\{ \left(-\frac{\Delta\Delta G}{RT} \right)_{Na^+} + \left(-\frac{\Delta\Delta G}{RT} \right)_{Cl^-} \right\} \quad (1)$$

All symbols are defined at the end of the paper. The numerical
values of $-\Delta\Delta G/RT$ for different ions applicable for cellulose
acetate (Eastman CA-398) membrane material and aqueous feed
solutions are known [4,15]. Therefore, the quantity $\ln C^*_{NaCl}$
representing the average pore size on the membrane surface can
be calculated from eq 1. Thus a cellulose acetate reverse osmosis
membrane can be specified in terms of its A and $\ln C^*_{NaCl}$ values
at any given operating pressure and temperature. Conversely,
for a membrane so specified, data on f and (PR) can be predicted
using the applicable value of the mass transfer coefficient k
and the basic reverse osmosis transport equations [14]. For
example, for very dilute NaCl-H$_2$O feed solutions (i.e., where
osmotic pressure effects may be assumed negligible), (PR) is
same as (PWP), and solute separation f is given by the relation

$$f = \left[1 + \frac{(D_{AM}/K\delta)}{v_s} \exp\left(\frac{v_s}{k} \right) \right]^{-1} \quad (2)$$

where

$$v_s = \frac{AP}{c} \quad (3)$$

and the value of $D_{AM}/K\delta$ is obtained from eq 1.

In a similar manner, one may also specify a cellulose acetate
ultrafiltration membrane in terms of A and $\ln C^*_{NaCl}$. While there
is no problem in obtaining the value of A for the ultrafiltration
membrane from direct experimental measurement, often it is not
possible to obtain reliable value for $\ln C^*_{NaCl}$ from direct
experimental data on NaCl separation, because ultrafiltration
membranes generally give very low or practically negligible
solute separation for NaCl from aqueous NaCl-H$_2$O feed solutions.

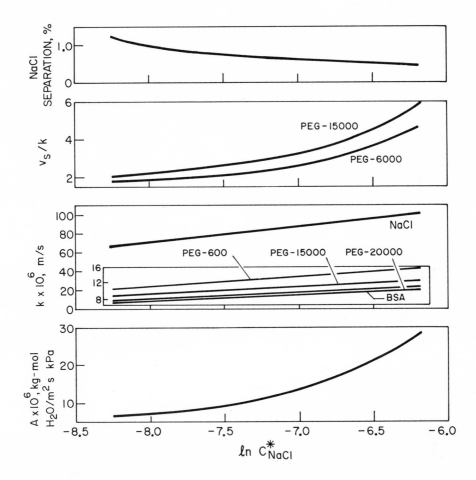

Fig. 3. Effect of porosity on membrane surface on pure water permeability constant, relative mass transfer co- efficients for NaCl, PEG-6000, PEG-15000, PEG-20000, and BSA, data on v_s/k, and solute separation for NaCl. All data are for dilute feed solutions at the operating pressure of 137.9 kPa gauge (20 psig), and cellulose acetate membranes.

This may be seen by a set of calculations on f values for various $\ln C^*_{NaCl}$ values corresponding to a set of given A and k_{NaCl} values using eq 1 to 3. For example, such calculated results given in Fig. 3 show that when $\ln C^*_{NaCl}$ values are greater than -8 (which is the case for many useful ultrafiltration membranes), NaCl separations are ∿1% or less which are too small for precise determinations of $(D_{AM}/K\delta)_{NaCl}$ and hence $\ln C^*_{NaCl}$ values. Consequently, for such membranes, alternate methods of

Table 2. Summation of Polar, Steric, and Nonpolar Parameters for
          Some Macromolecular Solutes Applicable to Cellulose
          Acetate Ultrafiltration Membranes and Aqueous Feed
          Solutions[a]

Solute	$\Sigma[(-\Delta\Delta G/RT) + \delta^*\Sigma E_s + \omega^*\Sigma s^*]$
PEG-200	0.89
PEG-300	0.53
PEG-400	0.21
PEG-600	-0.49
PEG-1000	-1.03
PEG-1500	-1.62
PEG-4000	-2.6
PEG-6000	-3.8
PEG-9000	-4.3
PEG-15000	-4.7
PEG-20000	-5.3
BSA	-5.9

[a]Data of Hsieh, Matsuura and Sourirajan [16,17,20].

determining reliable $\ln C_{NaCl}^*$ values are necessary.

Such methods applicable for $\ln C_{NaCl}^*$ values in the range
-11.2 to -6.6 (based on $D_{AM}/K\delta$ values in m/s) have been developed.
In these methods polyethylene glycols PEG-6000 and PEG-15000 are
used as reference solutes in dilute (50 ppm) aqueous feed solutions
in the ultrafiltration experiments. Experimental results have
shown [16,17] that at such low feed concentrations, there is
no pre-gel or gel formation on the membrane surface during
ultrafiltration, (PR) is essentially the same as (PWP), $D_{AM}/K\delta$
values for the above solutes are independent of feed concentration,
feed flow rate and operating pressures up to 690 kPa gauge (∿100
psig) or less, and the values of k for the above solutes are also
independent of such operating pressures. Further the $D_{AM}/K\delta$
value for each PEG-solute and its corresponding $\ln C_{NaCl}^*$ value
are related by the expression [4,15,18,19]:

$$\ln (D_{AM}/K\delta) = \ln C_{NaCl}^* + \ln \Delta^* + (- \frac{\Delta\Delta G}{RT}) + \delta^*\Sigma E_s + \omega^*\Sigma s^* \quad (4)$$

For the ultrafiltration (reverse osmosis) systems under discussion,
the sum of the polar, steric and nonpolar parameters
$(\Sigma(-\frac{\Delta\Delta G}{RT} + \delta^*\Sigma E_s + \omega^*\Sigma s^*))$ for the PEG-solutes has been determined
as given in Table 2, and the scale factor $\ln \Delta^*$ has a value of
-0.92. Based on these results, two methods [17,20] are described
below for obtaining the $\ln C_{NaCl}^*$ value for a cellulose acetate

ultrafiltration membrane.

Method 1. Determine experimentally (PWP), (PR), and f values
for a 50 ppm PEG-6000-H_2O feed solution at two operating pressures
in the range 137.9 to 689.5 kPa gauge (20 to 100 psig) at the
same feed flow rate. Since [4] at each pressure,

$$D_{AM}/K\delta = \frac{(PR)}{3600S\rho} \cdot \frac{1-f}{f} \left[\exp \left\{ \frac{(PR)}{3600Sk\rho} \right\} \right]^{-1} \tag{5}$$

$$k = \frac{(PR)_I-(PR)_{II}}{3600S\rho} \ell n \left[\left\{ \frac{(PR)_I}{(PR)_{II}} \cdot \frac{(1-f_I)f_{II}}{(1-f_{II})f_I} \right\} \right]^{-1} \tag{6}$$

where the subscripts I and II for (PR) and f refer to the
respective quantities at the chosen two operating pressures.
Using the experimental data, calculate k from eq 6, $D_{AM}/K\delta$ from
eq 5, and finally, $\ell n\ C_{NaCl}^*$ from eq 4 using the data given in
Table 2.

Method 2. Determine experimentally (PWP), (PR) and f for two feed
solutions 50 ppm PEG-6000-H_2O and 50 ppm PEG-15000-H_2O at the same
operating pressure of about 138 kPa gauge ($\sim$20 psig) and the same
feed flow rate. Equations 4 and 5 are also valid for the above
reverse osmosis systems; further, it has been shown that [4]

$$k = k_{ref}[D_{AB}/(D_{AB})_{ref}]^{2/3} \tag{7}$$

Let subscripts I and II represent the solutes PEG-6000 and
PEG-15000 respectively, and let

$$\Delta = [(-\frac{\Delta\Delta G}{RT}) + \delta^*\Sigma E_s + \omega^*\Sigma s^*]_I$$
$$- [(-\frac{\Delta\Delta G}{RT}) + \delta^*\Sigma E_s + \omega^*\Sigma s^*]_{II} \tag{8}$$

and

$$\alpha = [(D_{AB})_{II}/(D_{AB})_I]^{2/3} \tag{9}$$

Then, from eq 4, 5 and 7,

$$k_I = \frac{(PR)_I}{3600S\rho} \cdot \frac{1-\alpha}{\alpha} \left[\ell n \left(\frac{f_I}{1-f_I} \cdot \frac{1-f_{II}}{f_{II}} \right) + \Delta \right]^{-1} \tag{10}$$

From data already in the literature [17], $\Delta = 0.9$ and $\alpha = 0.788$.
Using the experimental data on $(PR)_I$, f_I and f_{II}, and the above
values of Δ and α, calculate k_I from eq 10, $(D_{AM}/K\delta)_I$ from eq 5,
and finally $\ell n\ C_{NaCl}^*$ from eq 4 using the data given in Table 2.
Both the above methods of calculating $\ell n\ C_{NaCl}^*$ for a cellulose
acetate ultrafiltration membrane have been successful [17,20].
Method 2 is particularly useful for specifying membranes whose
average pore sizes are relatively bigger, and for which the values
of v_s/k for PEG-solutes tend to become greater than unity

especially at higher operating pressures [16]. For dilute feed
solutions, it has already been shown [16] that when v_s/k is greater
than unity, solute separation decreases with increase in operating
pressure.

PREFERENTIAL SORPTION AT THE CELLULOSE ACETATE MEMBRANE-AQUEOUS
SOLUTION INTERFACE

Just as in reverse osmosis, in ultrafiltration also, the
preferential sorption characteristics at the membrane-solution
interface govern solute and solvent transport during the process.
The membrane specification techniques described above are based
on the assumption that in a PEG solute-H_2O-cellulose acetate
membrane reverse osmosis system, water is preferentially sorbed
at the membrane-solution interface. This assumption is based on
the chemical nature of the solute, solvent and membrane materials
involved; and, the detailed transport studies on such systems
[16,17] support the above assumption. In any ultrafiltration
system involving cellulose acetate membranes and aqueous feed
solutions, it is first necessary to ascertain whether water or
solute is preferentially sorbed at the membrane-solution interface,
in order to apply or develop appropriate transport equations for
subsequent system analysis and process design. The physicochemical
criteria developed earlier for preferential sorption at membrane-
solution interfaces in reverse osmosis [21] are applicable for
such preferential sorption in ultrafiltration as well. In
addition, data on adsorption experiments and data on liquid-solid
chromatography (LSC) offer more direct means of ascertaining the
preferential sorption situation at the membrane-solution interface
involved [20,22]. Such data, some of which are illustrated below,
are particularly useful in the study of ultrafiltration processes
where the nature of solutes involved and the factors governing
solute and solvent transport through the membrane are often even
more complex than in reverse osmosis.

By equilibrating a weighed quantity of dry cellulose acetate
membrane material in powder form with an aqueous solution of the
solute under consideration, and by determining the concentrations
of the solute in the solution before and after such equilibration,
one can ascertain whether solute or solvent is preferentially
adsorbed on the surface of the polymer material. When solute is
preferentially adsorbed on the surface of the polymer material,
one can expect that the concentration of the solute in the
solution at equilibrium will be less than its initial
concentration, whereas the opposite will be the case when water
is preferentially adsorbed on such surface. Thus the quantity
representing the difference between the initial and equilibrium
concentrations of the solute in the solution will be positive for
preferential adsorption of solute and negative for preferential
adsorption of solvent at the polymer material-solution interface.

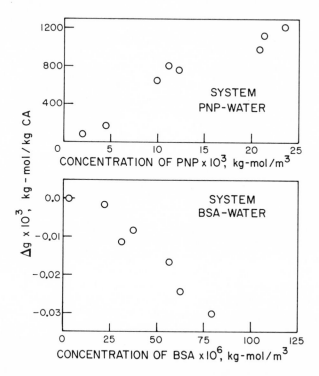

Fig. 4. Data on adsorption of p-nitrophenol (PNP) and bovine serum albumin (BSA) on cellulose acetate powder at 23-25°C. Data of Hsieh et al. [20].

This is illustrated by a set of adsorption data given in Fig. 4 obtained with a sample of cellulose acetate material in powder form equilibrated with paranitrophenol (PNP)-water and bovine serum albumin (BSA)-water solutions [20]; in the figure, the quantity Δg (in kg mol/kg CA) represents the amount of solute adsorbed per unit weight of cellulose acetate polymer material. These data show that in the PNP-water-cellulose acetate material system, solute is preferentially adsorbed at the polymer material-

solution interface, whereas in the BSA-water-cellulose acetate
material system solute is negatively adsorbed (i.e., water is
preferentially adsorbed) at the above interface. The former
result is consistent with the higher acidity of PNP relative to
that of water, and the net basic character of the cellulose
acetate polymer material [21]; and, the latter result is
consistent with the presence of significantly high proportion of
charged amino acids in the BSA molecule [23] and the consequent
repulsion of solute at the polymer material-solution interface.

It has been shown that LSC data on relative retention
volumes (V_R') (= retention time x flow rate) of different solutes
can offer some definitive indication of the preferential sorption
characteristics at polymer material-solution interfaces when the
polymer material is used in the chromatographic column [22]. For
example, considering that in LSC, the data on retention volume of
D_2O is essentially the same as that of ordinary water, it can be
expected that when water is preferentially sorbed at the polymer
material-solution interface, the retention volume of solute to
be relatively lower than that of D_2O, and when solute is
preferentially sorbed at the polymer material-solution interface,
the retention volume of solute to be relatively higher than that
of D_2O. With respect to cellulose acetate polymer material
(CA-398-3), the data on V_R' reported earlier [22] for raffinose,
D-sorbitol, sucrose, glycerol, magnesium sulfate and sodium
chloride correspond to the former case (preferential sorption of
water) and the V_R' value for phenol corresponds to the latter case
(preferential sorption of solute). These results are consistent
with the acidity-basicity criteria for preferential sorption
developed earlier for reverse osmosis separations [21].

In addition to data on retention volumes, the corresponding
data on peak areas with and without the polymer column in the
chromatography apparatus are also of interest, particularly in
ultrafiltration transport. If the solute is either rejected
or weakly adsorbed at the polymer material-solution interface
and if the solute does not undergo any physical change, then one
may expect that the peak area ratio (= peak area with column/
peak area without column) to be unity within experimental error.
But, if the solute is strongly adsorbed on the surface of the
polymer material, or if the solute tends to aggregate to form
either simple agglomerates or pre-gels or gels, the mobility of
the solute through the column will be prevented, retarded or
slowed down resulting in a definite or apparent peak area ratio
of significantly less than unity. Further, if the V_R' value for
the solute is relatively less than that of D_2O, and at the same
time, the peak area ratio is also less than unity, this could
indicate either that the solute is a mixture of several components
some of which are negatively adsorbed and some others of which
are positively adsorbed on the surface of the polymer material,

Table 3. LSC Data on Retention Volumes and Peak Area Ratios for
          Different Solutes with Cellulose Acetate and Polysulfone
          Columns

| Solute | Cellulose acetate | | Polysulfone |
	Retention volume, $V_R^!$ cm^3	Peak area ratio[a]	Peak area ratio[a]
D$_2$O	1.837	1.02	1.01
PEG-6000	1.286	0.90[b]	
BSA	1.068	0.84[b]	
Nonyl alcohol	1.891	0.261	0.03
Cutting oil	0.963	0.533	0.46

[a]Peak area with column/Peak area without column.
[b]There were significant tailings in both cases; data do not
  include area under tailings.

or that the solute, even though negatively adsorbed at the polymer
material-solution interface, has a tendency to form an agglomerate,
pre-gel or gel which lowers or retards its mobility through the
column. These possibilities are illustrated by the set of LSC
data given in Table 3.

    Table 3 gives data on retention volumes and the corresponding
peak area ratios for D$_2$O, PEG-6000, BSA, nonyl alcohol and a
commercial machine cutting-oil with cellulose acetate polymer
material (CA-398-3) in the chromatographic column; just for
comparison, similar data on peak area ratios for nonyl alcohol
and cutting-oil with a polysulfone polymer material in the
chromatographic column, are also given in the table. The
experimental details used in LSC experiments were the same as
those reported earlier [22,24,25].

    With respect to the cellulose acetate polymer material,
Table 3 shows that the $V_R^!$ values for PEG-6000 and BSA are less
than the $V_R^!$ value for D$_2$O, which indicates that water is preferentially adsorbed at the polymer-solution interface in both cases.
The results also show that for the above solutes the peak area
ratios are significantly less than unity. The latter results are
consistent with the possibility of pre-gel or gel formation at
the polymer material-solution boundary phase. This has already
been shown to be the case for the BSA-water-cellulose acetate
membrane system [20], and it is reasonable to expect such pre-gel
or gel formation in PEG-6000-H$_2$O-cellulose acetate membrane
system also at sufficiently high feed concentrations. For nonyl

alcohol, the V_R' value is higher than that of D_2O and the peak
area ratio is very much less than unity; both these results
indicate solute preferential sorption at the polymer material-
solution interface, which is consistent with the conclusion
derived earlier from the analysis of data on reverse osmosis
separations of higher alcohols in dilute aqueous solutions using
cellulose acetate membranes [19]. For cutting-oil, the V_R' value
is lower than that of D_2O and the peak area ratio is also much
less than unity; these results indicate that the cutting-oil used
is a mixture of several components some of which are preferentially
rejected and some others of which are preferentially adsorbed at
the polymer material-solution interface. Since such cutting-oils
usually contain both surfactants and hydrocarbon oils, the above
results are not surprising.

Comparing the data on peak area ratios for nonyl alcohol and
cutting-oil obtained with cellulose acetate and polysulfone
columns, the results show that (a) there is considerable preferential sorption of solute at the polymer-solution interface with
respect to both the solutes and both the polymers, (b) the
preferential sorption of nonyl alcohol is relatively more than the
overall preferential sorption of cutting-oil with respect to
each polymer, and (c) preferential sorption of solute is relatively
stronger for polysulfone than for cellulose acetate with respect
to each solute. It has been shown that preferential sorption
of solute at the membrane-solution interface results in pore-
blocking effect in reverse osmosis [26,27] and the quantity
$1-[(PR)/(PWP)]$ is a measure of such effect. For very dilute feed
solutions or where the osmotic pressure of the feed solution is
negligible, the quantity $1-[(PR)/(PWP)]$ is zero when there is no
pore blocking, and progressively increases towards unity as the
extent of pore blocking increases. A set of ultrafiltration
experiments were carried out with aqueous nonyl alcohol and
cutting-oil feed solutions using a few cellulose acetate and
polysulfone ultrafiltration membranes, and the results obtained
are given in Figs. 5 and 6. The results show that the quantity
$1-[(PR)/(PWP)]$ is (a) greater than zero in all experiments except
those involving cellulose acetate membranes and cutting-oil at
very low feed concentrations, (b) significantly higher for nonyl
alcohol system than for the cutting-oil system with respect to
both the membrane materials, and (c) significantly higher for the
polysulfone membranes than for cellulose acetate membranes with
respect to both the solution systems. These results are
consistent with the conclusions based on LSC data on retention
volumes and peak area ratios.

BASIC TRANSPORT EQUATIONS APPLICABLE FOR CELLULOSE ACETATE
ULTRAFILTRATION MEMBRANES

Just as in reverse osmosis, in ultrafiltration also, the

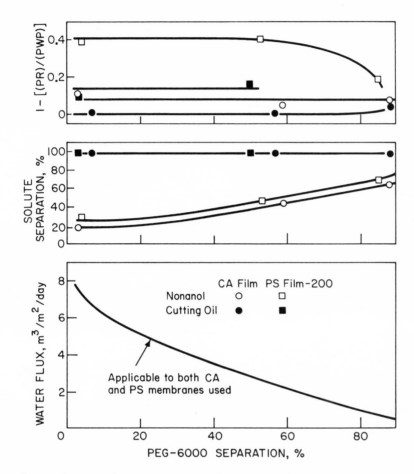

Fig. 5. Relative data on the performance of some cellulose
         acetate and polysulfone ultrafiltration membranes with
         60 ppm nonyl alcohol-water and 70 ppm machine cutting
         oil-water feed solutions. Operating pressure, 344.7 kPa
         gauge (50 psig).

basic equations governing solute and solvent transport through
the membrane must be applicable to all possible levels of solute
separation, and in addition, must satisfy the requirements of
the preferential sorption situation prevailing at the membrane-
solution interface in each case. While several variations in the
latter situation are possible in industrial ultrafiltration, and
all such variations have not yet been adequately studied, some

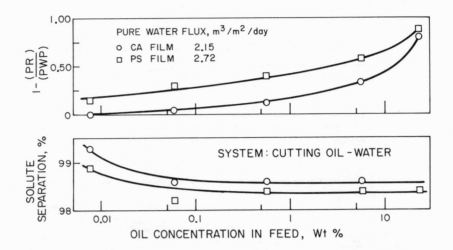

Fig. 6. Relative data on the effect of feed concentration on
         pore blocking effect in cellulose acetate and polysulfone
         membranes with machine cutting oil-water feed solutions.
         Operating pressure, 344.7 kPa gauge (50 psig).

definite statements can still be made with respect to the
following three cases.

    Case 1. When water is preferentially sorbed at the membrane-
solution interface, and there is no solute aggregation or pre-gel
or gel formation at the interface, then the transport of solvent
and solute through the membrane is governed by normal concentration
polarization effects only, and the following basic transport
equations which are applicable to reverse osmosis also apply to
ultrafiltration:

$$N_B = A[P - \pi(X_{A2}) + \pi(X_{A3})] \tag{11}$$

$$= \left(\frac{D_{AM}}{K\delta}\right)\left(\frac{1-X_{A3}}{X_{A3}}\right)(c_2 X_{A2} - c_3 X_{A3}) \tag{12}$$

$$= kc_1(1-X_{A3}) \ \ln\left(\frac{X_{A2}-X_{A3}}{X_{A1}-X_{A3}}\right) \tag{13}$$

From any given set of (PWP), (PR), and f data, one can calculate
the values of A and $(D_{AM}/K\delta)$ and k for the solute at the given
operating conditions; conversely, if the applicable values of
the latter three quantities are given, the values of (PR) and
f can be obtained from the above equations as illustrated
extensively in the literature [4,14,15]. PEG(-6000 to 15000)-

water-cellulose acetate membrane reverse osmosis systems involving low operating pressures (<690 kPa gauge) and low concentrations (<5000 ppm) of PEG solute are examples of ultrafiltration systems for which eq 11 to 13 are applicable [17].

Case 2. When preferential sorption of water and pre-gel or gel polarization occur simultaneously at the membrane-solution interface, both the mobility of water and that of solute through the membrane pores are retarded (because of the physical nature of pre-gels and gels [28]), resulting in a reduction in both water flux and solute flux. Such reduction is equivalent to the condition that the magnitudes of product rate and solute separation in ultrafiltration are controlled by reduced values of A, $D_{AM}/K\delta$ and k. Representing the latter reduced values as A_r, $(D_{AM}/K\delta)_r$ and k_r respectively, the basic transport equations applicable to Case 2 may be written as:

$$N_B = A_r[P-\pi(X_{A2}) + \pi(X_{A3})] \tag{14}$$

$$= \left(\frac{D_{AM}}{K\delta}\right)_r \left(\frac{1-X_{A3}}{X_{A3}}\right) (c_2 X_{A2}-c_3 X_{A3}) \tag{15}$$

$$= k_r c_1 (1-X_{A3}) \ln \left(\frac{X_{A2}-X_{A3}}{X_{A1}-X_{A3}}\right) \tag{16}$$

The values of A_r and A and those of $(D_{AM}/K\delta)_r$ and $D_{AM}/K\delta$ are related by the expressions:

$$1 -\left(\frac{A_r}{A}\right) = a\ X_{A2}^b \tag{17}$$

$$(D_{AM}/K\delta)_r/(D_{AM}/K\delta) = a'\ X_{A2}^{b'} \tag{18}$$

where a, b, a' and b' are constants which depend on the specific nature of the pre-gel or gel material and also the material of the membrane surface involved; in addition the constants a and b are functions of $\ln C_{NaCl}^*$, and the constants a' and b' are independent of $\ln C_{NaCl}^*$. Further,

$$k_r = f(A_r, X_{A1}) \tag{19}$$

which means that k_r is also a function of X_{A2}. This function is best established experimentally for the ultrafiltration apparatus used, using the physical properties (diffusivity and viscosity) of the solution systems involved, as illustrated in reference 17 and 20.

If the nature of pre-gel or gel material at the membrane-solution interface is such that X_{A3} is practically negligible (i.e., solute separation is essentially 100%), then the set of

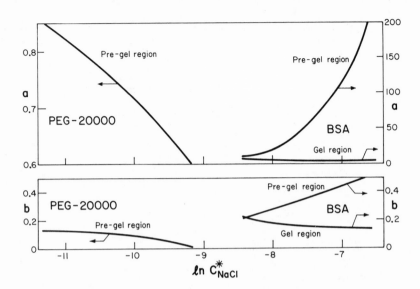

Fig. 7. Values of the constants a and b in eq 17 for PEG-20000-
water-cellulose acetate membrane and BSA-water-cellulose
acetate membrane systems. Data correspond to $X_{A2} \lessgtr$
1.5×10^{-5} for pre-gel region for both systems, and X_{A2}
values between $>1.5 \times 10^{-5}$ and 6.8×10^{-5} for gel region
for the BSA system. Data of Hsieh et al. [17,20].

eq 14 to 16 reduces to the following set of equations:

$$N_B = A_r[P-\pi(X_{A2})] \tag{20}$$

$$= k_r c_1 \; \ln\left(\frac{X_{A2}}{X_{A1}}\right) \tag{21}$$

PEG-20000-water-cellulose acetate membrane and BSA-water-
cellulose acetate membrane systems at low operating pressures
(<690 kPa gauge) and even at low feed concentrations are examples
of ultrafiltration systems for which eq 14 to 21 are applicable
[17,20]. Fig. 7 gives available data on a and b (in eq 17)
as a function of $\ln C^*_{NaCl}$ for the PEG-20000-water-cellulose
acetate membrane, and BSA-water-cellulose acetate membrane systems;
the values of a' and b' (in eq 18) are 0.0216 and -0.22 respectively for the former system, and 3.08×10^{-6} and -0.75 respectively for the latter system [17,20]. The effect of pre-gel
polarization on membrane performance is illustrated in Table 4

Table 4. Effect of Pre-gel Polarization on Membrane Performance

System: 500 ppm BSA-water-cellulose acetate membrane
Operating pressure: 137.9 kPa gauge (20 psig)

Film specifications: A kg-mol H_2O/m^2s kPa ℓn C_{NaCl}^*	Film 1 6.62×10^{-6} -8.25		Film 2 26.8×10^{-6} -6.5	
Basis for membrane performance data	Solute sepn,%	Product rate g/h cm^2	Solute sepn,%	Product rate g/h cm^2
Based on (no pre-gel) A, $D_{AM}/K\delta$, and k	83.0	5.92	3.0	24.
Based on (actual pre-gel) A_r, $(D_{AM}/K\delta)_r$ and k_r	98.1	3.80	93.7	11.7

by a set of calculated data on solute separation and product rate
for two cellulose acetate membranes specified in terms of A and
ℓn C_{NaCl}^*. For the purpose of this illustration, the feed solution
system chosen was 500 ppm BSA-water, the operating pressure chosen
was 137.9 kPa gauge (20 psig), and the values of a and b given
in Fig. 7, and those of k given in Fig. 3 were used in the
calculations. The data on solute separation and product rate
calculated on the basis of A, $D_{AM}/K\delta$ and k are for the hypothetical condition of no pre-gel polarization in the boundary
layer, whereas the corresponding data calculated on the basis of
A_r, $(D_{AM}/K\delta)_r$ and k_r are for the actual condition of pre-gel
polarization in the boundary layer under the experimental
conditions used. The calculated results given in Table 4
quantitatively illustrate the reduction in both solute flux and
solvent flux through the membranes as a result of pre-gel
polarization. Thus pre-gel polarization results in higher solute
separation and lower product rate, both of which are quantitatively
predictable from appropriate transport equations indicated above.

Case 3. This is the case of solute preferential sorption
at the membrane-solution interface. This case has not yet been
studied in detail with reference to any ultrafiltration system.
However limited studies have been made with respect to some
reverse osmosis systems [21,26,27] in which the following
observations have been made:
(a) Solute separation can be positive, negative or zero
     depending on the strength and magnitude of preferential

sorption and the mobility of the preferentially sorbed
species (relative to that of water) through the membrane
pores under the experimental conditions;

(b) At any given operating pressure, solute separation decreases
with increase in operating pressure;

(c) Both pore size and pore size-distribution on the membrane
surface affect solute and solvent flux through the membrane;

(d) Preferential sorption of solute at the membrane-solution
interface results in pore-blocking effect, and the quantity
$1-[(PR)/(PWP)]$ is a quantitative measure of such pore
blocking effect;

(e) The following transport equations are applicable to the
system at any given operating temperature and pressure:

$$X_{A2} = X_{A3} + (X_{A1}-X_{A3}) \exp\left(\frac{N_A+N_B}{ck}\right) \tag{22}$$

$$1 - \frac{(PR)}{(PWP)} = K_1\, X_{A2}^n \tag{23}$$

and

$$N_A = K_2\, X_{A2}^n \tag{24}$$

where K_1, K_2, and n are constants which are characteristics of
the system, and functions of $\ln C_{NaCl}^*$.

All the above observations may be expected to be valid for
the case of solute preferential sorption at the membrane-solution
interface in ultrafiltration also.

CONCLUSION

Though this paper is limited to cellulose acetate membranes
and aqueous feed solutions, neither reverse osmosis, nor ultrafiltration, nor the approach unfolded in this paper are limited
to such systems. The materials science of reverse osmosis
membranes [4] offers unlimited opportunities for developing
appropriate ultrafiltration membranes for a wide variety of
practical applications. With proper choice of reference solution
systems, the membrane specification techniques described in this
paper are extendable to still bigger pore ultrafiltration
membranes. Experimental adsorption and LSC data are valuable
aids for the development of both reverse osmosis and ultrafiltration membranes and processes. The engineering science of
ultrafiltration transport is still a virgin field calling for
extensive and systematic experimentation, correlation and analysis.
In spite of its long history, the field of ultrafiltration is
still underdeveloped. The reverse osmosis approach to ultrafiltration indicated in this paper can contribute significantly
to a fuller understanding and natural development of both reverse

osmosis and ultrafiltration membranes and processes.

NOMENCLATURE

A	= pure water permeability constant, kg-mol H_2O/m^2 s kPa
A_r	= reduced pure water permeability constant, kg-mol H_2O/m^2 s kPa
a, a'	= constants defined by eq 17 and eq 18 respectively
b, b'	= constants defined by eq 17 and eq 18 respectively
$\ln C^*_{NaCl}$	= constant representing pore structure of membrane surface as given by eq 4
c	= molar density of solution, kg-mol/m^3
D_{AB}	= diffusivity of solute in water, m^2/s
$(D_{AB})_{ref}$	= value of D_{AB} for reference solute (PEG-6000), m^2/s
$D_{AM}/K\delta$	= solute transport parameter (treated as a single quantity), m/s
$(D_{AM}/K\delta)_r$	= reduced solute transport parameter, m/s
f	= fraction solute separation
$-\Delta\Delta G/RT$	= polar free energy parameter
Δg	= amount of solute adsorbed per unit weight of cellulose acetate polymer, kg-mol/kg
K_1, K_2	= constants defined by eq 23 and eq 24 respectively
k	= mass transfer coefficient on the high pressure side of membrane, m/s
k_r	= reduced value of k, m/s
k_{ref}	= value of k for the reference solute (PEG-6000) at infinite dilution, m/s
N_A	= flux of solute through membrane, kg-mol/m^2 s
N_B	= flux of solvent water through membrane, kg-mol/m^2 s
n	= constant in eq 23 and eq 24
P	= operating pressure, kPa
(PR)	= membrane permeated product rate for given area of membrane surface, kg/h
(PWP)	= pure water permeation rate for given area of membrane surface, kg/h
S	= effective area of membrane surface, m^2
V'_R	= retention volume in chromatography experiment, cm^3
v_s	= permeation velocity of product solution, m/s
X_A	= mole fraction of solute

GREEK LETTERS

α	= quantity defined by eq 9
Δ	= quantity defined by eq 8
$\ln \Delta^*$	= quantity defined by eq 4 when polar, steric and nonpolar effect are each set equal to zero
$\delta^*\Sigma Es$	= steric parameter
$\pi(X_A)$	= osmotic pressure corresponding to mole fraction X_A of solute, kPa
ρ	= solution density, kg/m^3

$\omega^* \Sigma s^*$ =   nonpolar parameter

SUBSCRIPTS

1   = bulk feed solution on the high pressure side of
            membrane
2   = concentrated boundary solution on the high pressure
            side of membrane
3   = membrane permeated product solution on the atmospheric
            pressure side of membrane

REFERENCES

1. A.S. Michaels, Ultrafiltration, in: "Progress in Separation
    and Purification," Vol. 1, E.S. Perry, ed., Wiley-
    Interscience, New York (1968).
2. C.J. van Oss, Ultrafiltration Membranes, in: "Progress in
    Separation and Purification," Vol. 3, E.S. Perry and C.J.
    van Oss, eds., Wiley-Interscience, New York (1970).
3. W.F. Blatt, A. Dravid, A.S. Michaels and L. Nelson, Solute
    Polarization and Cake Formation in Membrane Ultrafiltration:
    Causes, Consequences, and Control Techniques, in:
    "Membrane Science and Technology," J.E. Flinn, ed., Plenum,
    New York (1970).
4. S. Sourirajan, Pure Appl. Chem., 50 (7), 593 (1978).
5. S. Sourirajan and B. Kunst, Cellulose Acetate and Other
    Cellulose Ester Membranes, in: "Reverse Osmosis and
    Synthetic Membranes," S. Sourirajan, ed., National Research
    Council Canada, Ottawa (1977).
6. B. Kunst and S. Sourirajan, J. Appl. Polym. Sci., 18, 3423
    (1974).
7. O. Kutowy and S. Sourirajan, J. Appl. Polym. Sci., 19, 1449
    (1975).
8. O. Kutowy, W.L. Thayer and S. Sourirajan, Desalination, 26,
    195 (1978).
9. T.A. Tweddle and S. Sourirajan, J. Appl. Polym. Sci., 22,
    2265 (1978).
10. G.R. Gildert, T. Matsuura and S. Sourirajan, J. Appl. Polym.
    Sci., 24, 305 (1979).
11. O. Kutowy, W.L. Thayer and S. Sourirajan, U.S. Patent 4, 145,
    295 dated March 20, 1979.
12. W.L. Thayer, L. Pageau and S. Sourirajan, Desalination, 21,
    209 (1977).
13. O. Kutowy, W.L. Thayer, C.E. Capes and S. Sourirajan, J. Sep.
    Process Tech. (submitted).
14. S. Sourirajan, "Reverse Osmosis," Chapter 3, Academic, New
    York (1970).
15. S. Sourirajan and T. Matsuura, A Fundamental Approach to
    Application of Reverse Osmosis for Water Pollution Control,
    in: Proc. EPA Symposium on Textile Industry Technology,

Williamsburg, Dec. 5-8, 1978. (EPA-600/2-79-104, May 1979)

16. F. Hsieh, T. Matsuura and S. Sourirajan, J. Appl. Polym. Sci., 23, 561 (1979).

17. F. Hsieh, T. Matsuura and S. Sourirajan, Ind. Eng. Chem. Process Des. Dev., 18, 414 (1979).

18. T. Matsuura, J.M. Dickson and S. Sourirajan, Ind. Eng. Chem. Process Des. Dev., 15, 149 (1976).

19. T. Matsuura, A.G. Baxter and S. Sourirajan, Ind. Eng. Chem. Process Des. Dev., 16, 82 (1977).

20. F. Hsieh, T. Matsuura and S. Sourirajan, J. Sep. Process Tech., (in press) (1979).

21. S. Sourirajan and T. Matsuura, Physicochemical Criteria for Reverse Osmosis Separations, in: "Reverse Osmosis and Synthetic Membranes," S. Sourirajan, ed., National Research Council Canada, Ottawa (1977).

22. T. Matsuura and S. Sourirajan, J. Coll. Int. Sci., 66, 589 (1978).

23. T. Peters, Jr., Serum Albumin, in: "The Plasma Proteins," 2nd ed., Vol. 1, F.W. Putman, ed., Academic, New York (1975).

24. T. Matsuura, P. Blais and S. Sourirajan, J. Appl. Polym. Sci., 20, 1515 (1976).

25. T. Matsuura and S. Sourirajan, Ind. Eng. Chem. Process Des. Dev., 17, 419 (1978).

26. T. Matsuura and S. Sourirajan, J. Appl. Polym. Sci., 16, 2531 (1972).

27. J.M. Dickson, T. Matsuura and S. Sourirajan, Ind. Eng. Chem. Process Des. Dev., (in press), (1979).

28. J.W. McBain, "Colloid Science," Chapter 11, D.C. Heath, Boston (1950).

CHEMICAL AND MORPHOLOGICAL EFFECTS ON SOLUTE DIFFUSION

THROUGH BLOCK COPOLYMER MEMBRANES

Yatin B. Thakore, Dien-Feng Shieh and Donald J. Lyman

Department of Materials Science and Engineering
University of Utah
Salt Lake City, Utah 84112

INTRODUCTION

The rate of aqueous diffusion of a solute through Cuprophane and other cellulosic membranes is controlled primarily on the basis of its molecular size, with the chemical structure of the solute having little influence on the diffusion rate. Stretching or chemical etching of cellulosic membranes can increase their permeabilities to solutes (1-3), but only at the expense of decreasing the limited selectivity of the transport of the smaller solutes (3). A study of block copolymer membranes was undertaken as a possible route to develop membranes which might show selectivity for both the molecular size and chemical structure of the diffusing solute. These block copolymer systems would be composed of suitable hydrophilic blocks to influence the aqueous permeability properties, and hydrophobic blocks to give the necessary mechanical strength to the membranes. Since solute diffusion would be through an aqueous solution of the hydrophilic block rather than just through water, one would expect that the interactions between the solute, water, and hydrophilic block would influence the rate of diffusion. Preliminary studies on one such block copolymer system indicated that the "partitioning" or selective solubility of a solute in the aqueous hydrophilic block solution could account for differences in diffusion rates of various solutes (4). However, the diffusion behavior of other solutes could not be adequately explained by this mechanism. In addition, the development of block copolyether-urethane membranes showed that the hydrophobic segment also appeared to contribute to selective diffusion (5,6). This report is part of a continuing study to determine the mechanism of aqueous diffusion of solutes through block copolymer membranes.

EXPERIMENTAL

The block copolyether-urethane membranes were synthesized by a modified solution polymerization technique described in detail previously (5,7). Polyoxyethylene glycol (mol wt 1570 daltons) was end-capped with methylene bis(4-phenyl isocyanate) in a dimethyl sulfoxide/4-methylpentanone-2 (50:50) mixed solvent system, then further reacted with additional methylene bis(4-phenyl isocyanate) and the appropriate monomeric diol to form the block copolyether-urethane. The polymer was isolated from its solution by precipitation in deionized water. The diols used in this study and the inherent viscosities of the block copolyether-urethanes are shown in Table I.

The block copolyether-ester was prepared by a melt poly-condensation technique (8). Polyoxyethylene glycol (30 mole per cent, mol wt 1401 daltons), ethylene glycol and dimethyl terephthalate with calcium acetate/antimony trioxide catalyst were reacted to give the copolymer. Inherent viscosity of the block copolyether-ester was 1.05, in 1,1,2,2-tetrachloroethane/phenol (40:60) at 30°C. (0.5% concn).

Table I.   Structure and Inherent Viscosities
           of Block Copolyether-urethanes

$$\{CNH\!\!-\!\!\bigcirc\!\!-\!\!CH_2\!\!-\!\!\bigcirc\ NHCO(CH_2CH_2O)_y\!\!-\!\!(CNH\!\!-\!\!\bigcirc\!\!-\!\!CH_2\!\!-\!\!\bigcirc\ NHCORO)_x\}_n^{a,b}$$

Diol (HOROH)	$\eta_{inh.}$ [c]
1,6 hexane diol	0.43
1,10 decane diol	0.33
α,α'-dihydroxy-p-xylene diol	0.39
Trans - 1,4 cyclohexane diol	0.37
Cis/Trans 1,4 cyclohexane diol	0.42

a.   The macroglycol was polyoxyethylene glycol
     (carbowax 1540, Union Carbide).

b.   Mole ratio of diol (HOROH) to polyoxyethylene
     glycol was 80:20.

c.   Inherent Viscosity in N,N-dimethyl formamide at
     30°C, concn 0.5 g/100 ml.

Films of the copolyether-ester and copolyether-urethanes were made by solution casting techniques. The solvents were dichloro-methane for the copolyether-ester and N,N-dimethyl formamide for the copolyether-urethane materials. Filtered solutions (15 to 20% solids by weight) were cast on glass plates with a calibrated casting knife. The copolyether-ester solutions were air dried; the copolyether-urethanes solutions were dried in a forced draft oven at 70°C for 45 minutes. The membranes were then soaked in deionized water to float them free from the glass casting plates.

Cuprophane membranes were obtained from the Sweden Freezer Co., (Artificial Kidney Division).

The hydrophilic block model compound was synthesized by end-capping polyoxyethylene glycol (mol wt 1000 daltons) with benzoyl chloride.

The relative half-time rates of transport, $T_{\frac{1}{2}}$, for different solutes through these membranes were determined using the Lyman Half-Time Dialysis Cell described elsewhere (5,8). The concentra-tions of the solutes were determined from concentration versus refractive index curves using a Brice-Phoenix Refractometer.

The viscosity measurements to determine the interactions between solute, water, and hydrophilic block were made at 30°C using an Ubbelohde viscometer. The solutions were made with the following concentrations: Solution I (10.0g block model compound, 20 ml water, 0 to 14.0g urea); Solution IA (5.0g block model compound, 20 ml water, and 0 to 14.0g urea); Solution IB (2.0g block model compound, 20 ml water, 0 to 5.0g urea); Solution II (10.0g block model compound, 20 ml water, and 0 to 5.0g sucrose).

Dynamic mechanical properties of the solution cast polyether-urethane and polyether-ester films were measured using a Rheovibron (Model DDV II, Toyo Measuring Co., Japan).

The glass transition temperatures were measured using a Perkin-Elmer Differential Scanning Calorimeter (Model II).

Urea and sucrose (Analytical Reagent grade, Mallinckrodt), α-alanine (Nutritional Biochemicals) and β-alanine (Eastman Kodak) were used as received.

RESULTS AND DISCUSSIONS

An important need exists for the development of membranes that control solute diffusion on the basis of both the size and the chemical structure of the solute. Our initial attempt at this was to design "partition membranes"; i.e., membranes which would

take advantage of differences in the partitioning, or solubility
of a solute between water and the aqueous/organic phase.   In
designing the molecular structure of a synthetic polymer to function
as a partition membrane, one is faced with the problem of combining
distinct sets of properties that are often incompatible with each
other: mechanical strength, water swellability, and chemical
stability.

     Block copolymers appeared to offer an ideal solution to this
problem.   One of the blocks would be hydrophilic to impart water
swelling and the other would be hydrophobic to impart mechanical
strength to the membrane.   The first block copolymer of this type
to be made was a block copolyether-ester based on polyoxyethylene
glycol and polyethylene terphthalate (8).   Membranes of this material
were characterized for half-time rates of transfer, $T_{1/2}$, and compared
to Cuprophane.   While Cuprophane membranes showed $T_{1/2}$ values that were
directly proportional to the molecular weight of the diffusing solute
(molecular weight being a rough measure of molecular size), there
was no apparent correlation between $T_{1/2}$ and molecular weight for the
block copolyether-ester membranes.   (Fig. 1 and Table II).   It was

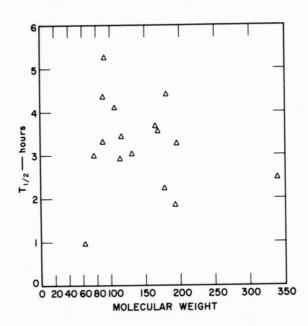

Fig. 1.   The effect of molecular weight on $T_{1/2}$ of various solutes
         through a copolyether-ester membrane.

Table II. $T_{1/2}$ for various solutes through Cuprophane and a block copolyether-ester membrane[a]

Compound	Mol Wt (daltons)	Membranes	
		Copolyether-ester[b] $T_{1/2}$(min)	Cuprophane $T_{1/2}$ (min)
Urea	60.8	58	68
α-Alanine	89.1	201	120
β-Alanine	89.1	317	120
Sarcosine	89.1	262	132
Glucose	180	268	223
Sucrose	342	397	270
Bacitracin	1411	568	720

a. Selected data from ref. 3.
b. Based on 30 mole percent polyoxyethylene glycol, molecular weight 1540.

of interest to note that the three solutes α-alanine, β-alanine and sarcosine (which have identical molecular weights but different chemical structure), while showing the similar $T_{1/2}$ values for the Cuprophane membrane, showed distinctly different $T_{1/2}$ values for the block copolyether-ester membranes. Determination of distribution coefficients for the various solutes between water and the water swollen block copolyether-ester membrane, indicated that in some cases, this unique separation possible with these membranes could be explained on the basis of partitioning. For example, α-alanine and β-alanine, which dialysed through Cuprophane in an identical manner could be separated using the block copolyether-ester membrane. α-Alanine ($T_{1/2}$ of 201 minutes) had a distribution coefficient of 0.197 while β-alanine ($T_{1/2}$ of 317 minutes) had a distribution coefficient of 0.099 (4). However, with other solutes the role of solubility in effecting diffusion was not as conclusive as with the alanines. For example, urea permeability in the block copolyether-ester was higher than in Cuprophane; in contrast, sucrose permeability in the block copolyether-ester was less than in Cuprophane. Yet, the solubilities of urea and sucrose were nearly equivalent in the copolyether-ester membrane.

To gain an insight into the nature of the solute movement through these water swollen hydrophilic block aggregations, the effect of added solutes on the viscosities of aqueous solutions of

a compound modeling the hydrophilic block were measured. Polyoxyethylene dibenzoate was the model hydrophilic block used in these
studies. The ratio of water to model block segment was similar to
that found within the membranes after equilibration with large
amounts of water. Since the polyoxyethylene glycol molecules show
little conformational change on dissolution in water (9), these
aqueous solutions of model segments should effectively represent the
channels through which the solute moves while diffusing through the
membrane. Varying amounts of urea or sucrose were added and the
viscosities determined at 30°C. The data is shown in Figure 2.
When sucrose is added to the aqueous model segment solution (see
curve II), the solution viscosity increases with increasing amounts
of sucrose. In contrast, when urea is added the solution viscosity
decreases (see curve I) until the molar amount of urea is nearly
equivalent to the molar number of repeat units of $\{CH_2CH_2O\}$ present
in the model segment. Further additions of urea result in a slight
increase in the solution viscosity.

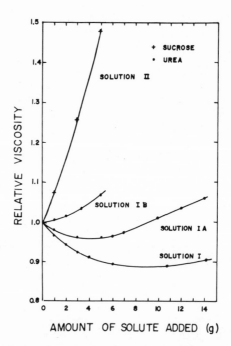

Fig. 2.   Relative viscosity of aqueous solutions of hydrophilic
          block model containing varying amounts of urea or sucrose
          (30°C). Solution compositions defined in Experimental.

These changes in solution viscosity of the membrane channel can be explained by changes in the structure of the water due to the presence of solute and hydrophilic block segments. Water is considered to form clusters of hydrogen bonded molecules in dynamic equilibrium with non-hydrogen bonded molecules (10,11). The degree of this clustering is affected by the presence of certain solutes. From the reported thermodynamic properties of aqueous polyoxyethylene glycol solutions (12), it would appear that an aqueous solution of the hydrophilic block model compound should show a highly ordered structure. When a water structure-making solute, such as sucrose (13) is added to the aqueous block model compound solution, the viscosity increases further. If this occurs in the channel through the membrane, the passage of sucrose through this membrane would be hindered and this is indeed observed. In contrast, urea is a known water structure-breaking type of solute (14,15) and additions of urea to the aqueous solution of the hydrophilic block model compound results in a decrease in solution viscosity. In this case, a reduced viscosity in the channel would result in an accelerated diffusion rate for urea through the membrane. Again, this is observed.

Further evidence for this water structure effect on viscosity is shown when the concentration of hydrophilic block model compound is decreased (Curves IA and IB). There is a progressively less pronounced effect on viscosity with the addition of the solute. Indeed, for Curve IB which represents a very dilute solution of the model compound, there is an almost normal increase of viscosity with increasing concentration when urea is added.

While these studies on the block copolyether-ester membranes were in progress, it was decided to prepare and characterize block copolyether-urethane membranes since they would have greater hydrolytic stability. A series of block copolyether-urethanes based on polyoxyethylene glycol were synthesized (5) in which the hydrophobic block was varied by using different coupling diols, to obtain varying mechanical properties. However, when the $T_{\frac{1}{2}}$ values were determined for various solutes through these membranes, striking differences in their permeability properties were observed (5,6) in which the nature of the hydrophobic block appeared to influence solute transport. To further study these effects we have resynthesized several of these copolyether-urethanes (Table I). The differences in the $T_{\frac{1}{2}}$ values for selected solutes are tabulated in Table III and are dramatically illustrated in Figure 3. As seen in Figure 3, changing the diol coupler in the hydrophobic segment from 1,6 hexane diol, to α, α'-dihydroxy-p-xylene diol, to 1,4 cyclohexane diol has only slight effect on urea transport through the corresponding membrane. However, the transport of α-alanine, β-alanine, and sucrose changes dramatically. For example, the diffusion of the alanines through membranes II or III are similar, yet by changing the diol from trans 1,4 cyclohexane diol to the

Table III.   T½ values for selected solutes through several
copolyether-urethane membranes.

Solute	Molecular Weight (Daltons)	Diol Segment			
		Cis/Trans 1,4 Cyclo hexane diol (I)	1,6 Hexane diol (II)	Trans 1,4 Cyclo hexane diol (III)	α,α'- Dihydroxy p-xylene diol (IV)
Urea	60.8	106	122	140	252
α-Alanine	89.1	811	606	364	791
β-Alanine	89.1	1095	592	398	1271
Sucrose	342.30	1640	1966	1282	1442

cis/trans mixture or the 1,6 hexane diol to α,α' dihydroxy p-
xylene, large differences in T½ times are observed between α- and
β-alanines. Thus it appears as though the hydrophobic segment can
contribute not only to the membrane's mechanical strength but to
its diffusion properties as well, thus giving added control over
the preparation of selective diffusion membranes.

These multiple block copolymers are known to form domain-matrix
structures because of the incompatibility of the two different block
segments making up the polymer chain (16). The completeness of
this microphase separation is dependent on the segment size and
the segment chemical structure. Thus one would expect that the
changes in diffusion noted above, as the hydrophobic segment is
changed, result from differences in the composition, or purity of
the hydrophilic phase.

The preliminary results of our dynamic modulus measurements
on these block copolyether-urethanes do indicate an incomplete
microphase separation (see Figures 4-6). Comparison of these
curves to those described by Bonart (17) for varying degrees of
completeness of phase separation, would lead one to estimate that:
(a) the two phases are highly mixed with the trans 1,4 cyclohexane
diol material; (b) there is slight separation of the phases with the
1,6 hexane diol material; and (c) there is somewhat more separation
with the cis/trans 1,4 cyclohexane diol material. Differential
Scanning Calorimetry data on these polymer gives support to this
phase mixing. This trend in phase separation does parallel the
diffusion data on the alanines, i.e. the two materials showing
greater phase mixing, show similar T½ values for α- and β-alanines
(Membranes II and III in Figure 3). When phase mixing decreases

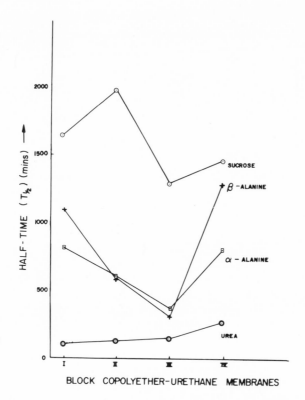

Fig. 3.   Effect of hydrophobic block structure on T½ of selected
          solutes for block copolyether-urethane membranes.
          Membrane I is based on cis/trans 1,4 cyclohexane diol,
          II on 1,6 hexane diol, III on trans 1,4 cyclohexane diol
          and IV on α,α' dihydroxy-p-xylene diol.

(membrane I) the T½ values for β-alanine is larger than for
α-alanine.

CONCLUSION

     Block copolymer membranes offer new potential for selective
diffusion of solutes on the basis of both molecular size and
chemical structure. The chemical nature of the hydrophilic block
and its association with water and the solute can influence solute
diffusion by (a) solute solubility effects; and (b) solution
viscosity effects. When the microphase separation is not complete,
the mixing of the hydrophobic block in the matrix phase can also
influence solute diffusion.

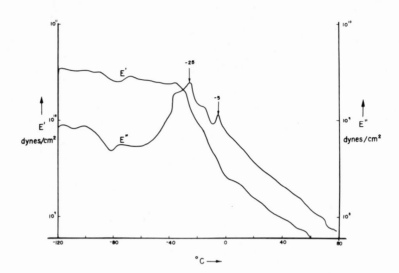

Fig. 4.   Dynamic Mechanical Modulus of Copolyether-urethane
          from 1,6 hexane diol (110 Hz).

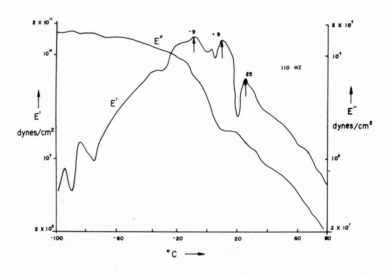

Fig. 5.   Dynamic Mechanical Modulus of Copolyether-urethane from
          cis/trans 1,4 cyclohexane diol (110 Hz).

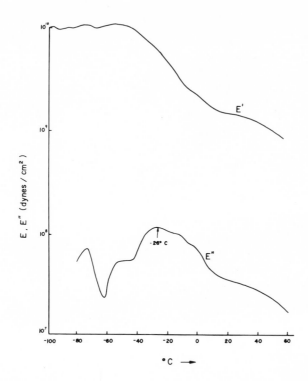

Fig. 6. Dynamic Mechanical Modulus of Copolyether-urethane
from trans 1,4 cyclohexane diol (110 Hz).

ACKNOWLEDGEMENT

This work was supported by the National Science Foundation,
Grant DMR 76-83681, Polymer Program.

REFERENCES

1. L. C. Craig and W. Konigsberg, J. Phys. Chem., 65,
    166 (1961).
2. C. E. Sachs and J. L. Funck-Brentano, Trans. Amer.
    Soc. Art. Int. Org., 9, 79 (1963).
3. D. J. Lyman, Proc. 3rd Int. Congr. Nephrol., 3, .
    265 (1966).
4. D. J. Lyman, B. H. Loo and W. M. Muir, Trans. Amer.
    Soc. Artif. Int. Organs, 11, 91 (1965).

5. D. J. Lyman and B. H. Loo, J. Biomed. Mater. Res.,
    1, 17 (1967).
6. D. J. Lyman, Proc. Fourth Conf. European Dialysis
    and Transplant Assoc., 4, 98 (1967).
7. D. J. Lyman, J. Polymer Sci., 45, 49 (1960).
8. D. J. Lyman, B. H. Loo and R. W. Crawford,
    Biochemistry, 3, 985 (1964).
9. J. L. Koenig and A. C. Angood, J. Polym. Sci: A-2,
    8, 1787 (1970).
10. H. Frank and W. Wen, Discuss. Farad. Soc., 24,
    133 (1957).
11. G. Nemethy and H. Scheraga, J. Chem. Phys., 36,
    3382 (1962).
12. M. L. Lakhanpal, K. S. Chhina and S. C. Sharma,
 • Indian J. Chem., 6, 505 (1968).
13. I. M. Klotz, Federation Proc., 24, S-24 (1965).
14. H. Frank, Ann. N. Y. Acad. Sci., 125, 737 (1965).
15. J. A. Rupley, J. Phys. Chem., 68, 2002 (1964).
16. D. C. Allport and W. H. Janes (eds.) "Block Copolymers",
    John Wiley and Sons, N. Y. (1973).
17. R. L. Bonart, L. Morbitzer and H. Rinke, Kolloid-Z,
    240, 807 (1970).

PRACTICAL ASPECTS IN THE DEVELOPMENT OF A

POLYMER MATRIX FOR ULTRAFILTRATION

Israel Cabasso

Gulf South Research Institute
P.O. Box 26518
New Orleans, LA   70186

Introduction

The wide acceptance of the ultrafiltration process as an integral mode of separation technology in the industrial and medical
complexes is a result of successive developments of the following
elements –
1) polymeric membranes (materials and morphology)
2) varied membrane configurations (flat sheet, tube and
   hollow fibers)
3) membrane devices (permeators)

The following discussion focuses on several aspects of the
first two elements, polymeric membranes and their configurations;
the third element related more to the basic engineering concepts
and fluid mechanics of specific ultrafiltration problems, and makes
practical use of the most suitable components available for specific
tasks.

In order to narrow the scope of the present discussion, we will
define an ultrafiltration membrane (UFM) as an anisotropic film
layer, having a tight (but porous) upper surface extending from
a highly porous, spongy substructure. The bottom face is also
highly porous (preferably a web-like structure), and is – unlike
the upper face – an integral extension of the membrane substructure.
The above definition of the surface of a UFM is illustrated in
Figure 1.

The polymer in this anisotropic structure (often termed "asymmetric") serves in two capacities: the upper surface ("skin") is
the permaselective barrier, and the substructure is its mechanical
support. Unlike the reverse osmosis membrane employed for water
desalination, the skin of a UFM permits viscous flow. Thus,
back-flushing a fouled UFM by reversing the direction of flow

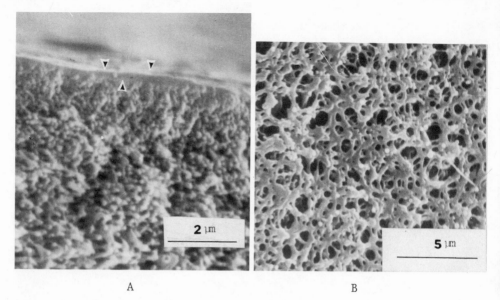

A   B

Figure 1: A modified PPO ultrafiltration membrane. A. Cross-
          section showing a tight skin resting on a spongy substruc-
          ture. B. Bottom face showing a highly porous, web-like
          structure.

through manipulation of hydraulic pressure is a common practice for
rejuvenation of plugged and fouled UFM units [as illustrated by
Breslau et al (1)].

     The properties of UFMs with respect to their structure should
be divided into: a) skin characteristics - pore density, dimensions, orientation and shape, and skin topography (e.g., roughness),
b) substructure - porosity, channel tortuosity, and macrovoids
(quantity, geometry and porosity), and c) bottom face - openings
(in the web-like structure or in a bottom porous skin).

     The chemical nature of a UFM should be identified with respect
to solvent and chemical resistance (pH, oxidizing agents, etc.),
surface activity (affinity for Lewis bases or acids) and its
wettability (hydrophilic/hydrophobic balance). Regarding the
membrane permaselectivity, the surface pore shape and chemical
entity are of major influence. For the UFM defined above, this
chemical entity may often be visualized as a sphere having the pore's
surface radius and center. The chemical entity of the skin pores
would exert decreased influence over membrane permaselectivity as
pore radii increase . Therefore, and for practical purposes, one
may conclude that permaselectivity in ultrafiltration membranes is
a result of pore/solute interaction, while permaselectivity of
reverse osmosis membranes (electrically neutral) may be assigned to
the (membrane) material/permeate interaction.

The above divisions provide simple criteria for the selection of materials from which UFMs may be constructed. As such, and by reviewing the family of commercially available polymers, one can conclude that initially, every material which may be fabricated to yield anisotropic morphology (Figure 1) having a neutral membrane surface with a pore size of $\sim$ 5 nm and more, would exhibit more or less the same permaselectivity when employed as a UFM for aqueous solution. Thus, cellulose esters, polysulfone, poly(vinylidene fluoride), poly(dimethylphenylene oxide), poly(acrylonitrile) and others which can be easily solution-cast (or spun) to provide a UFM via phase inversion process (commonly coagulated in water) can be "tailored" to yield the same transport and permaselectivity.

Hydrophilicity

The most practical division of these membrane materials is that which differentiates between the hydrophilic membranes and the hydro-phobic membranes. The precise division may be based on the material/ aqueous contact angle. Practical circumstances often dictate that dry UFMs not allowing convection flow of water at pressures above atmospheric pressure would be defined as hydrophobic.

The "break-through" pressure which produces a continuous liquid capillary (and hence, yields hydraulic convection flows whenever hydraulic pressures exceed the atmospheric pressure) is often far above the ultrafiltration operating pressure (Figure 2). There-fore, such membranes as polysulfone, PPO, and PVF_2, which exhibit superb mechanical strength and chemical stability, must either be stored wet, or must be manufactured with "wetting agents" (e.g., glycerine, "Triton," sodium dodecyl sulphate, etc.). Once in operation, these membranes cannot be subjected to wet-dry cycle if the break-through pressure exceeds the upper pressures allowed (or available) in the ultrafiltration system. Thus, for example, dry sterilization of permeators containing such membranes is not practiced. Unfortunately, the majority of the hydrophilic membranes are based on cellulose esters, which undergo a drastic, irreversible morphological alteration upon drying - generally, a collapse of the fine region of spongy substructure following a simultaneous change in the degree of crystalinity of the matrix.

The above problems, interestingly enough (and unfortunately), group the hydrophobic and hydrophilic membranes into the same class when installed in a unit operation, that is, membranes which should not be allowed to dry. This encompasses almost all the commercially available polymeric materials that are prepared via the phase inversion process (i.e., solution-cast and water coagulated to form a UFM). The solution to this problem has been recognized as utilizing a polymer consisting of a hydrophobic, stiff, skeleton accommodating pendant hydrophilic groups. The hydrophilicity of such polymers (a result of pendant group/water interaction) should be limited to 20 - 30 wt% water in a water-equilibrated polymer.

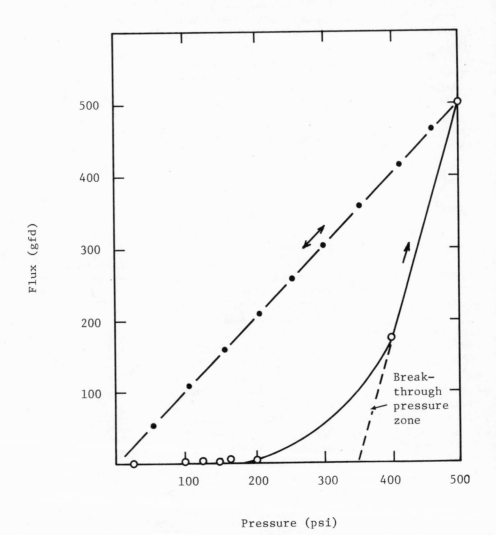

Figure 2:   Flux vs hydraulic pressure for a hydrophobic M-PPO
            ultrafiltration membrane (shown in Figure 1) with a
            cut-off of 800 - 1,200 MW. The dry membranes have a
            pressure break-through zone of 200 - 350 psi, and
            tolerate operating temperatures of up to 200°C.
            However, they should not be dried in an ultrafiltration
            system that is not equipped with a proper pumping
            capacity.

As such, one must define a hydrophilic pendant group in relation to the pertinent polymeric chain units, for example:

$$- CH - CH_2 -$$
$$\underset{X}{|}$$

In the present context, a hydroxyl group, for example, would not be
considered an adequate candidate, since it imparts hydrophilicity
to the polymeric backbone, that is, binding water molecules near
the main chain, thus shielding the ethylene and the phenylene
chains, which eventually causes gross plasticization of the matrix.
However, if the x unit is substituted, for example, by the following
phosphonic acid derivatives:

$$-CH_2-\underset{\underset{CH_3}{|}}{\overset{\overset{CH_3}{|}}{C}}-\overset{\overset{O}{\uparrow}}{P}-(OH)_2$$

and y is substituted by: $CH_2 - \overset{\overset{O}{\uparrow}}{P} - (OH)_2$,

the nature of the hydrophobic backbone will be retained, and both
the degree of swelling and the freedom of segmental motion would
be limited in the presence of imbibed water and could be controlled
by varying the extent of pendant group substitution. In order to
impart this type of hydrophilicity to a UFM, two approaches may be
taken: a) react the raw polymeric material with a hydrophilic pendant group, then cast the modified polymer into a membrane, and
b) graft onto the ready-made hydrophobic UFM a hydrophilic species.

Both methods deal with polymer modification; in the latter,
surface modification of the UFM matrix requires continuous bridging
(that is, the surface of the pore requires high density of the
hydrophilic moieties), as exposed hydrophobic domains are a hindrance
to the aqueous flow. In this case, depending on the nature of the
grafted moiety, the grafting on the hydrophobic pore should penetrate
into the polymer matrix as deep as 2 - 3 nm. Surface grafting (no
penetration at all) results in a very short durability. The groups,
in such instances, are either "swallowed" into the matrix, or are
hydrolyzed quickly as a result of the hydrophobic surface repulsion.

The degree of hydrophilic substitution onto a raw polymer - to
be cast as a UFM - is, in general, much less than the substitution
required for the ready-made hydrophobic membranes. Upon coagulation

of the modified polymer solution, the polymer chains rearrange so
as to have the polar and hydrogen bonding groups facing a water
phase while the more hydrophobic portions of macromolecules are
clustered under the surface.

In general, chemical modification of a hydrophobic UFM is hard
to achieve without damage to the matrix. Drastic measures, such as
radiation grafting have not yet been proven feasible and often result
in degradation products being continually leached into the ultrafiltration products. The treatment of ready-made UFMs is a preferred
method when a specific hydrophilic active ingredient with a strong
affinity for the polymer can interact with the matrix. As such,
gas phase sulfonation or halomethylation of polysulfone and PPO
membranes following hydrolyzation or surface reaction of the halo-
methylated moiety is very effective, and is partially discussed
below.   A method of instant sulfonation of the surface of a polysulfone membrane was described for polysulfone hollow fibers (2),
where the hydrophobic, water saturated fiber was rolled through a
sequence of baths containing sulphuric acid (which contains 7 - 8%
sulphur trioxide) and water, thus yielding a hydrophilic, fouling
resistant skin of PS - SO_3H .

Polymer Blends

Alteration of the hydrophilic/hydrophobic balance (as well as
other properties) of a membrane system can easily be accomplished
if the membrane is prepared from a multi-component polymer mixture
(i.e., a polymer blend). This method has not yet been fully
explored, due to the fact that most polymers are thermodynamically
incompatible. Most neutral polymers - in solution or in bulk -
when mixed, exhibit positive enthalpy of mixing with no significant
gain in the entropy of mixing. As such, no gain in the free energy
of mixing is obtained, and the components phase separate - i.e.,

$$\Delta G^m = \Delta H^m - T\Delta S^m \ , \ \Delta H > 0 \ , \ \text{and} \ T\Delta S \to 0$$, as increasingly high
molecular weight polymers are employed   (3).

In the case of polyelectrolyte complexes (i.e., ionic polymer
blends composed of negatively and positively charged macromolecules),
polymer miscibility is likely, due to ionic interaction (and entropy
gain due to the release of small counter ions to solution). Indeed,
such systems are utilized in the fabrication of UFMs with diverse
properties, and are well documented in the literature (4).

In recent years, a variety of nonionic, compatible polymer
blends suitable for membrane fabrication were identified. The
compatibility of these blends is usually due to specific interactions,
such as hydrogen bonding and polar-induced polar interaction. The

relevant examples hitherto are blend membranes containing cellulose acetate with: cellulose triacetate (5), polystyrene and polyphenylene oxide phosphonate esters (PPN) (6), and poly(4-vinylpyridine), recently reported by Aptel and Cabasso (7).

The formation of alloy membranes consisting of cellulose acetate and electron donor aromatic macromolecules was thoroughly investigated in our laboratories. The following macromolecules (compatible with cellulose acetate) were identified as candidates for the formation of an anisotropic ultrafiltration alloy membrane:

$$\sim CH-CH_2 \sim \qquad\qquad CH_2PO(OR)_2 \qquad\qquad \sim CH-CH_2 \sim$$

$$CH_2PO(OR)_2 \qquad\qquad Br \quad CH_3$$

The alloy membranes cast from solution mixtures of these hydrophilic polymers with cellulose acetate are prepared by the conventional phase inversion process, and are coagulated in an aqueous bath. The membranes can then be dried, since the spongy substructure of the alloy membrane will not collapse upon drying. However, all of these hydrophilic alloy membranes (which absorb 15 - 25 wt% water when equilibrated in an aqueous solution) exhibit contraction in the skin zone when first dried (Figure 3). This

Figure 3: Cross-section of the skin of a poly(4-vinylpyridine)/cellulose acetate alloy anisotropic membrane. Arrows designate the skin contraction zone after the first drying cycle (7).

contraction amounts to an insignificant loss in membrane surface
area (< 1%), which can be further controlled by changing the composition ratio of the components in the casting solution. The
membranes are durable in a wet-dry ultrafiltration operation; however,
permaselectivity increases (molecular weight cut-off decreases) upon
the first drying cycle, due to irreversible skin contraction at this
stage.

The chemical properties of such membranes may also be altered
by changing the polymer composition ratio. For example, an alloy
membrane composed of cellulose acetate/poly(phenylene oxide dimethyl
phosphonate) (1:1 w/w) can tolerate pH operating conditions in the
range of 10 - 10.7, which is higher than the tolerance level of
cellulose acetate (pH < 8) (8). In addition, chemical alteration
of at least one of the polymer components in the membrane may often
be possible. In an alloy membrane of poly(4-vinylpyridine)/cel-
lulose acetate, the poly(vinylpyridine) can readily be quaternized
by organic Lewis acids to yield an ion-exchange UFM (7).

Table 1: Ultrafiltration of PEG*-1500 with
Cellulose Acetate/Poly(4-vinylpyridine) Alloy Membranes
Aptel (9).

	CA	P_4VPy-CA 5/95	P_4VPy-CA 10/90
Flux (ℓ/d.m^2 atm)	750	115	50
Rejection (%)	35	90	99

* poly(ethylene glycol)

The introduction of an additional compatible polymer to a known
casting formulation (substituting a fraction of the original polymer
with the added polymer) produces a gross effect on the matrix formation and skin semipermeability. For example, the introduction of
a small fraction of poly(4-vinylpyridine) to a known UFM formulation
of cellulose acetate (cast from acetone/formamide into an aqueous
coagulation bath) has an extensive effect on the semipermeability
of the resulting nonannealed membrane, as shown in Table 1.

The manipulation of the membrane matrix via blending of hydrophobic and hydrophilic (water soluble) polymers was successfully
carried out at our laboratories, producing polysulfone UFMs in the
form of flat sheet and hollow fiber membranes (10). Polysulfone is
readily cast from water-miscible solvents into an aqueous coagulation
bath to form anisotropic membranes. The polymer is thus cast from

a DMF or DMA solution (15 - 18 wt%) into an aqueous solution to yield a flat sheet, macrovoid-free UFM. The major problems that have been encountered with this polymer are related to its hydrophobicity and to the fact that the available commercial polysulfones are not manufactured in the molecular weight range to yield the high viscosities required for spinning macrovoid-free ultrafiltration hollow fibers. Hence, the formulations for the casting of flat sheet membranes do not accommodate the proper viscosities for spinning hollow fibers; an increase in the polymer concentration (in order to obtain the necessary viscosity for spinning) results in low permeability membranes having dense skins and tight substructures.

Proper polysulfone spinning solutions have been obtained by blending this polymer with poly(vinylpyrrolidone) in DMF or DMA (10). Poly(vinylpyrrolidone) (a water soluble polymer) was found to be compatible with polysulfone when prepared with the above solvents. High viscosity spinning and casting dope mixtures, which exhibit a very long shelf life (several months for some solutions), were prepared for the fabrication of UFMs.

Flat sheet and hollow fiber UFMs were produced from a solution mixture displaying viscosities of up to 500,000 cP (50°C), with varied PVP/PS ratios (up to 3/2), employing grades of PVP ranging from 10,000 to 40,000 MW. Upon coagulation, the solvent and the PVP dissolve in the aqueous coagulating bath, and a polysulfone porous matrix is obtained. This method, if employed correctly, provides complete control of the final membrane morphology. Since this method combines a very hydrophilic polymer with a hydrophobic one, skin pore dimensions can be controlled by casting (or spinning) the dope mixture in atmospheres of varied humidities. A progressive increase in humidity at the stage prior to submersion in the coagulating bath induces micro-phase separation - between the two polymers - in the nascent membrane surface, yielding increasingly larger pore sizes. Conversely, casting the solution in a dry atmosphere yields a highly porous skin with pore dimensions exhibiting extremely small diameters; the membrane's molecular weight cut-off (of globular solutes) reaches values of less than 2,000 MW. Skin pores of a polysulfone hollow fiber spun from PVP/PS (2:3 w/w) are shown in Figure 4.

The ability to increase the dope viscosity while maintaining the membrane porosity was successfully employed in reducing and eliminating macrovoid formation in a nascent membrane, as will be discussed below.

An additional advantage provided by the fact that polysulfone and poly(vinylpyrrolidone) are a compatible pair is the ability to introduce a third polymer component into the dope mixture. In this instance, sulfonated polysulfone (containing a sulfonic acid group on the polysulfone phenyl ring), which is incompatible with poly-

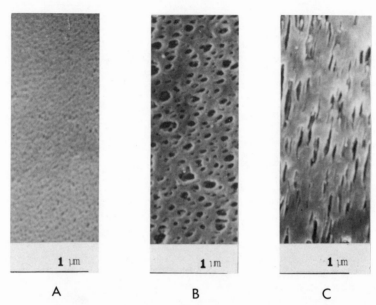

A   B   C

Figure 4:   Porous surfaces of polysulfone hollow fibers spun by the
            dry-jet wet spinning process. A. Spun in 20% relative
            humidity. B. Spun in 80% relative humidity. C. Oblong
            surface pores obtained by stretching the fiber between
            two godets in the last stages of solidification.
            Spinning dope: PS/PVP (3/2) in DMA.

sulfone (its parent polymer), can be mixed with the dope solution
containing PVP to yield a single phase dope composition. Thus, a
cation-exchange, polysulfone/sulfonated-PS hollow fiber may be spun
from a solution mixture containing PVP, consequently yielding a
membrane with a better fouling resistance and improved wettability
(11).

Macrovoids - Formation and Effect on UFM Performance

     Membranes which are solution cast or spun, and are subsequently
coagulated by a liquid medium frequently exhibit large voids in
conical, droplet, or lobe configurations. These voids sometimes
extend through the entire membrane cross-section (Figure 5). The
voids, in general, are the result of rapid coagulation of a solution
which is relatively low in either polymer concentration or viscosity. Exposing such a solution to a strong coagulant does not allow
slow and uniform exchange of solvent with nonsolvent at the nascent
membrane interface; rather, it results in immediate skin formation
and transient polymer aggregation beneath the skin. This induces
localized phase separations of areas having high and low polymer
concentrations, which often result in a subsequent intrusion of the

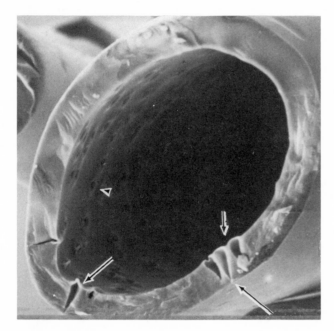

Figure 5:   Cross-section of an acrylic fiber UFM exhibiting conical
            macrovoids extending from the exterior skin to the bore
            surface.

liquid coagulant. Therefore, anisotropic membranes which are prepared by employing a strong coagulant often display large macrovoids and cavities; on the other hand, coagulating the membrane
with a relatively moderate quenching medium yields a macrovoid-free
membrane.

The above illustration somewhat oversimplifies a complex
phenomenon, which is partially discussed in the literature by various
investigators (12 - 14).   The origin of macrovoids and their
influence on UFMs can be traced by observing two types of macrovoids found in UFMs: 1) skinned voids (the most common), which
are a result of penetration and formation of a coagulant-rich
phase in the substructure, and 2) "open" macrovoids, whose walls
have, more or less, the same spongy morphology as the surrounding
substructure.

The first type of void is caused by an instant penetration
of the coagulant through the surface layer of the cast polymer
solution. This occurs simultaneously with the membrane coagulation
process, which is composed of three successive elements: 1) depletion of the solvent from the cast polymer surface (the solvent
diffuses rapidly outward to a lower chemical potential entity,

i.e., the coagulant), 2) contraction of the semicoagulated surface inward (toward the solvent-rich sublayer) and 3) diffusion of the coagulant into the cast layer.

The intrusion of the coagulant into the substructure - in a void formation mechanism - prevails simultaneously with the first coagulation stage upon exposing the cast layer to the coagulant (chiefly in droplet form, as is shown in Figures 5 and 6). The gross contraction of the nascent membrane inward - in the second stage of the mechanism - causes the structure to "close on the intruder," and a surface skin is then established.

Since the intrusion medium is progressively diluted by solvents, its effectiveness as a local coagulant within the sublayer decreases as it propagates. This is clearly illustrated in macrovoid wall morphologies (Figures 6 and 7) which exhibit dense, porous and web-like sponge structures along the penetrating cavity. The coagulation mechanism around the intruder prevails in the same fashion

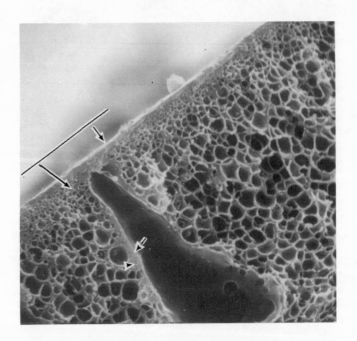

Figure 6: Cross-section of a UFM exhibiting a honeycombed substructure and tight-skinned macrovoid. Arrows designate the skin and show the extent of contraction. This membrane was cast from M-PPO (containing hydrophilic phosphonate ester groups).

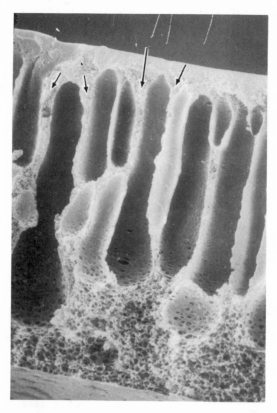

Figure 7: Cross-section of a polysulfone UFM exhibiting a large number of skinned macrovoids. Arrows designate the permeate flow channels detected by ultrafiltration of methylene blue.

described above for the membrane's formation, causing depletion of solvent and wall contraction (Figure 6). Cross-sections of membranes containing a small number of macrovoids may often lead to false observations when displayed in a scanning electron microscope micrograph. What is often interpreted as highly porous, spherical or conical macrovoids, in many instances, is a slice of the lower zone of a droplet (as can be seen in lower right of Figure 7).

A typical case is presented by Broens, et al (13), who, in discussing macrovoid formation, unfortunately displayed only the lower, porous zone of the macrovoid to support a claim that the boundaries of the voids do not solidify by gelation during their growth. The discovery of highly porous voids situated deep within the substructure, and having no clear links to the membrane surface (upper or lower), is, in most instances, also a result of a significant contraction of the nascent membrane during the phase inversion. The contraction of the membrane surface destroys the narrow neck of the droplet, isolating a pocket of the penetrated medium deep within the substructure (Figure 8).

The presence of macrovoids in UFMs is a serious drawback, since it increases the fragility of the membrane and limits its ability to withstand hydraulic pressures. Such membranes have lower elongation and tensile strength than macrovoid-free UFMs. The fact that most of the commercially available UFMs (utilized in medicine and industry) display large macrovoids points out the difficulty in achieving all the desired properties during the preparation of membranes; thus, mechanical quality is often sacrificed for transport properties.

Attempts to justify macrovoid formation in an anisotropic membrane as a necessary factor in the reduction of the sponge sublayer's resistance to permeation are, in most instances, irrelevant. The macrovoid skin exhibits a very high barrier to flow when it is

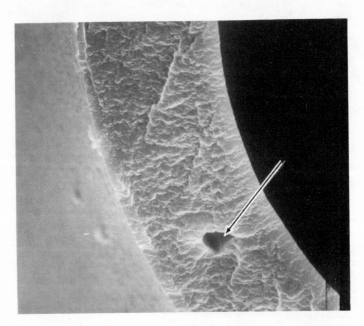

Figure 8:   Cross-section of a cellulose acetate hollow fiber membrane
            exhibiting a single macrovoid in the center of the struc-
            ture. This macrovoid was generated by the bore coagula-
            tion fluid. What is shown in the micrograph is the bottom
            section of a conical macrovoid which, at the initial stage
            of membrane formation, extended to the bore surface, and
            which was later eliminated by membrane contraction.

formed in a honeycombed sponge structure (Figure 6). The macrovoids in the open-cell foam structure (of polysulfone UFMs) shown in Figure 7 are also skinned, and the flow channels designated by arrows are occupying only a minute fraction of the membrane's substructure. (This flow channel may easily be traced by ultrafiltering an aqueous methylene blue solution, and subsequently examining the cross-section by light-microscopy.)

The major disadvantage of the UFM containing macrovoids is its vulnerability to elevated hydraulic pressures. Macrovoids that extend from the upper skin to the bottom surface of the membrane (Figure 5) are often seen in commercial ultrafiltration hollow fibers. These fibers have a rather low burst pressure, since the skin is not evenly supported.

Anisotropic membranes exposed to elevated hydraulic pressures undergo deformation, which may be equated with Young's Modulus and the yield-stress parameters of the materials used. The progressive yielding to the hydraulic pressure, and the resulting loss of pro- duction rate due to compaction (densification) of the matrix is of major concern in prolonging ultrafiltration and reverse osmosis operations.

The compaction of the anisotropic membrane is not a uniform contraction of the structure, because the membrane's anisotropic morphology displays a progressively increased porosity (or void vol- ume) from top to bottom. Therefore, the membrane's yield to hydrau- lic pressure commonly progresses from the bottom face of the membrane upward.

As long as the structure is laterally uniform, the lateral stress on the tight permaselective skin of the membrane is negli- gible. However, the existance of macrovoids in the structure commonly causes a nonuniform deformation of the membrane as a result of macrovoid collapse. This is clearly illustrated in Figure 9a, in which the collapse of the deeper and larger macrovoids effects the membrane structure and permaselectivity more than the many small macrovoids which are situated near the skin (which, at the stage shown in the micrograph, have not yet ruptured). The non- uniform compaction displayed in Figure 9a is a result of exposure of this particular membrane to a pressure of 200 psi (13.6 atm). In comparison, the macrovoid-free anisotropic membrane shown in Figure 9b exhibits a rather uniform lateral compaction which did not effect the membrane selectivity after the membrane was exposed to a pressure as high as 1,000 psi (68 atm).

Another important drawback of membranes having a large number of macrovoids is observed in in-vivo hemofiltration employing hollow fiber membranes. A certain degree of hemolysis is always

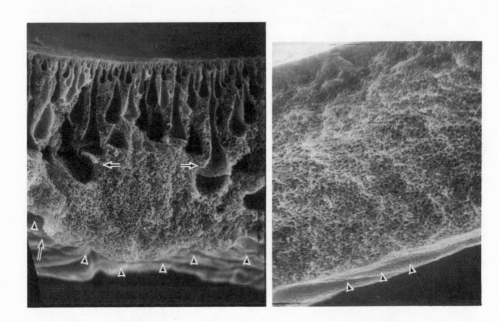

A   B

Figure 9:   Cross-section of two anisotropic membranes that were ex-
            posed to elevated hydraulic pressure for two weeks.
            A. A PPO-phosphonate membrane exhibiting a large number
            of intrusion voids was tested under pressures up to 200
            psi. The membrane's selectivity deteriorated at pressures
            greater than 200 psi. The arrows indicate nonuniform
            lateral compaction on the bottom face due to collapse of
            the large macrovoids, which subsequently increased the
            lateral stress on the permaselective skin. B. An alloy
            membrane composed of CA/PPO-phosphonate exhibiting uniform
            lateral compaction in the bottom surface after exposure
            to 1,000 psi.

caused by the rather sharp edges of each hollow fiber in the blood
entry port of hemodialysis and hemofiltration units. The existance
of large numbers of macrovoids (Figure 10) accelerates the rate of
hemolysis, as blood cells are damaged not only at the fibers' edges,
but also on the extremely rough macrovoid surfaces.

     Macrovoid formation is a more common phenomenon in wet spun
hollow fibers than in the analogous flat sheet membranes, partially
because, unlike flat sheet membrane casting, there is no casting
surface in the spinning process, thus forcing the nascent hollow
fiber membrane to establish mechanical integrity in a very short
time period. The rapid coagulation required and the relatively

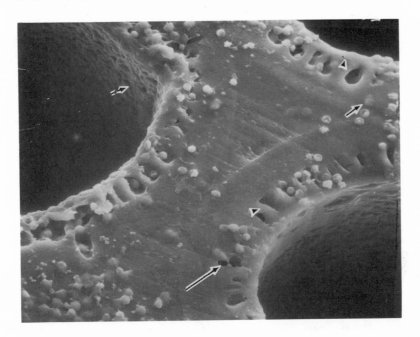

Figure 10: Cross-section of a commercial hemofiltration unit blood
            entry port. This unit was tested in-vivo for six hours
            and subsequently washed with saline. Arrows indicate
            heavy blood plasma deposits resulting from hemolysis on
            the sharp edges and macrovoids of the fibers. (The fibers
            are polyacrylonitrile potted with epoxy.)

low solution viscosities employed contribute to coagulant penetration into the nascent hollow fiber membrane, resulting in macrovoid
formation.

     There are several methods to reduce macrovoid formation in
the spinning process. The most convenient method is to increase
the dope mixture viscosity, thus increasing the resistance of the
emerging thread line to localized phase separation and coagulant
intrusion. Thus, spinning dope mixtures of high molecular weight
polymers yield fewer cavities than those of the same polymers, which
exhibit lower viscosities due to lower molecular weight. Also, as
reported in the preceding section, an increase in viscosity can be
achieved by blending the dope mixture with a coagulant-soluble
polymer. Figure 11 shows a cross-section of a polysulfone hollow
fiber spun from a blend of DMA-poly(vinylpyrrolidone)/polysulfone.

     Another method presently under investigation in our laboratories
is the controlled use of different coagulants for the exterior and

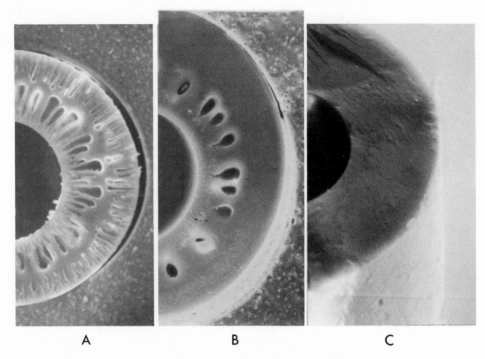

A   B   C

Figure 11:   Cross-sections of polysulfone hollow fibers spun by the
             dry-jet wet spinning process from a spinning mixture
             containing polysulfone/poly(vinylpyrrolidone)/DMA.   The
             viscosities (at 50°C) were: A. 3,000 cP, B. 10,000 cP
             and C. 180,000 cP   (14)

interior (bore) zones of the nascent fibers. In order to establish
anisotropic fibers with an exterior skin resting on a highly porous
macrovoid-free structure, the dry-jet wet spinning process is em‑
ployed (spinneret located above the exterior coagulant bath).   The
height of the jet is adjusted so that the viscosity of the emerging
polymer solution increases gradually outward before coagulation
prevails, and before the nascent fiber is submerged in the exterior
coagulant bath (Figure 12).

     The diffusion of the bore medium does not cause solidification
of the extrudate, but gradually increases the viscosity of the spun
solution. As its penetrating front propagates, the extrudate is
submerged in the exterior coagulating bath at the incipient moment
the bore medium reaches the exterior zone. This method is very
effective in impeding macrovoid formation. However, improper
spinning conditions, or submerging the nascent fiber in the exterior
coagulating bath prematurely may cause a fiber split (Figure 13).

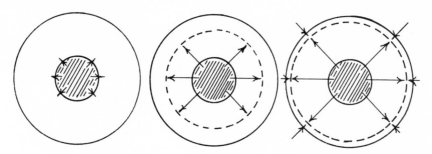

Figure 12: Schematic of a nascent hollow fiber membrane cross-section spun in the dry-jet wet spinning process in a fashion that impedes macrovoid formation. Arrows designate the propagating fronts of the interior and exterior coagulants. A. extrudate upon emergence from the spinneret. B. nascent fiber upon submersion in the exterior coagulant bath. C. exterior and interior coagulant merger zone.

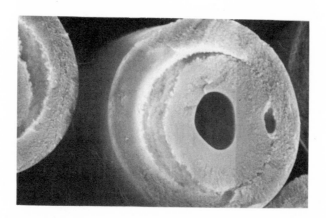

Figure 13: Split fiber matrix caused by improper use of coagulants and temperature gradient in spinning ultrafiltration hollow fibers by the technique described in Figure 12.

Modification of Skin Pores

Surface pore dimensions and density determine the basic properties of UFMs. In a high flux UFM, the skin is 10 - 20% pore area.
Typical UFM pores are round. However, if necessary, the nascent membrane may be stretched to yield oblong pores which will not allow
passage of globular species having a spatial cross-section less than
the total area occupied by the oblong pore. These two pore structures - round and oblong - are shown in Figure 4 for polysulfone UFMs.

In order to modify the pores' chemical environment, pores were
coated with active polymer ingredients such as poly(vinylpyridine),
or polyethyleneimine, which can then be cross-linked to yield a
quaternary surface. A method which has been developed recently is
the gas-phase bromination of polysulfone and poly(dimethylphenylene
oxide), which yields a bromomethylated pore surface. The bromomethylated groups are reacted with p-aminonaphthalene sulfonic acid
to yield high charge density in the vicinity and within the pore
area (15). Such membranes exhibit - as may be expected - relatively
high separation rejection of low molecular weight (< 1,000) organic
salt solutes, when compared to the separation of neutral solutes
having even higher molecular weights.

The separation via UFM of sodium dodecyl sulfate $(CH_3(CH_2)_{11}$
- OSO_3Na, 228 MW), a surfactant which exhibits a critical micelle
concentration of $\sim$ 2,000 ppm, was studied with the above treated
UFMs. The sigmoid shape of the rejection profile indicates formation of micelles below the CMC (Figure 14). The membranes exhibit
rather high separation of the SDS even at concentrations as low as
100 ppm. This is probably due to transient micelle formation on the
surface of this high flux, negatively charged UFM.

The rather high separation of SDS micelles ($\sim$ 23,000 MW)
was successfully employed to trap and separate low molecular weight
organic components, such as phenols, that otherwise could not be
effectively separated, even by conventional reverse osmosis membranes.

Acknowledgment

A major portion of the studies reported in this paper
was sponsored by the U.S. Department of Interior, Office
of Water Research and Technology.

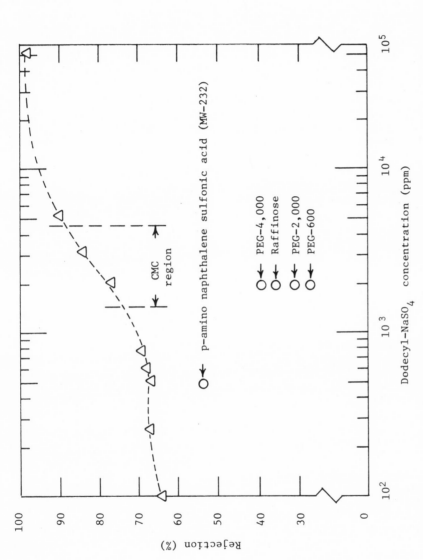

Figure 14: Rejection vs concentration of sodium dodecyl sulfate in water for an M-PPO membrane that was sulfonated with p-aminonaphthalene sulfonic acid. Product flux ranges from 70 - 40 gfd (50 psi) for the entire concentration profile. Rejection of other solutes by this membrane are designated by circles.

References

1.   B.R. Breslau, A.J. Testa and B.A. Milnes , "Advances in Hollow
      Fiber Ultrafiltration Technology" in this volume.

2.   I. Cabasso, E. Klein and J.K. Smith, NTIS, PB 248666 (1976).

3.   S. Krause, J. Macromol. Sci. Chem. 7, 251 (1972).

4.   A. Michaels, Ind. Eng. Chem. 57 (10) 32 (1965).

5.   C. Cannon, U.S. Patent 3,497,072 (1970).

6.   I. Cabasso, J. Jagur-Grodzinsky and D. Vofsi, in Polymer Alloys
      K.C. Frish and D. Klempner, Eds., Plenum, New York, (1977)

7.   P. Aptel and I. Cabasso, J. Appl. Polym. Sci. (in press).

8.   I. Cabasso and C.N. Tran, J. Appl. Polym. Sci. 23 2967 (1979).

9.   P. Aptel, private communication.

10.   I. Cabasso, E. Klein and J.K. Smith, J. Appl. Polym. Sci.
      20 2377 (1976).

11.   I. Cabasso and A.P. Tamvakis, J. Appl. Polym. Sci. 23, 1509
      (1979).

12.   M.A. Frommer and D. Lancet in Reverse Osmosis Membrane Research
      H.K. Lonsdale and H.E. Podall Eds., Plenum Press, New York (197

13.   L. Broens, F.W. Altena and C.A. Smolders, Desalination 32
      33 (1980).

14.   I. Cabasso, E. Klein and J.K. Smith, J. Appl. Polym. Sci.
      21 165 (1977).

15.   I. Cabasso, to be published.

PERMEABILITY PARAMETERS OF A NOVEL POLYAMIDE MEMBRANE

M.S. Lefebvre, C.J.D. Fell, A.G. Fane and A.G. Waters

School of Chemical Engineering
University of New South Wales
Australia, 2033

INTRODUCTION

The discovery by Loeb and Sourirajan (1) of a technique for preparing anisotropic cellulose acetate membranes having high solvent permeability and good solute rejection characteristics has undoubtedly led to the commercial exploitation of pressure driven membrane separation processes. However, as industrial applications have increased, it has been recognised that cellulose acetate membranes do have severe limitations. They are biodegradable, and subject to attack by strong acids and alkalis (3>pH>8), and creep under reverse osmosis pressures. Their applicability, in ultrafiltration processes in particular, is limited because of their poor resistance to temperature (typically T<50°C) and to sanitizing fluids, and to the relative ease with which they can be irrecoverably fouled by protein solutions.

These factors have initiated an active industrial research effort into alternative membranes. Successes include the du Pont aromatic polyamide Permasep series of membranes (2) which although having a low flux can be used in a multiple hollow fibre configuration for desalination, and the new industrial ultrafiltration membranes offered by companies such as Abcor, Der Danske Sukkerfabrikker, Dorr Oliver, Patterson-Candy and Rhone Poulenc. Full details of the constitutive polymers in these ultrafiltration membranes are not readily available, but it would appear that they are polysulphones, coagulated polyacrylonitrile, polyacrylonitrile-polyvinylchloride co-polymers, polyacrylonitrile-polysulphone composites, or other mixed polyelectrolyte species.

The success of the du Pont polyamide membrane and the known excellent resistance of polyamides to extremes of pH and temperature,

has lead to considerable interest in these particular polymers.
Recent developments in polyamide membrane research have been reviewed
by Blais (3). He comments, for reverse osmosis applications in particular, that polyamide membranes are "not very impressive in their
permeability and their selectivity to solutes of economic importance
is, at best, equal to current state-of-the-art cellulosic membranes".
Later reports by Riley et al (4) on poly (ether/amide) composite
membranes and by Endoh et al (5) on membranes fabricated from polyamides incorporating carboxylic groups, are more encouraging.

Existing polyamide membranes are prepared by a phase-inversion
process analogous to that proposed by Loeb and Sourirajan (1) for
the preparation of cellulose acetate membranes. The polyamide is
dissolved in a suitable solvent (dimethyl acetamide, dimethyl formamide or dimethyl sulphoxide) to form a dope. The dope is spread in
a thin film over a flamed glass sheet, and an initial diffusion
period is allowed to form the upper surface of the membrane and to
establish a concentration gradient in the film. The film is then
wholly coagulated by immersing it in a bath of a solvent diluent
(often water) in which the polymer is not soluble. The membranes
so produced have a thin skin (1 μm in thickness) backed up by a
cellular support formed in the coagulation step. The constitutive
water present in such polyamide membranes is generally lower than
that of equivalent cellulose acetate membranes (6), and it is believed that the low flux of polyamide membranes can in part be attributed
to this. A more hydrophilic character would be expected to increase
the water permeability.

Membranes suitable for ultrafiltration have been classified as
either diffusive or microporous (7), although most current generation
ultrafiltration membranes are probably microporous in character.
They are also highly anisotropic, relying on their thin skin (0.1 to
1.0 μm), 'active' surface to effect separation. It is the formation
of this thin skin which on the one hand makes the membranes viable
and on the other hand can limit their flux. This is because the film
forming techniques usually rely on surface evaporation of volatile
solvent prior to precipitation, and result in a membrane with a free
area (total area of pore openings per unit surface) significantly
less than the internal porosity of the membrane structure. The water
permeability of the membrane will depend, inter alia, on the available free area.

It is normally assumed that flux under gel-polarised conditions
of ultrafiltration is independent of the membrane water permeability
(8,9). However, this is arguable, particularly if the membrane is
viewed in terms of its free area, rather than its apparent resistance
to flow. The results of Lonsdale et al (10) on the permeability of
composite reverse osmosis membranes are pertinent since they show
how the free area of the 'low resistance' support significantly
influences the flux through the high resistance thin film. There

is a clear analogy between the gel-polarised ultrafiltration membrane and the composite reverse osmosis membrane. Thus it can be reasoned that the necessary anisotropy of an ultrafiltration membrane should not be achieved at the expense of the free area of the surface.

In the present paper we discuss the development and characteristics of a new type of polyamide membrane (11) with increased water affinity and an apparently high free area. This membrane is based on a chlorinated polyamide (typically nylon 6;6,6;4,11 or 12) and appears to have excellent potential as an ultrafiltration membrane, having high permeability, excellent thermal and chemical stability and good selectivity.

THE NEW MEMBRANE

In developing the new membrane it was reasoned that if the polymer could be modified by halogenation it would become more hydrophilic, and should exhibit higher water permeability. Moreover, if the film forming step could be modified to avoid densification of the surface, a membrane with higher flux capabilities should ensue.

The manufacturing technique for the new membrane is described below (11):

(i)   dissolution of the polymer in aqueous acid solution in the presence of a surface active agent to give a dope;

(ii)   maturation of the dope at constant temperature for a predetermined time or until the measured viscosity indicates the desired level of substitution and depolymerisation has occurred;

(iii)   knife coating of the dope on a non-porous support to give a predetermined thickness (typically 80-300 μm);

(iv)   immediate coagulation of the dope film by quenching in water to increase pH;

(v)   conditioning of the resultant membrane by prolonged rinsing at a preset temperature in water possibly containing ammonia or sodium hydroxide.

During dope maturation using hydrochloric acid, a fixation of approximately 1.5% w/w of chlorine occurs, with the evolution of hydrogen bubbles occurring during the first six hours of dope maturation.

The choice of polymer concentration in the dope is important as it is this which primarily determines the macrostructure of the

membrane. On immersing the dope film in the water bath, the pH chang
at the upper surface of the film is rapid and polymer precipitation
occurs, first as a film, and below this as a porous polymer net
enclosing cells or alveolae of increasing diameter. Cell formation
is a result of polymer diffusion to precipitation sites. The diamete
of the cells at any given depth into the membrane is determined by th
rate of pH change, the concentration of polymer in the dope, and the
diffusivity of the polymer. It should be noted that prior to coagulation there is no attempt to allow surface evaporation, and this
means that anisotropy results only from the differential coagulation.

By choice of polymer, concentration of acid and the surface
active agent in the dissolution medium, and period of maturation
before membrane formation, it is possible to exercise quite wide
control over the characteristics of the resultant membrane, and to
achieve molecular weight cut-offs in either the reverse osmosis or
ultrafiltration ranges.

For comparison of different membranes made by the above process,
we have adopted the codification as summarised in Table 1.

Table 1.   Coding System Used for Membranes

Example: Membrane CL6 40 50/25/5E 1BO

C   Type of mineral acid used

L   Type of polymer

6   Type of polyamide

40   Concentration (gms) of polymer in dope

50   Concentration (mls) of acid in dope

25   Water content (mls) of dope

5   Concentration (mls) of surfactant in dope

E   Type of surfactant used

1   Maturation time in days

B   Thickness code

O   Coagulation bath code

MEMBRANE STRUCTURE

A CL6 40 50/25/5E 1BO membrane has been chosen to illustrate the structural features of the new membrane. The polymer used for this particular membrane is polyamide 6 in the form of a highly oriented, finely divided textile yarn (44 dtex/13 filaments - bright MLB). The initial distribution of chain lengths present in this material means that the resulting dope contains polymer molecules of different molecular weight, and there is thus the opportunity for extensive cross linkage to occur during the coagulation step.

Figures 1 and 2, which are transmission electron microscope views of a freeze fractured section of the membrane (magnification 580x and 4200x respectively), give details of membrane macrostructure which is clearly anisotropic. The membrane is made up of many layers of cells, and in the enlarged view (Figure 2) it can be seen that a regular progression in cell sizes exists right up to the membrane surface, with cell diameter increasing with distance from the upper surface. We term this macrostructure 'geometrically divergent.'

By using a different dope formulation (e.g. CL6 25 50/25/5 E 1BO) it is possible to obtain a membrane structure in which cell size shows little change with distance from the surface. With even less polymer in the dope (CL6 17.5 50/25/5 E 1BO), cells in the interior of the membrane become open ended alveolae rather than spheres and the mechanical strength of the membrane deteriorates. (These membranes are only useful as biofilters).

At the surface and between the cells the membrane is microporous. For the CL6 40 50/25/5E 1BO membrane the micropores appear to increase in size from the upper (initially formed) surface to the bottom. This view is supported by the results of a series of tests in which successive layers of the membrane were removed by careful surface abrasion, and the membrane used to ultrafilter protein solutions. The results suggest a variation of molecular cut-off with depth as depicted in Figure 3; this characteristic is termed molecular divergence.

It is apparent from Figures 1-3 that the new membrane is more gradually anisotropic than other membranes. One potential advantage of this is that quality control during the membrane's manufacture need not be as stringent as for existing skinned membranes, as the gradient in molecular weight cut-off of successive cell layers means that imperfections in one layer will be compensated for by the succeeding layer without serious consequences on membrane rejection.

It is also apparent from water permeability measurements (to be discussed below) that the new membrane has a high free area, made up of more pores per unit surface than is normally achievable. However, confirmation of this awaits detailed examination by ultra-high

Fig. 1.   Transmission electron microscope view of freeze fractured
cross section of membrane.   Low magnification.

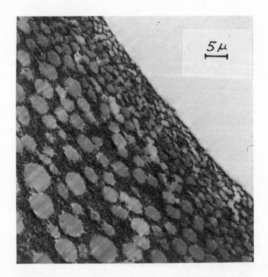

Fig. 2.   Transmission electron microscope view of freeze fractured
cross section of top surface of membrane.   High magnification.

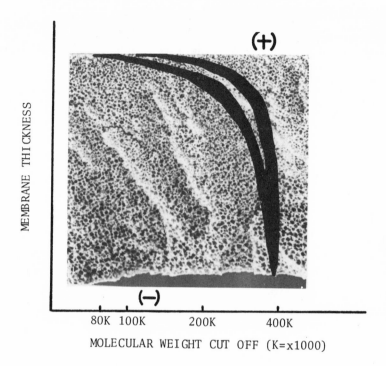

Fig. 3. Molecular weight cut-off of membrane as function of depth.
Successive cell layers removed by abrasion.

resolution (5 Å) electron microscopy.

MEMBRANE PERFORMANCE

The new membrane is unusual in that it can be used with either
surface facing the pressurised feed solution, although its performance
differs in these two situations. When used with the surface which
was water-side upwards during membrane formation in contact with the
feed solution, the membrane is said to be in the (+) configuration.
When reversed, the membrane is in the (−) configuration. Figure 3
indicates this terminology.

When used in the (+) configuration the membrane behaves as a conventional skinned membrane, although its relatively high water permeability appears to be reflected in rather higher than expected fluxes
under gel-polarised conditions. When used in the apparently anomalous
(−) configuration the membrane has unusual and attractive properties.
Details of performance of the membrane in both configurations are
discussed below.

Effect of Membrane Configuration on Water Permeability

The markedly different results obtained using the membrane in
(+) and (-) configurations are illustrated in Figure 4. Here the permeability of a CL6 40 50/25/5E 1BO membrane to distilled water at a
transmembrane pressure of 100 kPa has been determined in a 15 cm^2
stirred cell. Two runs have been performed - one with the membrane
initially in the (+) configuration (run B) and the other with the membrane initially in the (-) configuration (run A). After the passage of
several cell volumes of water, the membrane has in each case been
reversed and the run continued.

In both configurations the initial fluxes are similar, but in
the (+) configuration permeability rapidly declines with increased
water passage. Reversal of the membrane appears to substantially
restore the initial flux. The experimental results can be explained
by recognising that in the (+) configuration the bulk of resistance to
fluid passage occurs at the unsupported upper surface of the membrane,
and the membrane thus compacts. In the (-) configuration, stresses are
concentrated near the lower face of the membrane which is supported.

Further evidence of the ability of the membrane to recover from
the effect of transmembrane pressure is afforded by Figure 5. Here
a CL6 40 50/25/5E 1BO membrane in a stirred cell has been subjected
to water passage at a transmembrane pressure of 100 kPa. Flux is
observed to decline to 1/3rd of its initial value with the passage
of water. The transmembrane pressure has then been released and the
membrane allowed a period of 2-12 hours to recover before the experiment is continued. It is observed that with each depressurisation
the membrane partly recovers its initial flux, and this flux then
continues to decline with further water passage. All of these
observations are consistent with transmembrane pressure exerting a
semi-reversible structural change on polymer moieties comprising the
cell walls. It is of interest to note that the membrane reported in
Figure 5 had been annealed at 60°C for 1 hour before being placed in
service. This annealing slows the rate of flux decline as shown by
comparing the data with that in Figure 4.

Effect of Dope Preparation and Formulation on Permeability

Figure 6 gives a plot of the permeability of two membranes
(CL6 40 54/25/6 E 3H BO and 3 BO) to distilled water as a function
of the transmembrane pressure applied. These membranes have been
formed from dopes matured for 3 hours (3H BO) and 3 days (3 BO)
respectively, and used in the (+) configuration.

The permeabilities reported are taken after 15 mls of water have
passed, and may be considered to be effective initial permeabilities.
Both show a similar decline in membrane permeability as transmembrane
pressure is increased, although the fluxes for membranes made from

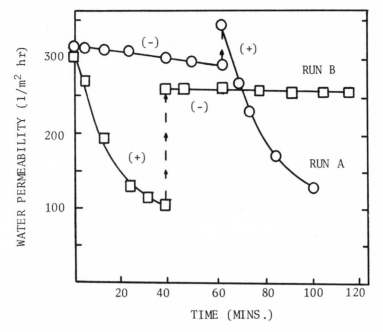

Fig. 4.   Membrane permeability in (+) and (-) configurations.
Membrane reversed and run continued at times indicated.
Temperature 20°C.

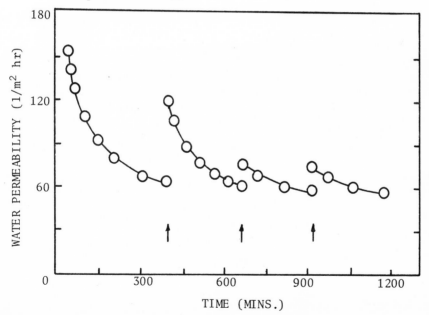

Fig. 5.   Membrane permeability for membrane in (+) configuration.
Transmembrane pressure released at times arrowed.
Temperature 20°C.

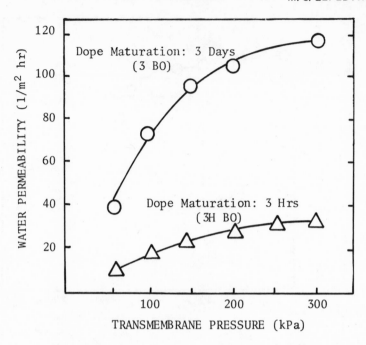

Fig. 6. Permeability of CL6 40 membranes of different dope
maturation time. Temperature 20°C.

dopes matured for 3 days are better than those for membranes with the
shorter dope maturation period. The improvement in performance as a
function of dope maturation reflects the shorter chain length of the
polymer in the more mature dope, and the tendency for the polymer
aggregates formed from this dope to have a more open structure. Flux
decline with increasing transmembrane pressure is a result of changes
to the nature of the polymer (amorphous to crystalline) as it is subject to increasing stress. Avenues for water passage through the
polymer net comprising the cell walls narrow with increasing stress,
and water permeability accordingly declines.

Both parent polymer chosen and solvent water content of the dope
exert profound effect on the permeability of the resulting membrane.

We define dope percentage solvent content as:

$$\% \text{ Solvent Content} = \frac{\text{Mls(Acid+water+surfactant)}}{\text{Gms polymer/2+Mls(Acid+water+surfactant)}}$$

$$\ldots (1)$$

In equation (1) the effective dissolved polymer density is taken as
2.0 gm/c.c.. Using equation (1) the % solvent content of a
CL6 40 50/25/5 E dope would be calculated as 80%.

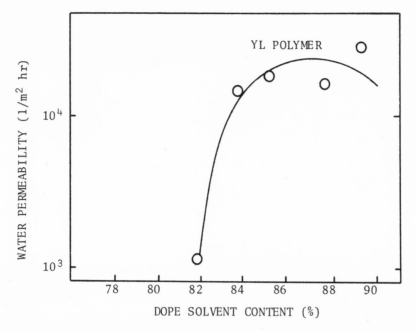

Fig. 7.   Water permeability as function of dope solvent
content for YL polymer. Temperature 20°C.

Initial water permeability for a transmembrane pressure of
100 kPa is plotted against % solvent content for a number of different
parent polymers in Figures 7 and 8. The acid used in membrane preparations in these examples is hydrochloric.

Figure 7 is for membranes formed from a medium molecular weight
highly crystalline polyamide 6 which we term YL. It is available
in Australia in the form of fibres of 77 dtex. The water permeability
of YL membranes increases steadily as % solvent content in the dope
increases until a solvent content of 86%, beyond which membrane
strength deteriorates rapidly.

Figure 8 gives equivalent curves for IP and M series polyamides.
Both are derived from nylon 6. IP is in the form of amorphous polymer
chips (3 mm dia., M.W. 22,000 n.a.) whilst M is the highly crystalline
monofilament of 0.4 mm in diameter formed by spinning and drawing of
these chips (drawing ratio 3.9). The monofilament (M), being of
lesser diameter than the chips takes less time to dissolve and undergoes more depolymerisation during dissolution. M dope contains a
narrower spectrum of molecular weights of chlorinated polyamides, and
when coagulated gives a membrane of superior strength at low concentration of polymer to that produced from IP. Neither polymer, however, gives a membrane having a permeability as good as that obtained
using YL as the parent polymer.

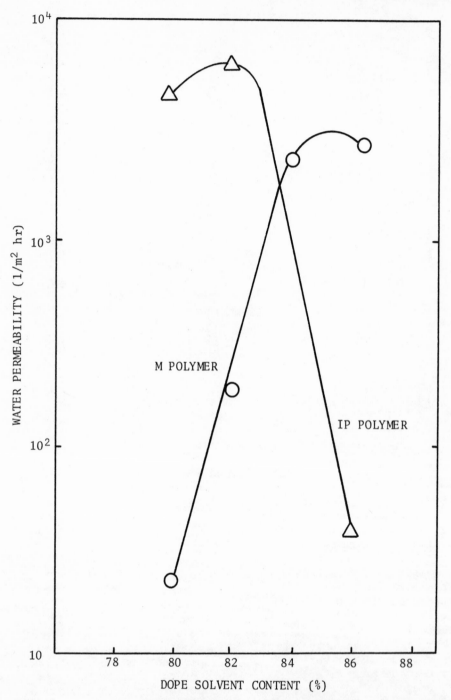

Fig. 8. Water permeability as function of dope solvent content
for IP and M polymer. Temperature 20°C.

Membrane Rejection Characteristics

It is experimentally observed that there is a direct relationship between membrane permeability and membrane rejection characteristic, with high water permeability membranes formed from low acid or low polymer content dopes typically having a rejection capability only for large molecular weight species. By choice of polymer starting material, dope solvent content and maturation time, membranes can be tailored to meet specific applications.

Table 2 gives examples of some different types of membranes produced in the present programme and an indication of their rejection characteristics.

SPECIFIC APPLICATIONS

Ultrafiltration of Oil-Water Emulsions

Figure 9(a) shows the flux-transmembrane pressure characteristic for the ultrafiltration of an oil-water emulsion (20% of mineral oil in water) at 20°C in a 45 cm^2 thin channel module fitted with a CL6 40 50/25/5E 1BO membrane in the (+) configuration. The figures on the curves are concentrate side cross flows through the (35 mm x 3 mm)

Table 2. Rejection Characteristics of Typical Membranes Produced

Description*	Characteristics
CL6 40 50/20/5 E 1BO	Rejects 80% bovine serum albumin (BSA)
CL6 35 50/20/5 E 1BO	Does not reject BSA but rejects 99% blue dextran
CL6 30 50/25/5 E 1BO	Passes blue dextran but rejects oil from oil/water emulsion
CT6 22.5 50/25/5 E 1BO	Biofilter rejecting 0.6 μm bacteria
CT6 28 50/25/5 E 1BO	Biofilter rejecting 0.2 μm bacteria
CT6 40 50/25/5 E 1BO	Rejects 96% of protein present in cattle blood serum but not non-protein N
CT6 40 80/0/1 E 4H BO	Rejects riboflavin, 35% NaCl$^{\triangle}$, and 35-40% bivalent ions$^{\triangle}$

* Both L and T polymers are polyamide 6 in the form of 44 dtex yarn.
$\triangle$ Tested at 4000 kPa under reverse osmosis conditions, all other tests at 200 kPa.

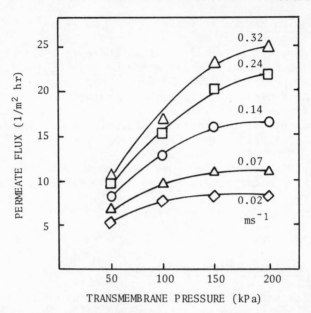

Fig. 9(a). Flux-transmembrane pressure characteristic for
ultrafiltration of oil-water emulsion. Cross flows
in ms⁻¹.

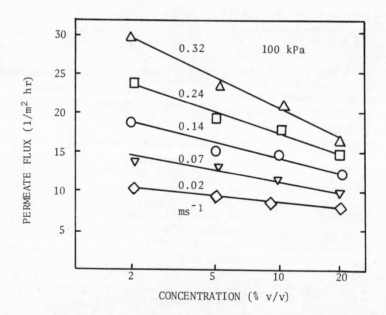

Fig. 9(b). Flux-concentration characteristic for ultrafiltration
of oil-water emulsion. Transmembrane pressure 100 kPa.
Cross flows in ms⁻¹.

channel in ms^{-1}. The curves show classic concentration polarisation
behaviour. Figure 9(b) shows the flux-concentration characteristics
at 100 kPa for the same membrane at different cross flows, with substantial improvement in flux at the higher cross flow velocities.
Fluxes are generally better than those reported previously (12).

Ultrafiltration of Protein Solutions

Figure 10 shows the flux-transmembrane pressure characteristics
for the ultrafiltration of a 0.5 wt % pepsin solution (Merck Index:
9:6942) with a CT 40 50/25/5E 1BO membrane in a 15 cm^2 stirred cell
at 50 to 150 kPa. In these experiments the water flux was first
stabilised and the membrane was then tested in the (+) position, and
then reversed to the (-) position. The membrane appears to perform
better in the (-) configuration. This is thought to be due to the
buildup of solute within the interstices of the membrane. Initially
the permeability to pepsin of the chosen membrane in its (-) configuration is high, but permeability rapidly drops and rejection rises
until a steady state flux is achieved, and pepsin becomes moderately
rejected as in the (+) configuration. We envisage that the membrane
becomes saturated with pepsin and commences to function as a skin
membrane, but the particular properties of the surface pepsin layer,
or the high free area of membrane surface mean that better than
expected fluxes are achieved.

Ultrafiltration of Acid Whey

Figure 11 shows flux-time curves for ultrafiltration of hydrochloric acid skim milk whey at 20°C by a CT6 35 50/25/5 E 1BO membrane. The equipment used is a Der Danske Sukkerfabrikker (D.D.S.)
0.2 m diameter model operated at an inlet transmembrane pressure of
150 kPa. Liquid cross flow is 8 1/min.

Curve A is for the membrane in its (+) configuration, whilst curve
B is for the membrane in its (-) configuration. Curve C is for a D.D.S.
GR60P membrane having a nominal molecular weight cut-off of 25,000.

The new membrane in its (+) configuration appears to perform
similarly to the D.D.S. GR60P membrane. However, when used in its
(-) configuration, the new membrane gives a much superior flux. Again,
we conjecture that in the (-) configuration an intermediate molecular
weight species is entrapped in the membrane at the commencement of
ultrafiltration and the ultimate rejection and flux characteristics
of the membrane are those of the composite of the original membrane
and the retained species. Notably, in the (-) configuration in skim
milk service, the membrane does not foul (24 hours trial), nor is
the rejection of protein (93%) significantly different from that
obtained with the D.D.S. GR60P membrane.

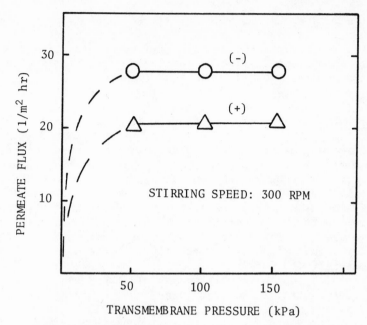

Fig. 10.   Flux-transmembrane pressure characteristic for ultrafiltration of pepsin solution in stirred cell at 20°C.

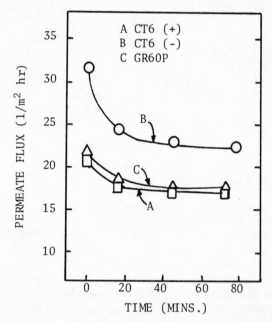

Fig. 11.   Flux-time curve for ultrafiltration of acid whey on DDS
ultrafilter. P_i = 200 kPa, P_o = 100 kPa.   Temperature
20°C.

Ultrafiltration of Cattle Blood Plasma

A CL6 40 50/25/5E 1BO membrane in the (-) configuration has been used for the ultrafiltration of cattle blood plasma at $25^{\circ}C$, pH = 7.5, transmembrane pressure 180 kPa, in a flat plate ultrafiltration unit with a gap of 1.5 mm. The achieved average ultrafiltration flux for a two-fold concentration of the solids initially present is approximately 18 $1/m^2hr$. Protein rejection is better than 92% measured by the Biuret method. This flux compares with a literature reported (13) value of initial flux of 9 $1/m^2hr$ at $30^{\circ}C$.

Again, it is considered that an intermediate molecular weight species is retained within the membrane and provides the basis for the membrane rejection characteristic. Indeed, when the feed is first applied to the membrane, protein rejection is poor. However, rejection quickly improves (c.a. 10 minutes) and settles, along with flux, to a relatively steady value.

Figure 12 is a photograph of a 1000 ℓ/24 hours Australian pilot plant for the ultrafiltration of cattle blood plasma. Fresh cattle blood collected at a local abattoir is first split into serum and cell fractions by centrifugation. The serum fraction is then fed to flat-plate-geometry ultrafilters incorporating the new membranes in which it is concentrated from 10% solids to 25% solids. The concentrate stream is dried to a free flowing white-to-cream coloured powder in a spouted bed dryer. The powder has excellent redissolution properties and is a sought-after bland protein additive for human food formulations.

DISCUSSION

Although basically of a preliminary nature, the results obtained to date with the new chlorinated polyamide membrane are sufficiently encouraging to warrant further extensive investigation of its capabilities, and such work is currently underway at the University of New South Wales.

There are a number of interesting points that arise from the work completed to date.

The first is that the new membrane appears to have properties that are desirable for industrial ultrafiltration membrane applications. It is resistant to pH in the range 2-14 and can be boiled (if necessary) for sterilisation purposes. Moreover, it is proof against concentrated sodium hydroxide solutions to $80^{\circ}C$. From a viewpoint of chemical stability it therefore appears attractive when compared with available cellulosic and polyelectrolyte membranes.

Fig. 12. Pilot scale unit for ultrafiltration of cattle blood
          plasma.

The second is that the new membrane, by reason of an inferred
higher free area, seems to give fluxes equivalent to or better than
commercial ultrafiltration membranes at lower transmembrane pressures.
This is an important observation as it means a lower energy requirement for a given ultrafiltration and also that the membrane structure
itself need not be designed to withstand as severe compactive forces.
Hence advantage can be taken of membranes having higher inherent
permeability.

The third is that the new membrane can profitably be used in a
(-) configuration. In this configuration the membrane gives better
fluxes for mixed protein solutions than do existing commercial membranes. As indicated earlier, we believe this to be a result of the
entrapment of solute within the membrane. The resultant membrane is
not a 'dynamic' membrane, nor is it fouled or plugged, but rather,
is an internal membrane of a specific protein supported by the underlying polyamide structure, and full advantage is taken of the openness
of the support structure. Indeed, following on from the success of
this approach there is the strong suggestion that the membranes for

a particular ultrafiltration should be specifically developed for the application so that the initial chlorinated polyamide membrane is compatible with the solutes in solution and allows the optimum entrapment of solute species to give a 'tight' resultant membrane.

It remains to be seen to what extent membranes of the type described in the present paper can replace both cellulosic and aromatic polyamide and polysulphone membranes in typical food plant ultrafiltration service.

CONCLUSION

Production of chlorinated polyamide membranes by a simple dissolution, re-coagulation procedure gives membranes which can have a range of permeabilities and molecular weight cut-offs. Both permeability and rejection are governed by the type of parent polymer used, and solvent (acid) level in the initial dope, with low polymer content dopes giving the most permeable membranes.

The new membranes can be used in either of two configurations. In the so-called (-) configuration, with the uppermost formed surface away from the process fluid, attainable flux for certain ultrafiltration applications appears better than can be achieved using conventional ultrafiltration membranes, and it is considered that this is a result of solute retention within the membrane to give a composite having good rejection characteristics. The approach used - that of tailoring a membrane for a specific process application rather than selecting from a limited range of available membranes with predetermined molecular weight cut-offs - would appear to be a profitable line for further investigation.

ACKNOWLEDGEMENTS

The authors wish to acknowledge support from Anderson Equipment Co-Operative Pty. Ltd., the Australian Meat Industry Authority, the Australian Research Grants Committee and T.R.S.A. Pty. Ltd. One of us (A.G.W.) is supported by an Australian Government Postgraduate Research Award.

REFERENCES

1. S. Loeb and S. Sourirajan, University of California at
    Los Angeles Engineering Report No. 60-60 (1961).
2. H. Ohya, H. Nakajima, K. Takagi, S. Kagawa and Y. Negishi.
    Desal. Vol.21, No. 3, p.257 (1977).

3. P. Blais in "Reverse Osmosis and Synthetic Membranes".
    S. Sourirajan Ed. National Research Council, Canada.
    Ottawa (1977).

4. R. L. Riley, R. L. Fox, C. R. Lyons, C. E. Milstead,
    M. W. Seroy and M. Tagami. Desal. Vol.19, p.113 (1976).

5. R. Endoh, T. Tanaka, M. Kurihara and K. Ikeda. Desal.
    Vol.21, No. 1, p.35 (1977).

6. R. E. Lacey and S. Loeb in "Industrial Processing with
    Membranes", p.135, Wiley - Interscience New York (1972).

7. A. S. Michaels in Prog. in Sep. and Purif. Wiley, p.297
    (1968).

8. M. C. Porter, Ind.Eng.Chem.Prod.Res.Develop., Vol.1, No.3,
    p.234 (1972).

9. W. F. Blatt, A. Dravid, A. S. Michaels and L. Nelsen in
    "Membrane Science and Technology". Plenum, New York
    (1970).

10. H. K. Lonsdale, R. L. Riley, C. R. Lyons and D. P. Carosella
    in "Membrane Processes in Industry and Biomedicine".
    M. Bier Ed. Plenum p.101 (1971).

11. M. S. Lefebvre and C. J. D. Fell, Australian Patent
    Application: 40897/78.

12. R. Matz, E. Zisner and G. Herscovici. Desal. Vol.24, p.113
    (1978).

13. I. von Bockelmann, P. Dejmek, G. Eriksson and B. Hallström
    in "Reverse Osmosis and Synthetic Membranes".
    S. Sourirajan Ed. National Research Council, Canada,
    Ottawa (1977).

FORMATION OF POLY(METHYL METHACRYLATE) MEMBRANES UTILIZING STEREOCOMPLEX PHENOMENON

Y.Sakai, H.Tsukamoto, Y.Fujii, and H.Tanzawa

Basic Research Laboratories, Toray Industries, Inc.
1111 Tebiro, Kamakura-shi, Kanagawa 248, Japan

INTRODUCTION

Sheet[1] and hollow fiber membranes with stabilized porous structure were formed from poly(methyl methacrylate) (PMMA) by making use of stereocomplex phenomenon, which occures in the mixture solution of isotactic and syndiotactic PMMA, and which is a stoichiometric and thermoreversible sol-gel phase transition phenomenon.[2)-5)] This report will show the formation processes, permeability characteristics, and medical applications of PMMA membranes.

EXPERIMENTAL

Membrane Preparation

The processes of membrane preparation are outlined in Figure 1. Isotactic PMMA anionically polymerized and syndiotactic PMMA radically polymerized were dissolved together in solvent such as DMSO above the gelation temperature. The tacticities in triad of these polymers measured by NMR[6)] were as follows :
Isotacticity of isotactic polymer was more than 95 %. Iso-, hetero-, and syndio-tacticity of syndiotactic polymer were 13, 32, and 55 %, respectively.

In case of sheet membranes, the dope solution was spread between glass plates and then was cooled to under the gelation temperature, and the gelled membrane, which can be handled with hands, was immersed into water. In this process, the solvent was replaced by water. The thickness of sheet membranes was between 50 to 150 microns. Hollow fiber membranes were made with a similar method.

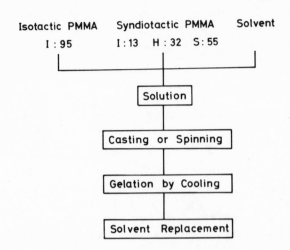

Fig. 1. Processes of PMMA sheet and hollow fiber
membranes preparation.

Figure 2 shows the transmission electron micrographs (TEM) of
a sheet membrane, water content of which was 65 wt%.   Samples for
TEM observation were embedded with water-soluble epoxide and stained
with osmium.   The dark portion of these photographs corresponds to
stained epoxide domain in the samples.   The sheet membranes are
almost uniform as the photographs show, which was expected from the
membrane preparation processes.   On the other hand, the structure
of hollow fiber membranes is heterogeneous ;   that is, a kind of
skin layer is formed at the both surfaces as Figure 3 shows.

<u>Measurement of Permeabilities</u>

The water permeation coefficient Lp is defined by

$$J_V = Lp\Delta P \qquad\qquad (1)$$

where J_V and ΔP are volume flux of water and transmembrane pressure,
respectively.   The solute diffusive permeability P_2 is defined by

$$J_S = P_2 \frac{\Delta C}{\lambda} \qquad\qquad (2)$$

where J_S, ΔC, and λ are solute flux, concentration difference across
the membrane, and membrane thickness, respectively.   Permeabilities
of sheet membranes are expressed in terms of λLp and P_2.   These
expressions reflect appropriately uniform membrane structure, as
fluxes J_V and J_S are reciprocal to the membrane thickness.   For

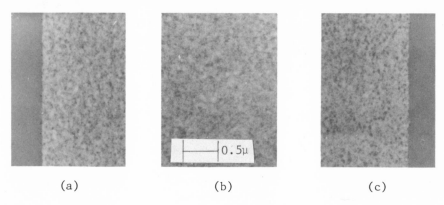

(a)   (b)   (c)

Fig. 2.   Transmission electron micrographs of a PMMA sheet
          membrane, water content of which was 65 wt% : (a) and
          (c) structure near both surfaces ; (b) internal
          structure.

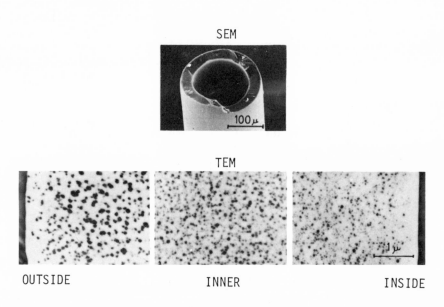

Fig. 3.   Scanning electron micrograph and transmission electron
          micrographs of a PMMA hollow fiber membrane.

hollow fiber membranes, λLp and P_2 are replaced with ultrafiltration
rate (UFR) and membrane resistance (R_M), respectively, as their
structure is not uniform.   UFR corresponds to Lp in eq. (1) and R_M
corresponds to λ/P_2 in eq. (2), and they express the overall permeability of membrane.   As for the apparatuses for measuring these

permeabilities, batch cells were used for sheet membranes and flow systems were used for hollow fiber membranes.

The rejection which is defined by eq. (3) was also measured for hollow fiber membranes.

$$R_j = 1 - \frac{C_U}{\bar{C}_B} \qquad\qquad (3)$$

C_U and $\bar{C}_B$ are solute concentrations in the filtrate and in the filtrant, respectively. As model substances, vitamin B_{12}, inulin, cytochrome C, myoglobin, pepsin, ovalbumin, albumin, and γ-globulin were used.

RESULTS AND DISCUSSION

The dependence of water permeability λL_p and NaCl permeability $P_2(NaCl)$ on water content for sheet membranes is shown in Figure 4. The ratio of isotactic to syndiotactic polymer was fixed at 1/5 and the water content was controlled by varying the polymer concentration of dope solution between 10 and 40 wt%. Water permeability is highly sensitive to water content, on the other hand, NaCl permeability is less influenced by water content. This fact can be explained by the difference of the driving forces for both permeabilities as eqs. (1) and (2) predict.

As a comparison, the data for a Cuprophane® sheet membrane, which is cellulosic and more hydrophilic than PMMA membranes, are also plotted in the same graph. The water content of Cuprophane® was 42 wt%, and at this water content λL_p of PMMA membrane was 3 to 4 times higher than that of Cuprophane®, whereas $P_2(NaCl)$ for both membranes were almost the same. This characteristic of PMMA membranes can be attributed to the hydrophobicity of the material.

Figure 5 shows the dependence of λL_p and $P_2(NaCl)$ on the content of isotactic polymer. The polymer concentration of the dope, determining factor of the water content of the membrane, was fixed at 20 wt%, and therefore the water content of the membranes thus obtained was practically the same, around 65 wt%. Note that λL_p reaches a minimum value at about 30 %, whereas $P_2(NaCl)$ remains almost constant. As already reported[3)-5)] the stereocomplex is stoichiometrically formed at the isotactic and syndiotactic polymer ratio of 1/2. Our results fit this ratio. It was found, furthermore, that the membranes with higher content of isotactic polymer, data of which were plotted extremely right in this Figure, were opaque, whereas other membranes were almost transparent. These observations suggest that the stereocomplex formation contributes to the membrane preparation. Figures 4 and 5 show that the water and solute permeabilities of PMMA membranes can be controlled

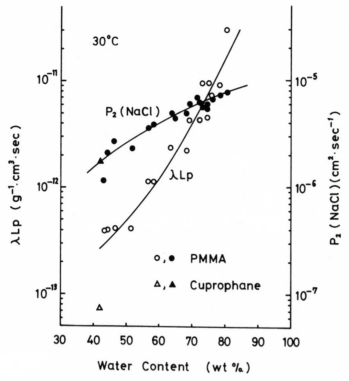

Fig. 4.   Dependence of water and NaCl permeabilities of
          PMMA sheet membranes (Iso/Syn = 1/5) and
          a cellulosic membrane on water content.

independently by varying the polymer concentration and the polymer
ratio of the dope solution.

     Figure 6 shows the dependence of UFR and R_M for urea of hollow
fiber membranes with the wall thickness of 40 microns on the polymer
concentration of the dope.   These data of hollow fiber membranes
essentially correspond to those of sheet membranes shown in Figure 4,
and well demonstrate high water permeability of PMMA hollow fiber
membranes, too.

     Figure 7 illustrates the relationship between the membrane
resistance of one kind of PMMA hollow fiber and the molecular weight
of various model solutes.   As a comparison, the data for a regenerated cellulosic hollow fiber were also plotted.   This figure shows
that the membrane resistance of the PMMA hollow fiber for the higher
molecular weight substances or so-called middle molecular substances,
such as vitamin B_{12} and inulin, is 1/3 to 1/7 of those observed
for the cellulosic hollow fiber, although the membrane thickness

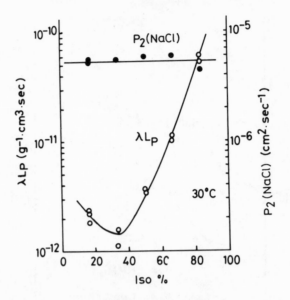

Fig. 5. Dependence of water permeability and NaCl permeability
         on the content of isotactic polymer :
         polymer concentration was fixed at 20 wt%.

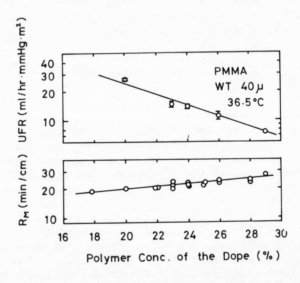

Fig. 6. Dependence of ultrafiltration rate and membrane
         resistance for urea of hollow fiber membranes on
         polymer concentration of the dope.

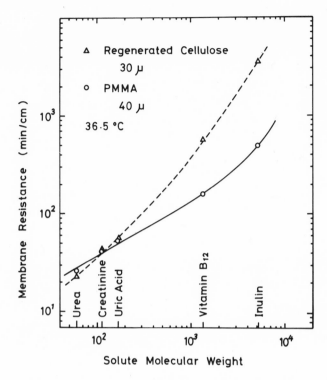

Fig. 7.   Dependence of the membrane resistance of a PMMA
          hollow fiber and a cellulosic hollow fiber on the
          molecular weight of various model substances.

of PMMA hollow fiber is larger than that of cellulosic hollow fiber.
As also shown in Figure 8, where sieving characteristics of two kind
of hollow fiber membranes are demonstrated, PMMA is more permeable
for so-called middle molecular substances than the cellulose.
Albumin, molecular weight of which is 69,000, however, is completely
rejected by the PMMA hollow fiber membrane.   Higher water permeability, lower membrane resistance and lower rejection for middle
molecular substances, suggest that PMMA hollow fiber has larger
pores than cellulosic fiber.

     Based upon these observations, clinical evaluation of PMMA
hollow fiber units not only as a conventional dialyzer but also as
a non-dialytic ultrafilter began around 4 years ago, and high UFR
has been proven.   In addition to regular type of PMMA hollow fiber
units which is now widely used for conventional dialysis, a new

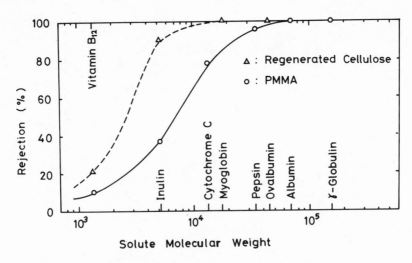

Fig. 8.   Sieving characteristics of a PMMA hollow fiber and
          a cellulosic hollow fiber.

version with higher flux has been developed for several new therapeutic modes, such as simultaneous hemofiltration/hemodialysis and
hemofiltration.

CONCLUSIONS

     PMMA sheet and hollow fiber membranes were formed utilizing
the stereocomplex phenomenon.   Their permeabilities for water and
solutes such as NaCl or urea can be independently controlled by
designing both the polymer concentration of and the ratio of isotactic PMMA to syndiotactic PMMA in the dope solution.

     Water permeability of the sheet membrane was 3 to 4 times
higher than that of cellulosic membrane, such as Cuprophane®, when
they were compared at the same water content.   Permeabilities of
a PMMA hollow fiber membrane for so-called middle molecular substances such as vitamin B_{12} and inulin were 3 to 7 times higher
than those of a cellulosic hollow fiber membrane.

     Their higher ultrafiltration characteristics have been proven
clinically, and units of PMMA hollow fiber are now clinically used
as a dialyzer and/or ultrafilter for hemodialysis, simultaneous
hemofiltration/hemodialysis, and hemofiltration therapies.

The authors wish to thank Mr. K.Yoshimura of Toray Research Center, Inc. for the electron microscopy study.

References

1) Y.Sakai and H.Tanzawa, J. Appl. Polym. Sci., 22, 1805 (1978).
2) W.H.Watanabe, C.F.Ryan, P.C.Fleischer, Jr., and B.S.Garrett,
     J. Phys. Chem., 65, 896 (1961).
3) A.M.Liquori, G.Anzuino, V.M.Coiro, M.D'Alagni, P.De Santis, and
     M.Savino, Nature, 206, 358 (1965).
4) H.Z.Liu and K.J.Liu, Macromolecules, 1, 157 (1968).
5) T.Miyamoto and H.Inagaki, Polym. J., 1, 46 (1970).
6) F.A.Bovey and G.V.D.Tiers, J.Polym. Sci., 44, 173 (1960).

ADVANCES IN HOLLOW FIBER

ULTRAFILTRATION TECHNOLOGY

Barry R. Breslau
Anthony J. Testa
Bradford A. Milnes
Gunars Medjanis

Romicon, Inc.
100 Cummings Park
Woburn, Massachusetts 01801

INTRODUCTION

The success of a membrane system in a given application is
often dependent on a recognition of the unique demands that the application places on processing as well as membrane technology. Often
success can only be achieved by combining advances in one form of
technology with those of the other. In this paper, examples of this
type are presented; they not only introduce new developments in membrane research but also demonstrate the diversity of hollow fiber
processing.

HOLLOW FIBER OPERATION

The operation of Romicon's wide diameter hollow fibers, illustrated in Figures 1 and 2, is analogous to that of permeable tubes
with the process stream flowing through the center of the fiber and
the permeate stream passing through the walls.

As in tube flow, the conventional laws of hydrodynamics apply
and the flow regime may be either laminar or turbulent depending
on the flow characteristics of the fluid and the pressure drop along
the fiber length (Table 1). Furthermore, just as a moving fluid
exerts a shear force along the tube wall in tube flow, the process
fluid exerts a similar force along the active membrane surface.
This force tends to sweep the surface thereby minimizing fouling.

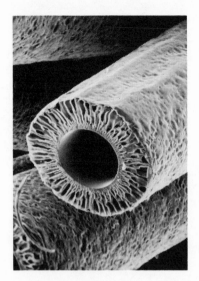 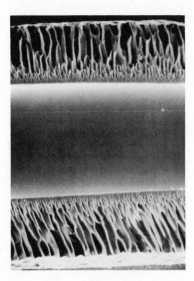

Fig. 1. Photomicrographs of a Romicon hollow fiber showing the
         inside and outside diameters, the inside active membrane
         surface and the outer support structure.

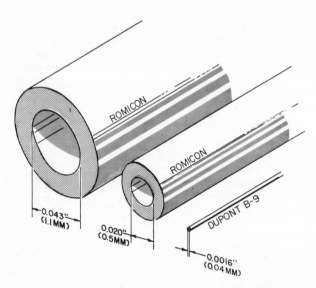

Fig. 2. Scale drawing showing the inside diameter of Romicon's
         wide diameter ultrafiltration fibers compared to DuPont's
         reverse osmosis fiber.

Table 1. Flow Characteristics for Water at 50°C as a
Function of Pressure Drop and Fiber Dimensions

ΔP, psi	Diameter, Inches	Length, Inches	Velocity, fps	Reynolds Number
10	0.020	43.0	2.7	740
10	0.020	25.0	5.5	1540
10	0.043	43.0	6.1	3630
10	0.043	25.0	8.0	4780
15	0.020	43.0	4.0	1110
15	0.020	25.0	8.8	2460
15	0.043	43.0	7.5	4450
15	0.043	25.0	9.8	5850
20	0.020	43.0	5.3	1480
20	0.020	25.0	10.2	2840
20	0.043	43.0	8.6	5140
20	0.043	25.0	11.3	6760

This is an important point and is one of the principal advantages of having the active membrane surface on the inside of the fiber. If fouling does occur, it can generally be removed by backflushing without any risk of damage to the fibers.[1]

The analogy to tube flow should always be considered when specifying Romicon hollow fibers. These fibers are available in different lengths and diameters and the effect of these dimensions on performance is often critical. For example, when a relatively viscous stream is to be processed, the widest diameter fiber should be specified in the shortest length. This will result in the highest velocity for a given pressure drop because the frictional drag will be minimized. This same fiber should also be chosen for streams where cleaning is expected to be difficult because it offers the capability of cleaning under conditions of maximum hydrodynamic shear. On the other hand, if the process fluid is relatively clean and can be processed under conditions of low flow, a long, small diameter fiber should be selected. Although this fiber is limited in its applications, it offers significant economic advantages as a result of its larger membrane area per unit volume and lower pumping requirements.

APPLICATIONS OF HOLLOW FIBER ULTRAFILTRATION

As the following examples illustrate, the properties of the fluid being ultrafiltered determine not only the choice of the fiber

but also the processing technique. In the instance of electrodeposition paint, the fibers are faced with a fluid that requires a
high flow to prevent fouling; whereas in the production of ultrapure water, the fibers are faced with a very low fouling stream
which can be processed under conditions of low flow and high recovery.

ELECTRODEPOSITION PAINT

The electrodeposition paint process is one of the oldest commercial applications of ultrafiltration and its general acceptance
is so widespread that ultrafiltration equipment is considered a
standard part of all electrodeposition systems.

In electrodeposition, the work piece to be coated functions as
an electrode and undergoes a typical set of electrode reactions.[2]
However, two methods of electrocoating are available, anodic and
cathodic. In anodic paint tanks the work piece becomes the anode
and the paint is deposited by an oxidation reaction, while in cathodic paint tanks the work piece becomes the cathode and the paint
is deposited by reduction. The inherent advantages of the cathodic
process are due to the fact that the piece being coated (the cathode)
is passive and is not affected by the reaction process. Conversely,
the work piece is active during the anodic process and results in
a small fraction of substrate dissolution. The inclusion of these
dissolved metal ions in the organic coating reportedly reduces the
effective corrosion protection of the anodically deposited film.[3,4]
As a result of the increased corrosion resistance, large scale
users, particularly automotive manufacturers, are rapidly converting
their anodic systems to the newly developed cathodic paint.[5]

The Role of Ultrafiltration in Electrocoating

When the coated work piece is withdrawn from the paint tank,
it carries with it a considerable quantity of excess paint referred
to as drag-out. This drag-out represents a direct loss of very expensive paint unless it can be recovered and returned to the tank
in a useable form. Obviously, rinsing the work piece with a spray
of deionized water would remove the paint, but the water dilution
would affect the stability of the paint emulsion making it unuseable.
The use of a solvent solution would overcome this, but the addition
of excess solvent to the tank would alter the delicate chemical
balance. Clearly, the best solution is to use an ultrafiltration
system operating directly on the paint to provide sufficient paint
permeate to rinse off the drag-out. In this application, the ultrafilter operates at low recovery and the paint solids are not
increased by more than 1%. The permeate stream is sent to a closed
loop, counter-current rinse system to wash the work piece and is

eventually returned to the paint tank with the recovered paint; the retained paint stream is returned directly to the paint tank. Since both permeate and retained streams are returned to the tank, the solids level and composition of the paint remain unaffected by this operation.

Anodic Paint and the XM50 Membrane

The XM50 membrane is a vinyl copolymer with a molecular weight cutoff of 50,000. This membrane is used extensively in waste and chemical treatment applications. The slight electronegativity of this membrane made it the logical choice for use with the negatively charged anodic paint and its excellent performance is attested to by over seventy systems operating in the field. In general, because of the high flow requirement, short, wide diameter (43 mil) fibers are preferred for anodic paint; however, long, wide diameter (43 mil) fibers have also been successfully employed. Typical data showing the flux versus time performance of the XM50 membrane on anodic paint is presented in Figure 3. While this membrane gives extremely stable fluxes over a long period of time, it can be cleaned with a solution of the solvents normally used in the paint makeup, such as butyl cellosolve.

The only processing requirements for anodic paint are: (1) rapid paint velocity, which is achieved with a 20 psi pressure drop across the fiber length, and (2) routine backflushing with permeate when the flux rate has declined to the minimum acceptable level. These procedures will generally insure stable fluxes over a considerable period of time.

Cathodic Paint and the CXM Membrane

Initial studies on cathodic paint began with the determination of the performance of the XM50 membrane. This membrane experienced severe flux decline, almost on contact with cathodic paint, and attempts to restore fluxes by frequent backflushing proved only partially successful. While work continued on the XM50, it soon became apparent that cathodic paint would require a completely new membrane. Reasoning that the rapid fouling of the XM50 membrane was due to electrostatic interaction between the positive functionality of the cathodic paint and the negative membrane surface, development of a membrane with a polymer matrix having positive functional groupings was undertaken. This was indeed the right direction to take as shown in Figure 4 which compares the flux performance of two early cathodic paint membranes (designated CXM) versus the XM50 membrane. Subsequent work carried out at automotive manufacturing facilities in Massachusetts and Michigan led to the optimization of the CXM membrane as indicated by the data presented

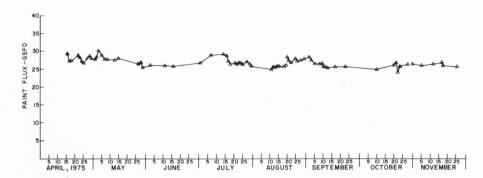

Fig. 3. Flux versus time performance data for the XM50 membrane
         on anodic paint (PPG).

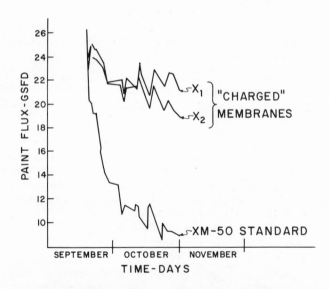

Fig. 4. Comparative data showing the flux versus time performance
         of the XM50 membrane versus that of the newly developed
         "charged" membranes on cathodic paint (PPG). The desig-
         nation X_1 and X_2 refer to different "charge density" levels.

in Table 2. This data shows consistent fluxes despite a number of periods where the system was shut down due to plant maintenance programs and holidays. Such shutdowns produced severe flux declines with the XM50 membrane. (While frequent mention is made in Table 2 of the use of an acetic acid soak during the weekends, this has since proved to be an unnecessary procedure. Scheduled shutdowns are now accomplished simply by flushing and soaking in deionized water.) To better show the stable flux performance of the CXM membrane, the data of Table 2 is presented graphically in Figure 5 for reference.

The CXM membrane is a permanently "charged" membrane and the extent of its "charge", or its "charge density", is critical in determining its performance on cathodic paint. To quantify the "charge", the membrane is characterized by its ability to reject divalent ions, a unique capability for an ultrafiltration membrane and the result of Donnan exclusion.

The CXM membrane is now a standard Romicon product and it is used in all cathodic paint applications. Its success has stimulated extensive research on developing a second generation of "charged" membranes for applications heretofore excluded from consideration by ultrafiltration membranes. The capabilities of this technology when extended beyond the "charge density" level required by cathodic paint is illustrated in Figure 6 which shows the rejection characteristics of a test membrane on simple divalent waters as a function of volume reduction. This type of membrane will open up new applications for ultrafiltration in the near future.

Processing Techniques for Cathodic Paint

The performance of the "charged" CXM membrane on cathodic paint is greatly enhanced by the self-cleaning capabilities of hollow fiber cartridges. These capabilities are reviewed below with reference to Figures 7 through 9.

Hollow fiber cartridges are normally mounted vertically as shown in Figure 7. Paint enters the cartridge from either the bottom or top at a pressure of 25 psi, and exits at a pressure of 5 psi. The 20 psi pressure drop across the cartridge provides a paint flow of 30 to 35 gpm. As the paint flows through the fibers there is sufficient pressure (15 psi average internal pressure) to generate permeate which, on filling the shell, exits through the upper permeate port en route to the rinse station.

In normal operation, the shell or permeate side of the cartridge is typically at a pressure of 1 to 2 psi due to frictional losses in the pipeline downstream to the rinse station. Periodically, however, the flow of permeate is intentionally interrupted

Table 2.   Performance Data for the CXM Membrane
on Cathodic Electrodeposition Paint*

Test Period, Days	Individual Cartridge Fluxes,			Average Fluxes, GSFD	Standard Deviation, GSFD
	#1 GSFD	#2 GSFD	#3 GSFD		
1	26.9	26.8	26.0	26.6	0.49
2	27.7	26.9	26.6	27.1	0.57
3	30.4	29.9	29.4	29.9	0.50
4	30.6	31.4	30.6	30.9	0.46
5	32.9	31.0	32.6	31.8	1.02
6-9**					
10	30.4	30.9	29.4	30.2	0.76
11	30.3	31.8	31.0	31.0	0.75
12	29.7	28.5	28.2	28.8	0.79
13	30.3	29.2	28.5	29.3	0.91
14	25.9	24.5	24.0	24.8	0.98
15-16**					
17	26.8	26.3	25.9	26.3	0.45
18	27.2	27.4	27.1	27.2	0.15
19	26.8	25.9	26.0	26.2	0.49
20	26.2	26.8	25.6	26.2	0.60
21	28.0	27.4	27.7	27.7	0.30
22-23**					
24	27.5	26.0	26.1	26.5	0.84
25	26.8	26.3	26.3	26.5	0.29
26	27.7	26.8	27.1	27.2	0.46
27	29.2	28.2	28.2	28.5	0.58
28	29.8	28.3	29.5	29.3	0.79
29-38**					
39	30.4	31.2	30.6	30.7	0.42
40	30.1	29.2	30.1	29.8	0.52
41	30.4	28.5	29.8	29.6	0.97
42	29.8	29.1	30.6	29.9	0.75
43-44**					
45	27.1	24.4	27.4	26.3	1.65
46	29.4	25.3	28.8	27.8	2.21
47	28.9	25.0	28.6	27.5	2.17
48	30.3	24.7	30.1	28.4	3.18
49	30.6	25.9	31.4	29.3	2.97
50-53**					
54	30.4	22.5	30.0	27.6	4.45
55	28.9	21.9	29.2	26.7	4.13
56-58**					
59	28.1	25.0	28.6	27.2	1.95
60	25.9	19.5	26.6	24.0	3.92
61	31.2	30.3	28.2	28.9	1.54
62-65**					
66	30.1	31.2	28.9	30.1	1.15
67	27.2	24.2	27.4	26.3	1.79
68	26.0	21.6	26.8	24.8	2.80
69	28.6	27.1	28.0	27.9	0.755

* PPG Cathodic Paint
** System generally stored in acetic acid during scheduled shutdowns

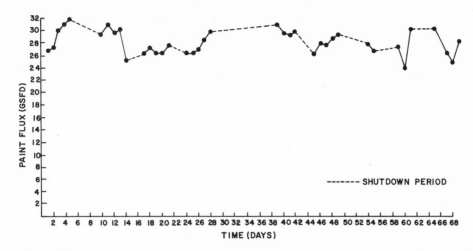

Fig. 5. Flux versus time performance data for the CXM membrane
on cathodic paint (PPG). Data represents average values
of three independent test cartridges.

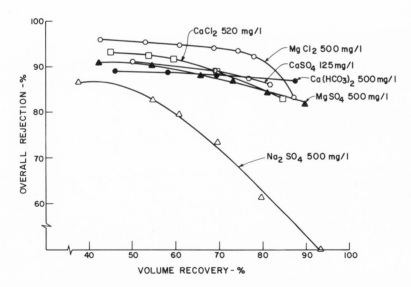

Fig. 6. Rejection versus volume reduction data for a highly
"charged" membrane operating on simple divalent ions.

by closing the upper permeate port (Figure 8). Whenever this occurs the cartridge is said to be in a recycling position as paint continues to flow but permeate does not leave the cartridge. In this condition the pressure within the shell of the cartridge will climb to approximately 15 psi which is the average of the inlet and outlet paint pressures. Since the internal pressure at the inlet of the cartridge (the high pressure side) is still greater than the shell side pressure, permeate will continue to be produced from this half of the cartridge (Figure 9). The shell, however, is at a higher pressure than the paint at the discharge or exit half of the cartridge, and hence there is a continuous flow of permeate from the shell back into the fiber. This "backflushing" action covers the entire discharge or low pressure side of the cartridge. The recycling operation therefore provides a simple route for backflushing one half of the cartridge without interrupting paint flow or introducing foreign chemicals. The backflushing action of recycling is extremely effective because it is coupled with the continued hydrodynamic sweeping action of the paint itself as it continues to flow. The entire cartridge can be backflushed by combining the recycling operation with a change in direction of paint flow as illustrated in Figures 8 and 9.

The reverse flow and recycling operations described above are flux maintenance procedures as they achieve a cleaning effect without

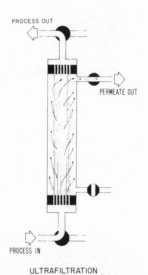

PROCESS OUT

PERMEATE OUT

PROCESS IN

ULTRAFILTRATION

Fig. 7.   Schematic drawing showing the operation of a hollow fiber cartridge in the ultrafiltration mode with the direction of flow upward.

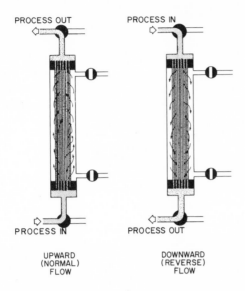

Fig. 8. Schematic drawings showing the operation of a hollow
         fiber cartridge in the recycle mode for both upward
         and downward flows.

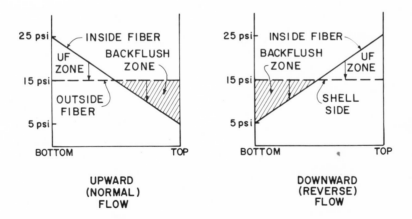

Fig. 9. Pressure profile in a hollow fiber cartridge during the
         recycling operation showing the ultrafiltration and
         backflushing zones.

a major interruption of the process. These operations are
routinely automated on all cathodic paint systems and they serve
to maximize the long stable performance of the CXM membrane.

Processing Techniques for Cleaning

The need to clean a cathodic paint system is usually dictated
by the pretreatment line. In all electrodeposition systems, the
surface to be coated must be thoroughly cleaned and prepared before
electrodeposition by exposing it to a number of chemical treatment
steps which normally end with a deionized water rinse. Depending
on the effectiveness of the final rinse, some of the pretreatment
chemicals may be carried into the paint tank. These contaminants
(typically phosphates) accumulate in the paint tank. In anodic
paint tanks no serious interaction has been reported between the
paint and dragged-in phosphates; in cathodic paint tanks, however,
the phosphates and paint have been found to react forming an insoluble lead phosphate complex which tends to foul the membrane and
reduce its flux. When this occurs, the system must undergo chemical
cleaning.

The reverse flow and recycling operations described above for
routine flux maintenance become powerful cleaning techniques when
used with a cleaning solution. In this instance "cleaning solution"
permeate is used to backflush the membrane while simultaneously
exposing it to the hydrodynamic sweeping action of the cleaning
solution itself. This is an extremely effective technique and the
typical cleaning time for a large scale cathodic paint system is
only twenty to thirty minutes.

The effectiveness of the cleaning cycle in restoring fluxes
is shown in Figure 10 which presents data obtained at a Georgia
automotive facility. This plant is typically cleaned every three
to four weeks.

It should be mentioned that cathodic paint systems do not
always require cleaning as evidenced by Figure 11 which was obtained at a metal fabricating plant in Illinois. This system has
operated continuously since December, 1978 without any cleaning
program other than the routine flux maintenance techniques of reverse flow and recycling. The performance of this system reflects
the effectiveness of the flux maintenance program.

WATER

Ultrafiltered water can be broadly characterized as being free
of colloidal and suspended solids, microorganisms and high molecular
weight contaminants. It is used in a wide range of applications

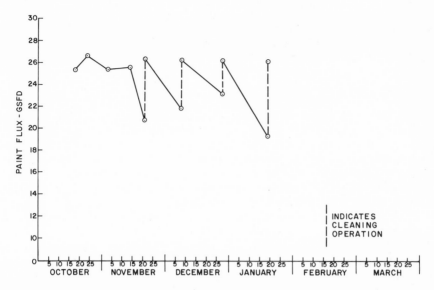

Fig. 10.   Flux versus time performance data showing the effect of
           cleaning a CXM membrane system at an automotive facility
           in Georgia.   The paint type is cathodic.

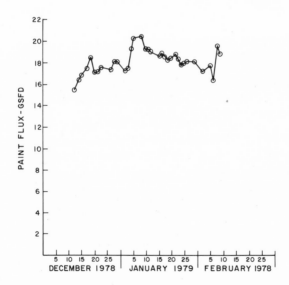

Fig. 11.   Flux versus time performance data for a CXM membrane
           system at a metal finishing facility.   The paint type
           is Glidden cathodic.

throughout industry but its most well known application is in the electronics industry which requires ultrapure, 18 megohm, water.[6] Other important applications are its use for pyrogen removal in the pharmaceutical industry and in the production of particle-free water for the chemical processing industry.

The ultrafiltration of water is distinctly different from that of paint because water is a relatively clean, non-fouling stream and can be processed under conditions of extremely low flow. Water therefore can be processed in single pass operation with the smaller 20 mil diameter fibers. While this method is not suitable for most industrial streams, significant economic advantages can be realized on those streams where it is feasible. The capability of operating in either a recirculating high flow domain or a single pass low flow domain is one of the most significant advantages of hollow fibers.

Water Processing and the New GM80 Membrane

The new GM80 membrane is a high flux acrylic copolymer with a molecular weight cutoff rating of 80,000. It was introduced in 1979 as a high flux, high strength replacement for the original GM80 membrane. While both membranes have the same molecular weight cutoff, the new GM80 membrane is approximately twice as strong with a burst strength of over 200 psi. It was developed specifically for water processing and is available only in the smaller 20 mil diameter fiber size.

Processing Techniques for Single Pass Operation

The single pass operation of a hollow fiber cartridge is shown schematically in Figure 12. The feed stream, water in this case, enters the process side of the cartridge typically at a pressure of 25 psi and a flow rate of 5 gpm where it produces permeate at a rate of 90 to 95% of the feed (90% recovery shown, 4.5 gpm). The difference between the feed and permeate, or product, streams represents the waste or "blowdown". While the "blowdown" is not generally recognized as being "dirty", it does contain concentrated amounts of those species present in the feed that were retained by the membrane. Thus, if the feed stream for the preceding example contained 100 ppm suspended solids, the waste stream would contain ten (10) times as much (5.0 gpd feed/0.5 gpd bleed) or 1000 ppm suspended solids. This increment in suspended solids occurs gradually along the cartridge length as the feed stream is progressively dewatered so that the "stream" being processed at the exit or discharge end of the cartridge is considerably "dirtier" than the initial feed stream. As a result of this concentrating effect, the membrane surface at the discharge end of the cartridge tends to foul significantly more rapidly than the lower or inlet end. This

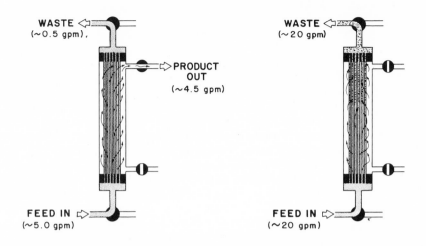

SINGLE PASS OPERATION   FAST FLUSH OPERATION

Fig. 12. Schematic drawings showing the operation of a hollow
          fiber cartridge in the single pass and fast flush mode.

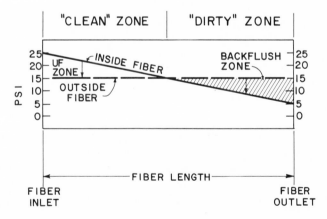

Fig. 13. Pressure profile in a hollow fiber cartridge during the
         fast flushing operation showing the ultrafiltration and
         backflushing zones.

effect is easily observed, even on water, by a gradual darkening of the discharge end if it is not exposed to regular cleaning.

The pressure of the waste stream is typically 23 to 24 psi as this stream must be throttled to get the desired permeate rate. Therefore, the pressure drop across the cartridges is only 1 to 2 psi and the hydrodynamic shear effect of the process fluid on the membrane surface, which is critical for flux maintenance in processing paint, is virtually negligible. The shear effect must be increased for flux maintenance and cleaning, and this is accomplished simply by opening the waste or "blowdown" valve to allow for a greater flow of feed (Figure 12). If this operation is coupled with the simultaneous closing of the permeate line, the single pass system will experience the same backflushing effect of the "recycling" operation described earlier for paint. However, in single pass systems, this "fast flushing" operation is especially effective because the backflushing action is directed specifically toward the "dirty" or discharge half of the cartridge as shown in Figure 13.

Performance Data for Water

Performance data on city water from Woburn, Massachusetts is presented in Figure 14. This data was obtained using GM80 membranes operating at a recovery of 90%, (i.e., 90% of the feed water was collected as permeate). The data shows that relatively stable fluxes were obtained by cleaning the system twice a day with a fast flush-backflush-fast flush cycle. The significant increase which resulted after approximately 220 hours of continuous operation was due to the addition of 50 ppm of sodium hypochlorite to the backflush tank once per day for sanitizing purposes. Once the sanitizing step was added, the fast flush-backflush-fast flush cycle became extremely effective in maintaining fluxes. This type of performance is typical of virtually all hollow fiber water systems.

Low flow, single pass, water processing systems offer favorable economics for both capital and operating costs. Since the high flux GM80 water processing cartridges contain the most membrane area per cartridge offered by Romicon (53 ft^2), the required number of cartridges and the size of associated equipment is minimized. In addition, power and labor costs are negligible, and the only major operating costs are depreciation and membrane replacement.

SUMMARY

Examples have been given of two diverse membrane applications which illustrate advances in membrane development and processing

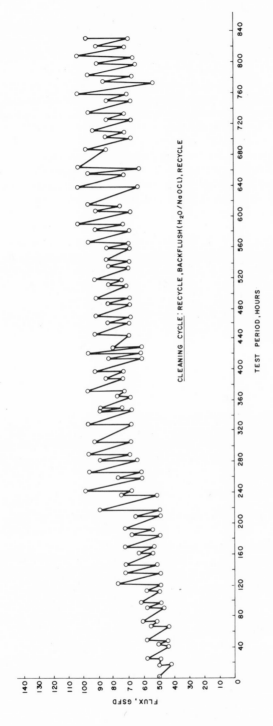

Fig. 14. Flux versus time performance data for the GM80 membrane operating on city water from Woburn, Massachusetts. Single pass operation was used with a permeate recovery rate of 90%.

technology. Cathodic electrodeposition paint is a highly fouling
stream which represented a challenge to all membrane systems because of its electrostatic interaction with conventional membrane
surfaces. To cope with this stream, the "charged" CXM membrane
was developed which has the ability to reject divalent ions. This
membrane resists surface fouling by preventing permanent bonding.
Processing techniques were then introduced for flux maintenance
which insure that the membrane surface is continuously cleansed
of any possible build-up of foulants. The "recycling" operation
combines the hydrodynamic shearing effect of the paint itself with
a backflushing action caused by the continuous flow of permeate.
Reversing the flow direction insures that the backflushing effect
covers the entire membrane surface. These operations coupled with
the natural tendency of the CXM membrane to resist fouling insure
long and stable performance.

Water is a completely different application in that it is a
relatively non-fouling stream and should be processed under conditions of low flow for economic reasons. However, even "non-fouling"
streams when operated under low flow conditions will result in
gradual surface fouling with time. Techniques were therefore introduced to show how water can be processed in the single pass mode
with the flux maintained by a periodic "recycle" or "fast flush"
operation.

ACKNOWLEDGEMENT

We would like to acknowledge the cooperation of the Ford Motor
Company, the General Motors Company and B-Line Industries for allowing Romicon to conduct pilot studies at their facilities.

REFERENCES

1. B. R. Breslau, E. A. Agranat, A. J. Testa, S. Messinger,
    and R. A. Cross, Hollow Fiber Ultrafiltration, Chem.
    Eng. Prog., 71:12 (December, 1975).
2. D. G. Anderson, E. J. Murphy, and J. Tucci III, Cathode
    Reactions and Metal Dissolution in Cationic Electrode-
    position, J. Coatings Tech., 50:648 (November, 1978).
3. E. J. Murphy, Cationic Electrocoating: Myth Versus
    Reality, Publication FC78-422 (1978), The Association
    for Finishing Processes of SME.
4. E. L. Jozwiak, Jr., Cathodic Rather than Anodic - Is
    There Justification, Publication FC79-697 (1979),
    The Association for Finishing Processes of SME.
5. C. W. Corray, Anodic Versus Cathodic Electrocoating, A
    User's Changeover and Experience, Technical Papers,
    Surface Coating 1979 (Chemical Coaters Association).

6. R. Lukacik, Ultrafiltration Applications in the Prepara-
    tion of Pure Water for Electronics Applications,
    Proceedings, 178th ACS National Meeting, Division of
    Colloid and Surface Chemistry.

TRANSPORT BEHAVIOR OF ASYMMETRIC POLYAMIDE FLAT SHEET MEMBRANES AND HOLLOW-FINE FIBERS IN DIALYSIS-OSMOSIS AND HYPERFILTRATION EXPERIMENTS

W. Pusch

Max-Planck-Institut für Biophysik, Kennedy-Allee 70
6000 Frankfurt am Main, W. Germany

ABSTRACT

Using 0.1 molar brine solutions of $NaCl$, Na_2SO_4, $MgSO_4$, $NaNO_3$, $CaCl_2$, Na_3PO_4, and $AlCl_3$, the transport parameters of the phenomen-ological relationships such as the the hydrodynamic, ℓ_p, and the os-motic, ℓ_π, permeability as well as the asymptotic salt rejection, r_∞, were determined for asymmetric Du Pont Aramid (B-9) flat sheet membranes by means of hyperfiltration experiments. Moreover, the same transport parameters were obtained for hollow-fine fiber mod-ules made from the same polymer (B-9) as well as the polymer employ-ed in seawater devices (B-10) using 0.1 molar and 0.5 molar $NaCl$ brine solutions. In addition, the transport parameter ℓ_p, and the reflection coefficient σ were measured in dialysis-osmosis experim-ents as functions of $NaCl$ concentration using flat sheet membranes (B-9) and applying different boundary conditions.

The asymptotic salt rejection, r_∞, is always beyond 0.9800 for all the salts used. Depending on the membrane sample, even values beyond 0.9900 are obtained for the different salts. The experiment-al findings clearly demonstrate that there is no qualitative nor quantitative difference between flat sheet membranes and hollow-fine fibers if the comparison is made on the basis of those transport parameters which do not depend on the thickness of the active layer such as the salt permeability, P_s, for instance. The hydrodynamic permeability of the polyamide membranes used varies drastically with the salt applied due to swelling effects. The lowest hydrodynamic permeability results for Na_3PO_4 while the largest value is obtained with $NaCl$ brine solutions. Furthermore, the flat sheet membranes ex-hibit strong asymmetry effects in dialysis-osmosis experiments under boundary conditions far from near-equilibrium. These asymmetry effects originate from concentration polarization effects in the porous matrix of the asymmetric membranes.

INTRODUCTION

During the early stages of brackish and seawater desalination by hyperfiltration (reverse osmosis') Loeb-Sourirajan type or asymmetric cellulose acetate (CA) membranes[1] were exclusively used because of their comparably larger water permeability and high salt rejection[2-4]. The excellent transport characteristics of these membranes relative to salt solutions are due to the asymmetric structure and a poor solubility of salts in the so-called active layer of these membranes. As shown by Merten et al.[5], the asymmetric CA membranes consist of a very dense layer (active layer) on top of a porous matrix acting as an excellent natural support for the active layer which is considered to be equivalent to a very thin dense homogeneous CA membrane. On the other hand, asymmetric CA membranes exhibit several disadvantages such as large compaction effects under pressure[6], degradation by hydrolysis[7], a low stability against chemical attack (acids, basis, organic solutes, oxidizing agents, etc.), and a comparably large flux decline with time[8]. Since the membrane compaction is mainly due to the lower strength of the porous matrix, Cadotte et al.[9] as well as Riley et al.[10] developed so-called composite membranes in order to minimize the compaction. A breakthrough in this respect was the discovery and development of polyamides as permeable film forming materials[11-13] which are superior to CA in many respects. Du Pont, for instance, commercialized the aramid (aromatic polyamide) membranes B-9 for brackish and B-10 for seawater desalination and these membranes combined the durability of material with the desirable characteristics of asymmetric membranes, namely a dense skin with a very porous matrix. In casting membranes from polyamide solutions, therefore, two different procedures were adopted. Asymmetric membranes, on the one hand, and composite ones, on the other hand, were cast and assembled in 'hollow-fine fiber modules'[14] and 'spiral wound modules'[15], respectively.

A great deal of research was devoted to the characterization of solute and solvent transport across homogeneous and asymmetric CA membranes by transport and partition coefficients employing various transport relationships[16-20]. Yet, only a very few papers deal with the transport properties of non-cellulosic membranes. Very recently, membranes prepared from poly(4-vinyl pyridine)-acrylnitrile copolymers[21] and polyamides[22,23] were characterized in hyperfiltration and partially in dialysis-osmosis experiments. No complete characterization of asymmetric polyamide membranes by appropriate transport parameters is available from the literature. Thus, the present measurements were performed to fill in this hiatus.

EXPERIMENTAL

The transport parameters of the phenomenological relationships of the thermodynamics of irreversible processes are determined using two different experimental setups, dialysis-osmosis and hyperfiltration equipment as described in detail elsewhere[20,24]. First, the hydrodynamic permeability, ℓ_p, and the reflection coefficient, σ, of a flat sheet asymmetric aramid (B-9) membrane are measured in dialysis-osmosis experiments as functions of the NaCl concentration of the external solutions applying the following three different boundary conditions (see Figure 1):

a: $0 \leq c_s' \leq 0.5$ mol/l NaCl, $c_s'' = 0$, and thus $\Delta c_s/c_s' \simeq 1$;

b: $0 \leq c_s', c_s'' \leq 0.5$ mol/l NaCl, $c_s' > c_s''$, and $\Delta c_s/c_s' \ll 1$;

c: $0 \leq c_s', c_s'' \leq 0.5$ mol/l NaCl and $c_s' = c_s''$.

The last two boundary conditions correspond to systems near equilibrium whereas the first boundary condition leads to systems far from equilibrium at least at larger concentrations c'. With these boundary conditions established, the volume flux, q (cm/s), is measured as a function of the pressure difference, ΔP, across the membrane. Employing the following linear relationship yields the hydrodynamic permeability and the reflection coefficient by means of a regression line analysis:

$$q = \ell_p (\Delta P - \sigma \Delta \Pi)$$ (1)

where $\Delta P = P' - P''$ and $\Delta \Pi = \Pi' - \Pi''$ are the pressure and osmotic pressure difference across the membrane in atm, for instance.

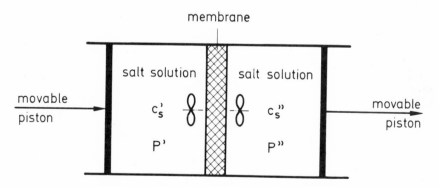

Figure 1: Diagrammatic presentation of an isothermal membrane system.

Second, the hydrodynamic, ℓ_p, and the osmotic, ℓ_π, permeability
as well as the asymptotic salt rejection, r_∞, are evaluated from
hyperfiltration data obtained with flat sheet asymmetric aramid (B-9)
membranes and hollow-fine fiber mini-modules of B-9 and B-10 by measuring the volume flux, q, and the salt rejection, $r = (c_s' - c_s'')/c_s'$,
as functions of the pressure difference, ΔP, across the membrane
maintaining the brine concentration constant at $c_s' = 0.1$ mol/l salt
in each case. Therewith, c_s'' is the salt concentration of the product (filtrate). The salt concentrations of the brine and product
solutions are analysed by atomic absorption spectroscopy (AAS) using
a Perkin-Elmer Atomic Absorption Spectrophotometer AAS 503 (Perkin &
Elmer, Bodenseewerk, Lindau, Germany). The volume flux and the salt
rejection were always measured at the steady-state of the system
which was reached when the concentrations c_s' and c_s'' no longer changed
with time. Six membrane samples taken from the same membrane sheet
were tested. Before the actual test measurements were performed, the
membranes were pre-pressurized for about three days at $P' = 60$ atm
using pure water as the feed solution to minimize compaction effects.
The transport parameters ℓ_p, ℓ_π, and r_∞ were then evaluated by means
of a regression line analysis employing Equation (1) and the follow-
ing relationship[25] using $\sigma = r_\infty$:

$$1/r = 1/r_\infty + (\ell_\pi/\ell_p - r_\infty^2) \cdot (\ell_p \Pi'/r_\infty) \cdot (1/q) \tag{2}$$

where Π' is the osmotic pressure of the brine.

RESULTS AND DISCUSSION

The hydrodynamic permeability, ℓ_p, and the reflection coefficient, σ, obtained from dialysis-osmosis experiments with a flat sheet
B-9 membrane are displayed in Figures 2 and 3 as functions of the
NaCl concentration of the external solutions. As is obvious from Figure 2, the hydrodynamic permeability, ℓ_p, depends strongly on both
the salt concentration c_s' and the boundary conditions used in measuring it. The substantial dependence of ℓ_p on c_s' as evident from the ℓ_p
vs. c_s' curve obtained under the boundary condition a) is due to swelling effects. The swelling or deswelling, respectively, affects mainly
the water content of the active layer of the asymmetric membrane and
thus leads to a variation of its hydrodynamic permeability. On the
other hand, measurements of ℓ_p, performed under the boundary conditions b) and c), yield a much stronger decay of ℓ_p with increasing
c_s' or $\tilde{c}_s = (c_s' + c_s'')/2$, respectively, than in case a). This stronger
decay with increasing c_s' is, in addition to swelling effects, due to
concentration polarization effects in the porous matrix of the asymmetric membrane as is well known from equivalent measurements with
asymmetric CA membranes[24].

In contrast with the experimental results relative to the variation of ℓ_p with the boundary conditions, there exists no such pro-

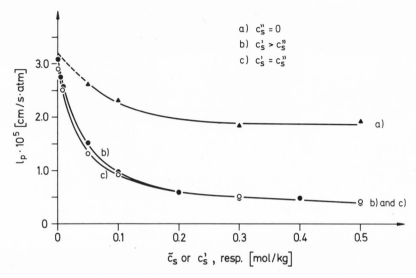

Figure 2: Hydrodynamic permeability, ℓ_p, as a function of mean salt concentration, $\tilde{c}_s$, or salt concentration, c'_s, respectively, at 25°C using NaCl solutions and a B-9 flat sheet membrane.

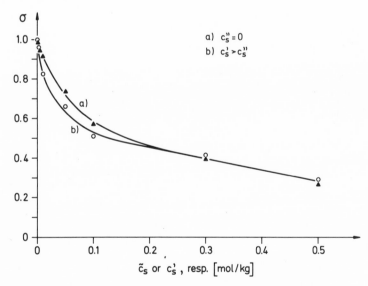

Figure 3: Reflection coefficient, σ, as a function of mean salt concentration, $\tilde{c}_s$, or salt concentration, c'_s, respectively, at 25°C using NaCl solutions and a B-9 flat sheet membrane. Reflection coefficient measurements require $\Delta c_s \neq 0$, thus no reflection coefficients are obtained from measurements under boundary condition c).

nounced difference in the reflection coefficients σ measured under the two boundary conditions a) and b) as is evident from Figure 3. These experimental findings indicate that the reflection coefficients evaluated from the experimental data by means of Equation (1) are nearly identical to the reflection coefficients σ' of the active layer of the asymmetric membrane at the corresponding salt concentrations. This conclusion implies that the reflection coefficient of the porous matrix is zero or nearly zero.

On the other hand, the reflection coefficient σ varies in the same manner with the salt concentration c_s' as the hydrodynamic permeability ℓ_p. At very small external salt concentrations, σ approaches the limiting value of 1. With increasing salt concentration, on the other hand, σ decreases drastically and possesses a value of about 0.3 at $c_s' = 0.5$ mol/1 NaCl. Since ℓ_p varies by about 35% of its value at $c_s' = 0$ when the concentration c_s' changes from zero to 0.5 mol/1 NaCl, the coefficient $A = \sigma \ell_p$ changes by more than a factor of ten over the concentration range used. Therefore, the strong change of A with c_s, as also found by Loeb et al.[23], is to a large extent a consequence of the variation of σ with c_s' or $\tilde{c}_s$, respectively. The substantial decay of σ with increasing salt concentration corresponds to a considerable increase of the salt partition coefficient K_s with increasing c_s.

TABLE 1

Transport Parameters of Asymmetric Polyamide B-9 Flat Sheet Membranes

(Brine Concentration $c_s^2 = 0.10$ mol / l)

Salt	r_∞		$l_p \cdot 10^5$ (cm/s·atm)		$l_\pi \cdot 10^5$ (cm/s·atm)		l_π / l_p	
	B-9/1	B-9/2	B-9/1	B-9/2	B-9/1	B-9/2	B-9/1	B-9/2
Na Cl	0.9980	0.9938	1.400	1.739	1.484	1.948	1.057	1.118
$Na_2 SO_4$	0.9996	0.9973	1.272	1.597	1.284	1.629	1.007	1.019
$Mg SO_4$	0.9989	0.9943	1.040	1.258	1.058	1.331	1.017	1.058
$Na NO_3$	0.9936	0.9868	1.099	1.346	1.348	1.766	1.227	1.312
$Ca Cl_2$	0.9943	0.9803	0.999	1.176	1.047	1.321	1.048	1.123
$Na_3 PO_4$	0.9912	0.9803	0.962	1.119	0.969	1.112	1.007	0.994
$Al Cl_3$	0.9921	0.9851	1.094	1.197	1.139	1.276	1.041	1.066

The transport parameters ℓ_p, ℓ_π, and r_∞ obtained from hyper-filtration experiments with a B-9 flat sheet membrane for a constant brine concentration of $c_s' = 0.1$ mol/l salt are collectively shown in Table 1. Therewith, the transport parameters of two different membrane samples, taken from the same batch, are exhibited. Sample 1 exhibited the best performance of the six sample membranes tested whereas sample 2 corresponds to the average performance of the six sample membranes used. As can be seen from Table 1, the hydrodynamic permeability of the B-9 membrane depends considerably on the salt used. The lowest permeability is obtained with Na_3PO_4 whereas the largest one results for NaCl. Since no concentration polarization exists in the porous matrix under hyperfiltration conditions, the hydrodynamic permeabilities measured in hyperfiltration experiments can only be compared to those values of ℓ_p obtained from dialysis-osmosis experiments under the boundary condition $c'' = 0$. Moreover, the variation of ℓ_p with the salt employed is mainly caused by swelling effects of the active layer as concentration polarization effects in the porous matrix of the asymmetric membrane are non-existent under hyperfiltration conditions.

Further the ratio ℓ_π/ℓ_p, an approximate measure of the salt permeability, possesses its largest value for a $NaNO_3$ brine solu-tion and its lowest one for a Na_2SO_4 solution in agreement with common experience regarding the salt permeabilities of synthetic membranes. The corresponding normalized salt permeabilities, $P_s/d = (\ell_\pi/\ell_p - r_\infty^2) \cdot \ell_p \Pi'$, are summarized in Table 2 where, in addition, also apparent normalized salt permeabilities of the hollow-fine fibers are displayed. As is obvious from this table, the normalized salt permeabilities vary much stronger with a variation of the anion

TABLE 2

Normalized Salt Permeabilities of Polyamid Flat Sheet Membranes and Hollow-Fine Fibers at 25°C

Membrane	c_s' (mol/l)	$(P_s/d) \cdot 10^6$ (cm/s)						
		Na Cl	Ca Cl$_2$	Al Cl$_3$	Na NO$_3$	Na$_2$ SO$_4$	Mg SO$_4$	Na$_3$ PO$_4$
B-9 flat sheet 1	0.1	3.89	3.71	4.98	11.87	0.577	0.592	1.56
B-9 flat sheet 2	0.1	10.34	11.94	9.17	20.52	2.27	2.59	2.45
B-9 hol-low fiber	0.1	0.35	---	---	---	---	---	---
B-10 hol-low fiber	0.1	0.28	---	---	---	---	---	---
B-10 hol-low fiber	0.5	0.38	---	---	---	---	---	---

than with a change of the cation and its valence. In accordance with
the salt permeabilities, the asymptotic salt rejection, r_∞, of the
sample membrane 1 is always beyond 0.99 whereas that of the sample
membrane 2 is always beyond 0.98. In addition, a broad range of pH
is covered with the salts used ranging from pH $\simeq$ 3.3 ($AlCl_3$ up to
pH $\simeq$ 13 (Na_3PO_4).

A considerable difference between the measured reflection coefficients, σ, and the asymptotic salt rejections, r_∞, is apparent.
This difference was also found with asymmetric CA membranes[22] and is
partially due to the fact that the measured values of σ characterize
the barrier effect to the salt of the entire membrane, active and
porous layer, whereas r_∞ essentially corresponds to the reflection
coefficient of the active layer. Apart from this difference, there
probably still exists a difference between the reflection coefficient
of the active layer, σ', and r_∞. This difference might originate to
some extent from the different experimental setups and the different
pressures employed[26]. However, a full explanation for this experimentally established difference is still not available[27].

The transport parameters received with the B-9 and B-10 hollow-
fine fiber mini-modules are summarized in Table 3. Since no pressure
drop, existing inside each individual fiber (product flow), was taken
into account in evaluating these transport parameters from the original experimental data, these transport parameters might be considered
as apparent ones. As is obvious from Table 3, the apparent hydrodynamic and osmotic permeabilities of the B-9 hollow-fine fibers are
smaller by a factor of about ten than those of the B-9 flat sheet
membrane. On the other hand, the asymptotic salt rejection, r_∞, and
the salt permeability, P_s, are similar to those of the flat sheet
membrane. The symptotic NaCl rejection of the B-10 fibers is larger
than the one of the B-9 fibers at c' = 0.1 mol/l NaCl. The apparent
hydrodynamic permeability of the B-10 hollow-fine fibers is much
smaller than the one of the B-9 hollow-fine fibers. Figure 4

TABLE 3

Apparent Hydrodynamic and Osmotic Permeabilities and Intrinsic
Asymptotic Salt Rejections of Hollow - Fine Fibers at 25° C

Hollow-Fine Fiber Module	c_s' (mol/l)	r_∞	$l_p \cdot 10^6$ (cm/s · atm)	$l_\pi \cdot 10^6$ (cm/s · atm)	l_π / l_p
B-9	0.1	0.9930	1.746	1.798	1.030
B-10	0.1	0.9983	1.277	1.334	1.045
B-10	0.5	0.9941	1.171	1.174	1.003

summarizes these experimental findings displaying the $1/r$ vs. $1/q$ straight lines for the different membrane systems. The slopes of these straight lines correlate to the normalized apparent salt permeabilities, P_s/d. It should be noted that extremely good straight lines were obtained with the B-9 and B-10 flat sheet membranes and hollow-fine fibers, respectively, indicating that compaction effects are negligible. Because of the comparably low compaction, the transport behavior of the membranes stays more constant with time than that of asymmetric cellulose acetate membranes, for instance.

Regarding the determination of intrinsic transport parameters of hollow fibers, the following should be said. In order to rigorously treat the hydrodynamics inside hollow fibers, the Navier-Stokes equations and deformation equations of the fibers must be solved simultaneously. No complete integral of this complex system of differential equations is available from the literature although there exist several approximate solutions for special limiting cases[28-32].

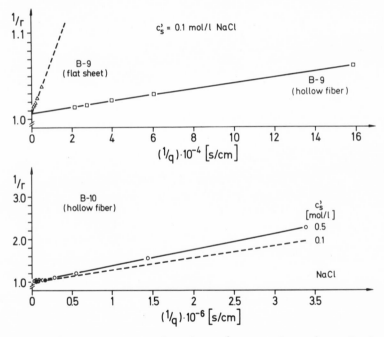

Figure 4: Reciprocal salt rejection $1/r$ as a function of the reciprocal volume flux $1/q$ using a flat sheet B-9 membrane, B-9 and B-10 hollow-fine fiber mini-modules and brine solutions of 0.1 mol/1 NaCl and 0.5 mol/1 NaCl, respectively, at T = 298 K. The corresponding brine pressures P' ranged from 10 atm to 60 atm for the flat sheet membrane and the B-10 mini-modules and from 10 atm to 30 atm for the B-9 mini-modules. The dashed lines are extrapolated as the volume fluxes of these membrane systems are much larger than the ones of those membrane systems used for comparison in each case.

Solving, for instance, a set of approximated Navier-Stokes equations[3] allows at least an estimation of the pressure drop present inside a hollow-fine fiber along its axis due to the product flow as discussed in the Appendix. Such an estimation results in a pressure drop not exceeding 3 atm at a brine pressure of 60 atm using an inner diameter d_i = 41 μm and a fiber length of 37.5 cm of a single fiber. Averaging then over the fiber length, yields a mean back pressure of about 2 atm inside the fiber thus reducing the hydrostatic pressure difference across the fiber wall by less than 3% of the applied pressure. Therefore, the intrinsic transport coefficients of hollow-fine fibers will always be larger than the apparent ones. It should be mentioned that the value of r_∞ is not affected by the pressure drop inside the fibers. Even after taking into account the pressure drop, there still exists a substantial difference between the intrinsic permeabilities of the B-9 flat sheet membranes and the corresponding hollow-fine fibers indicating the existance of a much thicker active layer in the hollow-fine fibers than in the flat sheet membranes.

CONCLUSIONS

The characterization of B-9 flat sheet membranes as well as B-9 and B-10 hollow-fine fibers by appropriate transport parameters for different salts demonstrates that the polyamide membranes exhibit excellent transport properties and resistivities against low and high pH values of the corresponding brine solutions. In addition, there exists little change of the transport properties with time at pressures up to about 60 atm or 30 atm, respectively.

APPENDIX

Following the example of Chen and Petty[33], the measured volumetric product rate $\dot{Q}$ from a fiber can be correlated with the fiber length, the inner and outer diameter of the fiber, and the mean hydrodynamic permeability of the fiber wall as outlined below. Assuming the validity of Hagen-Poisseuille's law at every cross-section of the fiber, the following differential equation results:

$$(d\dot{Q}/dz) = (\pi r_i^4/8\eta)(d^2p/dz^2) = -2\pi r_o \bar{\ell}_p (P' - r_\infty \Delta\Pi - p) \qquad (3)$$

where

$\dot{Q}$ = volumetric product rate from a single fiber (cm^3/s)
r_i = inner fiber radius (cm)
r_o = outer fiber radius (cm)
η = fluid viscosity (μbar·s)
$\bar{\ell}_p$ = mean hydrodynamic permeability of fiber wall (cm/s·atm)[34]
r_∞ = asymptotic salt rejection
$\Delta\Pi$ = $\Pi' - \Pi''$ = osmotic pressure difference across fiber wall (atm)
Π' = osmotic pressure of the brine solution (atm)
Π'' = osmotic pressure of the product solution (atm)

P' = hydrostatic brine pressure (atm)
p = local bore pressure at z (atm)
z = axial distance (cm)

Employing the boundary conditions $(dp/dz) = 0$ at $z = 0$ and $\Delta P = P'$, corresponding to $p = 0$, at $z = \ell$, this differential equation is then integrated to yield:

$$\Delta P(z)/P' = \{P' - p(z)\}/P' = \{\cosh(az)\}/\{\cosh(a\ell)\} \qquad (4)$$

where $a = (4/r_i^2)\sqrt{\eta r_o \overline{\ell}_p}$ (cm^{-1}) and ℓ = fiber length (cm). Further, the following expression results for the volumetric product rate from a single fiber:

$$\dot{Q} = \{\pi r_i^2 \sqrt{r_o \overline{\ell}_p} \cdot (P' - r_\infty \Delta\Pi)/2\sqrt{\eta}\} \cdot \tanh(a\ell) \qquad (5).$$

Using the following values of r_i, r_o, and ℓ of the mini-modules, the mean hydrodynamic permeabilities compiled in Table 3 are obtained, for B-9 mini-modules: $r_i = 20.5$ µm, $r_o = 40.5$ µm; and for B-10 mini-modules: $r_i = 20.5$ µm, $r_o = 50$ µm; $\ell = 37.5$ cm for both types of mini-modules.

ACKNOWLEDGEMENTS

The author appreciates the donation of aramid flat sheet and hollow-fine fiber mini-modules by PERMASEP ^{R} Products, a division of E.I. Du Pont de Nemours & Company, Inc., Wilmington, Delaware, and in addition is grateful to Mrs. B. Bäppler and Mr. H. Hildebrandt, Max-Planck-Institut für Biophysik in Frankfurt, for their assistance in carrying out the hyperfiltration experiments and the corresponding analyses. Mrs. H. Müller has to be thanked for measuring the transport coefficients of the B-9 flat sheet membrane in dialysis-osmosis experiments. The work was financially supported by the "Bundesminister für Forschung und Technologie", Bonn, Germany.

REFERENCES

1. S. Loeb and S. Sourirajan; Adv. Chem. Ser. 38, 117 (1962).
2. S. Loeb; in: *Desalination by Reverse Osmosis*, pp. 55-91, U. Merten, ed., The M.I.T. Press, Cambridge, Massachusetts, 1966.
3. H.K. Lonsdale, U. Merten, R. Riley, K.D. Vos, and J.C. Westmoreland; "Reverse Osmsosis for Water Desalination", OSW, Res. & Develop. Progr. Rept. No. 111, May 1964.
4. H.K. Lonsdale; in: *Desalination by Reverse Osmosis*, pp. 93-160, U. Merten, ed., The M.I.T. Press, Cambridge, Massachusetts, 1966.
5. R. Riley, U. Merten, and J.O. Gardner; Desalination 1, 30 (1966).
6. S. Sourirajan and T.S. Govindan; Proc. 1st Intl. Desalination Symposium, Paper SWD/41, Washington, D.C., Oct. 3-9, 1965.

7. E.J. Breton, Jr.; "Water and Ion Flow Through Imperfect Osmotic Membranes", OSW, Res. & Develop. Progr. Rept. No. 16, 1957.

8. S. Loeb and S. Manjikian; Ind. Eng. Chem. Proc. Res. Dev. 4, 207 (1965).

9. L.T. Rozelle, J.E. Cadotte, W.L. King, A.J. Senechal, and B.R. Nelson; "Development of Ultrathin Reverse Osmosis Membranes for Desalination", OSW, Res. & Develop. Progr. Rept. No. 659.

10. R.L. Riley, H.K. Lonsdale, and C.R. Lyons; Proc. 3rd Intl. Symposium on Fresh Water from the Sea, Vol. 2, p. 551, A. and E. Delyannis, eds., published by the editors, Athens, 1970.

11. J.W. Richter and H.H. Hoehn; US Patent 3,567,632 (1971).

12. R. McKinney, Jr.; in: Reverse Osmosis Membrane Research, H.K. Lonsdale and H.E. Podall, eds., Plenum Press, New York - London, 1972, p. 253.

13. R.L. Riley, G.R. Hightower, C.R. Lyons, and M. Tagami; Proc. 4th Intl. Symposium on Fresh Water from the Sea, Vol. 4, p. 333, A. and E. Delyannis, eds., published by the editors, Athens, 1973.

14. H.H. Hoehn and D.G. Pye; US. Patent 3,497,451 (1970); assigned to E.I. Du Pont de Nemours and Company, Wilmington, Delaware.

15. R.L. Riley, G. Hightower, and C.R. Lyons; in: Reverse Osmosis Membrane Research, H.K. Lonsdale and H.E. Podall, eds., Plenum Press, New York - London, 1972, p. 437.

16. C.E. Reid and E.J. Breton; J. Appl. Polymer Sci. 1, 133 (1959).

17. H.K. Lonsdale, U. Merten, and R.L. Riley; J. Appl. Polymer Sci. 9, 1341 (1965).

18. B. Keilin; "The Mechanism of Desalination by Reverse Osmosis", OSW, Res. & Develop. Progr. Rept. No. 117, Nov. 1963.

19. W. Banks and A. Sharpless; "The Mechanism of Desalination by Reverse Osmosis, and its Relation to Membrane Structure", Final Rept. to the OSW, 1965; and J. Appl. Chem. 16, 28 (1966).

20. W. Pusch; Ber. Bunsenges. physikal. Chem. 81, 854 (1977).

21. E. Oikawa and T. Ohsaki; Desalination 25, 187 (1978).

22. W. Pusch and R. Riley; Desalination 22, 191 (1977).

23. G.D. Metha and S. Loeb; J. Membrane Sci. 4, 335 (1979).

24. W. Pusch; Desalination 16, 65 (1975).

25. W. Pusch; Ber. Bunsenges. physikal. Chem. 81, 269 (1977).

26. W. Pusch and G. Mossa; Desalination 24, 39 (1978).

27. G. Jonsson; Desalination 24, 19 (1978).

28. Th.E. Tang and S.-T. Hwang; AIChE Journal 22, 1000 (1976).

29. J.M. Thorman and S.-T. Hwang; Chem. Eng. Sci. 33, 15 (1978).

30. O.H. Varga; Stress-Strain Behavior of Elastic Materials, Interscience, New York, 1966.

31. S.A. Stern, F.J. Onorato, and Ch. Libove; AIChE Journal 23 (4), 567 (1977).

32. J.J. Hermans; Desalination 26, 45 (1978).

33. C. Chen and C.A. Petty; Desalination 12, 281 (1973).

34. J. E. Sigdell; A Mathematical Theory for the Capillary Artificial Kidney, page 19 equations 48 and 51, Hippokrates Verlag, Stuttgart (1974)

SEPARATION OF MACROMOLECULES BY ULTRAFILTRATION: INFLUENCE
OF PROTEIN ADSORPTION, PROTEIN-PROTEIN INTERACTIONS, AND
CONCENTRATION POLARIZATION

Kenneth C. Ingham, Thomas F. Busby, Ylva Sahlestrom
and Franco Castino

American Red Cross Blood Services Labs.
Bethesda, Maryland 20014

INTRODUCTION

Ultrafiltration through microporous membranes is a well-established technique for concentrating dilute protein solutions and separating proteins from low molecular weight solutes such as salts or ethanol (Friedli et al., 1977; Guthörlein, 1977; Mercer, 1977), or from much larger particles such as cells (Colton et al., 1975). However, the use of this tool for fractionating proteins according to size has progressed less rapidly. Several difficulties can be identified:

- non-uniform pore size
- protein adsorption
- concentration polarization
- protein-protein interactions

The problem of heterogeneous pore size can only be solved through improvements in the manufacturing process. This requires a thorough understanding of the chemical mechanisms by which these polymeric networks are synthesized. Judging from the rate of appearance of new membranes on the market, and from the growing number of publications related to this subject, we can anticipate continued progress on this front. The use of nuclear bombardment followed by chemical etching is an example of a new approach which has led to some improvement (Quinn et al., 1972).

A solution to the problem of protein adsorption requires the availability of a synthetic material which has little affinity for proteins and which can also be fashioned into thin porous sheets capable of withstanding varying degrees of mechanical stress. Such a material should be hydrophilic and bear very little net charge. These are challenging criteria for the polymer chemist. An alternative approach to this problem might be to saturate the binding sites on a series of membranes with an appropriate reagent, perhaps even a protein, and then evaluate the sieving and flow characteristics with respect to the desired application.

Even in the absence of protein adsorption, there is a tendency for rejected proteins to accumulate at the membrane surface, modifying the filtration characteristics. The advent of cross-flow systems such as thin channels and hollow fibers, in which the feed solution is rapidly recirculated over the membrane surface, has significantly reduced this problem. However, in order to have finite flux, even with these systems, one must tolerate a finite amount of "concentration polarization", a term which refers to the increase in the concentration of rejected species with decreasing distance from the membrane (curve a, Fig. 1). Under certain conditions, the concentration at the wall, C_w is believed to approach the concentration, C_g, at which gel formation occurs. According to the gel polarization model (Michaels, 1968; Blatt, et al., 1970;

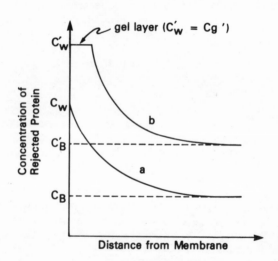

Fig. 1.   Schematic illustration of concentration polarization. C_B is the bulk concentration of rejected protein. C_w is the concentration adjacent to the membrane. Curves a and b refer to two different bulk concentrations.

Porter and Nelson, 1975), the gel layer which consequently covers the surface of the membrane, offers "hydraulic resistance" to solvent flux and would presumably alter the net sieving properties as well. Once a gel layer is formed, a further increase in the

bulk concentration is envisioned to simply increase the thickness
of that layer (curve b, Fig. 1), further lowering the flux according to the following equation:

$$J = k_s \ln (C_g/C_B) \qquad \text{Equation (1)}$$

where k_S, the so-called mass transport coefficient, is a measure of
the efficiency with which the rejected solute is transferred from
the membrane surface back into the bulk fluid. Its magnitude
depends on the channel dimensions, the diffusibility of the rejected solute, and the fluid shear rate at the membrane surface.

The popularity of the gel polarization model can be attributed
to the success with which it has been applied to the analysis of
flux versus concentration data. A number of workers (Blatt et al.,
1970; Porter and Nelson, 1975; Probstein et al., 1979) have documented the linearity of data plotted according to Equation (1).
However, much less attention has been given to evaluating the
effects of polarization on the sieving properties of the membrane.
A gel layer which is capable of providing mechanical resistance to
solvent flux should also increase the rejection of a second partially rejected solute, whose size is much larger than that of the
solvent molecules. We have data which indicate that this does not
occur, even under conditions where the gel polarization is expected
to apply. Furthermore, there is considerable variation in the
reported values of C_g determined by extrapolation of flux data
plotted according to Equation (1). Since C_g is a property of the
protein under a given set of conditions (pH, temperature, ionic
composition), it should be possible to systematically evaluate the
dependence of C_g on solution conditions for different proteins and
compare the results with solubility measurements; this has not been
done. Finally, a key assumption in the derivation of Equation (1)
is that the total flux over the entire channel length is small
relative to the flow through the channel (Probstein et al., 1979).
Since it is desirable to maximize flux with most applications, we
are faced with a situation in which the only available theory may
not apply under the most relevant conditions.

A fourth potential complication in the use of ultrafiltration
for fractionating proteins is related to the role of protein-protein
interactions. In a complex mixture such as human plasma, reversible
interactions between different proteins can interfere with attempts
to separate them by any method. In the case of ultrafiltration,
such interactions can be expected to lower the effective sieving
coefficients of the constituent proteins. While such interactions
are often viewed as a nuisance, in some cases it might be feasible
to exploit them by appropriate manipulation. To illustrate the
concept we will present some preliminary data obtained with a model
binary mixture of interacting proteins.

MIXTURES OF POLYETHYLENE GLYCOL AND ALBUMIN

Polyethylene glycol (PEG)* is a water-soluble synthetic polymer which has received increasing attention as a fractional precipitating agent for the purification of proteins (Polson et al., 1964; Janssen and Ruelius, 1968; Newman et al., 1971; Wickerhauser and Hao, 1972; Foster et al., 1973; Curling et al., 1977; Wickerhauser et al., 1979; Hao et al., 1980). Our interest in the separation of macromolecules by ultrafiltration originated from the need for a more efficient method of removing PEG-4000 from the various fractions produced. During the course of our investigations, it became apparent that PEG could serve as a useful probe of molecular events occurring at the membrane surface, especially those which modify the effective sieving properties (Busby and Ingham, 1980). Our ability to monitor the passage of PEG through the membranes was greatly facilitated by the availability of [3H]-PEG-4000, which was added as a tracer.

Aqueous mixtures of human albumin (Cohn fraction V) and/or PEG-4000, each at an initial concentration of 50 mg/ml, were ultrafiltered using an Amicon TCF-10 spiral-channel unit in the diafiltration mode, in which the sample volume, V_0, remains constant. The filtrate was collected using an LKB-1700 fraction collector operated in the time mode. Subsequent measurement of the volume of each fraction enabled the flow rate to be monitored throughout the experiment. Summation of the amount of PEG in the filtrate allowed the concentration remaining in the retentate, C, to be determined. For a partially rejected solute, this concentration usually decreases exponentially according to the following equation:

$$C = C_0 \exp \left(- k \ V/V_0 \right) \hspace{3cm} \text{Equation (2)}$$

where C_0 is the initial concentration. A value of the sieving coefficient, k, (or the rejection coefficient, R = 1 - k), can then be obtained from the slope of a semilog plot of % solute remaining versus V/V_0, the number of volumes of solvent exchanged. Occasionally, such plots exhibit curvature, reflecting either a dependence of k on solute concentration or changes in other variables, such as membrane porosity during the course of diafiltration (Busby and Ingham, 1980). This approach to determining k is thus superior to that of simply averaging several determinations of the ratio of the concentration in the filtrate to that in the retentate.

---

*PEG = poly(ethylene glycol), poly(ethylene oxide), polyoxyethylene. Chemical formula: $HOCH_2CH_2(CH_2CH_2O)_nCH_2CH_2OH$. PEG-4000 signifies a heterogeneous mixture having a nominal average molecular weight of 4000 daltons.

Several membranes which readily passed PEG in the absence of protein were found to exhibit increased rejection of the synthetic polymer when albumin was added to the solution. Representative data obtained with the Amicon PM-30 membrane are presented in Figure 2A, where the open circles pertain to PEG alone and the filled symbols to mixtures containing albumin at several concentrations. Figure 2B illustrates the large reduction in flux caused by the addition of protein as well as the gradual increase in flux which accompanied the removal of PEG during the course of a given experiment, presumably due to decreasing viscosity. Note that while the flux decreased steadily with increasing protein concentration, rejection of PEG was independent of protein concentration between 1 and 100 mg/ml, although greater than in the absence of protein. The amount of albumin lost from the cell during these experiments never exceeded 5% of the initial material.

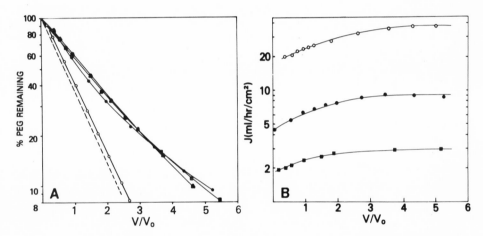

Fig. 2.   Diafiltration of aqueous solutions containing PEG-4000 in the absence (O) or in the presence of albumin at 1 mg/ml (●), 50 mg/ml (▲) and 100 mg/ml (■). The sample was recirculated at 210 ml/min under an applied pressure of 20 psi. The dashed line shows expected result for unrejected solute with k = 1. (Adapted from Busby and Ingham, 1980)

If these changes in filtration characteristics were due to concentration polarization, then one would expect them to be reversible. However, when the protein was removed and the system thoroughly flushed with water, the sieving coefficient for PEG remained essentially the same as in the presence of albumin. Furthermore, although the flux increased upon removal of albumin, it did not return to the value which prevailed prior to exposure to albumin. The suspicion that this lack of reversibility was due to protein ad-

sorption was confirmed by the observation that overnight treatment
of the membrane with trypsin restored the original flux and sieving
properties. It would appear that the increased rejection of PEG
in the presence of albumin is due entirely to adsorption effects,
whereas the decreased flux is due to a combination of adsorption
and polarization.

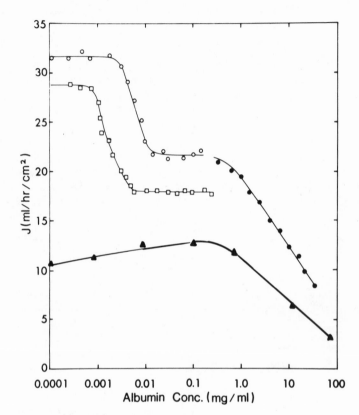

Fig. 3.   Dependence of flux on albumin concentration. The concen-
           tration of albumin shown is the theoretical concentration
           based on the volume of solution transferred from the
           reservoir to the cell (see text). Membranes were Amicon
           PM-30 (□,○,●) and YM-30 (▲).

In order to further resolve these two phenomena, the flux was
measured as a function of continuously increasing protein concentration over a 10^6-fold range. This was accomplished by including
albumin in the buffer reservoir, but not in the initial sample.
Since the protein is fully rejected by the membrane, it accumulates
in the ultrafiltration cell during diafiltration. Results obtained
with the PM-30 membrane are shown by the upper three curves in
Figure 3. The sharp decrease in flux, which occurred between 0.001

and 0.01 mg/ml in two separate experiments, is undoubtedly due to a
saturable process involving protein adsorption. The linear decrease
which occurred at high protein concentration is similar to that reported by others and is presumably due to concentration polarization
according to Equation (1). However, if this latter effect involves
the formation of a gel, we must conclude that the gel is capable of
lowering flux without influencing the sieving characteristics, since
the rejection of PEG did not change over this range of protein concentration (Fig. 2).

The lower curve in Figure 3 represents data from a similar
experiment with the new YM-30 membrane of Amicon, which is claimed
to have less tendency to adsorb proteins. Our flux data support
this claim since the sharp transition which occurs at low protein
concentration with the PM-30 membrane is absent here. However,
there was a slight decrease in the sieving coefficient for PEG from
approximately 0.8 to approximately 0.6 after exposure to albumin,
and this effect was reversed by trypsin treatment. This suggests
that the sieving characteristics are more sensitive to protein adsorption than are the flux properties. Since the flux values for
YM-30 were generally lower than PM-30, the new membrane did not
appear to offer any advantage for this particular application,
i.e., removal of PEG from albumin.

MIXTURES OF ALBUMIN AND LYSOZYME

Lysozyme is a basic protein having a large net positive charge
at neutral pH, where albumin has a large net negative charge. The
molecular weight of lysozyme is about 14,000, compared to about
69,000 for albumin. At low ionic strength and neutral pH, these
two proteins interact reversibly, forming electrostatic complexes
of undetermined stoichiometry which can be dissociated by addition
of salts (Steiner, 1953; Miekka and Ingham, 1979). Since both of
these proteins are readily available in appreciable quantities and
reasonable purity, they constitute an appropriate model system for
investigating the effects of such interactions on various separation
methods.

The feasibility of manipulating the rejection of lysozyme by
controlling its interaction with albumin is illustrated in Figure 4.
These experiments utilized an Amicon PM-30 membrane which had been
previously exposed to albumin. As mentioned in the previous section,
albumin is almost completely rejected by this membrane. At a concentration of 1 mg/ml in 0.01 M phosphate buffer, pH 7, lysozyme
alone passed freely through the membrane. The slight downward concavity in the diafiltration curve could reflect a small amount of
reversible self-association of lysozyme under the initial conditions, which would diminish as the concentration was lowered
(Imoto et al., 1972). When excess (50 mg/ml) albumin was added,
the lysozyme was almost completely rejected (k ~ 0.02) by virtue

$MW = 14,600 \pm 200$

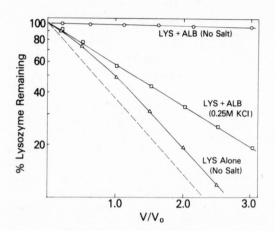

Fig. 4.   Influence of albumin (50 mg/ml) on the rejection of
           lysozyme at pH 7 in 0.01 M phosphate buffer with (□)
           and without (Δ,0) KCl. The initial concentration of
           lysozyme was 1 mg/ml. Dashed line same as Fig. 1.

of its hetero-association with albumin. This effect was strongly
reversed by the presence of 0.25 M KCl, which is known to partially
inhibit complex formation; higher concentrations are required for
complete inhibition (Miekka and Ingham, 1979). It should be noted
that KCl had essentially no effect on the sieving of lysozyme in
the absence of albumin (not shown).

     The effect of albumin and KCl concentration is further ex-
plored in Figure 5. The circles again show the retention of lysozyme in the presence of albumin at low ionic strength, and the open
squares show the removal of lysozyme when 0.25 M KCl is present.
In the experiment described by the triangles, albumin was initially
absent in the sample chamber, but was present in the buffer reservoir, both in the absence of KCl. Thus, as diafiltration proceeded,
the albumin concentration increased steadily according to the upper
scale, causing the development of upward concavity in the diafiltration curve as the remaining lysozyme was complexed. The filled
squares correspond to an analogous experiment in which KCl and
albumin were initially present in the sample chamber, but not in
the buffer reservoir. Thus, the concentration of KCl in the sample
steadily decreased, allowing complexation of the remaining lysozyme
and also causing upward curvature.

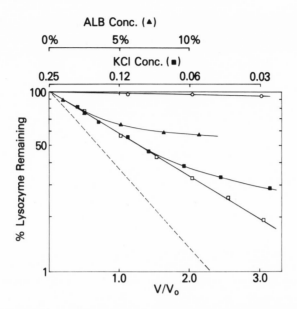

Fig. 5.   Influence of albumin on the rejection of lysozyme.
The open symbols are identical to those in Fig. 4.
The filled symbols illustrate the effect of increasing
albumin (▲) or decreasing KCl (■) according to the
upper scales, as described in the text. Dashed line
same as Fig. 1.

In the absence of knowledge regarding the reversible interaction between these two proteins, one might at first be inclined to
attribute the above observations to gel polarization by albumin at
low ionic strength, with consequent increase in rejection of lysozyme. In that context, the effect of KCl might be attributed to
an inhibition of gel formation. In order to further examine this
question, a similar set of experiments was performed with a mixture
of albumin and α-lactalbumin. The latter protein is strikingly
similar to lysozyme, having the same molecular weight and a similar
amino acid sequence (Brew and Campbell, 1967). However, its iso-
electric point is at pH 5, compared to pH 11 for lysozyme. Thus,
at neutral pH it has a net negative charge and is consequently not
expected to form electrostatic complexes with albumin. The diafiltration of this protein is illustrated in Figure 6. Although addition of albumin caused a small increase in the rejection of α-
lactalbumin, the effect was much smaller than with lysozyme.
Although it is conceivable that a gel layer could have different
effects on the rejection of different proteins of the same size,
one would expect the protein having the same net charge as albumin
to be more effectively rejected. The small increase in the rejection of α-lactalbumin in the presence of albumin could be due to

a slight interaction between these proteins, even though they have the same net charge.

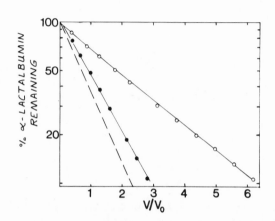

Fig. 6.   Diafiltration of α-lactalbumin (1 mg/ml) in the absence (●) or presence (O) of albumin (50 mg/ml) under conditions identical to Fig. 4.   Dashed line same as Fig. 1.

DIRECT MEASUREMENT OF ADSORPTION AND CONCENTRATION POLARIZATION

The results obtained above illustrate the importance of distinguishing the effects due to protein adsorption from those due to concentration polarization. Although both phenomena tend to decrease the efficiency of macromolecular separations, different approaches are required to minimize their effects. In this section, we describe some experiments in which an absorbance monitor was used to directly observe changes in the concentration of albumin in a closed-loop filtration circuit, illustrated schematically in Figure 7. The filtration units were Amicon In-Line® ultrafilters, containing a bundle of hollow fibers (total filtering area 2,000 cm²), made of material having a nominal molecular weight cutoff at 50,000 daltons (Silverstein et al., 1974).

The albumin used for these experiments was obtained by passing Cohn fraction V over a large column (2.5 x 160 cm) of Sephadex G-75

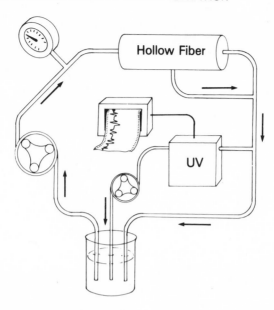

Fig. 7.   Schematic illustration of closed-loop ultrafiltration
           circuit with UV detector inserted to allow continuous
           monitoring of the bulk protein concentration. Trans-
           membrane pressure was controlled by partially occluding
           the outlet line of the Amicon XP-50 hollow-fiber unit.

(Pharmacia) which served to remove dimers, higher polymers and other
contaminants (Busby and Ingham, 1980). Solutions of albumin at
various concentrations were circulated through the filtration unit
at a controlled rate by means of a roller pump (Sarns model 5500).
Transmembrane pressure was adjusted by partially occluding the
outlet line between the filtration unit and the reservoir and was
measured by means of a mercury manometer. The outlet and filtrate
lines were combined and returned to the reservoir which was magnet-
ically stirred. A small part of the returning fluid was diverted
through a UV detector (LKB), which allowed the optical density (OD)
at 280 nm of the bulk solution, external to the filtration unit, to
be continuously monitored. Such measurements were confirmed by
withdrawing small samples from the reservoir and measuring the OD
on a Cary 118C spectrophotometer. Independent measurements confirmed the absence of albumin in the filtrate, indicating complete
rejection by the membrane.

     The upper trace in Figure 8 is an example of the kind of data
which were obtained with new filtration units not previously exposed to protein. Stepwise increases in the transmembrane pressure
($P_0 < P_1 < P_2...$) caused a progressive decrease in the OD of the
external solution as protein was "sequestered" inside the hollow

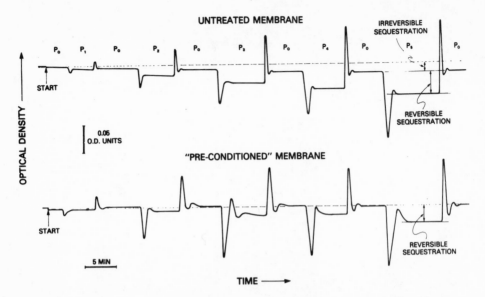

Fig. 8.   Influence of transmembrane pressure on optical density
           of external portion of closed-loop circuit described in
           Fig. 8. The albumin concentration was   5 mg/ml in
           0.05 M potassium phosphate buffer containing 0.1 M KCl.
           The solution was circulated at a constant rate of 325
           ml/min ($\dot{\gamma}_w$ = 2670 sec^{-1}) and pressure was varied up to
           600 mm Hg ($P_0 < P_1 < P_2 \ldots$).

fibers. This decrease was only partially reversed when the original
pressure was restored. Parallel measurements revealed irreversible
effects upon the filtration rate as well. This lack of reversibility is undoubtedly due to adsorption of protein onto the membrane
surface.

     The lower trace in Figure 8 shows that when the experiment was
repeated on a "conditioned" membrane which was previously exposed
to albumin under the highest pressures anticipated, only reversible
sequestration was observed. Restoration of the original pressure
at any point caused essentially complete recovery of the initial OD
of the external circuit. Parallel measurements indicated that the
filtration rate also returned to the original value. These reversible effects were attributed to concentration polarization.

     Using this approach, it is possible to quantitate the mass of
albumin retained within the filtration unit and to evaluate the dependence upon solution conditions and fluid dynamic parameters. For
example, Figure 9A illustrates the effect of transmembrane pressure
on the amount of albumin irreversibly adsorbed at several pH values.

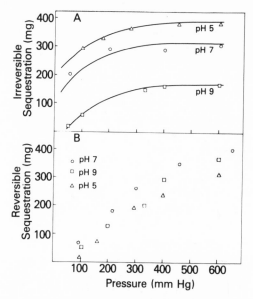

Fig. 9.   Influence of transmembrane pressure on the amount of
          albumin irreversibly (upper panel) or reversibly (lower
          panel) sequestered in the hollow-fiber unit at several
          pH values.

The amount of adsorption is seen to increase with increasing
pressure. Presumably, the higher filtration rates which occur at
higher pressures tend to push more proteins onto the filtering surface. Stretching of the membrane may also occur, allowing greater
penetration of the protein into the porous network. The amount of
adsorption appears to reach a plateau at pressures above 400 mm Hg,
presumably indicating a saturation of binding sites. The amount
adsorbed at any given pressure is seen to increase with decreasing
pH, in agreement with several reports in the literature (Dillman
and Miller, 1973, and references therein).

    The amount of protein reversibly sequestered via concentration
polarization appears to be less sensitive to pH, as shown in Figure
9B. The difference between the three curves is within the experimental variation observed between filtration units. In contrast to
adsorption, concentration polarization did not quite reach a plateau in the range of pressures employed. Parallel measurements of
the filtration rate indicated that it also continued to increase
with increasing pressure in this range. In general, when using a

preconditioned hollow-fiber unit, the filtration rate tended to
follow a pattern similar to that of reversible sequestration with
respect to its dependence on pressure.

Experiments performed at different shear rates indicated that
this parameter had very little influence on protein adsorption, but
had a significant effect on reversible sequestration. At a protein
concentration of 5.6 mg/ml and a constant transmembrane pressure of
600 mm Hg, increasing the wall shear rate, $\dot{\gamma}_w$, from 1480 sec^{-1} to
3330 sec^{-1} decreased the amount of protein reversibly sequestered
from ~400 mg to ~150 mg. This result is compatible with our conclusion that the reversible effects are due to concentration polarization, since that phenomenon is expected to decrease with increasing
$\dot{\gamma}_w$ (Michaels, 1968; Blatt et al., 1970; Porter and Nelson, 1975).

DISCUSSION

The use of ultrafiltration as a unit process in large-scale
fractionation of macromolecules has been confined primarily to the
concentration of dilute solutions and to removal of low molecular
weight solutes. Part of the lack of progress in the use of this
tool for fractionating proteins according to size can be attributed
to the discouraging tone of some of the articles which have been
written on the subject. The concept that rejected species inevit-
ably accumulate on the membrane surfaces, forming a gelatinous layer
whose properties then govern filtration efficiency, has, in our opinion, been overemphasized. Since albumin solutions are frequently
used to illustrate the validity of Equation (1), derived from the gel
polarization model, we were somewhat surprised to find that PEG-4000
could be easily separated from albumin by this method (Busby and
Ingham, 1980). It soon became apparent that the decreased sieving
coefficient for PEG which occurred in the presence of albumin, was
due not to the formation of a gel, but to irreversible adsorption of
protein. These results indicate that PEG can serve as a useful inert
probe of events which occur at the membrane surface and which modify
the sieving characteristics. In this particular example, PEG-4000
appeared to be right on the verge of being rejected by the virgin
PM-30 membrane. Thus, it was optimally suited for detecting such
alterations. With membranes of larger pore size, such alterations
might go unnoticed unless a higher molecular weight PEG was used.
In any case, it is clear that before attributing changes in flux or
rejection properties to concentration polarization, one must go back
and repeat measurements on the exposed membrane to determine the
appropriate base line.

Our observations with mixtures of albumin with lysozyme or α-lactalbumin, clearly violate the rather pessimistic
rule of thumb offered by Nelson (1977), that two proteins should
differ by a factor of ten in their molecular weights if they are to
be efficiently separated by ultrafiltration. Albumin, whose molecu-

lar weight differs from the others by only a factor of five, could be clearly separated, provided that conditions were chosen to minimize their interaction with each other. Since albumin is almost completely rejected by this membrane, whereas α-lactalbumin is only partially rejected, it follows that other proteins of intermediate size might also be separated from albumin in this manner.

A thorough knowledge of the reversible protein-protein interactions in the mixture whose fractionation is contemplated, opens the possibility of selectively manipulating the ultrafiltration of specific proteins in that mixture. The enhanced rejection of lysozyme in the presence of albumin at low ionic strength and its reversal by addition of salt, clearly demonstrates this important concept. In addition to hetero-interactions, such as between albumin and lysozyme, there are many examples of proteins which self-associate under various conditions. The oligomeric state of numerous enzymes can frequently be manipulated by addition or removal of specific effectors which bind to the subunits (Koshland, 1970; Frieden, 1971). Another approach might be to add fully rejected synthetic polymers containing covalently attached ligands which bind selectively and reversibly to the protein of interest. These approaches potentially allow the inherent biological specificity of the system to be exploited for purposes of fractionation. Their advantage over conventional affinity chromatography would be the potential for efficiently processing large volumes of source material.

The use of a UV detector to directly monitor changes in protein concentration, offers great potential for investigating in detail the factors which influence adsorption and concentration polarization. Although our initial attention has been confined to the steady-state portions of the recordings in Figure 8, analysis of the transient changes which accompany the changes in pressure may ultimately provide interesting clues regarding the mechanism by which the steady-state equilibria are established. For example, the rapid decrease in OD immediately following an increase in pressure is coincident with a corresponding overshoot in flux which has been discussed by others (de Filippi and Goldsmith, 1970). The relaxation back to the new steady-state level can be attributed to a shift of protein towards the filtering wall where it accumulates and provides additional resistance to flux, either by osmotic pressure which counteracts the applied pressure, or by increasing the thickness of the putative gel layer.

We are reluctant to postulate gel formation because of the lack of an effect of albumin on the rejection of PEG, and only a slight effect with α-lactalbumin (when using preconditioned membranes). Although the sieving data were obtained using a different membrane in a different device, the shear rates were similar and the concentration of albumin was at least ten-fold greater than that utilized with the hollow-fiber experiments. In our opinion, the effects on

flux, in both systems, are more reasonably explained by osmotic
pressure effects. Analyses of the non-ideality of concentrated
protein solutions indicate that osmotic pressures sufficient to
appreciably counteract the applied pressures, can be achieved at
concentrations substantially below those required for gel formation
(Ross and Minton, 1977). It remains to be seen whether such a model
can generate an equation similar to Equation (1) without invoking
gel formation. In this context, the extrapolated value usually referred to as "C_g" would be viewed as the bulk concentration at which
the concentration at the membrane surface becomes high enough to
produce an osmotic back-pressure sufficient to counteract the applied
pressure and prevent flux. Such a model would obviously not preclude
the possibility of gel formation at even higher concentrations, but
would offer an explanation for the fact that "C_g" is often substantially lower than the known solubility, especially with highly soluble proteins such as albumin.

The amount of albumin which is irreversibly adsorbed onto the
hollow-fiber membranes is substantial. Its measurement by this
relatively insensitive method is made possible by the large membrane surface area (2000 cm^2) relative to the volume (330 ml) of
dilute (5 mg/ml) protein solution. At higher protein concentrations,
irreversible sequestration is difficult to quantitate because the
amount adsorbed, even at saturation, is a trivial fraction of the
total. Using the data in Figure 9A, one calculates surface concentrations as high as 175 $\mu g/cm^2$, which is between two and three orders
of magnitude higher than values reported for adsorption of albumin
and other proteins to various nonporous synthetic surfaces (Brash
and Lyman, 1969; Dillman and Miller, 1973; Brynda et al., 1978).
This suggests that the effective surface area available for adsorption is much greater than the nominal area estimated from the internal diameter of the cylindrical fibers. It is likely that the
polymeric network contains numerous dead-end pores into which albumin can penetrate and can be adsorbed or trapped. Another possibility is that adsorption may not be limited to a monolayer. Adsorbed
molecules could undergo conformational changes which enable them to
serve as a surface for further adsorption.

It should be pointed out that the UV monitor can alternatively
be placed in the outlet line proximal to the point at which it is
joined to the filtrate line (Fig. 7). Under these conditions, an
increase in pressure causes an increase in OD reflecting the higher
concentration emerging from the unit. However, this is more than
compensated for by the decrease in flow rate as solvent is diverted
through the membranes such that the net change in OD of the bulk
external fluid is decreased (as observed in Fig. 8 with the UV
monitor in the original position). A more elaborate system, containing a UV monitor in both positions and a flow meter in the filtrate line, would allow all necessary parameters to be monitored in
real time. Such an approach may eventually provide a more thorough

understanding of the phenomenon of dynamic concentration polarization at the molecular level. Such studies should be done with a "conditioned" unit in order to avoid artifacts due to irreversible protein adsorption.

REFERENCES

Blatt, W. F., Dravid, A., Michaels, A. S., and Nelson, L., 1970, in: Membrane Science and Technology, J. E. Flinn, ed., Plenum Press, New York.

Brash, J. L., and Lyman, D. J., 1969, J. Biomed. Mater. Res., 3:175.

Brew, K., and Campbell, P. N., 1967, Biochem. J., 102:258.

Brynda, E., Houska, M., Pokorná, Z., Cepalova, N. A., Moiseev, Y. V., and Kálal, J., 1978, J. Bioeng., 2:411.

Busby, T. F., and Ingham, K. C., 1980, J. Biochem. Biophys. Methods, (in press).

Colton, C. K., Henderson, L. W., Ford, C. A., and Lysaght, M. J., 1975, J. Lab. Clin. Med., 85:356.

Curling, J. M., Berglöf, J., Lindquist, L.-O., and Eriksson, S., 1977, Vox Sang., 33:97.

de Filippi, R. P., and Goldsmith, R. L., 1970, in: Membrane Science and Technology, J. E. Flinn, ed., Plenum Press, New York.

Dillman, W. J. Jr., and Miller, I. F., 1973, J. Col. Interfac. Sci., 44:221.

Foster, P. R., Dunnill, P., and Lilly, M. D., 1973, Biochim. Biophys. Acta, 317:505.

Frieden, C., 1971, Ann. Rev. Biochem., 40:653.

Friedli, H.,, Fournier, E., Volk, T., and Kistler, P., 1977, Vox Sang., 33:93.

Guthörlein, G., 1977, in: Proc. Int. Workshop on Technology for Protein Separation and Improvement of Blood Plasma Fractionation, H. E. Sandberg, ed., U.S. DHEW Publication No. 78-1422.

Hao, Y. L., Ingham, K. C., and Wickerhauser, M., 1980, in: Methods in Plasma Fractionation, J. Curling, ed., Academic Press, New York and London.

Imoto, T., Johnson, L. N., North, A. C. T., Phillips, D. C., and Rupley, J. A., 1972, in: The Enzymes, P. D. Boyer, ed., 3rd Edn., Academic Press, New York.

Janssen, F. W., and Ruelius, H. W., 1968, Biochim. Biophys. Acta, 151:330.

Koshland, D. E., 1970, in: The Enzymes, P. D. Boyer, ed., 3rd Edn., Academic Press, New York.

Mercer, J., 1977, in: Proc. Int. Workshop on Technology for Protein Separation and Improvement of Blood Plasma Fractionation, H. E. Sandberg, ed., U.S. DHEW Publication No. 78-1422.

Michaels, A. S., 1968, Chem. Eng. Progr., 64:31.

Miekka, S. I., and Ingham, K. C., 1979, Fed. Proc., 38:519, and
        manuscript in preparation.
Nelson, L., 1977, in: Proc. Int. Workshop on Technology for
        Protein Separation and Improvement of Blood Plasma
        Fractionation, H. E. Sandberg, ed., U.S. DHEW Publication
        No. 78-1422.
Newman, J., Johnson, A. J., Karpatkin, M. H., and Puszkin, S.,
        1971, Brit. J. Haemat., 21:1.
Polson, A., Potgeiter, G. M., Largier, J. F., Mears, G. E. F.,
        and Joubert, F. J., 1964, Biochim. Biophys. Acta, 82:463.
Porter, M. C., and Nelson, L., 1975, in: Recent Developments in
        Separation Science, Vol. 2, N. N. Li, ed., Plenum Press,
        New York.
Probstein, R. F., Leung, W.-F., Alliance, Y., 1979,
        J. Phys. Chem., 83:1228.
Quinn, J. A., Anderson, J. L., Ho, W. S., and Petzny, W. J.,
        1972, Biophys. J., 12:990.
Ross, P. D., and Minton, A. P., 1977, J. Mol. Biol., 112:437.
Silverstein, M. E., Ford, C. A., Lysaght, M. J., and Henderson,
        L. W., 1974, N. Engl. J. Med., 291:747.
Steiner, R. F., 1953, Arch. Biochem. Biophys., 47:56.
Wickerhauser, M., and Hao, Y. L., 1972, Vox Sang., 23:119.
Wickerhauser, M., Williams, C., and Mercer, J. E., 1979,
        Vox Sang., 36:281.

ACKNOWLEDGEMENTS

        This report constitutes Publication No. 461 from the American
Red Cross Blood Services Laboratories. The work was also supported
by the National Institutes of Health (Grants RR 05737 and HL 21791),
and by a Career Development Award to K.C. Ingham (HL 00325). We are
indebted to Mr. Reginald Jenkins for technical assistance and to
Mrs. Marcia Robinson for her precise preparation of the manuscript.

ULTRAFILTRATION IN AN UNSTIRRED BATCH CELL

D. R. Trettin, M. R. Doshi

The Institute of Paper Chemistry
Environmental Sciences Division
P. O. Box 1039
Appleton, WI   54912

The ultrafiltration of bovine serum albumin (BSA) in .15 M NaCl (pH = 7.40) and .10 M sodium acetate (pH = 4.70) was investigated in an unstirred batch cell. Data were analyzed to minimize the effects of inter-membrane variations and other experimental errors. BSA systems were studied over a wide range of concentration (1–10 g/100cc) and at higher pressures than previously reported (150 psig). Results confirm that BSA does reach a pressure-independent flux region. A constant property integral method solution, which has the simplicity of the popular film theory model, compared quite well with the exact solution. Agreement between acquired BSA data and theoretical predictions is excellent, with the maximum error less than 3%. Not only is the assumption of constant diffusion coefficient shown to be reasonable for the unstirred batch cell, but also that the difference in the calculated value of the diffusion coefficient for BSA at pH = 4.70 and pH = 7.40 is small. This finding is further confirmed by separate diffusion coefficient determinations by ultracentrifuge technique. The values of C_{gel} as determined from ultrafiltration data are 37.0 g/100cc at pH 4.7 and 57.5 g/100cc at pH 7.4. These values correspond reasonably well to literature values.

Submitted for publication in Ind. and Eng. Chem. Fundamentals.

MORPHOLOGY OF SKINNED MEMBRANES:

A RATIONALE FROM PHASE SEPARATION PHENOMENA

C.A. Smolders

Department of Chemical Technology
Twente University of Technology
P.O. Box 217, 7500 AE Enschede, The Netherlands

INTRODUCTION

Asymmetric skinned membranes, first developed by Loeb and Sourirajan[1] for cellulose acetate as a polymer material, owe their practical value in ultrafiltration and reverse osmosis techniques to the minimization of the hydrodynamic barrier while the selectivity and the mechanical properties of the membranes are maintained. These membranes consist of a very thin dense skin (0.1 to 0.5 μm thickness) and a porous sublayer (0.1 to 0.2 mm thick) of the same polymeric material, see e.g. ref. 2. It has been shown a.o. by Frommer[3] that for the same polymer one can obtain different kinds of morphologies in the asymmetric membrane by changing the preparation conditions. Also it has become common knowledge[4-6] that many other polymers than cellulose acetate can be used to prepare asymmetric, skinned membranes.

The majority of the commercially available membranes are produced by the so-called phase inversion process[5]. In this process a homogeneous polymer solution is transformed into a two-phase system in which a solidified polymer phase forms the porous membrane structure, while a liquid phase, poor in polymer, fills the pores. Different techniques can be found in the literature which by phase inversion lead to quite a variety of membrane morphologies: skinned and unskinned membranes; porous sublayers with an even distribution of pore sizes or with increasing pore sizes over the membrane thickness; porous sublayers with cavities, sometimes extending over the full thickness of the membrane. The precipitation techniques are:

(a) Precipitation from the vapour phase[7,8].
In this very early developed technique membrane formation is accomplished by penetration of a precipitant for the polymer solution film from the vapour phase, which is saturated with the solvent

161

used. A porous membrane is produced without a skin and with an even
distribution of pores over the membrane thickness (bacteria filters
are produced this way).

(b) Precipitation by controlled evaporation[9-11].
The polymer is dissolved in a mixture of a good and a poor solvent,
of which the good solvent is the more volatile. The polymer precipitates when the solvent mixture shifts in composition to a higher
nonsolvent content. A skinned membrane can be the result.

(c) Immersion precipitation[8,12-15].
This technique which was successful for the preparation of a reverse
osmosis membrane for the first time in Loeb and Sourirajan's procedure has been studied and exploited most for the production of skinned membranes. The characteristic feature of course is the immersion of the cast film in a nonsolvent coagulation bath.

(d) Thermal precipitation[15].
Here a solution which is on the verge of precipitation is brought
to separation by a cooling step. When evaporation of the solvent
has not been excluded the membrane will have a skin.

In recent years we have gathered experimental evidence in our
laboratory[14,16-18] that the phase inversion process can be specified separately for the formation of the skin and the sublayer in
terms of two different types of phase separation which occur in
polymer solutions of medium and high concentrations. These types of
phase separation are: (micro)crystallization or gelation for the
skin formation and liquid-liquid phase separation, followed by gelation of the concentrated polymer phase, for the formation of the
porous sublayer.

In this survey we will describe the phase separation phenomena taking place and their consequences for the membrane morphology (skin, porous structure, cavities, etc.).

LIQUID-LIQUID PHASE SEPARATION AND GELATION IN CONCENTRATED POLY-
MER SOLUTIONS

Liquid-liquid (L-L) Phase Separation

When a homogeneous solution becomes thermodynamically unstable,
e.g. by the introduction of a nonsolvent, the original solution can
decrease its free enthalpy of mixing by splitting up into two liquid
phases of different composition. In Fig. 1 a schematic drawing shows
such a free enthalpy surface and following the line A B C D E one
sees that all compositions indicated by points between B' and D' of
the phase diagram will split up in two phases of composition B' and
D'. This process is called liquid-liquid phase separation. There
are two kinetic ways for L-L separation to occur: either by nucleation and growth of the second phase or by spinodal decomposition.
Entering the miscibility gap e.g. at the side of B' (this is a
rather concentrated polymer solution with a certain, low amount of

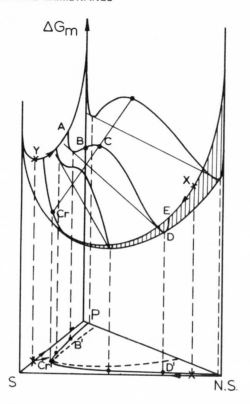

Fig. 1. Sketch of the ΔG_m surface and miscibility gap for the system polymer P, solvent S and nonsolvent NS.

nonsolvent) a nucleus of composition D' (this a tiny drop of high nonsolvent content with almost no polymer in it) will be formed in the polymer solution. This nucleus together with many other nuclei will grow in size until they touch each other and coalesce or until their growth and coalescence is stopped because of gelation of the surrounding polymer solution. The second kinetic way to L-L phase separation would occur if one could so quickly penetrate into the miscibility gap that one would pass the dotted line (the spinodal) before nucleation could take place. For those compositions in the spinodal region the solution is unstable with respect to infinitesimal small concentration fluctuations. The solution would then separate spontaneously[19,20] into interconnected regions of high and low polymer concentration ending up in intertwined networks of phases with compositions near B' and D'. As a matter of fact spinodal decomposition is a highly improbable phenomenon in polymer solutions, since nucleation and growth kinetics are too fast for the homogeneous solution to reach the required composition region.

Of course it is easy to draw the connection now between L-L

phase separation and the genesis of the porous sublayer in membrane formation[14]. The most interesting question here is, which of the polymer phases will be nucleated when starting with a normal membrane casting solution. In principle two possibilities exist, depending on the composition of the initial polymer solution with respect to the so-called critical point, point $C_r^!$ in Fig. 1. For concentrations higher in polymer concentration than $C_r^!$ (hence lying on the branch $C_r^!B'$) the nuclei will consist of the dilute phase. The reverse is true for initial polymer concentrations below $C_r^!$ (branch $C_r^!D'$). Since the critical point in polymer solutions is generally located at rather low polymer concentrations (<10 weight percent in polymer) and membrane forming solutions contain in general more than 10 weight percent of the polymer one should expect L-L phase separation with nucleation and growth of the dilute phase to be the rule.

Crystallization and Gelation

When the thermodynamic quality of a polymer solution is decreasing which may occur by loss of solvent, by lowering of the temperature or by the introduction of a nonsolvent, most polymers are able to form ordered agglomerates. In very dilute solutions the polymer molecules can form single crystals of lamellar type, being only a few hundred Angströms thick and often many microns in the lateral direction. From solutions of medium concentration more complex morphologies occur i.e. dendrytes or spherulites. These latter structures may contain, except for the ordered regions, appreciable amounts of amorphous polymeric material[21].

The formation of ordered structures is dependent not only on the thermodynamic quality of the solution but also on the ability of the macromolecules to crystallize in the time available. So if for a certain polymer solution at medium concentration (say about 10 to 20%) crystallization and L-L phase separation should both be possible thermodynamically, then the kinetically slower crystallization process would be surpassed by the fast process of L-L phase separation. The sequence of events might well be changed on increasing the polymer concentration. Then, by increase in polymer supersaturation the rate of nucleus formation for crystallization is increased substantially, while that for the L-L separation stays essentially the same, being more dependent on nonsolvent content for its nucleation process.

In Fig. 2 we have given a schematic representation of the free enthalpy behaviour in the region of high polymer/low nonsolvent content of the three component system. One sees that at higher concentrations the free enthalpy of mixing of the solution can be lowered by the formation of solid, crystalline polymer in equilibrium with a solution of a certain lower polymer content. At a lower (or higher) temperature the ΔG_m surface will change its shape and the crystalline polymer will be in equilibrium with a solution of lower (or higher) polymer concentration. The so-called melting point curve

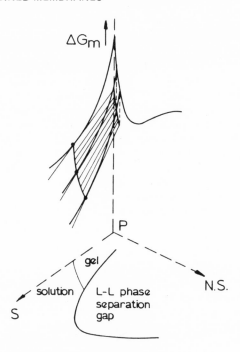

Fig. 2.   Sketch of the free enthalpy behaviour at high polymer/low
          nonsolvent content for the system polymer P, solvent S and
          nonsolvent NS, explaining the solution/gel transition.

for a polymer/solvent pair shows this dependance of melt temperature on composition.
       For crystallization at medium and especially at high concen-
tration, there is a diminished chance to find spherulites which have
grown out to microscopically observable size, if time is short. Instead, by the increased rate of nucleation and the limited growth
rate the solution will contain numerous submicroscopical ordered
regions, perhaps not larger than a nucleus. These microcrystalline
regions act as physical crosslinks in the polymer solution and a
thermoreversible gel is formed. Without excluding crosslinks of a
different kind to be operative in membrane formation, we think that
the type of gelation just described (nonsolvent or temperature induced nucleation at high polymer concentration) could be the leading principle for structure formation in the skin of membranes and
for the gelation of pore walls in the substructure after L-L separation. It is very hard to demonstrate that these gels contain
crystalline material, using standard diffraction techniques[22]. Cal-
orimetric measurements can be used, however, for studying these
thermoreversible gels.

EXPERIMENTAL EVIDENCE FOR TWO PHASE SEPARATION TYPES: GELATION AND
L-L SEPARATION

Experimental evidence has been gathered for the occurrence of
gelation *and* L-L separation in membrane forming systems in recent
studies from our laboratory. These systems comprised the polymers
polyurethane[14], polydimethylphenyleneoxide (PPO)[16], cellulose acetate[17] and polysulfone[18] in appropriate solvent/nonsolvent mixtures.
L-L phase separation was found to occur at low and medium polymer
concentrations and at variable nonsolvent content. This phase separation was detected by measuring cloud-points upon cooling homogeneous solution samples sealed in glass tubes, at different cooling
rates. Cloud points should not be dependent on cooling rate (as gelation does) and turbidity should disappear at the same temperature
on reheating. With proper care in the experimentation the L-L separation could be found for the polymers mentioned above. Gelation was
detected by measuring melting peaks with the help of differential
scanning calorimetry (DSC). Gelation occurs on cooling homogeneous
samples of higher polymer concentrations and relatively low nonsolvent content. For cellulose acetate the polymer concentration had
to exceed 40% and at ambient temperatures and low water content
(about 10%) one had to wait for a long period (one day) before the
melting peak showed up on reheating. Hence gelation often needs
very pronounced supersaturation conditions to proceed with proper
speed. In systems typically forming UF membranes (PPO, polysulfone,
polyacrylonitrile) the precipitation proceeded much faster than for
cellulose acetate. Also for these UF-type polymers the cloud points
(or the binodal curves in the phase diagram) show up at lower nonsolvent content than for cellulose acetate. For quantitative data
the reader is referred to the original papers mentioned.

PHASE SEPARATION AND MEMBRANE MORPHOLOGY

In this section we will discuss the conditions during phase
separation which bring about the most characteristic features of
asymmetric membranes: skin, sponge structure and cavities.

Skin Formation

The ultimately determining factor for the skin formation is
the local polymer concentration in the toplayer of the polymer solution film at the moment of precipitation. This is best illustrated by comparing the two different techniques of precipitation of
cast films mentioned earlier in the introduction: precipitation
from the vapour phase (a) and immersion precipitation (c).

In the first method the precipitation is accomplished at an
effectively unchanged polymer concentration in the toplayer, since
the vapour phase is saturated with the solvent. Then, by diffusion

of nonsolvent into the film the only type of phase separation which
can take place is L-L separation, giving a symmetric membrane without a skin. In the immersion technique the solvent depletion from
the toplayer of the solution film is extremely fast (diffusion aided
by stirring in the bath). An increase in polymer concentration in
the toplayer is the result. This increase improves the conditions
for gelation to occur. The gelation will be favoured by the penetration of nonsolvent. The higher the polymer concentration has become before nucleation in the skin sets in, the more numerous and
the smaller will be the nuclei because of higher supersaturation.

It will be clear now which factors favour the formation of a
more finely structured, i.e. denser skin and therefore of a reverse
osmosis type membrane:

- a higher initial polymer concentration of the solution will
favour the conditions for a large supersaturation in the toplayer
before nucleation sets in;

- a lower tendency of the nonsolvent to induce L-L separation
in the system or to penetrate the cast film will delay the onset of
gelation till sufficient solvent depletion has been obtained and
again supersaturation is favoured. A proper choice of nonsolvent
type and the use of certain additives to the coagulation bath (salt,
glycerin, etc.) will serve these purposes.

- lowering the temperature of the coagulation bath will increase supersaturation while decreasing growth kinetics for the nuclei.

The Porous Sublayer

The formation of the skin will increase the barrier for the
diffusion of solvent out and nonsolvent into the sublayer of the
polymer solution. This means that in the solution below the skin,
phase separation will take place at much lower polymer concentration compared to that in the skin. At concentrations of the original casting composition (or a little higher by solvent loss through
diffusion) this phase separation will be of the L-L separation type.
In many systems optical microscopy[14] allows one to follow the coagulation process closely. By doing so one often sees two phenomena
occurring simultaneously in the phase separation of the solution.

First, one observes the coagulation front which proceeds into
the polymer solution at a certain penetration rate. In some cases
one clearly sees the existance and coalescence of droplets[14]. This
is a direct proof of L-L separation taking place, where nuclei of
the dilute phase are formed and grow out, as soon as the nonsolvent
concentration has increased to values bringing the system within
the binodal region (Fig. 1). The sponge structure of the sublayer
in the final membrane is a result of this L-L separation; the walls
between the droplets have solidified by gelation of the second phase
in the L-L separation, i.e. the original solution phase which increased in polymer concentration. If the concentration of the poly-

mer at the point where L-L separation sets in does not increase
too much over the depth of the cast film, the nucleation density
will not change much over the film tickness and a uniform pore
structure will result. In cases where solvent loss from the film
remains important, the number of pores down the membrane thickness
decreases and the pore size increases.

The second phenomenon, which one very often observed in UF membrane forming systems (in immersion precipitation) is the growth of
large voids starting under the skin and extending their size over
appreciable distances (10-100 μm). Analogous phenomena can be found
in studies on wet-spun polymer fibres[23-27]. Earlier explanations for
its mechanism are: 1) the penetration of nonsolvent through defects
(cracks) in the surface of the fibre[23] or of the membrane[28] and 2)
the simultaneous diffusion of both solvent and nonsolvent to certain randomly distributed loci under the toplayer[18,25]. Some observations remain interesting for further consideration:

- through optical microscopy one observes that the voids grow
faster than the -diffusion controlled- coagulation front is proceeding.

- often the growth of the voids slows down and the coagulation
front for L-L separation may pass beyond the lower end of the cavity.

- the boundary of the voids with the surrounding solution remains fluid initially; no sign of direct skin formation can be detected. In those cases where L-L separation proceeds in the solution between the voids, coalescence with small droplets remains
possible, resulting in "open" walls of the voids, see Fig. 3. In at
least one case, i.e. for polyacrylonitrile, one observes a direct
gelation (without L-L separation) of the solution between the voids,
see Fig. 4.

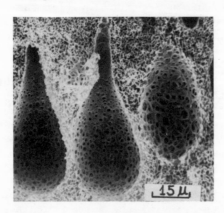
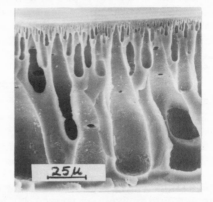

Fig. 3.   SEM-photograph of a   Fig. 4.   SEM-photograph of a cross-
          cross-section of a   section of a polyacryloni-
          polysulfone membrane,   trile membrane, showing
          showing voids with   voids with solid (gelled)
          open walls.   walls without detectable
                                                 pores.

For the interpretation of the phenomena observed, we think that two kinds of driving force are operative for the generation and growth of voids:

(1) The decrease of free enthalpy ΔG_m upon mixing in the voids solvent, originating from the polymer solution and nonsolvent, penetrating from the coagulation bath. The larger this decrease in ΔG_m is, the more pronounced the void formation can be; see the arrow at point X in Fig. 1.

(2) The change in ΔG_m when a polymer solution expells by a kind of syneresis process, the solvent it contains. This can be understood if one realizes that the polymer molecules try to change to a less expanded conformation in the solvent/nonsolvent mixture which increases in nonsolvent content before the L-L precipitation concentration is reached; see the arrow at point Y in Fig. 1.

The sum of both ΔG_m effects determines the kinetics of the void formation. The active driving force for solvent transport to the voids, together with the favourable diffusive transport in the voids make the voids grow faster in the early stages than the normal coagulation front proceeds. The overall mechanism of void formation can be described as follows:

- When the skin is formed, nonsolvent penetrates into the underlying polymer solution faster at certain spots in the skin, e.g. a thinner part of the skin or a local loose arrangement of the structural units in the skin, giving a more favourable pathway for diffusion. Only in systems with a large driving force for solvent/ nonsolvent mixing this heterogeneous type of nucleus is formed.

- Solvent is expelled from the surrounding polymer solution (syneresis) to these statistically spread loci and a gradient in nonsolvent concentration is set up in the void, ranging from a rather low value near the interface with the polymer solution to a high value near the skin surface. There is a fluid interface between the polymer solution and the void.

- Because of the syneresis effect and the facilitated diffusion in the void in comparison with the polymer solution phase, the void may grow faster initially than the coagulation front proceeds. When the syneresis becomes less effective the growth of the voids depend solely on diffusion of solvent over larger distances to the voids, so that the growth of the voids may slow down and the coagulation front may proceed beyond the void.

It is emphasized here that the content of the voids during their growth is not in equilibrium with the surrounding polymer solution in the sense of L-L separated binodal phases. Therefore the L-L separation may proceed separately in the solution between the voids, or a gelation may take place there. Since the content of the voids neither is in equilibrium with the coagulation bath, one often sees the result of a second type of L-L phase separation which occurs in the voids. Fig. 5 gives an example for the system cellulose acetate/dioxane/water. If the void contains a very low amount of polymer during its formation this polymer will be separated when the void content is gradually replaced by pure nonsolvent from

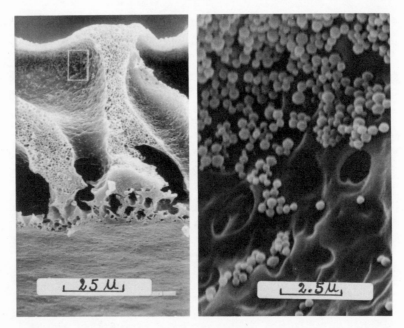

Fig. 5.   SEM-photograph of a polymer latex formed by phase separa-
          tion of a very dilute polymer solution in the voids of a
          cellulose acetate membrane (CA in dioxane precipitated by
          water).

the coagulation bath. Now in the very diluted polymer solution in
the void nucleation of the concentrated phase will initiate the
phase separation and a typical polymer latex as shown in Fig. 5 is
formed.

CONCLUSIONS

     In conclusion it has been shown that in membrane forming sys-
tems the skin is formed by gelation of the toplayer, at increased
polymer concentration by solvent loss, while liquid-liquid phase
separation is responsible for the formation of the porous sublayer.
Typical morphological features of skinned membranes (skin, sponge
structure, cavity formation) are discussed in the context of conditions during phase separation.

REFERENCES

1.   S. Loeb and S. Sourirajan, *Advan. Chem. Ser.*, 38:117 (1962).
2.   R.L. Riley, J.O. Gardner and U. Merten, *Desalination*, 1:30
        (1966).

3.   M.A. Frommer and D. Lancet *in*: "Reverse Osmosis Membrane Re-
     search", H.K. Lonsdale and H.E. Podall, eds., Plenum Press,
     New York (1972).
4.   A.S. Michaels *in*: "Advances in Separation and Purification",
     E.S. Perry, ed., Wiley, New York (1967).
5.   R.E. Kesting, "Synthetic Polymeric Membranes", McGraw Hill,
     New York (1971).
6.   M.T. So, F.R. Eirich, R.W. Baker and H. Strathman, *Polymer
     Letters*, 11:201 (1973).
7.   R. Zsigmondy and W. Bachman, *Z. Anorg. Allgem. Chem.*, 103:109
     (1918).
8.   H. Strathmann and K. Koch, *Desalination*, 21:241 (1977).
9.   J.D. Ferry, *Chem. Rev.*, 18:373 (1936).
10. K. Maier and E. Scheuermann, *Kolloid.Z.*, 171:122 (1960).
11. R.E. Kesting, *J. Appl. Polym. Sci.*, 17:1771 (1973).
12. a. S. Sourirajan, "Reverse Osmosis", Acad. Press, New York
         (1970).
     b. R.E. Kesting, "Synthetic Polymeric Membranes", McGraw Hill,
         New York (1971).
     c. H.K. Lonsdale and H.E. Podall, eds., "Reverse Osmosis Mem-
         brane Research", Plenum Press, New York (1972).
13. M. Guillotin, C. Lemoyne, C. Noel and L. Monnerie, *Desalina-
     tion*, 21:165 (1977).
14. D.M. Koenhen, M.H.V. Mulder and C.A. Smolders, *J. Appl. Polym.
     Sci.*, 21:199 (1977).
15. G.B. Tanny, *J. Appl. Polym. Sci.*, 18:2149 (1974).
16. L. Broens, D.M. Koenhen and C.A. Smolders, *Desalination*, 22:
     205 (1977).
17. F.W. Altena and C.A. Smolders, "Proc. Prague Microsymp. Calo-
     rimetry", to be published in *J. Pol. Sci.*, Part C (1980).
18. L. Broens, F.W. Altena, C.A. Smolders and D.M. Koenhen,
     Desalination, 32:33 (1980).
19. J.W. Cahn, *J. Chem. Phys.*, 42:93 (1965).
20. C.A. Smolders, J.J. van Aartsen and A. Steenbergen, *Kolloid
     Z. Z. Polym.*, 243:14 (1971).
21. B. Wunderlich, "Macromolecular Physics", Acad. Press, New
     York (1973).
22. J.H. Wendoff and E.W. Fischer, *Kolloid Z. Z. Polym·*, 251:884
     (1973).
23. J.P. Craig, J.P. Knudsen and U.F. Holland, *Textile Res. J.*,
     32:435 (1962).
24. V. Gröbe and K. Meijer, *Faserf. Textiltechn.*, 10:214 (1959).
25. V. Gröbe, G. Mann and G. Duve, *Faserf. Textiltechn.*, 17:142
     (1966).
26. J.P. Knudsen, *Textile Res. J.*, 33:435 (1962).
27. A. Ziabichi, "Fundamentals of Fibre Formation", Wiley, New
     York (1976).
28. H. Strathmann, K. Koch, P. Amer and R.W. Baker, *Desalination*,
     16:179 (1975).

CHARACTERIZATION TECHNIQUE OF STRAIGHT-THROUGH POROUS MEMBRANE

Kenji Kamide and Sei-ichi Manabe

Textile Research Laboratory,
Asahi Chemical Industry Company, Ltd.
Hacchonawate 11-7, Takatsuki, Osaka 569, Japan

INTRODUCTION

Recently the technology of separating materials dispersed or dissolved in liquid through use of polymeric porous membrane has attracted significant attention. Of course, the field to which porous membrane is successfully applicable, is exclusively determined by the radius (average and its distribution) of the pores existing in ex- and interior of the membrane. For example, the membrane with mean pore size less than 4 nm are available for desalination and those with 2-8 nm are for dialysis type artificial kidney and above 10 nm for ultrafiltration. Except the membrane used for the concentration of Uranium by the diffusion method, the majority of the porous membranes employed in both laboratory and industrial scales is principally polymeric materials, because, polymeric porous membrane has the following advantages: (1) Possibility of usage of various materials with different chemical nature, (2) attainment of high degree of porosity ($\simeq 70$ %) and wide range of average pore size, (3) readily formability of various forms (hollow fiber membrane, envelope membrane).

The porous polymeric membranes have unavoidably more or less the pore size distribution. Then, the magnitude of the mean pore size depends on the measuring method utilized. Nevertheless, this point has not unfortunately attracted attention in the literature. Up to now the numerous studies published hitherto were mainly concentrated to new procedure of the membrane production or the permeation characteristics of the membrane. And we can find no comprehensive study correlating the pore characteristics of polymeric membrane and performance including the permeability and permselectivity. In order to obtain the structure-performance rela-

tionships, it is highly desirable to establish the method of characterizing the pore characteristics of polymeric porous membranes,
including porosity, average pore size, and pore size distribution.

In this paper we will propose some methods for characterizing
the pore characteristics of polymer porous membranes with mean pore
size higher than 10 nm, which are based on our permeation theory[4]
for gas and liquid. In addition, we intend to discuss the application limits of these method.

THEORETICAL BACKGROUND

Fig.1 shows a schematic representation of porous membrane models used in this paper. It is assumed here that all pores have
circular cross section, penetrating from a surface to another one,
that is, the membrane has straight-through pores. We consider two
cases: The case where pore size is equal in both surfaces (cylindrical pore) and the case where pore size is not equal in both surfaces (i.e., truncated cone shape pore in Fig.1(c)). N is pore density (number of pores for a unit surface area), r is radius of pore,
d is membrane thickness. θ_R is defined by eq.(1)

$$\theta_R = \tan^{-1}\left\{(r_a-r_b)/d\right\} \tag{1}$$

where r_a and r_b are pore radii on the front side and back side,
respectively. For polymeric porous membranes, d lies in the range
100-500 μm and either r_a or r_b is in the range of 0.01-10 μm. Therefore, $\theta_R \ll 1$ holds here.

We denote the number of pore, whose radius of pore is between
r and r+dr, existing in unit surface area by N(r)dr. Here N(r) is
the distribution function of pore radius. If we use r_a in place of
r in Fig.1(c), N(r_a) by N(r). The various kinds of mean pore radius $\bar{r}_i$ are defined by eq.(2):

$$\bar{r}_i = \int_0^\infty r^i N(r)\,dr / \int_0^\infty r^{i-1} N(r)\,dr, \quad i=1,2,\cdots \tag{2}$$

Pore density N is given by

$$N = \int_0^\infty N(r)\,dr \tag{3}$$

The i-th moment of r, X_i, is also defined by eq.(4),

$$X_i = \int_0^\infty r^i N(r)\,dr \tag{4}$$

The porosity and the surface area of pore per unit volume of membrane are represented by Pr and Sr, respectively. When Pr $\geq$ 0.3,
Sr is approximated to the surface area of the membrane per unit
volume with high accuracy. Pr and Sr in the cases of Fig.1(b) and
1(c) are given by the following rather simple equations if $\tan\theta_R$
$\fallingdotseq \theta_R$ can be approximated.

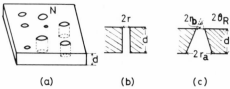

(a)   (b)   (c)

Fig. 1. Schematic representation of polymeric membrane[1].
(a), straight through porous membrane; (b), cylindrical
pore (cross section of membrane); (c), trancated cone
shape pore (cross section of membrane); r_a, pore radius
of front side; r_b, pore radius of back side.

$$Pr = \pi\{N\bar{r}_2 \cdot \bar{r}_1 + (\theta_R \cdot d)\bar{r}_1 + (\theta_R \cdot d)^2\}/3 \tag{5}$$

$$Sr = \pi N(2\bar{r}_1 + \theta_R \cdot d) \tag{6}$$

When $\theta_R = 0$, both eqs. (5) and (6) reduce to the corresponding equations for straight-through porous membrane, that is,

$$Pr = \int_0^\infty \pi r^2 N(r)\,dr = \pi X_2 = \pi N\bar{r}_2 \cdot \bar{r}_1 \tag{5'}$$

$$Sr = 2\pi\int_0^\infty rN(r)\,dr = 2\pi N\bar{r}_2 \tag{6'}$$

If we can estimate the values of $N(r)$, d and θ_R by using some adequate techniques, the characteristics (r_i, N, X_i, Pr and Sr) of the membrane as shown in Fig. 1 can be determined throughly by the use of eqs. (2)-(6). In the following sections we will demonstrate the methods of determing $N(r)$, $\bar{r}_i$, Pr and pore shape.

Methods for Evaluating $N(r)$

Mercury intrusion method (MI method). We denote summation of the apparent volumes of a membrane (v_s) and mercury (v_g) at pressure P by $v(p)$. The volume contraction due to the loading pressure P, $\Delta v(p)$, is given by

$$-\Delta v(p) = v(p) - v(0) = v_s \int_{f_2(p)}^{f_1(p)} (\pi/3\tan\theta_R)\{r_b^3 - (8\alpha^3/P^3)\cos^3(\theta + \theta_R)\}\times$$

$$N'(r_b)\,dr_b + \int_{f_1(p)}^\infty (\pi/3)\cdot d\cdot(r_b^2 + r_b r_a + r_a^2)N'(r_b)\,dr_b + v_g\beta_g P$$

$$\cdots \tag{7}$$

with $f_1(p) = (2\alpha\cos(\theta + \theta_R)/P) + (\tan\theta_R)\cdot d$, and $f_2(p) = 2\alpha\cos(\theta + \theta_R)/P$, and

$$r_b^2 + r_b r_a + r_a^2 = 3r_b^2 - 3r_b d\cdot\tan\theta_R + (d\cdot\tan\theta_R)^2 \tag{8}$$

Where α is the surface tension of mercury, θ the contact angle between mercury and membrane material, $v(0)$ the $v(p)$ value at $P=0$, β the compressibility of mercury. $N'(r_b)$ is the distribution function of r_b and $N'(r_b)dr_b=N(r_a)dr_a$ (that is $N(r)dr$) holds.

Differentiation of eq.(7) with respect to r_j, which is given by eq.(9), and rearrangement of the equation obtained by use of eq.(8) yields eq.(10):

$$r_j = 2\sigma\cos(\theta+\theta_R)/P \qquad (9)$$

$$N(r-d\cdot\theta_R/2) = \frac{-P^4\{(dv(p)/dP)-v_g\beta_g\}}{v_s\cdot\pi\cdot d\cdot 2\sigma\cos(\theta+\theta_R)\,[\{2\sigma\cos(\theta+\theta_R)\}^2-2/3(d\cdot\theta_R)^2 P^2]}$$
$$\cdots (10)$$

Since v_g, θ_R, and d can be obtained experimentally and β_g and θ are given by a literature, $N(r)$ can be calculated by using eq.(10) from experimentally obtained $\Delta v(p)$ vs P curve.

<u>Fluid permeability & bubble pressure method (BP method)</u>. Fig. 2 shows the schematic representation of cross sectional view of membrane contacting with two kinds of fluids of S_2 and S_1. In this figure, these two kinds of fluids should not be miscible with each other and $\theta_R=0$ is assumed. The pressure P_1 is loaded on the side 1 and S_1 is liquid or gas (gas in general) and S_2 should be liquid. When the pressure difference $\Delta P(=P_1-P_2)$ becomes larger than a certain critical value $\Delta P'_1$, the fluid of S_1 passes as a bubble through a pore with radius r to side 2, $\Delta P'_1$ which is derived from the principle of balance of force and given by eq.(11),

$$\Delta P'_1 = 2\sigma_2/r \qquad (11)$$

where, σ_2 is the surface tension of the fluid S_2. According to eq.(11), the minimum pressure difference ΔP_{min} which originates the first bubble gives a maximum pore radius, r_{max}, that is $r_{max}=2\sigma_2/\Delta P_{min}$.

The permeation rate of fluid S_1 per unit area of a membrane designated by J, is given by eq.(12) or (13) depending on whether S_1 is incompressible or compressible fluids.

$$J = k\Delta P\int_{2\sigma/\Delta P}^{\infty} r^4 N(r)dr \qquad \text{for incompressible fluid} \quad (12)$$

$$J/\bar{P} = k\Delta P\int_{2\sigma/\Delta P}^{\infty} r^4 N(r)dr \qquad \text{for compressible fluid} \quad (13)$$

where, $\bar{P}=(P_1+P_2)/2$, $k=\pi/8\eta_1 d$, and η_1 is viscosity of fluid S_1.

On the basis of eqs.(12) and (13), $N(r)$ can be estimated from the permeation rate by eqs.(14) and (15), which are applicable in the case of $\theta_R=0$.

$$dJ/d\ln\Delta P = J+2\sigma kN(r)(2\sigma/\Delta P)^4 \quad \text{or} \quad N(r)=(dJ/d\ln\Delta P-J)/2\sigma k(2\sigma/\Delta P)^4$$

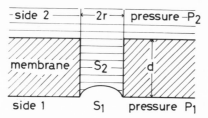

Fig. 2. Cross sectional view of membrane with fluid.
P_2, pressure loaded on front surface of membrane; P_1, pressure loaded on back surface, $P_1 > P_2$; side 2, front side; side 1, back side; m, membrane; S_2, solvent 2; S_1, solvent 1; r, pore radius; d, thickness of membrane.

$$\cdots (14)$$

$$d(J/\bar{P})/d\ln\Delta P = J/\bar{P} + 2\pi k N(r)(2\pi/\Delta P)^4 \text{ or } N(r) = \{d(J/\bar{P})/d\ln\Delta P - J/\bar{P}\}/2\pi k (2\pi/\Delta P)^4 \cdots (15)$$

In the case of $\theta_R \neq 0$, the correction function $F(\theta_R)$ derived for incompressible fluid by Oka[2] is assumed to be applicable to compressible fluid.

By applying similar procedure, eqs. (16) and (17) are derived in the case of $\theta_R \neq 0$.

$$N(r) = \{dJ/d\ln\Delta P - J\}/\{2\pi k F(\theta_R)(2\pi/\Delta P)^4\} \text{ for incompressible fluid} \cdots (16)$$

$$N(r) = \{d(J/\bar{P})/d\ln\Delta P - J/\bar{P}\}/\{2\pi k F(\theta_R)(2\pi/\Delta P)^4\} \text{ for compressible fluid} \cdots (17)$$

where, $F(\theta_R)$ is the correction function originally given by Oka for incompressible fluid by eq. (18)

$$F(\theta_R) = (16/3)\{(1-\cos\theta_R)/\theta_R\}(1+2\cos\theta_R)/(2+\cos\theta_R+\cos^2\theta_R) \qquad (18)$$

<u>Ultrafiltration-permselectivity method (UP method)</u>. When mean pore size is less than 30 nm, above two method are not applicable because of large absolute value of $\bar{P}$. According to our permselectivity theory[3], the ratio of the concentration between filtrate and filtrand when particles with radius a are dispersed or dissolved in the filtrand, $\mathcal{G}(a)$, can be expressed in terms of r and $N(r)$ by eq. (19),

$$\mathcal{G}(a) = \int_0^\infty \mathcal{G}(a,\delta) r^4 N(r) dr / \int_0^\infty r^4 N(r) dr \qquad (19)$$

where, $\mathcal{G}(a,\delta)$ is $\mathcal{G}(a)$ in the case when $N(r)$ is represented by δ function being theoretically given by us[3]. $\mathcal{G}(a)$ vs $a/(\bar{r}_3 \cdot \bar{r}_4)^{1/2}$ curves depends significantly on $N(r)$ and this relation for the model membrane with various types of $N(r)$ have been numerically calculated. By the application of a curve fitting technique to the experimental $\mathcal{Y}$ vs $a/(\bar{r}_3 \cdot \bar{r}_4)^{1/2}$ curve, the best fitting $N(r)$ function can be selected.

Gas permeability method (GP method). It has been shown in the previous paper[4] that the permeability coefficient $\mathbb{P}(P_1,P_1)$ $(=\lim_{P_2 \to P_1}$ $\mathbb{P}(P_1,P_2)$, P_1 and P_2; gas pressure loaded on membrane, $P_1 > P_2$) of inorganic gas for the membrane of $\bar{r}_4 > 15$ nm can be approximately expressed by eqs.(20)-(23).

$$\mathbb{P}(P_1,P_1) = \int_{\lambda_1/2}^{\infty} \mathbb{P}_v N(r)\,dr + \int_0^{\lambda_1/2} \mathbb{P}_f N(r)\,dr \qquad (20)$$

with

$$\mathbb{P}_v = (\hbar Z/8\eta)r^4 P_1 (T_s/P_s \cdot T) + \left\{(2-f_1)/f_1\right\}(\pi/4)r^3(2\pi RT/M)^{1/2}(T_s/P_s T) \cdots (21)$$

$$\mathbb{P}_f = \left\{(2-f_o)/f_o\right\}(4/3)r^3(2\pi RT/M)^{1/2}(T_s/P_s T) \qquad (22)$$

$$\lambda_1 = (\eta/ZP_1)(\pi RT/2M)^{1/2} \qquad (23)$$

where, P_s is the standard pressure of 76 cmHg, T_s the standard temperature of 273.15 K, f_1 the Maxwell reflection coefficient for slip flow, f_o the Maxwell reflection coefficient for free molecular flow, λ_1 the mean free path of the gas at $P=P_1$ and at temperature T, η the viscosity of the gas, M the molecular weight of gas, Z a constant $(=1.01325 \times 10^6/76.0)$. Eq.(20) indicates that in this case gas flow in a given pore is consisted of either viscous flow (V flow) or free molecular flow (F flow) and which flow is dominated is determined by the ratio of r to λ_1. We can evaluate $N(r)$ from eq.(20) by employing the following approximation given by eq.(24).

$$\int_0^{\lambda_1/2} r^3 N(r)\,dr = (1/2)(\lambda_1/2)^4 N(\lambda_1/2) \qquad (24)$$

The results obtained is shown by eq.(25).

$$N(\lambda_1/2) = A/B \qquad (25)$$

$$A = (-2P_1/\lambda_1)d(\mathbb{P}(P_1,P_1) - a_h X_4 P_1 - a_s X_3)/dP_1 + (2/\lambda_1)(\mathbb{P}(P_1,P_1) - a_h X_4 P_1 - a_s X_3$$

$$B = (3/2)\left[(a_f - a_s)(\lambda_1/2)^3 - a_h(\lambda_1/2)^4 P_1\right] - \left[0.52 a P_1/\left\{0.87(P_1+b) - a\right\}^2 \lambda_1\right] \left\{f_1/(2-f_1)\right\} a_s(\lambda_1/2)^3(\lambda_1/4)$$

where,

$$a_h = (Z\pi/8\eta) \cdot (T_s/P_s \cdot T) \qquad (26)$$

$$a_s = \left\{(2-f_1)/f_1\right\}(\pi/4)\cdot(2\pi RT/M)^{1/2}(T_s/P_s T) \qquad (27)$$

$$f_1 = 0.87 - a/(P_1+b) \qquad (28)$$

a and b are constants for a given gas species, and are approximated to be nearly equal to 5 (cmHg) and 18(cmHg),respectively,
from the experimental f_1 vs P_1 curves for hydrogen, helium, nitrogen, oxygen, carbon dioxide, and argon[9]. X_3 and X_4 defined before
by eq.(4) are determined by the method mentioned in the following
section of GP method. When the experimental value of $P(P_1,P_1)$ and
theoretical value of λ_1 in eq.(23) are substituted into eq.(25), we
can calculate $N(r)$ value at $r=\lambda_1/2$.

Electron microscopic (EM) method. In the case when pore is
the straight-through cylindrical pore as shown in Fig.1, $N(r)$ is
directly determined from observation of the membrane surface by EM
method. Scanning electron microscopic method (SEM method) is applicable to the membrane with larger pore size than ca.10 nm because
of its resolving power limit. When the pore size is less than 10
nm and probably, larger than 1 nm at least, we can utilize a transmission electron microscopic method (TEM method) combined with the
ultra-thin cutting technique and stereological analysis.

Method for Evaluating Mean Pore Size $\bar{r}_i$
_____

Most general method. The i-th order of mean pore radius $\bar{r}_i$
defined by eq.(2) is , in the strict sence, calculated from $N(r)$.
Accordingly, the method for evaluating $N(r)$ is also applicable to
determine r_i. However, the conventional methods proposed hetherto
(MI method and flow rate method), in which knowledge of $N(r)$ is not
necessary, affords also mean pore size,but its order has not been
clarified and therefore, mean pore size thus determined is change-
able according with the determination method employed. In this se-
section the strict meaning of the mean pore radii determined by conventional methods will be given and new methods specialized for determination or r_i will be proposed.

MI method. When the $v(p)-p$ curve is experimentally given in
advance, the mean pore radius $\bar{r}_{pp}$ can be estimated conventionally
by eq.(29)

$$\bar{r}_{pp} = \frac{\int_0^\infty (2a\cos\theta/P)(dv(p)/dp-v_g\beta_g)\,dp}{\int_0^\infty (dv(p)/dp-v_g\beta_g)\,dp} \qquad (29)$$

$\bar{r}_{pp}$ has been widely employed hitherto.
Substitution of eq.(10) into eq.(29) under the condition of
$\theta_R \ll 1$, leads us to the following equation:

$$\bar{r}_{pp} = \frac{3\bar{r}_3 \cdot \bar{r}_2 \cdot \bar{r}_1 - 9\bar{r}_2 \cdot \bar{r}_1 \cdot \theta_R \cdot d + 7\bar{r}_1 \cdot (\theta_R \cdot d)^2 - (\theta_R \cdot d)^3}{3\bar{r}_2 \cdot \bar{r}_1 - 6\bar{r}_1 \cdot \theta_R + (\theta_R \cdot d)^2} \tag{30}$$

Then, $\bar{r}_{pp}$ can be expressed in terms of $\bar{r}_i$, θ_R and d. The mean pore radius $\bar{r}_{pp}$ defined in eq.(29) is obviously equal to $\bar{r}_3$ at $\theta_R=0$ and nearly equal to $\bar{r}_3$ when $\theta_R \neq 0$.

Flow rate method. The filtration flux J' of solvent (the volumetric flow rate of solvent for unit surface area of the membrane) whose viscosity is η_s, can be readily obtained at a given pressure difference ΔP. From the observed J' and Pr values, a mean pore radius $\bar{r}_f$ is calculated by using Poiseuille equation (eq.(31)). Eq. (31) is stricltly valid only in the case when the pore size distribution is extremely narrow, but this equation has been, for long years, employed for the membranes whose pore size distribution is wide.

$$\bar{r}_f = (8\eta_s dJ'/\pi Pr \cdot \Delta P)^{1/2} \tag{31}$$

J' is generally expressed theoretically by eq.(32).

$$J' = \int_0^\infty (\pi r^4/8\eta_s) \varrho \Delta P / \partial d) F(\theta_R) N(r) dr \tag{32}$$

Eq.(31) can be rearranged by using eqs.(5) and (32) into

$$\bar{r}_f^2 = \bar{r}_4 \cdot \bar{r}_3 \cdot \bar{r}_2 \cdot \bar{r}_1 F(\theta_R)/(\bar{r}_2 \cdot \bar{r}_1 + (\theta_R \cdot d)\bar{r} + (\theta_R \cdot d)^2) \tag{33}$$

In deriving eq.(33), $\partial P/\partial d = \Delta P/d$ is assumed. Eq.(33) indicates that $\bar{r}_f$ is equal or nearly equal to $(\bar{r}_3 \cdot \bar{r}_4)^{1/2}$ when $\theta_R=0$ or $\theta_R \neq 0$, respectively.

GP method. The gas permeation in the range of $\lambda_1/2 < r_{min}$ (r_{min}; minimum pore radius) is mainly originated by viscous flow (V flow) and $\mathbb{P}(P_1,P_1)$ in this case can be expressed by[4]

$$\mathbb{P}(P_1,P_1) = C_1 P_1 + C_2 \tag{34}$$

where C_1 and C_2 are constants which can be determined experimentally. These two values (C_1 and C_2) change according with measuring temperature, gas species and pore characteristics.

It should be noted that in the case of $\lambda_1/2 < r_{min}$, the second term of the right hand side of eq.(20) is always zero. Comparison of eq.(34) with the equation derived by substituting eq.(21) into eq.(20) (not shown here) leads to

$$X_4 = C_1/\alpha_h \tag{35}$$

$$X_3 = C_2/\alpha_s \tag{36}$$

X_4 and X_3 are obtained from the experimental C_1 and C_2 values because parameters α_h and α_s can be calculated in advance. From X_4 and X_3 thus obtained, r_4 (referred to as r_{4v}) is calculated according to the following equation.

$$\bar{r}_{4v} = X_4/X_3 = (C_1/C_2)(\mathbf{a}_s/\mathbf{a}_h) \tag{37}$$

If Pr is known in advance, $\bar{r}_3$ (referred to as $\bar{r}_{3v}$) and $(\bar{r}_3 \cdot \bar{r}_4)^{1/2}$ (referred to as $(\bar{r}_3 \cdot \bar{r}_4)^{1/2}{}_v$) are given by eqs.(38) and (39), respectively[5].

$$\bar{r}_{3v} = (C_2/Pr)(\mathbf{\pi}/\mathbf{a}_s) \tag{38}$$

$$(\bar{r}_3 \cdot \bar{r}_4)^{1/2}{}_v = (C_1/Pr)^{1/2}(\mathbf{\pi}/\mathbf{a}_h)^{1/2} \tag{39}$$

Knudsen flow

The free molecular flow (F flow) is a dominant flow in the range of $\lambda_1/2 > r_{max}$, in which $P(P_1,P_1)$ is nearly constant $(C_0=P(0,0))$ independent of P_1 value. When Pr and C_0 are known, $\bar{r}_3$ (referred to as r3f) is calculated by using the following equation[5],

$$\bar{r}_{3f} = (C_0/Pr)(\mathbf{\pi}/\mathbf{a}_f) \tag{40}$$

where, $\mathbf{a}_f = (4/3)(2\mathbf{\pi}RT/M)^{1/2}(T_s/P_s \cdot T) \tag{41}$

Ultrafiltration method.   value at $\mathbf{y}=0.5$ in the $\mathbf{y}-\mathbf{a}$ curve has been regarded as a mean pore radius (denoted hereafter by r_a) without sufficient theoretical background except for recent our theory[3]. According to our permselectivity theory, the $\mathbf{y}$ value should be taken at the linear flow rate of filtrand u=0. If $\mathbf{a}$ value is to be determined, this method is applicable only for the case where the gradient of $\mathbf{y}-\mathbf{a}$ curve at $\mathbf{y}=0.5$ is less than $3(\bar{r}_3 \cdot \bar{r}_4)^{1/2}$. $\bar{r}_a$ value obtained under the above conditions, is nearly equal to $(\bar{r}_3 \cdot \bar{r}_4)^{1/2}$ as is deduced from the theoretical calculation of $\mathbf{y}-\mathbf{a}$ curve for model membrane with various type of $N(r)$[3].

Surface area method. When Pr and Sr can be measured in advance by some adequate technique, the mean pore radius $\bar{r}_s$ is defined by eq.(42).

$$\bar{r}_s = 2Pr/Sr \tag{42}$$

Pr in eq.(5) and Sr in eq.(6) are substituted into eq.(42), then we obtain

$$\bar{r}_s = \left\{ r_2 \cdot r_1 + (\mathbf{\theta}_R \cdot d)r_1 + (\mathbf{\theta}_R \cdot d)^2/3 \right\}/(r_1 + \mathbf{\theta}_R \cdot d/2) \tag{43}$$

Eq.(43) shows that $\bar{r}_s$ is equal or nearly equal to $\bar{r}_2$ when $\mathbf{\theta}_R=0$ or $\mathbf{\theta}_R \neq 0$, respectively.

Method for Determinating Porosity

Porosity Pr can be also calculated from $N(r)$ by using eq.(5).

In addition to this, other methods for estimating will be described below.

    Apparent density method. The apparent density of a membrane ρ_f is evaluated from its experimental weight and volume. On the other hand, the density of material constituting the membrane ρ_p is given in advance. Then, porosity (=Prρ) is calculated by the use of eq.(44) regardless of pore shape.

$$Pr\rho = 1 - \rho_f/\rho_p \tag{44}$$

Eq.(44) is valid for both closed and open type pores.

    Swelling method. When a membrane is kept being immersed in a liquid such as water or non-solvent for membrane materials, the liquid penetrates into pore. After stationary equilibrium is attained, the weight of membrane contained with liquid is measured and then total amount of the liquid penetrated into membrane Ws is evaluated. The porosity (=Prw) is easily calculated by using eq.(45),

$$Prw = (Ws/\rho s)/(Wp/\rho p + Ws/\rho s) \tag{45}$$

where, ρ_s is density of the liquid, Wp is dry weight of the membrane. This method is suitable when $Pr \lesssim 0.20$ from practical view point.

    MI method. When the total amount of volume loss resulted from penetration of mercury into pore, which can be regarded as limiting value of $(\Delta v(p) - v_g \beta_g P)$ at infinite P, is obtained, then porosity Prr is given by

$$Prr = (\Delta v(p) - v_g \beta_g P)/v_s \tag{46}$$

    EM method. Pr evaluated from N(r) obtained by SEM are referred to as Pre, and Pr obtained without aid of N(r) by test line method in the stereology (Prs) for an electron microphotograph is given by eq.(47)

$$Prs = \sum l_{a,i}/\sum L_i \tag{47}$$

where, L_i is the length of i-th test line, and $l_{a,i}$ is the length of line crossing pores of i-th test line.

Pore Shape

    Pore shape is an important characteristic parameter which identifies a membrane. It is very difficult to describe the shape quantitatively though this shape itself is easily observed by electron microscopy as far as pore size is larger than 10 nm.

    We newly introduce the polymer efficiency for constituting a pore P_E in order to express pore shape quantitatively[6]. P_E is defined as the volume fraction of polymer chain component constituting a pore in the total amount of polymer chain. The components which do not constitute a pore, that is the component having free end without connecting each other, do not deform when the tensile stress is loaded on the membrane. Consequently, by measuring the modulus of membrane, we can evaluate the amount of fraction of polymer chains deformed, that is P_E. Porous membrane is regarded as the bicomponent body constituting polymer material and air. According to Uemura and Takayanagi's theory[7] the storage modulus of the membrane having $P_E=1.0$, $E'_{1.0}$, is given by

$$E'_{1.0} = (4.5-0.045Pr)E'_p/(4.5+0.03Pr) \qquad (48)$$

where, E'_p is E' of polymer material constituting the membrane when $Pr=0$. By putting the observed dynamic tensile modulus E'_s of porous membrane into eq.(49), P_E can be calculated.

$$P_E = E'_s/E'_{1.0} = (4.5+0.03Pr)E'_s/(4.5-0.045Pr)E'_p \qquad (49)$$

EXPERIMENTAL

Sample Preparation

We adopted a commercially available polycarbonate membrane " nuclepore" (N series) with various mean pore size, from 15 to 800 nm manufactured by General Electric Comp.(USA) as the actual membrane having straight-through cylindrical pore. Hereafter we denote for example the nuclepore with 800 nm as Nu0.8. This membrane was washed with diethylether in order to leave out parafin on the membrane.

A commercially available membrane, having a trade name of "Millipore Filter" MF0.4 manufactured by Millipore corp. Ltd(USA), and the cellulose acetate membrane prepared by micro phase separation method, proposed by us(sample code number of SF series such as SF0.5 and so on) were used as the model membrane with complicated shape or trancated cone shape pore.

Measurement

MI method. Fig.3 shows the apparatus, by which v(p)-P curve can be obtained. v(p) is indirectly obtained by transformation of the experimental value of electrical resistance between A and B.

BP method. The apparatus shown in Fig.4 was designed and constructed for measurement of the J-P curve by the bubble pressure method combined with fluid permeability method. A membrane to be measured are immersed in cyclohexanol in advance of the measurements so as to make the solvent penetration into a pore.

UP method. New apparatus as shown in Fig.5 was constructed for carring out the measurement of filtration rate under various pressure and various filtrand flow rate[8]. Dilute aq. solutions of conc-

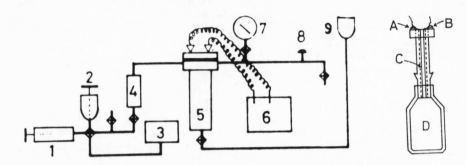

Fig. 3. Apparatus for the mercury intrusion measurement [1].
1, pressure control hundle; 2, oil tank; 3, weight; 4. oil
trap tank; 5, autoclave; 6, galvanometer; 7, pressure gauge
8, safety valve; 9, tank for water injection; A, B, C, wire
of P_t/I_r for measurement of electrical resistance; D, dila-
tometer placed in autoclave 5.

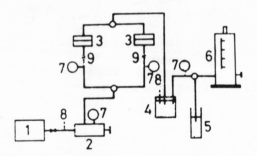

Fig. 4. Apparatus for measurement of the J–ΔP curve by the combined bubble pressure and fluid permeability method [1].
1, pressure source (compressor); 2, pressure control hundle; 3, filter holder; 4, trap for solvent and oil; 5, bubble tester; 6, flow meter; 7, pressure gauge; 8, thermometer; 9, safety valve.

entration from 0.005 to 0.1 wt % of model substances were employed.
In this case the concentration of the filtrand was chosen low in order that the osmotic pressure, caused by the concentration difference
between the filtrand and the filtrate, has negligible effect to
ultrafiltration rate observed. The concentration was measured by
spectroscopy (Hitachi spectrophotometer type 100-50, manufactured
by Hitachi Ltd, Japan),differential refractometry (ALC/GPC 201, manufactured by Waters Associate In. USA) and chemical analysis. The
φ value at zero flow rate u=0 was extrapolated from the experimental φ-u relation.

   GP method. In order to evaluate the gas permeability coefficient, the apparatus shown in Fig. 4 was used when $P_2 \geqslant 76$ cmHg, and
when $P_2 < 76$ cmHg the apparatus newly designed was used[9] and is schematically represented in Fig. 6. In this apparatus the fine particles
dispersed in gas are filtered out by PF2 having the same type membrane with that to be measured, and the contacting point of the support against membrane (F) was large enough to neglect additional
resistance of gas flow.
   EM method. N(r) is directly obtained by using scanning electron microscope (JSM-U3 type, manufactured by Japan Electron Optics
Laboratory Co. Ltd, Japan).

RESULTS AND DISCUSSION

N(r) Obtained by Using Various Methods

   Figure 7 shows the N(r)-r curve for a cellulose diacetate membrane SF0.5 estimated by MI method. The numbers in this figure denote the value of θ_R (in radian) assumed in the calculation of Eq.
(10). With an increase in θ_R, N(r) shifts to smaller r value, increasing its peak. The curve for $\theta_R = 1.7 \times 10^{-4}$ radian is the true N(r)
curve for this membrane, because this value is obtained by SEM.
   The N(r)-r curves for NuO.8 evaluated by various methods are
shown in Fig.8(a) and (b), in which two different N(r) curves by SEM
are demonstrated. This difference may have its origin in the
difference on production batches. Although the shape of N(r) curve
is different in its detailed structure depending on the method used,
the pore density N is nearly same in all methods such as 3.75×10^7(
(Number/cm^2) in MI method and 3.81×10^7 (Number/cm^2 in SEM method and
4.4-4.5×10^7 (Number/cm^2) in GP method. It is also clear that N(r)
obtained by GP method is independent of the kind of permeation gas
used.
   Figure 9 compares two N(r) curves for SF0.5 estimated by using
MI and SEM methods. Both N(r) curves cover nearly the same r range
having same peak location. The N values evaluated from N(r) curve
are found to be 1.0×10^8 (Number/cm^2) and 5.5×10^7 (Number/cm^2) for MI
and SEM methods, respectively.
   Figure 10 shows N(r)-r curves for MF0.4 estimated by MI(full
line) and BP(broken line) methods. The coincidence of both curves

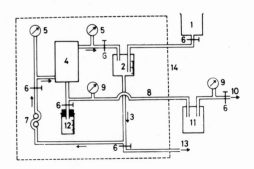

Fig. 5. Apparatus for carring out of filtration of blood (fluid)[8].
1, storage tank of blood; 2, blood bottle; 3, circular line
of blood; 4, artificial kidney; 5, pressure gauge; 6, bulb;
7, blood pump; 8, vacuum line; 9, pressure gauge for vacuum; 10, vacuum pump; 11, thermoregulated room; 12, recovery tube.

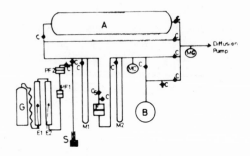

Fig. 6. Schematic representation of measuring instrument of gas
permeability coefficient[8].
A, storage tank of gas of high pressure side; B, storage
tank of gas of low pressure side; C and C_5, switch cock;
E1 and E2, drying tubes containing $CaCl_2$ and P_2O_5, respectively; F, filter holder; G, gas bomb; M1 and M2, mercury
manometer; PF1, prefilter with ceramic filter; PF2, prefilter with ceramic filter; PF2, prefilter with Nu series
membrane; MC, Macleod gauge.

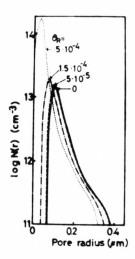

Fig. 7. Distribution function of pore radius N(r) of SF0.5 determined by MI method [1].
parameter denote θ_R (radian). $\theta_R=1.7\times10^{-4}$ was obtained by SEM method.

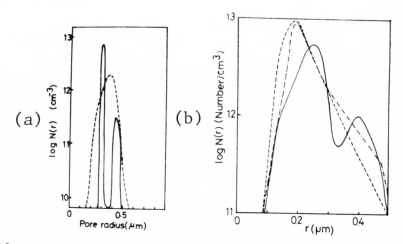

Fig.8. Distribution function of pore radius N(r) of Nu0.8.
(a): Full line, N(r) determined by mercury intrusion method; broken line, N(r) determined by SEM method; r, pore radius.
(b): Full line, N(r) evaluated by using SEM method; broken line, N(r) evaluated by substituting $\mathbb{P}(P_1,P_1)$ of Ar into eq.(25); chain line, N(r) evaluated by substituting $\mathbb{P}(P_1,P_1)$ of N_2 into eq.(25)

is fairly good. The values of N valculated via N(r) are estimated
to be 1.4×10^8 and 1.2×10^8 (Number/cm^2) for MI and BP methods, respectively. Irregular shape of pore in membrane MF0.4 makes SEM meth-
od inapplicable.

In summary, the N(r) values determined by using MI, BP, GP,
and EM methods, for the membrane having straight-through cylindrical
pore like Nu0.8, coincide with each other. In contrast, this is not
true for the membrane having pore of complicated irregular shape like
SF0.5 and MF0.4. Even in this case, the peak value of N(r) and
r value at its peak and N estimated are almost independent of the
methods employed.

In Table 1 are summarized the pore form to be measured and sample weight needed and characteristic values obtained for the various characterization methods. Table 2 demonstrates the pore size
range applicable with respects to the various characterization methods. As for MI method, the pore of open and semi-open types is measured, and total amounts of sample needed for this method is about
a few gram. The accuracy is comparatively high among the methods
used except small r range. It is very difficult to dicide the N(r)
value at small r less than 10 nm, because the contribution of $v_g \beta_g P$
at small r (tant is large P) in eq.(7) increases resulting in the
increase in the relative error of $(dv(p)/dP-v_g\beta_g)$value. SEM method
gives us very impressible feature about pore size and pore shape.
The sample needed is only a few milli-gram and this originates low
reliability of the data, because the fine structure of a membrane
easily changes in place to place. In addition to this fault, the
limitation that the pore must have regular shape in order to analyze quantitatively is another drawback in practice.

Mean Pore Radius $\bar{r}_i$

Table 3 represents the ratio $\bar{r}_4/\bar{r}_3$ evaluated for various membranes. It is evident that MI, SEM and BP methods give almost the
same value of this ratio for a membrane, and polycarbonate membrane
Nu0.8 has the sharpest pore size distribution.
Figure 11(a) and (b) show the change of $\bar{r}_i$ for Nu0.8, SF0.5 and
MF0.4 with an increase in mean order (i). In Fig.11(a) and (b) $\bar{r}_i$
is calculated by putting N(r) obtained by MI and SEM methods into
eq. (2), respectively. Full and broken lines with arrow in these
figures denote $(\bar{r}_4 \cdot \bar{r}_3)^{1/2}$ value estimated from N(r), and the
experimental value of $\bar{r}_f$, respectively. The theory indicates
that when N(r) - r curve is very sharp, the i dependence of $\bar{r}_i$
is very small. Then, the width of N(r) - r curve can be represent-
ed by $\bar{r}_i/\bar{r}_{i-1}$ reasonably. $(\bar{r}_3 \cdot \bar{r}_4)^{1/2}$ coincides fairly well with
r_f except for Nu0.8 as shown in these figures. Low measuring accu-
racy of Pr for Nu0.8 might explain the disagreement between
$(\bar{r}_3 \cdot \bar{r}_4)^{1/2}$and $\bar{r}_f$.
Table 4 summarizes mean pore radius, estimated by MI method
$\bar{r}_{pp}$ (eq.(29)) and flow rate method $\bar{r}_f$ for SF0.2, SF0.3, and SF0.4.

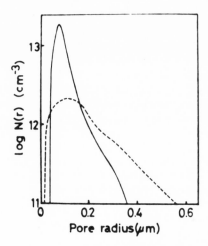

Fig. 9. Distribution function of pore radius N(r) of SF0.5 [1].
Full line, N(r)-r curve determined by mercury intrusion
method (θ_R=1.5x10^{-4} radian); broden line, N(r)-r curve determined by SEM method.

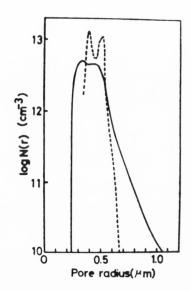

Fig. 10. Distribution function of pore radius N(r) of MF0.4 [1].
Full line, N(r)-r curve determined by mercury intrusion
method; broken line, N(r)-r curve determined by combination method of bubble pressure and fluid permeability.

Since, from the theoretical view point, the following relations $\bar{r}_{pp}=\bar{r}_3$, and $\bar{r}_f=(\bar{r}_4\cdot\bar{r}_3)^{1/2}$ hold, $\bar{r}_f$ should be equal to or larger than $\bar{r}_{pp}$. Table 4 confirms the above expectation.

Table 5 collects the parameters evaluated from $\mathbb{P}(P_1,P_1)$ for various polycarbonate membranes and η of gas. In the table, C_1' and C_2' correspond C_1 and C_2 in eq.(34),respectively, evaluated by the method proposed by Yasuda&Tsai[10], who ignored the distribution of pore size, assuming that the observed gas permeability coeffi- cient can be represented with the simple summation of the permeabi- lity coefficient due to free molecular flow and viscous flow $\mathbb{P}_f$ and $\mathbb{P}_v$. That is, they assumed that both V and F flows occur in a given pore coccurently. This sharply contradicts with the experimental results[4]. The mean pore radius obtained by Yasuda and Tsai method is denoted as $\bar{r}_{yt}$ and given by

$$\bar{r}_{yt} = (C_1'/C_2')(32\eta/3\pi Z)(2\pi RT/M)^{1/2} \qquad (50)$$

In Table 5, Prρ, Prw and Pre shown in the parentheses mean the Pr value estimated by apparent density method, swelling method and SEM method, respectively. By using the data in Table 5, we obtain $\bar{r}_{3f}$, $\bar{r}_{yt}$, $\bar{r}_{3v}$, $\bar{r}_{4v}$, and $(\bar{r}_3\cdot\bar{r}_4)_v^{1/2}$, respectively, from eqs.(40), (50), (38), (37) and (39), and are shown in Fig.12.

Figure 12(a) and (b) show the dependence of $\bar{r}_{3f}$ and $\bar{r}_{yt}$ on mo- lecular weight of permeating gas (M), respectively. $\bar{r}_{3f}$ remains al- most constant independent of M. In contrast to this, the value of $\bar{r}_{yt}$ does not show significant tendency with M due to rather wide scattering of data. Comparison of Fig.12(a) with Fig.12(b) suggests the high reliability of GP method (eq.(40)), which is used for de- terming $\bar{r}_{3f}$.

We obtain $\bar{r}_{3f}$, $\bar{r}_{3v}$, $\bar{r}_{4v}$ and $(\bar{r}_3\cdot\bar{r}_4)_v^{1/2}$ from the gas permeation data, corresponding to the viscous flow region, by using a specific gas, including H_2, He, CO_2, and N_2. Then, the above parameters can be averaged over various kinds of gas used and are denoted as $\bar{r}_{3fa}$, $\bar{r}_{3va}$, $\bar{r}_{4va}$, and $(\bar{r}_3\cdot\bar{r}_4)_{va}^{1/2}$, respectively. The relationship between $\bar{r}_{3va}$ and $\bar{r}_{3fa}$ for polycarbonate membranes is shown in Fig.13, from which the relation $\bar{r}_{3va}=\bar{r}_{3fa}$ is obtained, indicating that the mean pore size evaluated from the gas permeability data obtained in the free molecular flow range agrees well with that obtained in the vi- scous flow range.

In Fig.14(a)-(c) are shown the plots of various mean pore ra- dii evaluated by the gas permeation method against those by other methods ($\bar{r}_3$ and $\bar{r}_4$ by SEM method, designated as $\bar{r}_{3e}$ and $\bar{r}_{4e}$). In this figure $(\bar{r}_3\cdot\bar{r}_4)_p^{1/2}$ corresponds to $\bar{r}_f$. The relations $\bar{r}_{3f}=\bar{r}_{3e}$, $\bar{r}_{4v}=\bar{r}_{4e}$, and $(\bar{r}_3\cdot\bar{r}_4)_v^{1/2}/(\bar{r}_3\cdot\bar{r}_4)_p^{1/2}=1.2-2.2$ hold as is obvious from Fig.14 with some exceptions. In other words, the values of $\bar{r}_i$ obtained by using GP and SEM methods are consistent with each other as far as the membranes having straight-through cylindrical pore are concerned. It should be pointed out here that $\bar{r}_{4v}$ can be obtained only from the data of $\mathbb{P}(P_1,P_1)$ without using Pr.

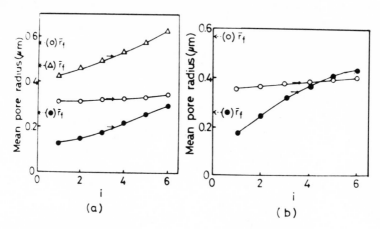

Fig. 11. Mean pore radii of Nu0.8, SF0.5, and MF0.4 [1].
(a), $\bar{r}_i$ determined by mercury intrusion method; (b), $\bar{r}_i$
determined by SEM method; O, Nu0.8; $\bullet$, SF0.5; $\triangle$, MF0.4;
$\longrightarrow$ indicates the value of $(r_3 \cdot r_4)^{1/2}$; $--\rightarrow$ indicates the
value of $\bar{r}_f$

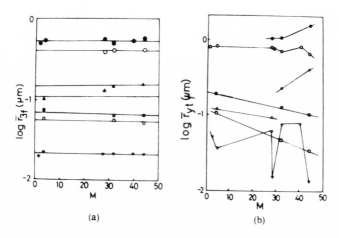

Fig. 12. Relationships between mean pore radii of polycarbonate
membrane and molecular weight of gas [5]. (a) $\bar{r}_{3f}$, third order mean pore radius evaluated by using $\mathbb{P}(P_1, P_1)$ in the range of free molecular flow of gas; (b) $\bar{r}_{yt}$, mean pore radius evaluated by the method of Yasuda-Tsai[10]; M, molecular
weight of gas; $\bullet$, Nu0.8; O, Nu0.6; $\blacktriangle$, Nu0.2; $\triangle$, Nu0.1;
$\blacksquare$, Nu0.08; $\square$, Nu0.05; $\blacktriangledown$, Nu0.03

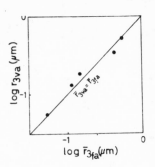

Fig. 13. Relatioship between $\bar{r}_{3va}$ and $\bar{r}_{3fa}$ of polycarbonate membrane[5].
$\bar{r}_{3va}$, average value of third order mean pore radius $\bar{r}_{3v}$
evaluated by using $\mathbb{P}(P_{1},P_{1})$ in the range of viscous flow;
$\bar{r}_{3fa}$, average value of r_{3f}

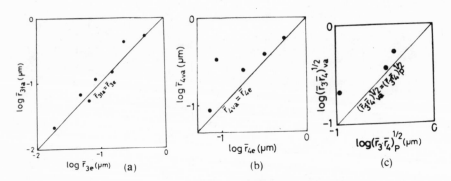

Fig. 14. Plots of mean pore radii evaluated by using gas permeation method proposed here against those by other methods [5].
(a), plots of $\bar{r}_{3fa}$ against $\bar{r}_{3e}$; (b), plots of $\bar{r}_{4va}$ against
$\bar{r}_{4e}$; (c), plots of $(\bar{r}_{3}\cdot\bar{r}_{4})_{va}^{1/2}$ against $(\bar{r}_{3}\cdot\bar{r}_{4})_{p}^{1/2}$; $\bar{r}_{ie}($
i=3 or 4), $\bar{r}_{i}$ values evaluated by SEM method; $\bar{r}_{iva}$(i=3 or
4), average value of $\bar{r}_{iv}$ evaluated by using $\mathbb{P}(P_{1},P_{1})$ in
the range of viscous flow; $(\bar{r}_{3}\cdot\bar{r}_{4})_{p}^{1/2}$, evaluated by using water flow rate

Table 1 Characterization method of pore

Methods	Pore form	Sample weight	area	Characteristic values
Mercury intrusion (MI method)	open, semi-	g	10^7	$N(r)$, $\bar{r}_1$, $\bar{r}_{pp}$, P_{rr}
Combination of bubble pressure and fluid permeability (BP method)	open	100 mg	10^6	$N(r)$, $\bar{r}_1$, r_{max}
Gas permeability (GP method)	open	100 mg	10^6	$N(r)$, $\bar{r}_1$, $\bar{r}_{3v}$, $\bar{r}_{4v}$ $(\bar{r}_3 \cdot \bar{r}_4)^{1/2}_v$, $\bar{r}_{3f}$
Ultrafiltration & permselectivity (UP method)	open	100 mg	10^6	$N(r)$, $\bar{r}_1$, $\bar{r}_a$
Electron microscopy (SEM method)	open, semi-	1 mg	1	$N(r)$, $\bar{r}_1$, P_{rs}
(TEM method)	open, semi-close	1 mg	1	$N(r)$, $\bar{r}_1$, P_{rs}
Flow rate	open	100 mg	10^6	$\bar{r}_f$
Surface area	open, semi-	g	10^7	$\bar{r}_s$

Table 2 Applicable pore size range of various characterization methods

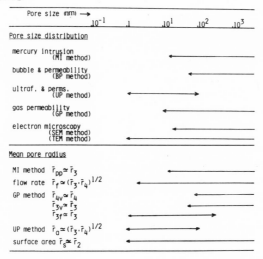

Table 3 Pore size ratio $\bar{r}_4/\bar{r}_3$ [1]

Sample code No.	Method	$\bar{r}_4/\bar{r}_3$
Nu0.8	MI*	1.02
	SEM**	1.02
SF0.5	MI	1.27
	SEM	1.20
MF0.4	MI	1.07
	BP***	1.04

* Mercury intrusion method
** Scanning electron microscopy
*** Combination method of bubble pressure and fluid permeability

Table 4 Values of $\bar{r}_{pp}$ and $\bar{r}_f$ [1]

Sample code No.	$\bar{r}_{pp}$ (nm)	$\bar{r}_f$ (nm)
SF0.2	100	240
SF0.3	320	380
SF0.4	500	460

Porosity Pr Obtained by Using Various Methods

The Pr values estimated for Nu0.8, SF0.5, and MF0.4 are summarized in Table 6. Prm in this table stands for the Pr obtained by putting $N(r)$ value evaluated by MI method into eqs.(2) and (3) and using eq.(5). For SF0.5 data are analyzed by assuming $\theta_R=0$, 5×10^{-5}, 1.5×10^{-4}, and 5×10^{-4} radian for the sake of comparison. The true Pr value corresponds to $\theta_R=1.7 \times 10^{-4}$, which is determined by SEM method. The values of Pre and Prs for MF0.4 could not be evaluated due to its complicated pore shape. The relation Prm=Pre holds for Nu0.8. If we assume that this relation holds even in the case of SF0.5, then θ_R must be 1.2×10^{-4} radian and this value is nearly equal to the value determined by SEM method (1.7×10^{-4} radian). In other words, as for Pr, the morphology of pore in a cellulose diacetate membrane SF0.5 can be approximately represented by a truncated cone shape model with $\theta_R=1.2 \times 10^{-4}$ radian.

Inspection of Table 6 leads us to the conclusion that (1) the density (eq.(44)), MI (eq.(46)) and swelling methods(eq.(45)) are at least recommended for evaluating Pr of membranes with complicated irregular shape pore (for example, MF0.45), and $Pr\rho$, Prr and Prw estimated by these three methods coincide with each other, (2) Prw changes depending on the interaction of solvent with materials constituting a membrane when water is employed as solvent. Cellulose acetate and cellulose acetate/nitrate membranes, such as SF0.5 and MF0.4, are examples and for these, the relation Prw $>Pr\rho$ holds, (3) in the case when a membrane having truncated cone shape pore or complicated shape pore, $Pr\rho$, Prr, and Prw can be nearly equal to each other within $\pm 7\%$ if an adequate θ_R value is chosen.

Pore Shape and Its Related Parameter P_E

Figure 15(a)-(e) demonstrate the scanning electron photomicrographs of various membranes. The pore shape changes from a membrane to a membrane. P_E values of these membranes are 0.21, 0.34, 0.49, 0.71, 0.91 for the membranes demonstrated in Fig.15(a),(b), (c), (d), and (e), respectively. With an increase in P_E, the shape of pore tends to change from irregular shape to circular shape.

Applicability Limit of Characterization Methods

It is noteworthy that $\bar{r}_i$ and $N(r)$ can be strictly defined for straight-through cylindrical pore membrane and these physical meanings may become to some what extent vague if the pore shape deviates from the above-mentioned ideal form. The characterization methods have their limit of applicability with respect to size and shape of pores in membrane. When pore size decreases down to less than 10 nm, BP, GP, and SEM methods are not applicable. For example, BP method loses the measurement accuracy as mentioned before, and in GP method,

Table 5   Characteristic values of $P(P_1, P_1)$ of various polycarbonate membrane [5]

Membrane	Pr	Gas	$\gamma^{a)}$	$C_0^{b)}$	$C_1^{c)}$	$C_2^{b)}$	$C_1'^{c)}$	$C_2'^{b)}$
Nu 0.8	0.22	H_2	0.88	15.1	14.5	13.0	14.5	13.0
	(Pr_ρ)	He	1.96	11.7	8.0	9.5	8.0	9.5
		N_2	1.75	4.6	7.7	3.3	7.7	3.3
		O_2	2.03	3.8	6.2	3.2	6.2	3.2
		Ar	2.27	3.7	6.1	2.7	6.5	2.7
		CO_2	1.47	3.8	8.2	2.8	8.2	2.8
		air	1.82	$—^{f)}$	6.6	3.0	6.6	3.0
Nu 0.6	0.14	N_2	1.75	2.0	2.2	1.5	4.57	1.5
	(Pr_ρ)	O_2	2.03	2.0	1.9	1.3	3.78	1.3
		CO_2	1.47	2.2	2.4	1.12	7.3	1.1
Nu 0.2	0.13	O_2	2.03	0.65	0.35	0.50	0.39	0.60
	(Pr_ρ)	CO_2	1.47	0.60	0.80	0.45	0.90	0.50
Nu 0.1	0.10	He	1.96	0.99	$—^{f)}$	$—^{f)}$	0.120	0.99
	(Pr_ρ)	N_2	1.75	0.49	0.50	0.40	0.141	0.48
Nu 0.08	0.072	He	1.96	0.41	$—^{f)}$	$—^{f)}$	0.066	0.34
	(Prw)	O_2	2.03	0.140	0.039	0.11	0.039	0.110
		CO_2	1.47	0.110	0.043	0.09	0.043	0.094
Nu 0.05	0.026	He	1.96	0.190	$—^{f)}$	$—^{f)}$	0.0140	0.127
	(Prw)	O_2	2.03	0.057	0.063	0.037	0.0060	0.045
		CO_2	1.47	0.050	0.063	0.027	0.0060	0.038
Nu 0.03	1.5×10^{-4}	H_2	0.88	$4.1^{d)}$	$—^{f)}$	$—^{f)}$	$3.33^{e)}$	$4.1^{d)}$
	(Pre)	He	1.96	$3.2^{d)}$	$—^{f)}$	$—^{f)}$	$1.22^{e)}$	$3.2^{d)}$
		N_2	1.75	$1.22^{d)}$	$—^{f)}$	$—^{f)}$	$2.24^{e)}$	$1.22^{f)}$
		CO	1.75	$1.14^{d)}$	$—^{f)}$	$—^{f)}$	$2.40^{e)}$	$1.14^{d)}$
		O_2	2.03	$0.98^{d)}$	$—^{f)}$	$—^{f)}$	$1.53^{e)}$	$0.98^{d)}$
		Ar	2.27	$0.97^{d)}$	$—^{f)}$	$—^{f)}$	$0.58^{e)}$	$0.97^{d)}$
		CO_2	1.47	$1.09^{d)}$	$—^{f)}$	$—^{f)}$	$0.50^{e)}$	$1.09^{d)}$

a) viscosity at 295 K in 10^{-4} poise.   b) 10^{-3} (PU).   c) 10^{-5} (PU/cmHg).   d) 10^{-7} (PU).   e) 10^{-10} (PU/cmHg).   f) not evaluated, PU=cm³(STP)/cm·sec·cmHg.

Table 6   Porosity $Pr(\%)$ determined by the various methods [1]

P_r	Nu 0.8	SF 0.5				MF 0.4
		$\theta_R = 0^{a)}$	$5 \cdot 10^{-5\,a)}$	$1.5 \cdot 10^{-4\,a)}$	$5 \cdot 10^{-4\,a)}$	
$P_{r\rho}$	22	68	68	68	68	72
P_{rm}	11	64	65	62	124	120
P_{rr}	18	61	61	61	61	73
P_{rw}	12	73	73	73	73	74
P_{re}	12	51	56	68	117	—
P_{rs}	10	48	53	64	110	—

a) Radian.

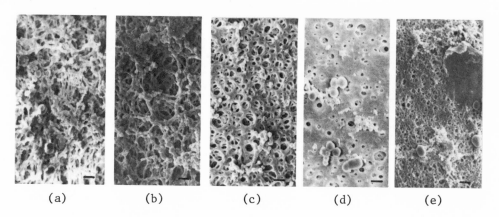

(a)   (b)   (c)   (d)   (e)

Fig. 15. Electron scanning micro-photographies for various cell-
     ulose acetate membrane.
     (a), SF0.70; (b), SF0.67; (c), SF0.62; (d), SF0.20; (e),
     SF0.18 ; scale bar indicates 1 μm.

other permeation mechanisms than F and V flows such as surface diffusion and diffusion become dominant in small pore size region.
And also in SEM method, insufficient resolving power of SEM make it
impossible to measure the diameter of pores clearly.

Then we discuss an applicability of GP method to cellulose diacetate membranes (SF series) having a rather complicated irregular
shape pore.

Figure 16 shows the ratio of $\bar{r}_{4v}$ or $\bar{r}_{3v}$ estimated by GP method
of O_2 to that ($\bar{r}_{4v}$ or $\bar{r}_{3v}$) of CO_2 for cellulose diacetate membranes,
as a function of r_{3v} determined by using O_2 ($_o r_{3v}$).

Figure 17 shows the relationships between r_{3v}, calculated data
on Prρ and $\mathbb{P}$(P_1, P_1) through use of eq.(38), and r_f. In the range
$r_{3v} \geq 30$ nm, the relation $\bar{r}_{3v} = 0.9 \bar{r}_f$ holds. Correlation between $\bar{r}_{4e}$
or r_{4v} and r_f is shown in Fig.18. Both $\bar{r}_{4v}$ and $\bar{r}_{4e}$ are evaluated
independently from the value of Pr. $\bar{r}_{4v}$ is nearly equal to $\bar{r}_{4e}$, being larger than twice of $\bar{r}_f$. This indicates that the value of Pr is
an important parameter for determining accurate values of $\bar{r}_f$ and $\bar{r}_{3v}$
Consequently, what kinds of Pr should be taken in the case of irregular pore shape is a problem now open for further study.

Figure 19 shows the relationship between Pre and Prρ. Open
marks are the values for polycarbonate membranes (Nu series) with
straight-through cylindrical pore and filled marks are those for
cellulose diacetate membranes (SF series). Full line is the theoretical curve calculated by eq.(51),

$$Pr\rho = \pi/6 + (Pre/\pi)(1/\bar{r}_2 \cdot N^{1/2}) \qquad (51)$$

and broken line is Pre=Prρ curve, which is valid for straight-through
gh cylindrical pore model. Eq.(51) has been derived by us for model
of membrane with spherical pore[11]. Evidently, the polycarbonate mem-

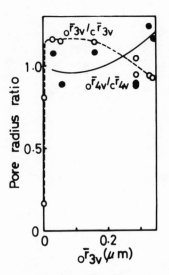

Fig. 16. Dependences of pore radius ratio $_o\bar{r}_{3v}/_c\bar{r}_{3v}$ and $_o\bar{r}_{4v}/_c\bar{r}_{4v}$ on $_o\bar{r}_{3v}$ [11].
● , $_o\bar{r}_{4v}/_c\bar{r}_{4v}$; ○, $_o\bar{r}_{3v}/_c\bar{r}_{3v}$; where, $_o\bar{r}_{3v}$ and $_c\bar{r}_{3v}$ are 3rd order mean pore radius obtained from the permeability data of O_2 and CO_2, respectively, and $_o\bar{r}_{4v}$ and $_c\bar{r}_{4v}$ are 4th order mean pore radius obtained from the permeability data of O_2 and CO_2, respectively.

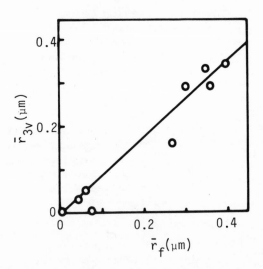

Fig. 17. Relationship between $\bar{r}_{3v}$ and $\bar{r}_f$ [11].

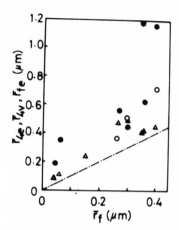

Fig. 18.   Plots of $\bar{r}_{4e}$, $\bar{r}_{4v}$, and $\bar{r}_{fe}$ against $\bar{r}_f$ [11].
●, $\bar{r}_{4v}$ obtained by using eq.(37); ○, $\bar{r}_{4e}$ obtained by
using SEM method; △, $\bar{r}_{fe}$ obtained by putting the relation $Pr=Pre$ into eq.(31); chain line indicates the case
when $\bar{r}_i=\bar{r}_f$ holds

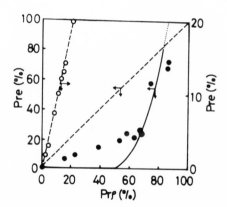

Fig. 19.   Plots of Pre against Pr𝑓 for polycarbonate membrane
and cellulose acetate porous membrane [11].
●, cellulose acetate porous membrane; ○, polycarbonate
membrane; ——, calculated value of Pre by using eq.(51)
derived for spherical pore model; ---, calculated value
of Pre by using the relation Pre=Pr𝑓 derived for straight
through cylindrical pore model

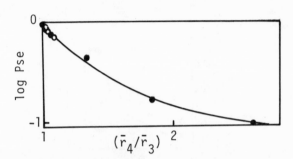

Fig. 20. Dependence of permselectivity Pse on $\bar{r}_4/\bar{r}_3$.
●, model membranes with $(\bar{r}_3 \cdot \bar{r}_4)^{1/2}$=776 nm and b=-4 - 5
in $N(r)=k_b r^b$; ○, actual membranes of Nu0.8, Nu0.6 and
SF0.3

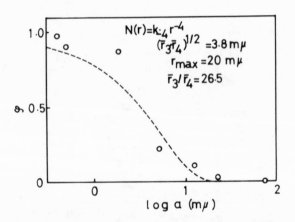

Fig. 21. Comparison between observed and theoretical φ values
of aq._solution of model substances with particle size
2a at u=0 for Cuprophan.
Open circle, observed value; broken line, theoretical
value

brane (Nu series) is succesfully represented by straight-through
cylindrical pore model and cellulose diacetate membrane(SF series)
is by spherical pore model. N(r) for the latter model determined
by using GP method does not coincide with that by using SEM or MI
method.

Application of UP Method to Membrane Having $\bar{r}_3 < 10$ nm

The methods mentioned above can not be applied to the membrane
of $\bar{r}_3 < 10$ nm. In this case UP method are strongly recommended. Since
nce to obtain the $\varphi - a$ curve by UP method is very time-consuming
and in addition, until recently this relation has no theoretical
background, this method is not popularized as a characterization
technique of membrane. According to theoretical basis of UP method,
given by us, this method can be applied for characterization of
pore by using the data of $\varphi - a$ exterpolated to $\bar{u}=0$, as well as
the transmission electron microscopy method combined with the ultra-
thin section/swelling technique.

Slope of $\varphi - a$ curve at $\varphi=0.5$, that is $(d\varphi/da)_{\varphi=0.5}$, gives
us the value of $\bar{r}_4/\bar{r}_3$ as shown in Fig.20, where, Pse is defined by
Pse$=(\bar{r}_3 \cdot \bar{r}_4)^{1/2} (d\varphi/da)_{\varphi=0.5}$. Then, by comparing the shape of this
$\varphi - a$ curve with the one of theoretical $\varphi - a$ curve we obtain the
function of N(r) having above $\bar{r}_4/\bar{r}_3$ value.

Figure 21 shows the $\varphi - a$ curve for the regenerated cellulose
membrane "Cuprophan" manufactured by Y.P.Bemberg Co. Ltd. Open
marks are observed values and broken line is the theoretical $\varphi - a$
curve for the membrane having N(r)=$k_{-4}r^{-4}$, $(\bar{r}_3 \cdot \bar{r}_4)^{1/2}=3.8$nm, $r_{max}=$
20 nm and $\bar{r}_4/\bar{r}_3=26.5$. The effective porosity for permeation of liquid is also evaluated from the data of permeability coefficient Pe
by using eq.(52) derived from the Hagen Poiseuille law.

$$Pr_s = 8\eta_s Pe/(\bar{r}_3 \cdot \bar{r}_4) \qquad\qquad (52)$$

Pr value thus obtained is 0.038 and is a little smaller than 0.11
of Prw and 0.03 - 0.07 of Prφ.

CONCLUSIONS

In order to characterize the pore structures many methods for
evaluating N(r), r_i, Pr and pore shape were established according
to our theories about permeation mechanisms of liquid and gas. When
the mean pore size is larger than 30 nm, N(r) and $\bar{r}_i$, and Pr values
evaluated are independent of the measuring method employed for the
straight-through cylindrical pore, but are dependent for the irregular shape pore. In latter case, what kinds of Pr should be taken
is unsolved problem, though, this is most important factor to evaluate N(r) and $\bar{r}_i$. On the other hand when the mean pore size is
less than 30 nm, UP method and EM (especially TEM) method combined
with ultra-thin cutting and swelling techniques are recommended.

REFERENCES

1. K. Kamide, S. Manabe, T. Matsui, Kobunshi Ronbunshu, 34, 299(19
   77).
2. S. Oka, Rep. Progr. Polymer Phys. Jpn., 6, 71(1963).
3. K. Kamide, S. Manabe, to be published
4. T. Nohmi, H. Makino, S. Manabe, K. Kamide, T. Kawai, Kobunshi
   Ronbunshu, 35, 253(1978).
5. T. Nohmi, S. Manabe, K. Kamide, T. Kawai, Kobunshi Ronbunshu,
   35, 509(1978).
6. K. Kamide, S. Manabe, T. Matsui, T. Sakamoto, S. Kajita, Kobunshi
   Ronbunshu, 34, 205(1977).
7. S. Uemura, M. Takayanage, J. Appl. Polym. Sci., 10, 113(1966).
8. K. Kamide, S. Manabe, K. Hamada, H. Okunishi, S. Miyazaki, Artificial Organ, 5, supplement, 139(1976).
9. T. Nohmi, S. Manabe, K. Kamide, T. Kawai, Kobunshi Ronbunshu,
   34, 729(1977).
10. H. Yasuda, J. T. Tsai, J. Appl. Polymer Sci., 18, 805(1974).
11. S. Manabe, K. Kamide, T. Nohmi, T. Kawai, Kobunshi Ronbunshu,
    to be published

FLOW RATES OF SOLUTIONS THROUGH ULTRAFILTRATION MEMBRANES

MONITORED BY THE STRUCTURE OF ADSORBED FLEXIBLE POLYMERS

P. Dejardin, C. Toledo[+] E. Pefferkorn and R. Varoqui

Centre de Recherches sur les Macromolécules, CNRS
6 rue Boussingault
67083 Strasbourg-Cedex, France

[+]Department of Physics
University Complutense of Madrid
Madrid, Spain

INTRODUCTION

Flexible macromolecules, due to their large dimensions compared with low-molecular weight substances, adhere well to interfaces and an adsorbed polymer layer may strongly influence the rheology of fluid flow into membranes in processes such as ultrafiltration of polymer solutions, osmometry, etc... From a practical point of view it is therefore of importance to determine the mechanism by which a polymer layer at the boundary between a flow system and a solid wall affects the flow pattern of the flowing phase (Silberberg, 1975; Myard, 1979). Systems in which transport processes are monitored by structural modifications of adsorbed chains might also be of importance in biological situations, and the study of the role of polymer adsorption in modifying the membrane permeability is of great interest (Pefferkorn et al., 1978; Singer et al., 1968; Silberberg, 1968).

In the present paper, we present results on the influence of adsorption of flexible polymer on the permeability of various ultrafiltration membranes. An experimental set up was realised by which flow rates of fluids through porous filters, with and without an adsorbed surface film, could be determined with high precision and reliability. Permeability measurements are first reported for porous sintered glass filter discs of $0.23\mu m$ average pore radius into which polystyrene was adsorbed from trans-decalin. The influence of molecular weight, temperature and pressure gradient was examined. A polyacid was also adsorbed into a Nuclepore membrane — a polycarbonate ultrafiltration membrane – of 0.1μm pore radius, and the permeability towards aqueous salt solutions ($CaCl_2$, KCl) was analysed as a function of ionic strength.

The permeability change is expressed via Poiseulle's law in terms of a hydrodynamic length of the adsorbed polymer and this parameter is compared to the average dimension of the polymer coil in free solution. A model is proposed to explain the mechanism by which a flexible polymer in the adsorbed state modifies the permeability of a membrane.

EXPERIMENTAL

Ultrafiltration membranes: Two kinds of porous discs whose specifications are given in Table 1 were used. The value of 0.23μm was determined from the rate of flow J_v of pure solvent through the filter according to Poiseulle's law:

$$J_v = aPS \rho^2 \tag{1}$$

P being the porosity, S the exposed total surface and ρ the pore radius. Taking into account the non-uniformity of the conduit radius in the porous glass discs, one finds that ρ, defined by Eq. 1, corresponds to $(\overline{\rho^4}/\overline{\rho^2})^{1/2}$, the root square of the ratio of the mean-fourth radius over the mean-square radius.

Polymers: Polystyrene fractions No. 1,2,3 and 4 were prepared by radical polymerisation followed by fractionation with a cyclohexane/methanol solvent/precipitant system. Fractions No. 5 and 6 were prepared by anionic polymerization. The polyacid was obtained by copolymerizing maleic anhydride with ethyl vinylether.[+] Radical copolymerization yields an alternated 1-1 copolymer[+] which after

Table 1. Membrane Specifications

	Pore Radius (average, in μm)	Thickness (cm)	Surface Area (cm^2)	Open Area
Pyrex glass discs (Sovirel product)	0.23	0.2	9.6	75%
Polycarbonate membrane (N005 Nuclepore Corp.)	0.10	0.001	4.9	12%

[+] The copolymer has the structure of an homopolymer provided one maleic acid and one ether residue are included in the definition of the monomer.

Table 2.   Molecular Weight of Polystyrene Samples

Fraction No.	$\overline{M}_w$(GPC)	$\overline{M}_w/\overline{M}_n$ (GPC)
1	9.5×10^5	1.50
2	8.43×10^5	1.24
3	5.91×10^5	1.24
4	3.98×10^5	1.21
5	2.10×10^5	–
6	0.918×10^5	1.10

hydrolysis has the following structure

$$\left(CH - CH - CH_2 - CH \right)_n \tag{2}$$
$$\underset{COOH}{|} \quad \underset{COOH}{|} \quad \quad \underset{OC_2H_5}{|}$$

Preparation, hydrolysis, fractionation and characterization have been described elsewhere (Pefferkorn et al., 1978).   A sample of molecular weight 1.7×10^5 has been used in the present work.

Solvents: Trans-decahydronaphthalene (Decalin) of analytical grade (Merck, 96% content by GC) was used for polystyrene solutions.   KCl, $CaCl_2$ of analytical grade and water of 18 MΩ cm purified by a 0.45μm Millipore filter was used in preparing the aqueous solutions.

Viscosity measurements were carried out with an automatic recording capillary viscosimeter (Gramain et al., 1970).

Method for measuring the flow rates through the filters: An improved device shown in Figure 1a was designed by which the pressure difference of the fluid between the membrane faces could be varied precisely step by step.   The rate of flow was determined by recording in front of two photodiodes the passage of the meniscus in a precision glass capillary.   Two capillaries of known radii were part of a system consisting of two toothed metallic wheels rigidly locked to a central horizontal axis.   The capillaries were clamped to the wheels a few cm. from the edge in diametrically opposite positions. The rotation and locking of the wheels at fixed positions permits a precise step by step variation of the pressure difference of the fluid between the membrane faces.   The photoelectric detection is

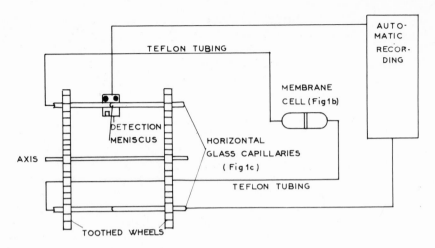

Fig. 1a.   Schematic drawing of flow cell system with automatic
recording and pressure devices.

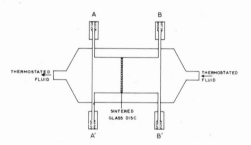

Fig. 1b.   Flow cell with thermostatic control, A and B openings
connected to capillaries, A' and B' openings for filling
and removal of liquid.

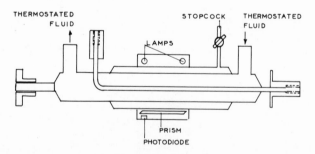

Fig. 1c.   Horizontal capillary with thermostatic control and
detection system.

of routine use in automatic capillary viscosimetry and its mode of
operation has been given elsewhere (Gramain et al., 1970). Measurements were automatically repeated on reversing the fluid flow in the
capillary by means of a small air pump and a pneumatic valve (not
shown in Figure 1a) connected to the lower capillary; the number of
runs was selected by the operator.

A sectional drawing of the flow cell and of the capillary is
shown in Figures 1b and 1c respectively. The flow cell consisted
of two glass compartments; each reservoir of about 20 ml was fitted
with two openings, two of which were connected via teflon tubing to
the capillaries, and the two others were for filling and removal of
the liquid from the cell. Efficient thermostatic control was
achieved by circulating a fluid at constant temperature through a
glass jacket which surrounded the cell. Thermal fluctuations
recorded with a Hewlett Packard quartz thermometer were less than
5×10^{-3}°C at 20°C. In the cell design of Figure 1b, the pyrex glass
disc was sealed to the glass envelope. In using Nucleopore membranes,
the membrane was clamped between the two glass half-cells by means
of a stainless steel holder and tightness was ensured by silicon
rubber sleeves. The cell was also directly immersed into the
constant temperature bath. The glass capillaries shown in Figure
1c were surrounded by an external evacuated glass envelope, which
prevented blurring from condensation at low temperatures (below
10°C) and ensured the proper working of the photodiodes.

The experiments were performed according to the following
procedure. The flow rate of pure solvent was first determined with
a given pressure difference. Solvent was then replaced by the
solution which was circulated through the wet disc under a pressure
difference in the range of 760 to 23 mm Hg. The decrease in time
of the flow rate of the polymer solution due to adsorption was
recorded during a period of several days. A constant value was
reached normally after a period of about ten days. This long time
seems to indicate that adsorbed polymers reach their equilibrium
conformation only slowly. The final value of flow rates was also
found to depend somewhat on the pressure difference under which
adsorption equilibria were obtained. After adsorption was completed
the solution was withdrawn from the cell and replaced by pure
solvent, the flow rate of which was subsequently measured. Most
adsorbed layers were found to be very stable and could not be
desorbed by pure solvent. A similar pattern was also most often
reported elsewhere (Lipatov et al., 1974). All results reported here
were obtained under conditions when desorption did not occur during
a period greatly extending the measurement time.

From Poiseulle's law it is possible to express the ratio of
volume flows J_v^o and J_v of the solvent before and after adsorption
in terms of a pore size reduction L_H:

$$L_H = \langle \rho \rangle \left[1 - \left(\frac{J_v}{J_v^\circ} \right)^{0.25} \right] \tag{3}$$

$\langle \rho \rangle$ being an average of the conduit radius in the case of a pore size heterogeneity. It can be shown that for small $L_H / \langle \rho \rangle$,

$$\langle \rho \rangle = \frac{\overline{\rho^4}}{\overline{\rho^3}} \tag{4}$$

In computing L_H, we have taken $\langle \rho \rangle = 0.23$ μm, which as shown before, is the value of the ratio $(\overline{\rho_4}/\overline{\rho_2})^{1/2}$. The absolute value of L_H is therefore probably somewhat in error. However, this point need not trouble us when the relative variations of L_H are discussed.[+]

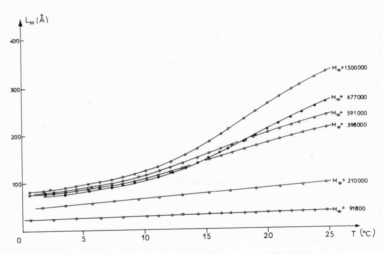

Fig. 2. Variation of L_H as a function of temperature and molecular weight for polystyrene adsorbed in porous glass discs.

+ Rowland and Eirich (1966) in their work, have presented an extensive characterisation of the pore size distribution in sintered
   glass discs. In the context of our discussion, this aspect
   would be only of borderline interest and we have not judged
   it necessary to carry out a complete characterization of the
   glass discs. The pore size variation in the Nuclepore membrane
   is small.

RESULTS AND DISCUSSION

Polystyrene Adsorbed into Glass Filter Disc

In Figure 2, L_H(Å) computed from Eq. 3 is represented as a
function of temperature and molecular weight. Adsorption equilibrium
was reached from a 0.1% wt solution in trans-decalin at 25°C. The
flow rate of pure solvent was recorded at 25°C and at lower temperatures.

The reduction of the pore dimension is seen to be quite large
especially for the high molecular weight polymer. This behaviour
clearly demonstrates that the polymer is not adsorbed in an extended
conformation, but rather suggests that the surface layer has much of
the polymer protruding deeply into the solution phase in the form of
loops and tails at the ends (Silberberg, 1967-1968; Hoeve et al.,
1965). The average loop dimension is a function of molecular weight
and solvent power, the latter being monitored by a temperature
change. The reversibility of the deformation of the part of the
polymer which extends into the solvent phase was tested by running
several increasing and decreasing temperature runs and the resulting
curves were always found to be identical. Moreover, the hydrodynamic
length did not display any change as the pressure difference applied
in measuring the flow rates before and after adsorption was changed.
The linear behaviour of the flow rate as a function of applied
pressure difference is illustrated in Figure 3 for a 1.5×10^6
molecular weight.

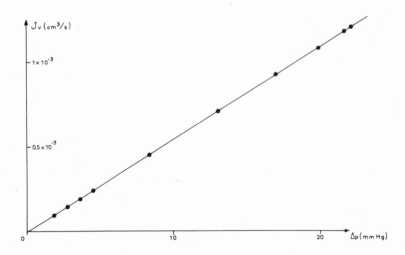

Fig. 3.   Volume flow as a function of applied pressure difference
in porous glass discs in the presence of adsorbed polystyrene.

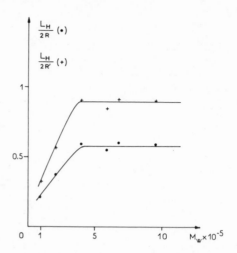

Fig. 4.   Ratio of L_H over geometrical (R) and hydrodynamic (R') coil
          dimensions as a function of molecular weight.

     In Figure 4, we have plotted $\dfrac{L_H}{2R}$ as a function of molecular
weight, the quantity R being the average dimension of the polymer
coil in free solution which was obtained from viscosity determinations according to the chemical Flory-Fox equation (Tanford, 1961),

$$R = 0.648 \, (Mw\,[\eta])^{1/3} \tag{5}$$

$[\eta]$ is the intrinsic viscosity. The values of $[\eta]$ with temperature
are reported in the Appendix.

     According to Rowland and Eirich (1966) the ratio $\dfrac{L_H}{2R}$ is a conven-
ient measure of the deformation or compression of the coil after
adsorption. A close inspection of the problem shows however, that a
better test would be the ratio $L_H/2R'$ with $R' = 0.67R$; R' is the
radius of a solid sphere with a translational motion through the
solvent and which experiences a similar frictional effect as the
translating coil (Flory, 1953). In Figure 4, $L_H/2R'$ is an increasing
function at low molecular weight, but the dominating plateau value
around one could signify, at a first glance, that the PS coils fit
almost undeformed on the surface. This conjecture must however be
altered in consideration of the relative variations of L_H and R with
the temperature. The ratio

$$\frac{dL_H/L_H}{dT} \Bigg/ \frac{dR/R}{dT}$$

i.e. the rate of change of L_H with temperature over the rate of change

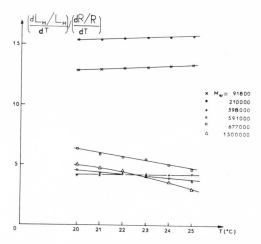

Fig. 5.   Relative variation of L_H over relative variation of R as a
function of temperature for polystyrene adsorbed in porous
glass discs.

of R with temperature is represented as a function of temperature in
Figure 5 and it is seen that in any case it largely exceeds unity.

From the result of Figure 5, and taking into account that in the
diffuse layer, the segment-segment and solvent-segment interactions
are in essence of the same thermodynamic character as in a free
solution, one must assume that the spatial distribution of monomers
in the diffuse part of the adsorbed layer presumably departs from
that which exists in the free coil. An exponential distribution,
rather than a gaussian one, was in fact advocated in many theoretical
papers (Hoeve, 1966; De Gennes, 1969). The authors have themselves
presented in two earlier papers a fluid mechanical model to explain
the permeability change of porous media caused by flexible adsorbed
polymer chains (Varoqui et al., 1977; Pefferkorn et al., 1978). The
theory was based on a modified Navier-Stokes equation,

$$\eta \nabla^2 v - \tau(r) f v \;\; = \;\; - \left| \text{grad } p \right| \tag{6}$$

v is the velocity of the fluid at radial position r in the pore, f is
a friction coefficient and $\tau(r)$ is the local spatial density of
monomers. $\tau(r)fv$ is in fact the supplementary friction force
experienced by the fluid from the monomers in the loops. If $\tau(r)$ is
known, Eq. 6 can be solved by the flow pattern v(r). J_v is obtained
by integration and L_H is obtained from Eq. 3. Eq. 6 was solved using
an exponential distribution:

$$\tau(r) \;\; = \;\; n_o \exp\left[- (\rho - r)/b \right] \tag{7}$$

n_0 is the monomer density at $r = \rho$ and b is the average distance of monomers from the pore wall.

L_H was found to be linear with b; moreover, it was also found that L_H can be very large, for instance, much larger than the mean distance b. More important perhaps in the context of membrane permeability seems to be the fact that a slight variation in b yielded a rather large change in L_H (Varoqui et al., 1977, Figure 2). The large variations of L_H observed in a small temperature domain are in agreement with the theoretical deduction.

Adsorption of Polyacids from Aqueous Solutions into Polycarbonate Membranes

The polyacid adsorption in the polycarbonate membrane was obtained from aqueous solutions, 0.1N in KCl, pH = 3 and 0.14% wt polymer concentration. After adsorption equilibrium was reached the solution was withdrawn and the flow rate of aqueous electrolyte solutions containing various amounts of $CaCl_2$ or KCl salts were determined at pH = 6. L_H, as previously, was obtained via Eq. 3 from the ratio J_V/J_V^o of the volume flows of the aqueous electrolyte solutions as a function of the normality of the solutions. An example of this behaviour is illustrated in Figure 6, where the pore radius increment is represented as a function of the salt normality.

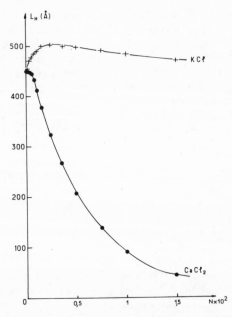

Fig. 6.   Variation of L_H as a function of electrolyte concentration, KCl and $CaCl_2$ of the aqueous solutions for the polyacid adsorbed in a Nuclepore membrane.

It is observed that the effect of adding small amounts of $CaCl_2$ is a large increase of the pore radius. A more than tenfold increase in the flow value was recorded as the concentration of $CaCl_2$ was varied from zero to 0.015N. The effect of adding a mono-monovalent salt KCl is displayed by the top curve in Figure 6. A different behaviour is then observed since L_H first reaches a maximum at 2.5×10^{-3}N KCl and then steadily decreases.

The spectacular effect of calcium ions in triggering the membrane permeability certainly originates from a large conformational change of the polyelectrolyte in the adsorbed layer. It is indeed well known from polyelectrolyte studies, that bivalent ions interact strongly in the form of ion-pairs with negatively charged polyion sites (Morawetz et al., 1954). The lowering of the thermodynamic affinity of the solvent as chelation by alkaline-earth salts occurs, results in a dehydration and a reduction of the polyion dimension. The reduced viscosity η_{sp}/C of the polyacid solution was determined at 0.14% wt concentration, pH = 6, for various added amounts of $CaCl_2$ (values of η_{sp}/C are given in the Appendix). In Figure 7, we have reported the ratio

$$\frac{d\ln L_H}{dN} \bigg/ \frac{d\ln(\eta_{sp}/C)^{1/3}}{dN}$$

as a function of the normality N of $CaCl_2$. Since, as a general rule, $(M_w\eta_{sp}/C)^{1/3}$ approximately follows the variation of the polymer dimension the ordinate values reported in Figure 7, indicate the rate of change of $\ln L_H$ over the rate of change of the logarithm of

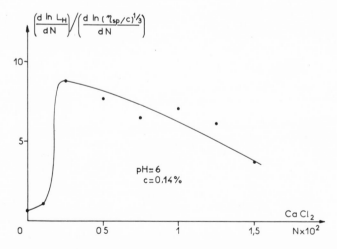

Fig. 7. Relative variation of L_H over relative variations of coil dimension as a function of electrolyte normality for polyacid adsorbed in a Nuclepore membrane.

polymer dimension with the salt concentration. These values display
a maximum at very low $CaCl_2$ concentration. Outside the small domain
at the origin, the ratio of the rate of variation again greatly
departs from unity at any salt concentration. This demonstrates
again that the mechanism by which adsorption impedes the flow through
a membrane cannot be described in terms of a stoppage of flow in a
region as large as the hydrodynamic thickness of the free coil.

CONCLUSION

Our aim was to present in this paper, some typical features of
flow variations in membranes caused by polymer adsorption. From the
examination of the experimental data, a certain number of meaningful
points were made apparent: Polymers are probably adsorbed with a
large change of their conformational topology from the random coil
solution state to the surface state. The permeability of membranes
in the presence of adsorption might be very sensitive to a small
change in the fluid parameters owing to conformational changes.

ACKNOWLEDGMENTS

This work was sponsored by the A.T.P., Project 3500 "Surfaces"
of the Centre National de la Recherche Scientifique. The authors
are indebted to Mr. C. Durand and G. Maennel for technical
assistance.

APPENDIX

Table A1. Variation of the Intrinsic Viscosity of
           Polystyrene in Trans-decalin with Temperature
           for the domain 20 - 25 $^\circ$C

$M_w x 10^{-5}$	$[\eta]$ (ml/g)
0.918	$20,84 + 0,118T(^\circ C)$
2.10	$25,26 + 0,396T(^\circ C)$
3.98	$23,57 + 1,091T(^\circ C)$
5.91	$26,48 + 1,301T(^\circ C)$
8.43	$23,32 + 1,899T(^\circ C)$
9.52	$20,02 + 2,222T(^\circ C)$

Table A2. Variation of the Reduced Viscosity η_{sp}/C of the Polyacid with the Calcium Chloride Normality of Aqueous Solutions.

$Nx10^2$ $CaCl_2$	(η_{sp}/C) (ml/g)
0	384
0,2	349
0,4	312
0,5	271
1	204
1,5	140
2,5	70

REFERENCES

Flory, P.J., 1953, "Principles of Polymer Chemistry," Cornell University Press, Ithaca, New York.
De Gennes, P.G., 1969, Rep. Prog. Phys., 32:187.
Gramain, P., and Libeyre, R., 1970, J. Appl. Polymer Sci., 14:383.
Hoeve, C.A.J., Di Marzio, E.A., and Peyser, P., 1965, J. Chem. Phys., 42:2558.
Hoeve, C.A.J., 1966, J. Chem. Phys., 44:1505.
Lipatov, Y.S., and Sergeeva, L.M., 1974, "Absorption of Polymers," John Wiley, New York - Toronto.
Morawetz, H., Kotliar, A.M., and Mark, H., 1954, J. Phys. Chem., 58:619.
Myard, P., 1979, Thesis, University Louis Pasteur, Strasbourg.
Pefferkorn, E., Schmitt, A., and Varoqui, R., 1978, J. Membrane Sci., 4:17.
Pefferkorn, E., Dejardin, P., and Varoqui, R., 1978, J. Colloid and Interface Sci., 63:353.
Rowland, F.W., and Eirich, F.R., 1966, J. Polymer Sci., 4:2033; ibid., 4:2401.
Silberberg, A., 1967, J. Chem. Phys., 46:1105; 1968, ibid., 48:2835.
Silberberg, A., 1968, p. 225 in "Hemorheology," A.L. Copley, ed., Pergamon, Oxford.
Silberberg, A., 1975, Coll. Int. CNRS, No. 233:81.
Singer, J., and Tasaki, T., 1968, in "Biological Membranes, Physical Fact and Functions," Chap. 8, D. Chapman, ed., Academic Press, New York.
Tanford, C., 1961, "Physical Chemistry of Macromolecules," John Wiley, New York.
Varoqui, R., and Dejardin, P., 1977, J. of Chem. Phys., 66:4395.

PROTEIN ULTRAFILTRATION: THEORY OF MEMBRANE FOULING AND ITS

TREATMENT WITH IMMOBILIZED PROTEASES

John A. Howell and Ö Velicangil

Department of Chemical Engineering
University College of Swansea
Swansea, United Kingdom

SUMMARY

To circumvent the severe flux losses encountered during
ultrafiltration of macromolecular solutions, enzymes were
immobilized on the membranes to hydrolyze the deposited solute
molecules.   This resulted in 26 to 78% flux enhancements,
averaged over a 20 hr period, during a daily run.   A
mathematical model was developed to explain gel formation and the
action of the enzyme precoat on the membrane surface.

INTRODUCTION

Separation and concentration of macromolecular solutions and
colloidal suspensions by ultrafiltration suffers from a major
drawback - significant flux losses due to the formation of a gel
layer. The effect of cleaning of the membrane surface with
conventional detergents or with dilute bases wears off after a
relatively short period and around two thirds of the total
subsequent flux drop occurs within the first 5 hours of a routine
20-hour run when ultrafiltering cheese whey (Velicangil and
Howell, 1977) or other macromolecular solutions of similar
concentration (Blatt et al., 1970).

To provide further insight on the formation of the gel layer
a comprehensive mathematical model was developed.   A parallel
investigation has been concerned with the preparation of self-
cleaning membranes by attaching proteolytic enzymes on the surface
of UF membranes so that clogging proteins are hydrolysed as they
are deposited.   Consequently, the hydraulic permeability of the
gel layer is enhanced and its rate of formation is retarded.

217

Among the vast number of studies on ultrafiltration, reported over the past decennium, only a few were concerned with designing alternative procedures to combat the declining permeation rates (Lee and Merson, 1976; Hayes et al., 1974). They mainly consisted of chemical treatments of the protein source by pH change, preheating or addition of various reagents to affect the polymer forming components in solution. To the authors' knowledge, the only reported attempt at making use of immobilized proteases to enhance the permeation rates of a membrane is that of Fisher and Lowell (1970) who were interested in using it for sewage treatment. They attached trypsin to a membrane cast from N-hydroxysuccinimide ester of cellulose acetate hydrogen succinate, but did not report any flux test of the membrane with sewage.

MATERIALS AND METHODS

Ultrafiltration membranes were obtained from Amicon Ltd., High Wycombe (types PM-10, PM-30, XM-50). Papain (E.C.No.3.4.4.10), two times crystallized from papaya latex, Bovine Albumin (Cohn Fraction V) and Haemoglobin (crude type) were purchased from Sigma London Chemical Co., Poole, Dorset. Corolase S100 (industrial papain product from carica papaya, purified and activated by a special process) and Proteinase P (a neutral protease complex, from Aspergillus cultures) were kindly donated by Röhm GmbH, Darmstadt, West Germany. Cheddar cheese whey was regularly obtained from Unigate Ltd., Johnstown, Carmarthen.

The ultrafiltration cell was a flat thin-channel spiral type with 1.5 mm channel depth and 150 mm in diameter. It was connected to a centrifugal pump which delivered the feed solution up to 4 atm. pressure at 5 lit./min. The system was operated at total recycle with permeate and retentate returned and thoroughly mixed in the feed reservoir. Temperature and pH were controlled, the probes being installed in the feed tank.

To simulate plant operational conditions, experiments with Cheddar cheese whey were carried out at 50°C and in the turbulent region ($R_e = 15200$) with PM-10 membranes. Those experiments with other protein sources were conducted at 25 or 30°C and at $R_e = 6000$ employing PM-10, PM-30 and XM-50 membranes.

Two separate attachment procedures were used. The general procedure adopted to attach refined papain onto the membrane was as follows: Prior to immobilization the membrane surfaces were etched in 3N HCl at 45°C stirring for 3 hours. Then they were perfused with the protease in buffer containing 0.15M KCl for about 1 hr. at 4°C. This was followed by circulating 0.1% glutaraldehyde solution (pH 8.5) in the module at room temperature for 50 minutes. The membranes were then treated further with ice-cold 0.05 M $NaBH_4$ for 20 minutes to reduce the remaining

aldehydic groups and thus to prevent their covalent cross-linking
to whey proteins in subsequent experiments.   Finally the membranes
were washed free of non-crosslinked enzyme with 1N NaCl.

For those experiments with crude proteases a very simple
immobilization procedure was employed.   The enzyme was dissolved
in buffer to a concentration close to the protein concentration of
the protein source to be separated (0.5%-0.75%).   This solution
was ultrafiltered through the membrane until the permeate flux
reached a steady value over the short term (usually 6 to 12 minutes).
Then for 4 to 5 minutes water was recirculated to purge the free
enzyme from the system.   For some experiments a glutaraldehyde
solution was subsequently circulated to cross-link and fix the
enzyme, whereupon the excess glutaraldehyde was reduced by
perfusion with $NaBH_4$.   At the end of a 20 to 22 hr UF run the
exhausted enzyme was removed from the membrane by detergent
cleaning.

The amount of bound protein was determined by difference by
the method of Lowry et al.(1951).   The proteolytic activities of
the free and immobilized proteases were assayed by the standard
Anson method with heamoglobin as substrate (Anson, 1938).

MATHEMATICAL MODEL

Ordinary Membranes

Existing models of ultrafiltration (Blatt et al, 1970;
Porter, 1972) attribute membrane fouling to the buildup of a
uniform gel layer on the membrane surface by a concentration
polarization mechanism.   Where the surface concentration is below
that of the gel it is postulated that there is a resistance to
solvent flow relative to the solute which affects flux as though
it were a solid layer.   When cheese whey is filtered the gel layer
is complex being made up of components of widely differing
molecular weights ranging from lactalbumin at around 10 Kdaltons to
residual caseinate complexes at around 300 Kdaltons.

Analysis of the concentration polarization mechanism is simply
done by means of the conservation equation for solute across the
boundary layer, ignoring flow and gradients in the plane of the
membrane

$$\frac{\partial C}{\partial t} = \mathcal{D}\frac{\partial^2 C}{\partial x^2} - u\frac{\partial C}{\partial x} \tag{1}$$

with the initial and boundary conditions $t = 0 \quad C = C_b$

$$x = 0 \quad C = C_b$$
$$x = L \quad \frac{\partial C}{\partial x} = uC_b$$

C_w is the solution concentration at the wall which is given at $x=L$ by

$$\frac{C_w}{C_b} = e^{\frac{uL}{\mathscr{D}}} + \sum a_n\, e^{-\lambda_n^2 t}\, X_n\left(\frac{\lambda_n^2 \mathscr{D}}{u^2}, \frac{uL}{\mathscr{D}}\right) \tag{2}$$

where λ_n and X_n are the eigenvalues and eigenfunctions respectively of the Sturm-Liouville problem associated with equation (1).

The flux across the membrane is given by

$$u = \frac{\Delta P}{\mu\left(R_m + \frac{\ell}{P_g}\right)} \tag{3}$$

The permeability of the gel layer P_g is calculated from the Carman-Kozeny equation

$$P_g = \frac{d^2}{180}\, \frac{\varepsilon^3}{(1-\varepsilon)^2} \tag{4}$$

Reasonable values for the physical parameters were assumed as given in the Nomenclature section at the end of this paper. These quantities relate specifically to our experiments on the ultrafiltration of cheese whey. As the flux decreased rapidly over the first minute and as the solution quoted assumed a constant velocity the time constants of the solution were first established. The first and smallest value of λ^2 for a transverse velocity of 3.72×10^{-5} ms^{-1} (which was reached at the end of the first minute) was 0.22s^{-1}.

The eigenvalue was larger at higher fluxes and thus the maximum time constant for the full effect of concentration polarisation to occur would be 5 seconds. The experimental technique employed did not permit any flux measurement at times shorter than six seconds and thus the flux decline due to the concentration polarisation effect could not be observed. In fact there is likely to be an even faster effect in view of the limited area occupied by the pores on the upper membrane surface. This space was shown by Preusser (1972) to be less than 0.05% of the total membrane area and thus velocities through the pores are 2000 times or more higher than those towards the membrane. Concentration polarisation could be expected to occur in the pores within a small fraction of a second. If one then calculates the thickness of gel layer required to cause the flux through the pores to drop from that observed with distilled water to that observed

after 10 seconds ultrafiltration of protein, assuming that the
Carman-Kozeny law holds, the resulting thickness is of the order of
a few Angstroms, in other words of molecular dimensions (and thus
the Carman-Kozeny law is unlikely to be valid).   The result does
suggest, however, that initial flux drops could be caused by
blocking of a percentage of the pores by a single protein molecule
convected there during the first moments of ultrafiltration.   If
one takes the distribution of pore sizes in an XM50 membrane as
measured by Preusser (1972), and assumes that the pores of sizes
close to the diameter of the solute molecule will be blocked but
that the remainder will still pass solvent, then the resultant
flux drop matches closely that observed over the first stages of
flux decline.

To determine whether the gel concentration is actually reached
at the membrane the steady-state transverse velocity u_s was
evaluated from the steady-state solution of Equation (1) with a
wall concentration equal to the gel concentration

$$u_s = \frac{\mathcal{D}}{L} \ln \frac{c_w}{c_b} \tag{5}$$

where $c_w = c_g$.

The value for the mass transfer coefficient k_m (= $\mathcal{D}/L$), was
calculated from the Dittus-Boelter correlation in thin channels.
The predicted value of u_s is very close to the water flux, but
significantly higher than the earliest measured flux with cheese
whey.   Thus gel polarisation based on the total membrane surface
cannot account for flux drops below this.

Further calculation then shows that the flux has declined
below that which will cause a gel concentration to be present at
the membrane surface by concentration polarisation and yet the
calculated gel layer is much smaller than has been observed at
later stages.   It is thus postulated that further build up of the
gel can occur by adsorption of protein molecules on to the membrane
surface to form a monolayer.   The mechanism then becomes
chemisorption continuing for several hours and building up an
increasing thickness of gel layer.   The kinetics of such
chemisorption should be observable and this work assumed that it
occurred as an nth order reaction and attempted to fit the data
with the best value of the index n.

The analysis of various flux data of the UF process yielded
a second order relationship between the rate of gel layer growth
and the wall concentration, c_w:

$$\frac{d\ell}{dt} = k_r c_w^{\ 2} \tag{6}$$

Combining equations (3) and (5) with equation (6) yields:

$$\frac{d\ell}{dt} = k_r c_b^{\ 2} \exp\left[2\Delta P/k_m \mu (R_m + \frac{\ell}{P_g})\right] \tag{7}$$

Self Cleaning Membranes

To explain the enhanced fluxes when using membranes pretreated with enzymes, an enzyme activity term is incorporated into equation (6):

$$\frac{d\ell}{dt} = k_r c_w^{\ 2} - k_1 \exp(-k_2 t) \tag{8}$$

By substituting equation (5) and equation (3) into equation (8) we obtain the following expression for the rate of gel layer growth:

$$\frac{d\ell}{dt} = k_r c_b^{\ 2} \exp\left[2\Delta P/k_m \mu (R_m + \frac{\ell}{P_g})\right] - k_1 \exp(-k_2 t) \tag{9}$$

Using a computer program (non-linear optimization by least squares), the parameters k_1 and k_2 of the equation (9) were evaluated. They compared well with those for free solution activity of the enzyme. Although the overall activity of Proteinase P (k_1) was decreased by 90% upon immobilization the activity decay constant, k_2, was also decreased by 40% (from 0.1 to 0.06), thus nearly doubling the enzyme half-life. A digital computer simulation of equations (3), (7) and (9) is presented in Figure 1, together with the experimental data from ultrafiltration of 0.5% bovine albumin through ordinary and pretreated (Proteinase P) PM-10 membranes. The run with papain-pretreated membrane (represented by the highest flux data in Figure 1) requires a more complex model as activators (cysteine and EDTA) are involved in the system.

ULTRAFILTRATION WITH IMMOBILIZED ENZYME MEMBRANES

Separation of Bovine Albumin and Haemoglobin

In these experiments either the same membrane was used in an alternating way as prototype and then as control in consecutive runs, or two different membranes with identical history were

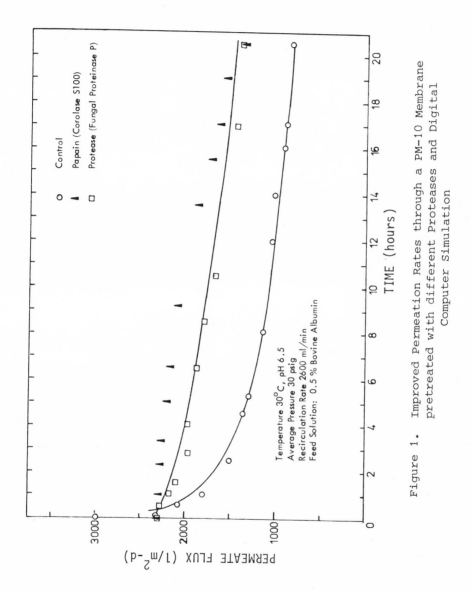

Figure 1. Improved Permeation Rates through a PM-10 Membrane
pretreated with different Proteases and Digital
Computer Simulation

employed as prototype and as control respectively.   The flux
enhancements for both cases were the same.   In one such set of
experiments, consisting of 21 hr runs, 26% flux improvement was
obtained with an enzyme membrane processing 0.5% native albumin and
23% improvement when processing heat denatured albumin.   In
another attempt where papain was fixed onto the membrane with
glutaraldehyde and no cleaning was performed between the two 24 hr
runs, by the end of the second day cumulative permeate was 214%
higher than that of the control.

        The net protein loss through the membrane due to the cleavage
of filtered albumin by the active enzyme was found to be between
5 to 7% of the processed protein.

        The results of similar experiments employing the same membrane
as prototype and as control are presented in Figure 2.   A papain
adsorbed membrane (activated by adding to the albumin solution,
0.005 M cysteine and 0.002 M EDTA) exhibited 44.4% improvement in
cumulative permeate over the control.   The flux loss for the
control over a 22 hr period was 55%, but the prototype showed only
17.6% flux decline over the same period.   In another run without
activators a 31.2% increase in permeate yield was obtained,
although after the first 6 hrs the flux decline accelerated,
presumably due to the rapid inactivation of papain, and the final
flux value matched the 22 hr value of the control.   The fourth
experiment was designed to distinguish whether the flux enhancements
with various prototypes were due to the physical effect of the
adsorbed papain layer as a prefilter coat or due to its biochemical
action.   To this purpose, $0.002M$ H_2O_2 instead of the activators
was introduced into the feed solution to inactivate papain completely.
After 6 hours the feed solution was replaced by a fresh batch
containing activators and no H_2O_2.   Although up to this point the
rate of flux decline was identical with the control a sudden
recovery was observed and the cumulative permeate was 13.4% higher
within the next 16 hours.   A similar experiment, but reversing
addition of H_2O_2/activators order was performed with 0.5% haemoglobin
ultrafiltration (Figure 3, Runs 3 and 4).   After 3 hours of
operation with activators, where the flux was constant and had
identical behaviour with that of Run 1 (papain-adsorbed and
activated membrane), 0.06 M H_2O_2 was introduced into the system.
A sudden drop in the flux was observed and it showed similar
behaviour to that of the control (Run 2) for the rest of the test.
When 3 portions of 0.006 M H_2O_2 was added at hourly intervals there
were corresponding flux decreases.

        Also, the most striking flux difference between the prototype
and the control (same membrane) was achieved with a haemoglobin
solution.   Over a 21 hrs run the cumulative permeate showed 77.8%
improvement.   Results of the ultrafiltration experiments with
various substrates and immobilized proteases are summarised in

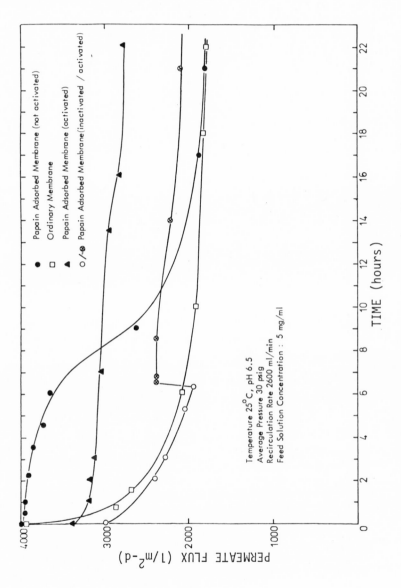

Figure 2. Ultrafiltration Rates on Ordinary and Pretreated
PM-30 Membranes for BSA Solution.

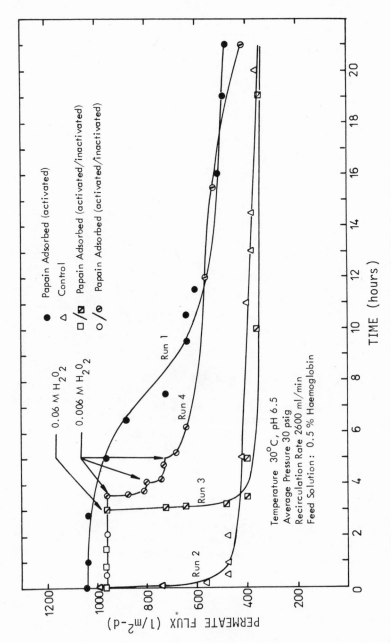

Figure 3.   Enhancement of Permeation Rate with Papain–Adsorbed
PM-10 Membrane.   (Data of 4 consecutive runs presented).

Table 1. When a neutral protease is employed the highest flux
improvement (49.6%) was well below the level observed with papain
(77.8%). This is in accordance with the difference between the
free solution activities of these two enzymes.

Cheese Whey Processing

UF of whey with PM-10 membranes was carried out in the presence
of the activators for papain (crystalline grade). The control
was operated for 20 hours every day. The remaining 4 hours were
spent cleaning with 0.1N NaOH, H_2O, and then with 2% Dymex or 0.3%
Tergazyme solution at 50°C. In contrast, immobilized enzyme
membranes were used continuously during the five days of a run.
PM-10 control membrane suffered a sharp loss of flux over the first
5 hours of a 20 hr run. It had then to be actively cleaned to
restore a reasonable flux. In comparison with the control, enzyme-
modified membranes exhibited 60% and 278% higher permeate yields.
Initially flux loss due to immobilization of papain amounted to 40%
for the enzyme loading of 0.0048 mg/cm . The flux of this membrane
matched the continuously dropping flux of the control in the 7th
hour.

CONCLUSIONS

Initial flux drop in ultrafiltration of cheese whey or other
proteins is probably due to local convective deposition of molecules
of protein close to or in the pores. This process, complete in
less than a second, caused the flux to drop below the level which
would cause gel polarisation at the surface by convection. Flux
drop in the succeeding minute occurs too slowly to be due to gel
polarisation but is much faster than the subsequent flux decay

Table 1. Percent Flux Improvements in Various Systems

Enzyme	Albumin 0.5%,25°C-30°C	Haemoglobin 0.5%,30°C	Cheddar Cheese Whey 50°C
Crude Papain (activated)	26%-70.7%	77.8%	0%
Crude Papain (not activated)	31.2%	–	–
Refined Papain (activated)	–	–	60%-278%
Neutral Protease	49.6%	–	–

over a period of hours.   The former decay is attributed to
adsorption of a monolayer at the membrane surface and the latter
decay is attributed to the reversible polymerisation of protein to
gel, or chemisorption.   The process is second order with respect
to the wall concentration c_w.   In the case of enzyme adsorption
a retardation of gel layer growth occurs.

   A reproducible and inexpensive method was developed to
introduce proteases onto the UF membranes and 26% to 78% flux
improvement was obtained during a standard 22 hr run when
processing 0.5% albumin or haemoglobin substrate.   The same
enzyme solution can be used up to 5 times (5 successive days) when
stored at 4°C in between the runs.   When papain is replaced by a
neutral protease the observed flux enhancements were still
satisfactory considering this enzyme does not require any
activation.   The engineering calculations related to the most
feasible system are underway.   The presented technique is
completely compatible with the requirements of the dairy industry
and does not interfere with the routine in a whey processing plant.
On the other hand, simple adsorption technique of crude enzyme did
not prove to be effective when processing cheese whey, probably due
to the high polymers of lactoglobulin, residual caseinate complexes
etc., building a gel layer of different nature than single protein
solutions (Lee and Merson, 1976a).   Cheese whey ultrafiltration was
effectively improved when recrystallised enzyme was covalently
cross-linked onto the membrane.

NOMENCLATURE

$\mathcal{D}_{50^\circ C}$	Diffusivity of proteins in whey	$16.1 \times 10^{-11} m^2 s^{-1}$
d	Average diameter	3×10^{-9} m
c_b	Solute concentration in the bulk solution	0.5%–0.6%
c_g	Gel concentration	60%
c_w	Concentration at the membrane surface	
k_m	Mass transfer coefficient	
k_r, k_1, k_2	Rate constants (equations (6,7,8))	
L	Boundary layer thickness	
ℓ	Gel layer thickness	
ΔP	Transmembrane pressure	$1.58 \times 10^5 N\ m^{-2}$
P_g	Gel permeability	$2 \times 10^{-20} m^2$
R_m	Membrane resistivity	$13 \times 10^{-11} m^{-1}$

Greek Symbols

ε	Voidage of gel layer	0.5
μ	Viscosity	$7.06 \times 10^{-4} kgm^{-1} s$

References

M. L. Anson, J. Gen. Physiology 22:79 (1938).

W. F. Blatt, A. Dravaid, A. S. Michaels and L. Nelsen, in "Membrane Science and Technology," J. E. Flinn, Ed., Plenum Press, New York (1970), 47-97.

B. S. Fisher and J. R. Lowell, Jr., "New Technology for Treatment of Wastewater by Reverse Osmosis," report to EPA on FWQA Program, 77 pp., September 1970.

J. F. Hayes, J. A. Dunkerley, L. L. Muller and A. T. Griffin, Austr. J. Dairy Tech. 29(3):132 (1974).

D. N. Lee and R. L. Merson, J. Food Sci. 41(2):403 (1976a).

D. N. Lee and R. L. Merson, J. Food Sci. 41(4):778 (1976b).

O. H. Lowry, N. J. Rosebrough, A. L. Farr and R. J. Randall, J. Biol. Chem. 193:265 (1951).

M. C. Porter, Ind. Eng. Chem. Prod. Res. Develop. 11(3):234 (1972).

H. J. Preusser, Kolloid-Z.u.Z. Polymere 250:133 (1972).

O. Velicangil and J. A. Howell, Biotech. Bioeng. 19:1891 (1977).

BOUNDARY LAYER REMOVAL IN ULTRAFILTRATION

Mahendra R. Doshi

Environmental Sciences Division
The Institute of Paper Chemistry
Appleton, WI 54912

INTRODUCTION

A large number of processes involving a transport of heat and/or mass between a solid and a fluid are characterized by a thin boundary layer in the vicinity of the wall. The boundary layer offers major resistance to heat and mass transfer and is often a rate limiting factor. Any attempt to reduce the boundary layer thickness, for example by operating under turbulent flow conditions, will improve transport rates. Another way of improving the performance is to remove the boundary layer by using the device described here. In this report, application of the boundary layer removal device to ultrafiltration will be discussed.

Membrane separation processes, e.g., reverse osmosis and ultrafiltration, are quite useful in concentration, separation and purification[1-6]. An attractive feature of membrane processes is that they are simple in operation and do not require a phase change. One of the problems associated with membrane processes is the phenomenon of concentration polarization due to the resistance offered by the boundary layer. As a selective membrane allows solvent to pass through but rejects most of the dissolved solute or suspended particles, the rejected constituents accumulate in the vicinity of the membrane, giving rise to concentration polarization. The concentration polarization reduces the permeation rate and increases the solute concentration in the permeate, both of which are undesirable. Additionally, the rejected constituent may form a gel layer (in ultrafiltration) or a precipitate (in reverse osmosis). It is, therefore, evident that the removal of the boundary layer will greatly increase the efficiency of membrane processes.

THEORY AND DESIGN CONSIDERATIONS

In a large number of cases, the concentration distribution in tubular or rectangular membrane ducts can be described by film theory[2,3,5,7-9]. In this section, results of film theory will be used to estimate the boundary layer thickness and concentration.

In film theory it is assumed that radial convection is balanced by radial diffusion in the boundary layer:

$$-V_w \frac{\partial C}{\partial y} = D \frac{\partial^2 C}{\partial y^2} \tag{1}$$

The permeation velocity, V_w, is taken to be positive; the coordinate system is shown in Fig. 1.

Solute flux at the membrane surface is given by:

$$[D \frac{\partial C}{\partial y} + V_w C]_{y=o} = V_w C_p \tag{2}$$

Since all the concentration changes take place in the boundary layer, concentration at the edge of the boundary layer thickness must be equal to C_o:

$$(C)_{y=\delta} = C_o \tag{3}$$

The solution of Equation (1) subject to the boundary conditions, Equations (2) and (3), is:

$$\frac{C - C_p}{C_o - C_p} = \exp\left(\frac{V_w}{D}(\delta - y)\right) \tag{4}$$

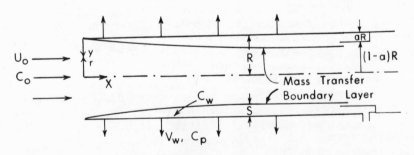

Fig. 1. Schematic diagram of tubular membrane duct.

Since, at $y = o$, $C = C_w$, we have

$$\frac{C_w - C_p}{C_o - C_p} = \exp\left(\frac{V_w \delta}{D}\right) \qquad (5)$$

The boundary layer thickness, δ, can be estimated from the corresponding heat or mass transfer problem, without transverse convection[10]. It has been found that the Leveque solution for heat transfer in a tube with constant wall flux can be used to estimate δ with reasonable accuracy:

$$k = \frac{D}{\delta} = 1.03 \left(\frac{U_o D^2}{X R}\right)^{1/3} \qquad (6)$$

In terms of the boundary layer thickness, the above equation can be rewritten as:

$$\frac{\delta}{R} = \frac{1}{1.03}\left(\frac{X D}{U_o R^2}\right)^{1/3} = (1.54)\left[\frac{X}{(2R) \cdot Pe}\right]^{1/3} \qquad (7)$$

Equation (7) for the boundary layer thickness is applicable for the case of laminar flow, (Re < 2100). Some typical values of the boundary layer thickness are given in Table 1. As can be seen from Equation (7) and Table 1, the dimensionless boundary layer thickness, δ/R, increases as Reynolds number or Schmidt number decreases or as the length of the tube, (X/2R), increases. On the other hand, the dimensional boundary layer thickness, δ, increases as the length of the tube, (X), the tube radius, (R), and diffusion coefficient (D) increases, and as the inlet velocity decreases.

The volumetric flow rate through the boundary layer can be calculated, by assuming a Hagen-Poiseuille velocity profile:

$$U = 2 U_o \left(1 - \frac{r^2}{R^2}\right) \qquad (8)$$

In writing the above equation we have assumed that the permeation velocity is too small to have a significant effect on the velocity profile. This is a reasonable assumption since V_w/U_o is generally less than 10^{-3}.

The ratio of volumetric flow rate through the boundary layer to the volumetric feed flow rate, f, can be calculated by integration:

$$f = \frac{\int_{R-aR}^{R} r \, U \, dr}{\int_{o}^{R} r \, U \, dr} = 4a^2 - 4a^3 + a^4 \qquad (9)$$

When a<<1, Equation (9) simplifies to

$$f \cong 4a^2 \qquad (10)$$

It can be seen from Table 2 that for some typical values of "a," relatively small fraction of the feed flows through the boundary layer. However, concentration in this small fraction will be high compared to the feed concentration.

TABLE 1

BOUNDARY LAYER THICKNESS, EQUATION (7)

$(X/2R) = 50$

	Sc	Re	$\frac{\delta}{R}$
1.	NaCl–H_2O	100	0.145
	Sc $\cong$ 600	500	0.0847
		1000	0.0673
		2000	0.0534
2.	Yellow Dextran — H_2O	100	0.0263
	Sc $\cong$ 10^5	500	0.0154
		1000	0.0122
		2000	0.00970
3.	TiO$_2$ or clay suspension	100	0.0122
		500	0.00715
	Sc $\cong$ 10^6	1000	0.00567
		2000	0.00450

TABLE 2

$$f = \frac{\text{Volumetric Flow Rate in Boundary Layer}}{\text{Volumetric Feed Flow Rate}}$$

a	f, %
$\frac{1}{32} = 0.03125$	0.4
$\frac{1}{16} = 0.0625$	1.5
$\frac{1}{8} = 0.125$	5.5

The boundary layer concentration, C_{BL}, is defined as:

$$C_{BL} = \frac{\int\limits_{R-aR}^{R} rUC\, dr}{\int\limits_{R-aR}^{R} rU\, dr} \simeq \frac{\int\limits_{o}^{a} yC\, dy}{\int\limits_{o}^{a} y\, dy}, \quad (a<<1) \tag{11}$$

where $y = R-r$.

If $a<<1$, Equation (11) can be simplified as indicated. By substituting for C from Equation (4) and integrating, one obtains:

$a \leq \delta/R,$

$$\frac{C_{BL}}{C_o} = \frac{C_p}{C_o} + \left(1 - \frac{C_p}{C_o}\right)\left(\frac{2}{a^2P^2}\right)e^{\frac{P\delta}{R}}\left[1 - (1 + aP)e^{-aP}\right]; \tag{12a}$$

$a \geq \delta/R,$

$$\frac{C_{BL}}{C_o} = \frac{C_p}{C_o} + \left(1 - \frac{C_p}{C_o}\right)\left(\frac{2R^2}{\delta^2P^2}\right)e^{\frac{P\delta}{R}}\left[1 - \left(1 + \frac{P\delta}{R}\right)e^{\frac{-P\delta}{R}}\right] + 1 - \frac{\delta^2}{a^2R^2} \tag{12b}$$

Under the special case of

$a = \frac{\delta}{R}$ (Perfect boundary layer removal)

$C_p = 0$ (Ideal membrane)

Equation (12) simplifies to

$$\frac{C_{BL}}{C_o} = \frac{2e^{aP}}{a^2P^2} [1 - (1 + aP)e^{-aP}] \tag{13}$$

Some values of the dimensionless boundary layer concentration, C_{BL}/C_o, and concentration polarization, C_w/C_o, are shown in Fig. 2. The concentration polarization rises much more rapidly than the boundary layer concentration, as the parameter aP is increased. An increase in concentration polarization increases osmotic pressure and induces the formation of a precipitate or a gel layer. It is, therefore, desirable to limit the concentration polarization to, say, less than 6. From Fig. 2, the boundary layer concentration (C_{BL}/C_o) will then be less than 2 and the corresponding value of aP will have to be less than 1.8.

In plotting Fig. 2, we have assumed that the boundary layer removal is perfect ($a = \delta/R$). The boundary layer concentration will be closer to the corresponding polarization value if $a < \delta/R$ and vice versa if $a > \delta/R$.

In the design of a tubular reverse osmosis or ultrafiltration (pregel region) unit, for a given feed concentration, C_w/C_o is fixed from practical considerations. Figure 2 can then be used to determine the appropriate value of aP. The phenomenological solvent transport equation can be used to estimate P:

$$V_w = A (\Delta P - \Delta \pi)$$

or in terms of dimensionless quantities:

$$P = P_o (1 - B_2(C_w/C_o))$$

Thus, P can be calculated for the various values of the applied pressure, ΔP. Since aP is known, corresponding values of a can be obtained.

For the case of a real membrane, Equations (5) and (12) can be used to calculate polarization and the boundary layer concentration as shown in Fig. 3. A rejection parameter, R_1 is defined on the basis of wall concentration,

$$R_1 = 1 - \frac{C_p}{C_w}$$

For a given value of C_w/C_o, the boundary layer concentration increases with an increase in the rejection parameter, R_1. Thus

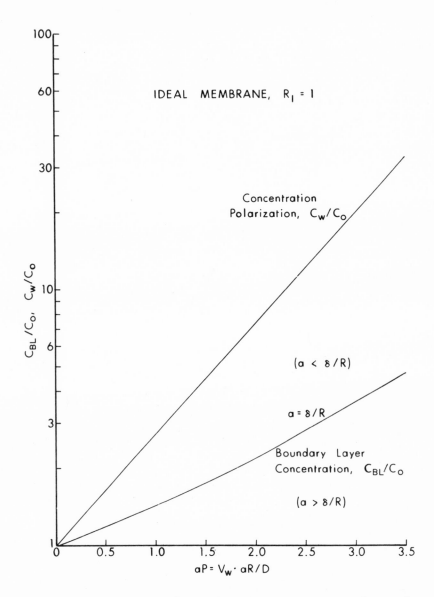

Fig. 2.   Variations of boundary layer concentration and
concentration polarization

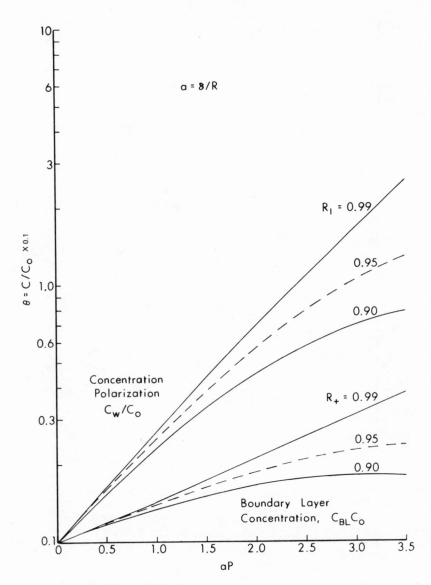

Fig. 3. Variation of boundary layer concentration and concentration
polarization for different values of rejection parameter, R_1

boundary layer performance and permeate quality will improve as tighter membranes are used.

EXPERIMENTAL

The experimental set up is shown in Fig. 4. Feed solution was pumped into a one-inch diameter by fifty one inches long Abcor ultrafiltration tubular unit. Permeate and core streams (from the central portion of the tube) were recycled back to the feed tank. The flow rate in the boundary layer was too small to be recycled. The back pressure valve at the outlet of the boundary layer removal device was adjusted so as to get the flow rate indicated in Table 2. That is, the volumetric flow rate in the boundary layer was maintained at its theoretical value. This was done to assure minimum flow disturbance due to the boundary layer removal device. Greater or smaller flow rates could result in the bending of the streamlines and consequent disruption of the boundary layer.

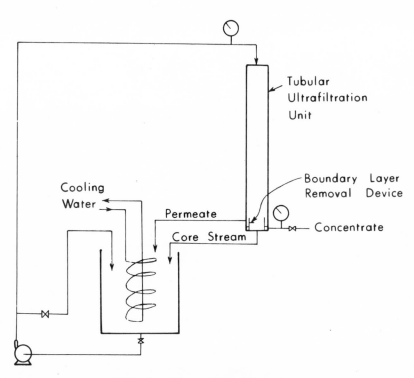

Fig. 4. Experimental set up

The feed flow rate was monitored periodically by stopwatch
and graduated cylinder. Pressure in the feed stream and the
boundary layer stream were measured by pressure gages. Temperature
of the feed stream was maintained constant by adjusting the cooling
water flow rate. Permeate, boundary layer and core stream samples
were collected at the beginning and end of a run.

As shown in Fig. 5a, the end connection was made by folding
the rubber gasket and compressing it by a stainless steel sleeve.
The purpose of this connection is to prevent the mixing of permeate
with the outgoing concentrated boundary layer. The thickness of
this rubber stainless steel sleeve seal was of the same order of
magnitude as the thickness of the boundary layer (see Table 2).
This type of end connection stagnates the boundary layer and
consequently deteriorates the membrane performance. It was,
therefore, decided to modify the end connection.

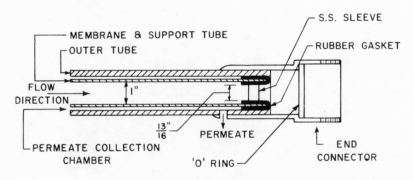

Fig. 5a. Original tubular membrane end connection

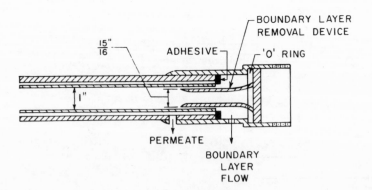

Fig. 5b. Modified end connection in tubular membrane and
          boundary layer removal device

In the modified end connection, Fig. 5b, the seal between permeate and boundary layer was achieved by applying epoxy adhesive to the membrane support edge, as shown in the figure. The boundary layer could thus be removed easily.

Abcor HFD, polyamide membrane (molecular weight cut off, 15,000) was used for the boundary layer removal experiment. Experiments were carried out with solution of Dextran and with a clay suspension. The concentration of Dextran was measured by analyzing total carbon. The amount of clay in suspension was determined by filtering with a 0.45 micron filter.

Since the shower water from Green Bay Packaging Inc. (NSSC mill) was in our laboratory, an experiment was performed to determine the efficiency of the boundary layer removal device utilizing this effluent.

RESULTS AND DISCUSSION

Experimental results with Dextran are summarized in Table 3. Experiments were carried out with Dextran 2000 (molecular weight = 2,000,000). Rejection of Dextran 2000 is about sixty percent.

It can be seen from the results that good boundary layer separation was achieved when the flow rate was 10 ml/sec (Reynolds number $\sim$ 500) or less. For higher flow rates, the boundary layer is relatively thin and consequently its removal is neither necessary nor practical. Tubular membranes are generally operated

TABLE 3

ULTRAFILTRATION OF DEXTRAN (M.W. 200,000)
ΔP = 30 psi

Tube Orientation	Flow Rate, mL/sec			Total Carbon mg/L			% Conc.
	Feed	Perm.	Bound Layer	Feed	Perm.	Bound Layer	
Horizontal	10.3	0.20	0.04	830	200	1010	22
	9.1	0.10	0.05	830	420	1310	58
Vertical (downward flow)	8.2	0.13	0.02	720	390	990	38

% Conc. = 100 x $\dfrac{\text{(Concentration in Boundary Layer)}-\text{(Feed Concentration)}}{\text{Feed Concentration}}$

under high flow rates to minimize concentration polarization. However, when the boundary layer removal device is used, we want to have maximum polarization and consequently lower flow rates.

The boundary layer removal was effective when the feed flow was downward in a vertical tube. In this case it was found that the concentration in the boundary layer was thirty-eight to fifty-eight percent greater than the feed concentration. The boundary layer flow rate was only 0.2 to 0.5 percent of the feed flow rate.

Results with clay suspension are shown in Table 4. Experiments were carried out with the upflow and downflow in a vertical tube. In these experiments, clay particles adhere to the membrane, thereby making the boundary layer somewhat immobile. As a result, the concentration of clay particles in the feed and boundary layer streams were not much different (Run 9, Table 4). In one experiment, it was noted that the feed concentration decreased from 58 mg/L to 6 mg/L during the four-hour operation. This phenomenon was observed even when the charge on clay particles was reduced by lowering the pH.

It appears that in the case of colloidal suspension, particles arrive at the membrane surface by following a streamline. Once in contact with the membrane, they are hard to remove. Thus, even though the boundary layer removal device is effective in the case of colloidal suspension, adsorption of the particles or cake formation does occur. Cake formation can be minimized by reducing the charge on colloidal particles and thereby improving the efficiency of the boundary layer removal device.

Finally, one experiment was carried out with the shower water from the Green Bay Packaging Inc. Results are shown in Table 5. As the flow rate increases, concentration polarization decreases

TABLE 4
ULTRAFILTRATION OF CLAY SUSPENSIONS

Pressure: 30 psi
Tube Orientation: Vertical

Run No.	Flow Direction and pH		Flow Rate, mL/sec			Solids (0.45 μ Filter) mg/L			% Conc.
			Feed	Perm.	Bound Layer	Feed	Perm.	Bound Layer	
9	Upward pH = 5		9.7	0.4	0.05	707	17.5	696	−2
			5.8	0.24	0.37	696	11.0	672	−3
10	Downward pH = 5	Initial after 4 hr	10.0	1.4	0.048	404	0	496	.23
			9.2	0.95	0.033	127	0	149	17
11	Downward pH = 2	Initial after 4 hr	13.3	4.4	0.032	57.9	1.5	112.3	94
			11.0	4.4	0.032	6.3	0.25	15.2	141

TABLE 5

ULTRAFILTRATION OF SHOWER WATER FROM
GREEN BAY PACKAGING INC.

Pressure: 100 psi

Flow Rate, mL/sec			Color, mg/L			% Conc.	% Rej.
Feed	Perm.	Bound Layer	Feed	Perm.	Bound Layer		
8.2	0.4	0.042	7,080	6,860	34,700	390	3
9.1	0.38	0.05	7,620	3,360	29,100	282	56
12.8	0.54	0.05	7,380	806	8,200	11	89

$$\% \text{ Conc.} = \frac{(\text{Conc. in Bound Layer}) - (\text{Feed Conc.})}{\text{Feed Conc.}} \times 100$$

$$\% \text{ Rej.} = \frac{(\text{Conc. in Feed}) - (\text{Conc. in Perm.})}{\text{Conc. in Feed}} \times 100$$

$$\% \text{ R} = \frac{(\text{Membrane Surface Conc.}^n) - (\text{Permeate Conc.}^n)}{(\text{Membrane Surface Conc.}^n)} \times 100 \approx 90$$

and consequently permeate concentration as well as boundary layer concentration decreases. Thus, at very low flow rates, the boundary layer removal device is very effective in producing a concentrate stream. But in this case permeate concentration is not much different from that of feed. On the other extreme, when the flow rate is very high, the membrane is effective in rejecting most of the solute or particulate matter, but the boundary layer device is of little benefit. However, if a membrane is very "tight" (i.e., highly rejecting to solute and particulate material), low flow rate operations will produce a concentrated boundary layer and quite dilute permeate.

The membrane rejection characteristic can be defined on the basis of what the membrane "sees" or in other words, in terms of membrane surface concentration, as shown in the bottom equation in Table 5. However, this quantity cannot be determined accurately, since the membrane surface concentration is not known. From the boundary layer data it appears that rejection constant of the membrane under study is about 90 percent. If the membrane rejection constant is higher, say 98 percent or greater, permeate concentration will decrease and boundary layer concentration will increase, both of which are desirable.

APPLICATIONS

Membrane processes are useful in concentration, purification and fractionation. The boundary layer device can be useful in all three applications.

In the pulp and paper industry, for example, fractionation of lignin components can be achieved since large molecular weight compounds will be in the boundary layer, medium molecular weight components will be in the core section and small molecular weight compounds will be in the permeate. Thus, three fractions are obtained, instead of just two fractions, by using the boundary layer removal device.

CONCLUSION

In principle and in practice a boundary layer removal device is simple to construct and install. The boundary layer device appears to work well at low feed rates. At higher feed rates, the boundary layer is too thin to make its removal practical.

In the case of colloidal suspension, boundary layer removal is effective, but the formation of a cake layer on the membrane surface does occur. In ultrafiltration of macromolecular solutions the boundary layer device works (at low flow rates) at the expense of membrane rejection. In this case optimum operating conditions must be determined. If the aim is concentration of feed, a boundary layer device will be quite valuable.

ACKNOWLEDGMENT

The author would like to thank member companies for supporting this project.

LIST OF SYMBOLS

A Solvent permeability constant

a Fraction of tube radius occupied by boundary layer removal device

B_2 Ratio of feed osmotic pressure to the applied pressure

C Solute concentration

C_o Feed concentration

C_p Permeate concentration

C_w Wall concentration

C_{BL} Boundary layer concentration

D   Solute diffusion coefficient

f   Fraction of feed recovered through boundary layer removal

k   Muss transfer coefficient

P   Wall Peclet number, $V_w R/D$

P_o $A\Delta PR/D$

ΔP Pressure drop across the membrane

Pe   Peclet number, $= Re. Sc.$

R   Tube radius

r   Radial coordinate

R_1 $1 - C_p/C_w$, rejection coefficient

Re $2 R V_o/\nu$, Reynolds number

Sc ν/D, Schmidt number

U   Feed velocity

U_o Feed velocity at inlet

V_w Permeate velocity

X   Axial coordinate

Y   Radial coordinate measured from the wall

δ Boundary layer thickness

ν Kinematic viscosity

π Osmotic pressure

θ C/C_o, dimensionless concentration

θp C_p/C_o

θw C_w/C_o

θ_{BL} C_{BL}/C_o

REFERENCES

1. Michaels, A. S., "New separation technique for the CPI,"
Chem. Eng. Progr. <u>64</u> (12), 31-43 (1968).

2. Blatt, W. F., A. Dravid, A. S. Michaels and L. Nelson,
"Solute polarization and cake formation in membrane ultrafiltration: causes, consequences and control techniques,"
in <u>Membrane Science and Technology</u>, J. J. E. Flinn, ed.,
pp. 47-97, Plenum Press, N.Y. (1970).

3. Sourirajan, S., <u>Reverse Osmosis</u>, Academic Press, N.Y. (1970).

4. Sourirajan, S., ed., <u>Reverse Osmosis and Synthetic Membranes</u>
— <u>Theory, Technology, Engineering</u>, National Research Council
Canada (1977).

5. Gill, W. N., L. J. Derzansky and M. R. Doshi, "Convective Diffusion in Laminar and Turbulent Hyperfiltration (Reverse
Osmosis) Systems," in <u>Surface and Colloid Science</u>, <u>Vol. IV</u>,
pp. 261-360, E. Matijevic (ed.), John Wiley, New York (1971).

6. Doshi, M. R., A. K. Dewan and W. N. Gill, "The Effect of Concentration Dependent Viscosity and Diffusivity on Concentration Polarization in Reverse Osmosis Systems," in <u>Water</u>

1971, pp. 323–339, L. K. Cecil (ed.), A.I.Ch.E. Symposium
Series No. 124, <u>Vol</u>. <u>68</u> (1972).

7. Brian, P. L. T., "Mass transport in reverse osmosis in <u>Desal-
ination</u> <u>by</u> <u>Reverse</u> <u>Osmosis</u>," U. Merten, ed., pp. 161-202, MIT
Press, MA (1966).

8. Derzansky, L. J. and W. N. Gill, "Mechanisms of brine-side
mass transfer in a horizontal reverse osmosis tubular membrane," <u>AIChE</u> <u>Journal</u> <u>20</u> 751-761 (1974).

9. Goldsmith, R. L., "Macromolecular ultrafiltration with micro-
porous membranes," <u>Ind</u>. <u>Eng</u>. <u>Chem</u>. <u>Fundam</u>. <u>10</u>, 113-120 (1971).

10. Knudsen, J. G. and D. L. Katz, "Fluid Dynamics and Heat Transfer," McGraw Hill, New York (1958).

INITIAL TIME STIRRED PROTEIN ULTRAFILTRATION STUDIES

WITH PARTIALLY PERMEABLE MEMBRANES

T. Swaminathan, M. Chaudhury[*] and K. K. Sirkar[**]

Industrial Wastes Division, NEERI
Nehru Marg, Nagpur, 440020, India

[*]Department of Civil Engineering
Indian Institute of Technology
Kanpur 208016, U.P., India

[**]Department of Chemistry and Chemical Engineering
Stevens Institute of Technology
Hoboken, NJ 07030

Initial time unsteady stirred ultrafiltration studies of solvent flux and solute retention have been carried out with two proteins, BSA and ovalbumin, and membranes XM 100 A and XM 300, both being partially permeable to the solutes. The time to attain a steady state solvent flux depends primarily on the applied pressure and solute bulk concentration with minor contributions from membrane permeability and solute size. The time to attain a steady state solute retention is much larger and is dependent of solute size, pressure, bulk solute concentration, and membrane permeability for the solvent. For the experimental conditions of bulk concentration varying between 0.05% to 0.5% and pressure between 10-35 psig, the steady state solvent flux is achieved in less than 270 seconds, whereas more than 10 minutes are required for a steady solute retention. Initial time solute retentions are considerably lower than the steady state values reported in membrane manufacturers' catalogs. The steady state solvent flux behavior of both systems conform to those of total retention membranes with respect to the gel-polarized region as well as the pre-gel-polarized region. The solute retentions depend on applied pressure and bulk concentration in a manner suggested by Blatt, et al.

ELECTROPHORETIC TECHNIQUES FOR CONTROLLING

CONCENTRATION POLARIZATION IN ULTRAFILTRATION

John M. Radovich

School of Chemical Engineering & Materials Science
University of Oklahoma
Norman, Oklahoma 73019

Robert E. Sparks

Biological Transport Laboratory
Department of Chemical Engineering
Washington University
St. Louis, Missouri 63130

ABSTRACT

An electrophoretic technique for controlling concentration polarization has been successfully applied to the ultrafiltration of charged macrosolutes-proteins and colloidal electrodeposition paints. The electric field controlled the build-up of retained macrosolutes at the membrane surface. Resultant fluxes were 75% to 500% higher than normal ultrafiltration fluxes. Selectivity was also improved as indicated by a 2 to 6 fold increase in the separation factor for solutions of proteins. These preliminary results are analyzed in terms of the important process variables.

INTRODUCTION

Membrane processes, such as ultrafiltration, are being used on an ever increasing scale to concentrate and purify macrosolutes.[1-5] Ultrafiltration, a pressure-driven process, separates molecules on the basis of size and shape. Molecules larger than the membrane pores are retained at the surface, while smaller solutes are convectively transported through the membrane with the solvent. The main disadvantages of ultrafiltration are the lack of membranes with sharp molecular weight cut-offs which would give good selectivity, and concentration polarization - the build-up of retained

molecules at the membrane surface. While improvement of the
selective permeability of ultrafiltration membranes requires
attention, the concentration polarization problem is much more
troublesome, because of its deleterious effects on membrane flux
and selectivity.

The modelling of the concentration polarization problem as
applied to macrosolutes has been covered extensively in the
literature.[6,7,8] Briefly, the concentration polarization model
states that the steady-state membrane flux, J, will be independent
of the transmembrane pressure drop. The concentration polarization
layer acts as a resistance in series with the polymer membrane.
This resistance varies such that a balance between convective
transport of macrosolutes toward the membrane (due to ΔP) and
diffusive back-transport (due to concentration gradients) is
maintained. Under these conditions, J can be calculated on the
basis of mass transfer of solute from the membrane surface
(concentration C_m) into the bulk solution (concentration C_b):
$J = K \ln (C_m/C_b)$ where K is an overall mass transfer coefficient.
The design problem to optimize the flux has thus been reduced to
maximizing K by adjusting the membrane geometry and fluid flow
conditions of the ultrafiltration system.[7,8,9]

In our process, electroultrafiltration, an electric field
which is perpendicular to the fluid flow across the membrane,
exerts an electrophoretic force opposite to that due to the
transmembrane pressure drop. Retained macrosolutes are kept clear
of the membrane by the electric field. Separation is accomplished
by interaction of the ultrafiltration membrane, which discriminates
on the basis of size/shape and the electric field which controls
the number of molecules of a given charge and mobility that reach
the membrane.

The potential for increasing flow in membrane processes by
using an electric field was recognized many years ago by Bechold[10]
and Mannegold.[11] Forced flow electrophoresis (FFE) developed by
Bier[12], uses an electric field to move charged particles away from
a filter. Applications of FFE, in which only the isoelectric
component in the mixture is purified, have been used in processing
of proteins, Bier[13], and clay-water solutions in normal[14] and
cross flow-filtration.[15]

Extension of the one-dimensional gel-polarization model[7] for
flux to the case where an electrophoretic force acts on the
macrosolute starts with the transport equation. At steady state,
for an impermeable membrane:

$$JC = -D \frac{dC}{dx} + \mu EC \qquad\qquad (1)$$

where C is the macrosolute concentration, μ is the electrophoretic mobility and E is the field strength. Integration gives:

$$J = k \ln C_m/C_B + \mu E \tag{2}$$

If operating under gel-polarization conditions, C_m is constant and the flux, while still independent of ΔP, will be increased by an amount μE when the electric field is applied. Under pre-gel polarization conditions, C_m will vary and the increase in flux due to E will not be directly proportional to μE.

Equation (2) is valid only when concentration polarization exists. At some electric field strength, $C_m/C_B \rightarrow 1$. Under these conditions, the convective transport of solute toward the membrane will be balanced by the electrophoretic transport of solute away from the membrane. The electric field strength at which this occurs is E_{cr}. A solute balance then gives

$$E_{cr} = J_o/\mu \tag{3}$$

where J_o is the flux in the absence of concentration polarization, i.e. the flux of pure buffer.

In addition to the electrophoretic transport of the macro-solute in the polarization boundary layer, the electroosmotic contributions to the total flux must be considered. Electro-osmosis is the flux thru the membrane due to the electric field in the absence of a transmembrane pressure drop. Electroosmosis in the membrane and in the gel layer were taken into account.

EXPERIMENTAL APPARATUS & PROCEDURES

The continuous flow cell which was used for the electroultra-filtration experiments is sketched in Figure 1. The process flow system is shown schematically in Figure 2. The cells were made of ultraviolet transmitting Plexiglas [R] and contained a membrane of about 75.0 cm^2. The separation of the electrode compartments from the ultrafiltration compartments was necessary to prevent contamination of the process solutions by electrolysis products and to prevent coating of the electrodes by the migrating macro-solutes.

Amicon Diaflo [R] membranes, PM and XM series, were used only once in each ultrafiltration experiment. These experiments were conducted in the pH range of 4.7 to 8.1. The proteins were dissolved in phosphate or acetate buffer solutions with minimal agitation. Bovine albumin, γ-globulin and fibrinogen were obtained from Sigma Chemical (St. Louis, Catalogue numbers A-4503, BG-11,

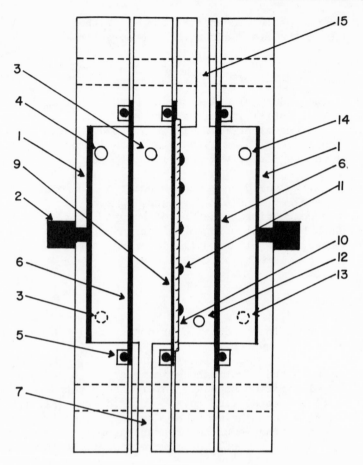

1.   electrodes
2.   electrode binding post
3.   high pressure buffer solution inlet
4.   high pressure buffer solution outlet
5.   O-ring
6.   dialysis membrane
7.   high pressure protein solution inlet
8.   high pressure retentate outlet
9.   ultrafiltration membrane
10.   membrane support
11.   membrane support grid
12.   low pressure ultrafiltrate outlet
13.   low pressure buffer solution inlet
14.   low pressure buffer solution outlet
15.   ultrafiltrate recycle inlet

Fig. 1.   Cross Section of Electroultrafiltration Cell

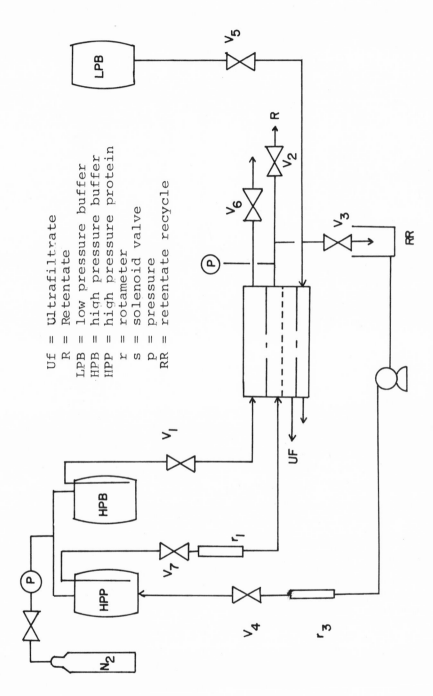

Uf = Ultrafiltrate
R = Retentate
LPB = low pressure buffer
HPB = high pressure buffer
HPP = high pressure protein
r = rotameter
s = solenoid valve
p = pressure
RR = retentate recycle

Fig. 2. Flow Diagram for Electroultrafiltration Experiments

F-4753). I^{125} labeled albumin was purchased from Mallinckrodt
Nuclear (St. Louis, Catalogue number 350). Solutions of I^{131}
labeled bovine γ-globulin were prepared[16] with isotopic iodine
in the form of sodium iodine in 0.1N NaOH, which was obtained from
Industrial Nuclear Corp. (St. Louis).

Protein concentration was measured by the adsorption of
ultraviolet light at a wavelength of 280 nm. For solutions of
single proteins, a Gilford Spectrophotometer, Model 240 was used
for all absorbance measurements with matched, 1.0 cm quartz
Pyrocells ^{R} . The concentrations of albumin and γ-globulin in
binary solutions were determined by scintillation counting of
solutions containing I^{125} labelled albumin and I^{131} labeled γ-
globulin.

A cationic, electrodeposition primer manufactured by PPG
Industries, Inc., ED-3002, was the model paint solution. All
paint solutions were at pH 6.5. Solids concentration was varied
by dilution with acetic acid. Solids concentration was measured
by gravimetric methods involving evaporation of solvent.

RESULTS

Our goal was to demonstrate that an electric field could be
used to control concentration polarization when ultrafiltering
macrosolutes. Thus, the dependence of flux on electric field
strength was determined for albumin (BSA), γ-globulin (BγG),
fibrinogen (BF) and ED 3002 paint. The effects of ΔP, ionic
strength (Γ/2), and pH were considered to a lesser extent. Figure
3 shows the effect of E on the steady state flux of 1 wt.% BSA
in phosphate buffer, pH 7.4, Γ/2 = 0.052M, for XM 50 membranes at
ΔP = 260 mm Hg. The curves represent the average of a number of
experimental runs, with the bars (I) indicating the standard
deviations.

Figure 4 illustrates the increase in flux due to the electric
field for a 0.3 wt.% BGG solution at pH 4.7 using an XM100A
membrane. The curve for the flux of buffer alone is also included
for comparison.

Figure 5 is plot of flux versus time with varying electric
field strength for ultrafiltering 0.3 wt.% BγG with a more open
XM300 membrane in acetate buffer at pH 4.7. A short pulse of
52.2V/cm quickly caused excessive joule heating in the ultrafiltrate compartment. This was partially controlled by cooling
the circulating buffer and protein solution.

The dependence of flux on electric field strength for
fibrinogen is shown in Figure 6. An XM100A membrane was used to

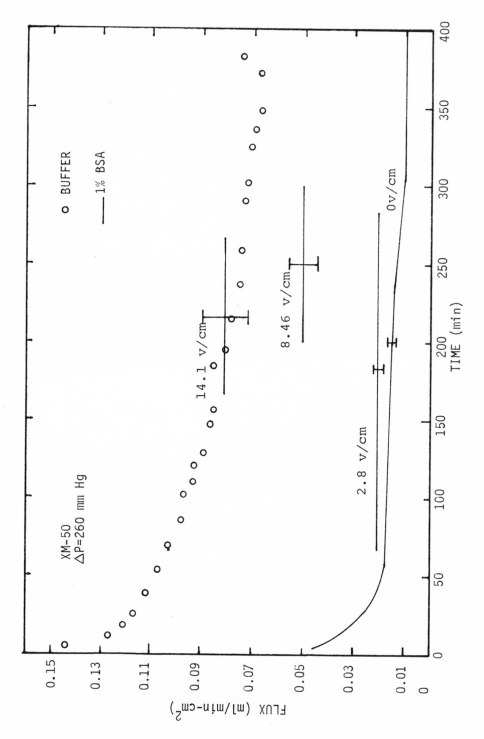

Fig. 3. Effect of Electric Field on the Flux of 1% BSA

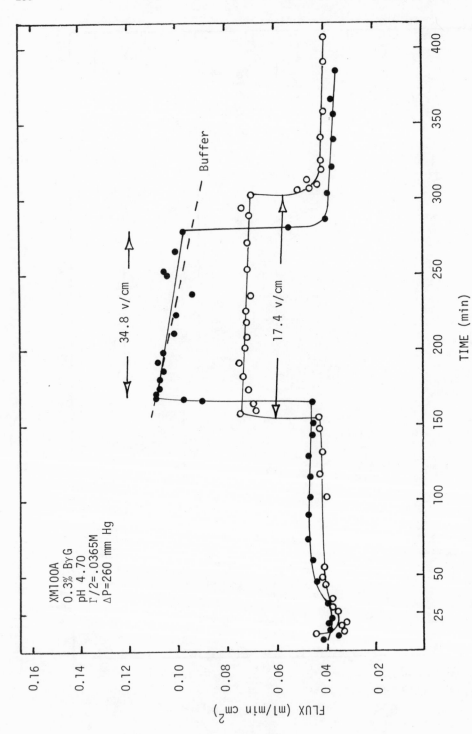

Fig. 4. Effect of Electric Field on the Flux of 0.3% BYG, XM100A Membrane

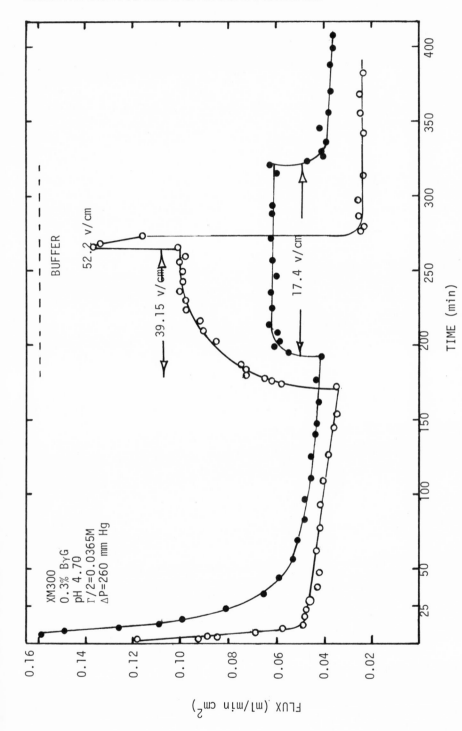

Fig. 5. Effect of Electric Field on the Flux of 0.3% BγG, XM300 Membrane

ultrafilter 0.1 wt.% fibrinogen in phosphate buffer at pH 8.2,
$\Gamma/2 = 0.058$ M.

Figure 7 shows the dependence of flux on field strength and
transmembrane pressure drop for 3 wt.% solution of ED-3002 paint
at pH 6.50. Above an overall field strength of about 2.5 V/cm,
the paint particles coated the cellophane membrane separating the
buffer and retentate compartments. There was a large voltage
drop across this deposited paint film which greatly decreased the
field strength in the retentate compartment.

Table 1 summarizes the results for ultrafiltering a mixture
of 1 wt.% BSA and 0.3 wt.% BYG at $\Delta P = 260$ mm Hg. The normal
flux (E = 0) was 0.031 ml/min cm^2 for XM100A and XM300 membranes
at pH 4.7, and 0.025 ml/min cm^2 at pH 8.1. As with the single
solute solutions, application of the electric field gives significant increases in flux.

Table 1.　Flux Dependence on Field Strength
for a Mixture of BSA and BYG

XM300	4.7	34.8	0.065
		39.1	0.060
		49.1	0.070
	8.1	30.5	0.104
		34.8	0.115
XM100A	4.7	34.8	0.065
		47.9	0.075
	8.1	17.4	0.062
		30.5	0.085

Electroosmosis

Figure 8 shows the effect of the electric field on the flux
of acetate buffer using an XM300 membrane for $\Delta P = 260$ mm Hg.
There is no perceptible change in flux indicating that membrane
electroosmosis was negligible for these conditions. Similar
experiments for the other membranes, at pH 4.6 to pH 8.1 gave
identical results for each electric field strength for paint and
protein solutions.

The magnitude of electroosmosis through any gel layer formed
during ultrafiltration was determined by applying an electric
field for $\Delta P = 0$ after the steady state flux was reached. Again,
the change in flux (< 5%) was insignificant indicating that gel-
layer electroosmosis was negligible.

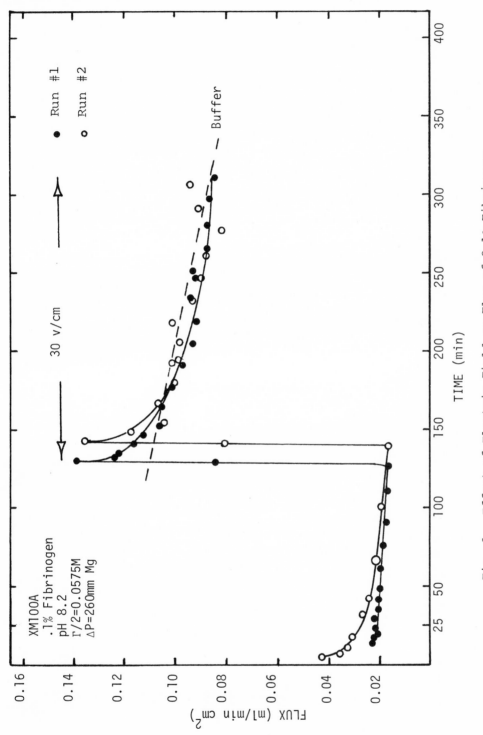

Fig. 6. Effect of Electric Field on Flux of 0.1% Fibrinogen

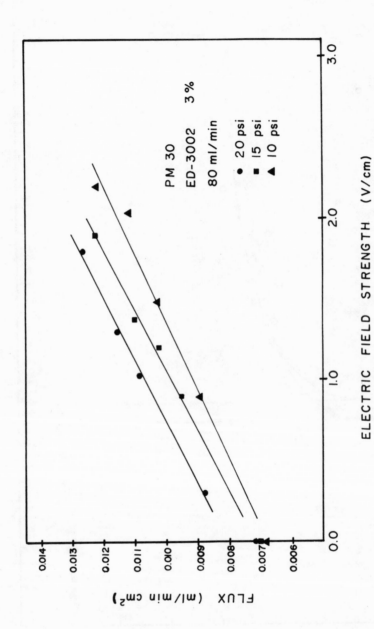

Fig. 7. Flux Dependence on Electric Field for 3.% ED-3002

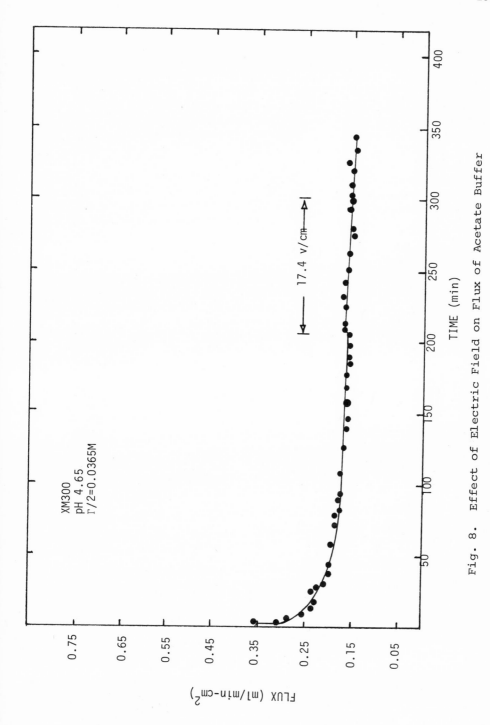

Fig. 8. Effect of Electric Field on Flux of Acetate Buffer

Flux Dependence on Other Operating Variables

The effects of changing the pH and ionic strength were minor compared to the effect of the electric field on flux. For protein solutions, as the difference between the process pH and the iso-electric point increased, the flux increased. Decreasing the ionic strength of the protein solutions increased the flux slightly when the electric field was applied. More importantly, a lower ionic strength gave less current which made it easier to control Joule heating in the cell.

The effect of changing ΔP is shown in Figure 7 for the paint solution. When the electric field was applied, the flux was no longer independent of the transmembrane pressure drop. This was also true for electroultrafiltration of the plasma protein solutions.

Prediction of E_{cr}

Once the dependence of flux on electric field strength is known, it is desirable to predict E_{cr}, the field strength required to eliminate the effects of concentration polarization. E_{cr} was obtained by plotting experimental data, J vs E. When $J = J_0$, $E = E_{cr}$. The results of this analysis are shown in Table 2 for albumin and gamma globulin solutions for $\Delta P = 260$ mm Hg. The critical voltage can also be calculated from equation (3). Those results are also shown in Table 2. They are in good agreement with the experimental data for membranes with high retention (XM50 and XM100A) but not for the more open XM300 membrane. There was only one data point for the fibrinogen solution. Coating of the cello-phane membrane by the electrophoretically migrating paint particles prevented a similar analysis in that case.

Retention and Separation for Electroultrafiltration

Retention is defined by the equation

$$R = 1 - C_u/C_R \tag{4}$$

where C_u and C_R are the bulk concentrations in the ultrafiltrate and retentate, respectively. The retention of the charged macro-solute in a single solute solution always increased when the electric field was applied. The magnitude of the increase depended on the operating variables.

For mixtures, the membrane's ability to discriminate between the different solutes is of great importance for achieving satisfactory separations. A measure of selectivity is the separa-tion factor, α, defined as:

$$\alpha = \frac{(1 - R_1)}{(1 - R_2)} \tag{5}$$

where R_1 and R_2 are the retentions for solute 1 and solute 2.

Table 3 lists the retention and separation factors for albumin and gamma-globulin solutions when ultrafiltering at $\Delta P = 260$ mm Hg with and without the electric field. It is evident that the electric field increases the retention of the charged solute. This is more graphically illustrated in Figure 9 which shows that the retention of uncharged albumin (isoelectric point: pH = 4.7) decreases. This is contrary to the well-known "solute retention by concentration polarization" in normal ultrafiltration.[7] The results in Table 3 can be summarized as follows:

1. At pH 8, the XM300 and XM100A membranes show high retention for both proteins, so the rate of separation would be slow. This process arrangement would work well for just protein concentration. The retention for both BSA and BγG increases when the electric field is on.

2. For the XM300 membrane at pH 4.7, the field exerts a strong effect, holding back γ-globulin while uncharged BSA is forced through the membrane. This leads to a 2 to 6-fold increase in α.

3. The XM100A membrane is too retentive for albumin leading to poor separation and low flux even when the electric field is on. The ultrafiltrate would contain less γ-globulin than albumin, but the retenate concentrations would change minimally.

SUMMARY

Under the certain conditions, an electric field can partially eliminate the effects of concentration polarization when ultrafiltering charged macrosolutes. For the limited range of voltages tested, it appears that flux is linearly related to electric field strength for single solute solutions such as albumin, gamma-globulin, and ED-3002. The data indicate that application of the electric field has the effect of changing the ultrafiltration process from a gel-layer-controlled to a pre-gel polarization-controlled operation. Flux is no longer independent of the transmembrane pressure drop when the electric field is on. The exact relationship between ΔP, flux and E (especially at small values of E), has yet to be determined. The effect of varying the fluid hydrodynamic conditions also must still be examined.

Table 2. Prediction of E_{cr} from Experimental Data

Membrane	Protein	pH	Slope ml/min v cm	Intercept ml/min cm^2	J_o^+ ml/min cm^2	r^2	E_{cr}^* (v/cm)	$E_{cr} = \dfrac{J_o}{\mu}$
XM50	BSA	7.4	0.00453	0.010	0.08	.941	15.5	19.6
XM100A	BYG	4.7	0.00146	0.0426	0.10	.994	34.7	37.9
SM300	BYG	4.7	0.00146	0.0398	0.16	.991	82.3	60.6

+ experimental data

* calculated by linear regression analysis

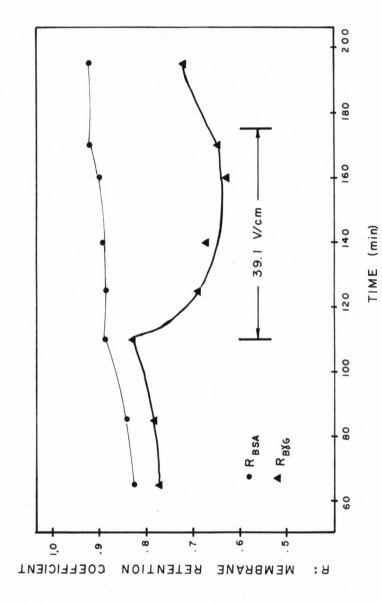

Fig. 9. Effect of Electric Field on Protein Retention in BSA & BγG Mixture at pH 4.7, XM300 Membrane

Table 3. Retention and Separation Factors for BSA and BγG Mixtures (α)

Membrane	pH	E(v/cm)	R_{BSA}	$R_{B\gamma G}$	α
XM300	4.72	0	.370	.792	3.03
		39.15	.487	.975	20.52
XM300	4.70	0	.569	.900	4.31
		39.15	.543	.946	8.46
XM300	4.70	0	.641	.927	4.92
		39.45	.59	.976	17.08
XM300	4.70	34.8	.642	.978	16.27
XM300	8.18	0	.923	.887	0.68
		30.45	.984	.972	0.57
XM100A	8.00	0	.964	.937	0.57
		30.45	.987	.987	1.0
XM100A	8.08	25.9	.976	.972	0.86
		0	.970	.963	0.81
XM100A	4.70	34.8	.958	.997	14.0
XM100A	4.70	34.8	.888	.987	8.61
		47.85	.987	.991	5.89
		21.75	.949	.994	8.50

Electrophoretic migration of the charged macrosolutes in the polarized boundary layer is the controlled transport mechanism rather than the electroosmotic contributions of the membrane and gel layer. Increased retention of charged species and deposition of paint on the opposite cellophane membrane are evidence that the charged particles are pulled away from the ultrafiltration membrane by the electric field in electroultrafiltration. This selective retention by the electric field gives increases in the separation of albumin from gamma globulin only in membranes which originally have a low retention for albumin. In electro-ultrafiltration, the build-up of a layer of retained gamma globulin is prevented (or the layer's permeability is altered) while albumin, at its isoelectric, point passes through the membrane.

REFERENCES

1. W.F. Blatt, Ultrafiltration for Enzyme Concentration, in: Methods in Enzymology, Vol. 22, M.W. Jakoby, ed., Plenum Press, New York (1971).

2. P.R. Klinowski, Ultrafiltration: An Energing Unit Operation, Chem. Engr. 85(11):169 (1978).

3. M.C. Porter and L. Nelson, Ultrafiltration in the Chemical, Food Processing, Pharmaceutical and Medical Industries, in: "Recent Developments in Separation Science," Vol. 2, N.N. Li, ed., CRC Press, Cleveland (1972).

4. P. Mears, ed., Membrane Separation Processes, Elsevier Scientific, New York (1976).

5. U.R. Lebras and R.M. Christenson, Ultrafiltration - A Process Aid in Electrocoating, J. Paint Tech., 44:63 (1972).

6. M.C. Porter, Ultrafiltration of Colloidal Suspensions, AIChE Symp. Ser., 68:21 (1972).

7. W.F. Blatt, A. Dravid, A.S. Michaels, and L. Nelson, Solute Polarization and Cake Formation in Membrane Ultrafiltration: Causes, Consequences and Control Techniques, in: Membrane Science and Technology, J.E. Flinn, ed., Plenum Press, New York (1970).

8. R. P. de Fillippi and R. L. Goldsmith, Application and Theory and Theory of Membrane Processes for Biological and Other Macromolecular Solutions, ibid.

9. M.C. Porter, Concentration Polarization with Membrane Ultra-
    filtration, Ind. Engr. Chem. Prod. Res. Develop. 11:235 (1972).
    (1972).

10. H. Bechold, Ultrafiltration and Electro-Ultrafiltration, in:
    "Colloid Chemistry", Vol. 1, J. Alexander, ed., Chemical
    Catalogue Company, Inc., New York (1926).

11. E. Mannegold, The Effectiveness of Filtration, Dialysis,
    Electrolysis and Their Intercombinations as Purification
    Processes, Trans. Fara. Soc. 33:1088 (1937).

12. M. Bier, Preparative Electrophoresis without Supporting Media,
    in: "Electrophoresis," M. Bier, ed., Academic Press, New
    York (1959).

13. M. Bier, Electrokinetic Membrane Processes, in: "Membrane
    Processes in Industry and Biomedicine," M. Bier, ed., Plenum
    Press, New York (1971).

14. S.P. Moulik, Physical Aspects of Electrofiltration,
    Environ. Sci. Tech., 5:771 (1971).

15. J.D. Henry, Jr., L.F. Lawler, C.H.A. Kuo, A Solid/Liquid
    Separation Process Based on Cross Flow and Electrofiltration,
    AIChE J., 23:851 (1977).

16. J.M. Radovich, "Electrophoretic Control of Concentration
    Polarization in the Ultrafiltration of Plasma Proteins,"
    D.Sc. Thesis, Washington University, St. Louis (1976).

PREDICTION OF PERMEATE FLUXES IN UF/RO SYSTEMS.

Miguel López-Leiva

Division of Food Engineering University of Lund
S-230 53 Alnarp, Sweden

INTRODUCTION

A membrane plant (Ultrafiltration or Reverse Osmosis) can be operated in one of two ways: batch or continuous. The continuous arrangement in its turn can be with partial recirculation (recirculating
system), or without recirculation at all (single-pass). In fig. 1 an
UF/RO circuit has been sketched. Depending on the value of the ratio
R/C(= recirculation flow/final product flow) this circuit can represent each of the three basic arrangements seen: When R/C = ∞ (C=0),
all the retentate is recirculated and the circuit corresponds to a
batch system. In the other extreme when R/C = 0 (R=0), we have the
case of a single pass-system. A value of R/C between these two boundaries represents a continuous system with partial recirculation.

The batch system is the most simple one. It needs little instru-
mentation and there is theoretically no limit for the minimum area
required, but it has the disadvantage of large residence times (order
of hours). This is of course an obvious drawback in the field of
food technology were the control of microbial growth is of paramount
importance.

The recirculating system is probably the most widely used layout in the food industry. It works at much shorter residence times
than the batch system (order of minutes), but still requires rather
extreme temperatures ($10°C > T > 48°C$) to avoid spoilage of the processed liquid. Due to the recirculation of part of the retentate this
system works at a lower relative capacity than the batch system,
presenting also a very spread age distribution curve.

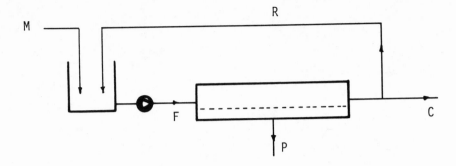

Fig. 1. Membrane separation layout.

Relative capacity of a UF/RO system is defined here as the ratio between the average flux the system gives at a certain degree of concentration and that it gives at the beginning of the process.

$$RC = \frac{\bar{J}(x)}{J(o)} \tag{1}$$

RC has been chosen in this work to measure the performances of the different UF/RO layouts.

The single-pass arrangement is the best from the point of view of control of microbial growth, since it has the shortest residence time with a sharp age distribution curve.

From other points of view the single-pass system seems also to be the most convenient one. The mechanical stress produced by this system upon the liquid is much smaller than in any other system. For instance in a recirculating process for the concentration of skim milk (x 5.5) a single particle, as an average, has to pass through the pump more than 300 times (1).

Even the amount of energy required in a single-pass system is smaller than in a recirculating system. To concentrate whey two times in a single-pass RO system the energy consumption amounts to 5 kWh/m^3 while the same concentration in a recirculating system is obtained with 11 kWh/m^3. (2).

The big problem with a usual single-pass layout is its low flexibility. It is very difficult to influence the concentration of the final product. A decrease of the flow through the equipment in order to obtain a more concentrated product will produce only a temporary change, since the permeate flux, being a function of the Reynolds number of the flow will also decrease. Alternatively an increase of the velocity pass the membrane to improve the flux of permeate it means at the same time a decrease of the residence time of the solution and consecuently the final concentration will vary in a way which is difficult to predict.

This disadvantage is avoided in a module of the rotary type. In this case the membrane itself (or a surface parallel to it) is put in motion and so the hydrodynamic of the flow is made independent of the passage of the product through the module. A laboratory scale module of this type have been developed in our department (3,4). It consists of a pair of concentric cylinders with the inner one rotatable and carrying the semipermeable membrane. The outer fixed cylinder works as a carcasse forming a narrow annulus where the solution to be treated flows . (Fig. 2).

In this module, changing the longitudinal velocity of the solution influences only the residence time of it since the mass transfer coefficient is independent of this axial velocity. In this way an effective change in the degree of concentration of the final product is obtained.

A   Rotary membrane

B   Mechanical seals

D   Annular gap

E   Shell

T   Thermocouple

F   Feed

C   Concentrate

P   Permeate

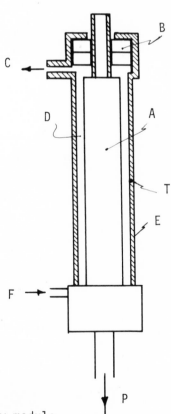

Fig. 2. Rotary module.

This work deals with the theoretical analysis of the permeation fluxes of single-pass and batch systems and their relationship with the degree of concentration. The analysis have been done for different membrane modules: stirred cell, tubular, hollow fiber, rotary. The theoretical results have been experimentally verified in the cases of a stirred cell and a rotary module.

THEORY

In Fig. 3 a general picture of the systems to be analyzed is depicted. It consists of a membrane of length L arranged in a certain geometrical configuration, with a transversal characteristic dimension d. S is the membrane area per unit of length, q_0 and q_k are the initial and final retentate flow respectively. At z=z, the retentate flow is denoted by q and the permeate flux by J.

The pricipial assumptions made in this study are:
1) The system is working in all its extension in the complete polarized region. This means that the concentration of solute on the membrane surface is constant and equal to the concentration of gelation (c_g).

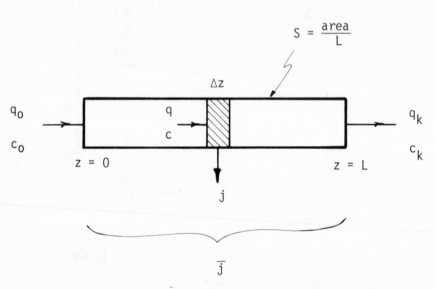

Fig. 3. Nomenclature used in the theoretical analysis.

2) The physical properties of the solution are dependent on concentration only when variations in the z-direction are considered. In the r-direction, no concentration dependence is taken into account.

3) The rejection of the membrane is 100 %.

4) In accordance with 1) variations of pressure do not influence the permeate flux. (?)

5) Constant temperature and constant S (membrane area/length).

With the first assumption and using the approach of Blatt et al., (5), we can write:

$$J = k \ln \left(\frac{c_g}{c}\right) \tag{2}$$

In this equation, $k = \frac{D}{\delta}$ is defined as a mass transfer coefficient.

Experimental data for k in eq. 2 are usually correlated according to:

$$Sh = \frac{k\ d}{D} = A\ Re^a\ Sc^{1/3} \tag{3}$$

Where d is an adequate dimension of the apparatus (diameter in the case of a tubular system) and $\underline{A}$ and $\underline{a}$ are parameters which depend on the geometry of the module.

A total mass balance over the element Δz in Fig. 3 gives:

$$\frac{dq}{dz} = -\ SJ \tag{4}$$

while a solute balance gives:

$$\frac{d(qc)}{dz} = 0 \tag{5}$$

Combining these two last equations we get:

$$SJ = \frac{q}{c}\ \frac{dc}{dz} \tag{6}$$

Introducing dimensionless variables:

$$X = \frac{c}{c_0} \qquad Q = \frac{q}{q_0} \qquad Z = \frac{z}{L}$$

Equation 6 becomes

$$SJ = q_0\ \frac{Q}{XL}\ \frac{dX}{dZ} \tag{7}$$

But for 100% rejection we can write:

$$Q = X^{-1} \tag{8}$$

we get:

$$SJ = \frac{q_0}{X^2 L} \frac{dX}{dZ} \tag{9}$$

The dependent variable J - permeate flux -, can be considered to be a function of either the longitudinal distance Z or the concentration X. In the first case equation 9 can be readily integrated to give:

$$\int_0^1 J dZ = \bar{J} = \frac{q_0}{SL} (1 - \frac{1}{F_k}) \tag{10}$$

In this equation $\bar{J}$ is the average permeate flux (as measured in the apparatus), and F_k is the degree of concentration of the system ($= c_k/c_0$).

The other possibility is to solve equation 9 considering J as a function of the dimensionless concentration X. To do this we make use of equation 2. Using dimensionless variables and with the help of assumption 3 we have:

$$\frac{J}{J_0} = \frac{k}{k_0} (1 - \frac{\ln X}{\ln X_g}) \tag{11}$$

k/k_0 is obtained from the empirical equation 3 as a function of the linear velocity of the liquid through the module and the physical properties of the solute in solution:

$$\frac{k}{k_0} = F_1(Q) \cdot F_2(X) = Q^b \frac{(D/D_0)^{2/3}}{(v/v_0)^{a-1/3}} \tag{12}$$

The function F_2 can be simplified making use of the following relationship which accounts for the dependence of D on concentration (9):

$$D = D^0 \left[1 + C \frac{d\ln y}{dC} \right] (v^0 \rho) / (\eta/\eta^0) \tag{13}$$

Here the superscript o represents infinite dilution, y is the activity coefficient, C is the concentration of solute in moles/l, v^0 is the partial specific volume of solvent and ρ is the density of the solution. From this expression for low concentrations or for y independent of C, we get

$$\frac{\mathbb{D}}{\mathbb{D}_0} = \frac{\nu_0}{\nu} \qquad (14)$$

and

$$F_2(X) = \left(\frac{\nu}{\nu_0}\right)^{-(a+1/3)} \qquad (15)$$

With this approximation it is possible to predict the relative capacity of a UF/RO system without the knowledge of the coefficient of diffusivity of the solute. Only the kinematic viscosity is involved.

For any stationary system b ≡ a in equation 12. For a rotary module and for axial Reynolds numbers larger than 100, the mass transfer coefficient becomes independent of the longitudinal component of the velocity (4) and hence b=0. The function F_2 (X) is characteristic for each solute in solution and must be determined separately for each case. Combining eqs. 9 to 12 we finally get:

$$\text{Relative capacity} = RC = \frac{\overline{J}}{J_0} = \frac{1 - 1/F_k}{\phi (F_k)} \qquad (16)$$

where

$$\phi(F_k) = \int_1^{F_k} \frac{dX}{F_2(X) \cdot X^{(2-b)} \left(1 - \frac{\ln X_g}{\ln X}\right)} \qquad (17)$$

Observing this function we see that the influence of the physical characteristics of the solution upon the relative capacity of a system is expressed by the kinematic viscosity of the solution (function F_2) and the concentration of gelation (X_g). The layout itself influences the value of the relative capacity by means of the parameters a (in F_2) and b.

In general we can conclude that the relative capacity of a given layout will decrease with F_k more sharply under the following circumstances:

1) Smaller concentration of gelation (X_g)

2) Larger values of a (and b) (turbulent flows)

3) More strongly is the dependence of viscosity and diffusivity with concentration.

The expression 16 was numerically solved as function of F_k for different values of the parameters a and c_0. All the results presented here are for Bovine Serum Albumin (BSA) solutions, since this is a macromolecule of well known physical characteristics. The concentration of gelation of BSA was taken equal to 58.5 g/100 cc in accordance with the work of Kosinski and Lightfoot (6). From the same authors we take the correlation for viscosity:

$$\mu = \exp (0.00244 \; c^2) \tag{18}$$

And for the coefficient of diffusion we use the relationship found by Keller et al., (7):

$$\mathbb{D} = \mathbb{D}_\infty \; \frac{\tanh (0.159 \; c)}{0.159 \; c} \tag{19}$$

These relationship together with our measurements of the density of BSA solutions were used to calculate the function F_2. This was done for both, the exact form of F_2 (eq 12) and the simplified form (eq. 15).

The function F_2 has two parameters: a, exponent of Reynolds in equation 3 and c_0, the initial concentration of the solution. The exponent a can acquire the values given in table 1 for the different systems shown.

Table 1. Values of a for different systems

a	System	Reference
0.42	Stirred vessel, laminar	(5)
0.47	Hollow Fiber, laminar	(8)
0.50	Channel, laminar	(10)
0.50	Rotating disk	(6)
0.64	Rotary module, vortex flow	(4)
0.69-0.80	Channel, turbulent	(5)
0.70	Stirred vessel, turbulent	(5)
0.80-1.20	Tubular, turbulent	(10)

RESULTS AND DISCUSSION

In Fig. 4 the approximate and exact forms of the function F_2 have been plotted. We see that both curves deviate considerable each other in all the range of concentrations studied. Examining equation 13 we can postulate that the difference between these functions can be ascribed to the term

$$\frac{d \ln y}{d \ln C}$$

which was taken equal to zero when the expression 14 was derived.

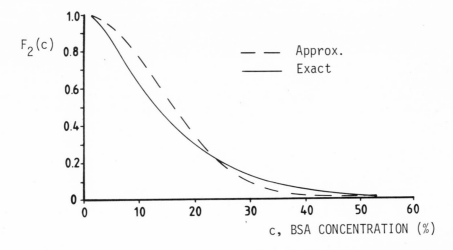

Fig. 4. Function F_2 for BSA solutions in a tubular turbulent system (a = 0.8) --- Approximate (eq. 15) —— Exact (eq. 12).

In Fig. 5 the relative capacity curves for batch and single-pass systems have been plotted. This was done for different modules (rotary, tubular and hollow fiber) for a 2 % solution of BSA. 1 represent the curve for a rotary module while 2 which coincide with 1 in all the range plotted is a time averaged batch tubular system. 3 and 4 represent single-pass tubular systems; 3 for laminar flow (hollow fiber) and 4 for a turbulent flow.

In general we conclude that the systems that give the best relative capacity are the rotary and the batch tubular, while the tubular single-pass shows the lower capacity in all the range studied, being worse when the flow is turbulent.

It is necessary to remember at this point that the flux of reference, J_0, is not the same for every system and so the results shown in Fig. 5 do not necessary mean that in a given situation the absolute values of the permeate flux will be ordered in the same way as the relative capacities are in Fig. 5.

Besides we must take into account that no consideration of the time of operation (for a batch system) or the amount of area (single-pass system) necessary to perform a given separation, is done in this analysis. From a practical point of view we can say that due to limitations in the size of the equipment required a single-pass system cannot concentrate more than 2-3 times while the batch system is limited by the time of operation which for a liquid food shouldn't be more than 2-5 hours depending on the working temperature and the final quality required for the product.

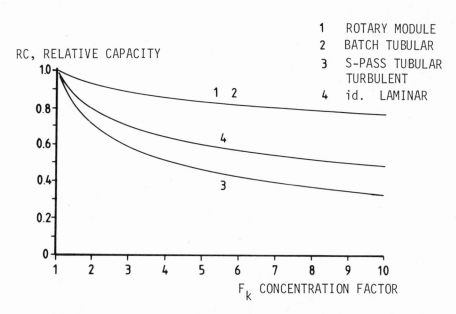

Fig. 5. Relative Capacity versus Concentration Factor for different
systems. 1. Rotary module. 2. Batch tubular. 3. Single-pass
tubular turbulent. 4. Hollow fiber.

In Fig. 6 the experimental results with a stirred cell are plotted together with the theoretical curves obtained when the function
F_2 is calculated out of the original definition (eq. 12) and when it
is out of the simplified expression (eq. 15). In this case we see
that the experimental values follow very close the exact solution,
while the simplified solution overestimates RC at the worse in
around 8 %.

This experiment was done in a lab-scale Amicon cell of 25.5 cm^2
of membrane area (DDS 600), with an initial volume of solution of
about 150 cm^3. The initial solution was 3.1 % of BSA at 25°C.

In Fig. 7 similar experimental results are presented for a
rotary module. The experimental points fall in this case between the
exact and the simplified theoretical curves. The values of RC calculated out of these curves differ no more than 6 % in all the range
studied. This experiment was performed in the rotary module shown
in Fig. 2, which used a 380 cm^2 DDS 600 membrane. The feed solution
was BSA 2 %, and the module was rotated at a constant velocity of
2100 rpm, and 0.3 MPa of pressure.

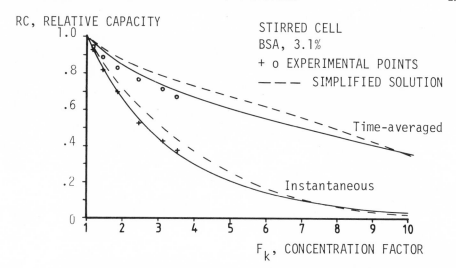

Fig. 6. Experimental verification of eq. 16. Stirred cell (a=b=0.75) and a 3.1 % solution of BSA. --- approximate solution. ——— exact solution. o experimental points (time-averaged). + experimental points (instantaneous).

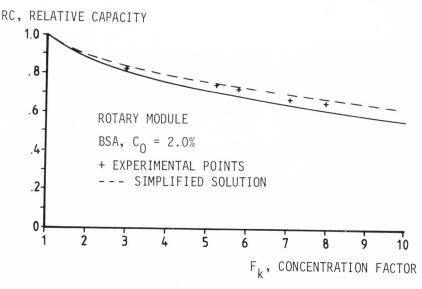

Fig. 7. Experimental verification of eq. 16. Rotary module (a=0.64; b=0) and a 2.0 % solution of BSA. --- approximate solution. ——— exact solution. + experimental points.

CONCLUSIONS

The detemination of the Relative Capacity curve for a given UF/RO system is of great practical interest since it allows the pre-diction of the flux this system will give in certain conditions, as-suming that its performance at a given reference state is known. The precision of the method is strongly related to how accurate the function F_2 represents the reality. A simplified method to calculate F_2 presented in this work has the advantage that only the knowledge of the kinematic viscosity of the solution is needed but it gives values of RC that are up to 10 % higher than the experimental ones.

We have applied this analysis to different layouts and found that both the batch and the rotary systems show the same and best RC-curve among those cases studied. This means that both systems are similar. The variation of concentration with the axial distance in the case of a rotary module is equivalent to the variation of concentration with time in the case of a batch system.

The common single-pass system (tubular for instance) shows a low RC-curve meaning that the membrane area in this case is poorly utilized. This is due to the decreasing of Reynolds number with axial distance. A remedy for this is of course the use of a Christmas tree-type of arrangement in order to mantain a certain minimal velocity inside the tubes.

NOMENCLATURE

a, b	=	Exponent of Reynolds number	(-)
c	=	Solute concentration	(% = g/100 cc)
c_g	=	Concentration of gelation	(%)
c_p	=	Concentration of permeate	(%)
c_0	=	Initial concentration	(%)
c_k	=	Final concentration	(%)
d	=	Characteristic length	(m)
	=	Diameter (tubes)	
	=	Annular gap (rotary module)	
F_k	=	Concentration factor	(-)
J	=	Permeate flux	$(m^3/m^2 \ s)$
J_0	=	Permeate flux at initial conditions	$(m^3/m^2 \ s)$
$\overline{J}$	=	Average permeate flux	$(m^3/m^2 \ s)$
k	=	Mass transfer coefficient	(m/s)
L	=	Length of UF/RO module	(m)
q	=	Retentate flow	(m^3/s)
q_0	=	Feed flow	(m^3/s)
Q	=	Dimensionless flow	(-)
Re	=	Reynolds number	(-)
S	=	Membrane area per unit length	(m^2/m)
Sh	=	Sherwood number = $k \ d/\mathbb{D}$	(-)

Sc = Schmidt number $\frac{\mathbb{D}}{\nu}$ (-)
X = Dimensionless concentration (-)
z = Coordinate (m)
Z = Dimensionless coordinate (-)
δ = Boundary layer thickness (m)
$\mathbb{D}$ = Coefficient of diffusivity (m^2/s)
μ = Viscosity (Kg/m s)
ν = Kinematic viscosity (m^2/s)

REFERENCES

1. L. Kiviniemi; Points of view in the calculation of membrane processes, in: "Industrial membrane technique", Edited by B. Hallström and G. Eriksson, Ingenjörsförlaget, (1977). (in Swedish).
2. J. Hiddink, Netherlands Institute for Dairy Research, private communication.
3. B. Hallström and M. López-Leiva, Description of a rotating ultrafiltration module. Desalination, 24 (1978) 273-279.
4. M. López-Leiva, Ultrafiltration in a Taylor vortex flow. AIChE 87th. National Meeting, Boston, August 19-22, 1979.
5. W.F. Blatt; A. Dravid; A.S. Michaels and L. Nelson. Solute polarization and cake formation in membrane ultrafiltration. Causes, consequences and control techniques, in "Membrane Science and Technology". Edited by James E Flinn, Plenum Press, 1970.
6. A.A. Kozinski and E.N. Lightfoot, Protein Layer Ultrafiltration: A general example of boundary layer filtration, AIChE Journal, Vol. 18, no 5, (1972).
7. K.H. Keller; E.R. Canales and S.I. Yurn. Tracer and mutual diffusion coefficients of proteins. The Journal of Phys. Chem. Vol. 75, no 3, (1971).
8. M. Cheryan, "Mass transfer characteristics of hollow fiber UF of soy protein systems". J. of Food Proc. Eng., 1, (1977), 269-287.
9. L.J. Gosting, Measurements and Interpretation of Diffusion coefficients of Proteins in: "Advances in Protein Chemistry" Vol. 11, 429 (1956).
10. R.P. de Filippi and R.L. Goldsmith, Application and theory of membrane processes for biological and other macromolecular solutions, in "Membrane Science and Technology", Edited by James E Flinn, Plenum Press, (1970).

DEMETALLATION OF CHELATING POLYMERS BY DIAFILTRATION IN

THE PRESENCE OF A PERMEABLE COMPLEXING AGENT

Anthony R. Cooper[*] and Robin G. Booth[+]

[*]Dynapol
1454 Page Mill Road
Palo Alto, CA 94304

[+]331 High Street, Palo Alto, CA 94301

ABSTRACT

Polymeric chelating solutes are retained by ultrafiltration membranes of appropriate porosity. The factors influencing the transport of a metallic impurity by diafiltration in the presence of a permeable complexing agent have been determined. This process may find applications where difficulties are encountered with conventional metal removal processes.

INTRODUCTION

During synthesis of polymeric dyes (1,2), metallic contamination of the product occurs by the use of either impure starting materials or a metal catalyst in its production. The polymeric dyes are good complexing agents, so they are effective scavengers for metallic ions. Often these metallic ion impurities may be removed by ion-exchange resins, provided the resin and the polymeric dye, if charged, have charges of the same sign. However, we have used ultrafiltration as a method for the purification of polymeric dyes (3). It became of interest, therefore, to determine whether metallic ions could be removed during the ultrafiltration step. The polymeric dyes are capable of complexing many metals such that no transport of these metal ions occurs during aqueous ultrafiltration. The use of a complexing agent to promote transport of metal ions in the presence of polymeric dyes has therefore been investigated.

EXPERIMENTAL

The polymeric dyes was prepared by coupling tartrazine to poly(aminoethylene), and contained 0.5 wt% unattached tartrazine and 544 ppm zinc. The complexing agent employed was ethylene dinitrilo-tetraacetic acid (EDTA) from J. T. Baker (Lot No. 415688).

The ultrafiltration unit was capable of recycling the retentate through a hollow fiber module (Romicon Inc., MA) at the following con-ditions: inlet pressure, 25 psig; outlet pressure, 15 psig; tempera-ture, 38°C. The hollow fiber module was a PM10 membrane with 1.8 ft^2 surface area, and an internal channel diameter of 20 mil. By recycl-ing the retentate and ultrafiltrate to the retentate tank, steady state conditions at various solution pH values and EDTA concentrations were established, and samples were taken from the retentate and ultra-filtrate streams. In all experiments, the polymeric dye concentration was 1.5 g/dl. Zinc analyses were performed by elemental analysis.

RESULTS AND DISCUSSION

The transport of a given species in ultrafiltration is generally defined by the rejection coefficient, σ, calculated from:

$$\sigma = 1 - C_U/C_R \tag{i}$$

where C_U is the species concentration in the ultrafiltrate,

C_R is the species concentration in the retentate.

The rejection coefficient was determined at a series of pH values without the addition of EDTA, and at 1000 and 10,000 ppm of EDTA. The results are shown in Table I and plotted in Figure 1. The data show poor transport of zinc in the absence of EDTA, or at pH values higher than 9 in the presence of EDTA at concentrations up to 10,000 ppm. At a fixed concentration of EDTA, the rejection coefficient of zinc is essentially insensitive to pH. As would be expected for the case of competitive complexation, an increase in the EDTA promotes transport of the zinc. The values of the rejection coefficient at 10,000 ppm EDTA are acceptable for the use of diafiltration as a method for zinc removal. In diafiltration mode, the retentate is maintained at con-stant volume by metering in an appropriate make-up stream equal to the ultrafiltrate flux. One diavolume is defined as the collection of a volume of ultrafiltrate equal to that of the retentate. The concen-tration of zinc remaining after N diavolumes, C_R^N, may be calculated from equation (ii):

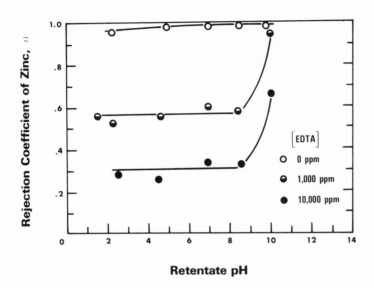

Figure 1. Transport of Zinc Across a Hollow Fiber PM10 Membrane as a Function of Retentate pH and EDTA Concentration in the Presence of a Polymeric Yellow Dye, 1.5 g/dl.

$$\frac{C_R^{\,N}}{C_R^{\,0}} = \exp\ (\sigma-1)N \tag{ii}$$

where $C_R^{\,0}$ is the initial zinc concentration. For ideal transport, $\sigma = 0$ and a one-decade reduction of zinc concentration may be achieved in 2.303 diavolumes. If $\sigma \approx 0.3$, as is the case of 10,000 ppm concentrations, then a one-decade reduction in zinc level would require 3.289 diavolumes. The rejection coefficient of zinc is plotted against EDTA concentration in Figure 2. An empirical fit of the data was obtained by equation (iii):

$$\sigma_{Zn} = 0.226\ \frac{([EDTA]+0.335)}{([EDTA]+0.079)} \tag{iii}$$

Instead of maintaining a constant level of EDTA during diafiltration, an initial charge of EDTA may be used. The mass balance for zinc and EDTA for performing N diavolumes with water make-up is as follows:

Table I

HOLLOW FIBER PM10 MEMBRANE TRANSPORT DATA FOR ZINC IN THE PRESENCE OF 1.5 g/dl
POLYMERIC YELLOW DYE AS A FUNCTION OF EDTA CONCENTRATION AND pH OF THE RETENTATE

RUN NUMBER	EDTA CONCENTRATION ppm	RETENTATE pH	ZINC CONCENTRATION[a] IN ULTRAFILTRATE, C_U ppm	ZINC REJECTION COEFFICIENT $\sigma = 1 - C_U/C_R$
Run 1				
a	0	2.3	.28	.966
b	0	4.9	.09	.988
c	0	7.0	.05	.993
d	0	8.5	.03	.996
e	0	9.8	.27	.967
Run 2				
a	1,000	1.5	3.62	.556
b	1,000	2.3	3.85	.527
c	1,000	4.6	3.53	.566
d	1,000	6.9	3.24	.602
e	1,000	8.4	3.42	.581
f	1,000	10.0	0.39	.953
Run 3				
a	10,000	2.5	5.92	.274
b	10,000	4.5	6.13	.248
c	10,000	6.9	5.37	.341
d	10,000	8.5	5.46	.331
e	10,000	10.0	2.73	.665

[a] Retentate concentration, C_R, of zinc was held constant at 8.15 ppm.

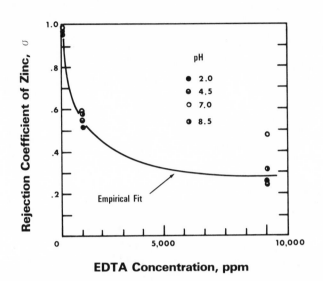

Figure 2.   Transport of Zinc Across a Hollow Fiber PM10 Membrane at Various EDTA Concentrations and pH Values in the Presence of a Polymeric Yellow Dye, 1.5 g/dl.

For zinc,

$$-\frac{d(\ln \theta_{Zn})}{dN} = (1-\sigma_{Zn})$$ (iv)

where θ_{Zn} is the concentration reduction ratio for zinc.

For EDTA,

$$-\frac{d(\ln \theta_{EDTA})}{dN} = (1-\sigma_{EDTA})$$ (v)

where θ_{EDTA} is the concentration reduction ratio for EDTA.

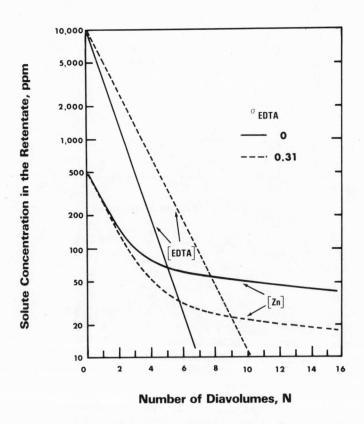

Figure 3. Calculated Curves for Reduction of Zinc and EDTA Concentrations During Diafiltration with a Hollow Fiber PM10 Membrane in the Presence of a Polymeric Yellow Dye, 1.5 g/dl.

Assuming σ_{EDTA} is a constant, equation (v) becomes

$$[EDTA] = [EDTA]^0 \exp (\sigma_{EDTA}-1)N \qquad (vi)$$

where $[EDTA]^0$ is the initial concentration.

For zinc, however, the rejection coefficient, σ_{Zn}, varies with the EDTA concentration. Substitution of equation (iii) in (iv) yields

$$\frac{d(\ln \theta_{Zn})}{dN} = -0.774 \frac{([EDTA]+0.335)}{([EDTA]+0.079)} \qquad (vii)$$

Substitution of equation (vi) in (vii) yields

$$\frac{d(\ln \theta_{Zn})}{dN} = -0.774 \left\{ \frac{[EDTA]^0 \exp (\sigma_{EDTA}-1)N + 0.335}{[EDTA]^0 \exp (\sigma_{EDTA}-1)N + 0.079} \right\} \qquad (viii)$$

Equation (viii) was evaluated using a Tektronix TEK31 programmable calculator for the condition: $[EDTA]^0$ = 10,000 ppm, $[Zn]^0$ = 544 ppm and N = 16. When σ_{EDTA} = 0, a final zinc concentration of 41 ppm was predicted. Experimentally, a final zinc concentration of 17 ppm was observed from this experiment. If the reverse calculation is performed to yield 17 ppm of zinc after 16 diavolumes, then σ_{EDTA} is required to be 0.31. The concentration profiles of EDTA for σ_{EDTA} of 0 and 0.31 are shown in Figure 3. Also plotted are the transport curves for zinc, based on these two values. The transport of zinc is enhanced if EDTA is partially rejected. For the efficient removal of zinc, the concentration of EDTA should be maintained above 100 ppm.

Several other metals were removed during the 16-diavolume ultra-filtration, viz., nickel, copper, lead and cadmium. The method may therefore be applicable as a general method for the removal of metallic impurities. Its application will depend on an economic comparison with other methods, such as ion exchange or solvent extraction. One may envision circumstances where ultrafiltration in the presence of a chelating agent may be the preferred method for removing metals from soluble polymers.

REFERENCES

1. D. J. Dawson, R. D. Gless and R. E. Wingard, Jr., Chem-Tech 6:724 (1976).
2. D. J. Dawson, R. D. Gless and R. E. Wingard, Jr., J. Am. Chem. Soc. 98:5996 (1976).
3. A. R. Cooper and R. G. Booth, J. Appl. Polym. Sci. 23:1373 (1979).

PROGRESS IN THE INDUSTRIAL REALIZATIONS

OF ULTRAFILTRATION PROCESSES

Enrico Drioli

Istituto di Principi di Ingegneria Chimica, Facoltà di
Ingegneria, University of Naples
Piazzale Tecchio, 80125 NAPLES, ITALY

INTRODUCTION

Pressure driven membrane processes show interesting properties
which make them very attractive for chemical reuse and energy
saving in various industrial fields[1]. Ultrafiltration in particular
appeared from the early 70's a very promising process in a large
number of different industrial separation steps in the pharmaceu
tical and food industry, in the treatment of industrial water
containing macromolecules and colloids etc. Large scale industrial
applications of ultrafiltration are, however, still in the early
stages of development and acceptance, largely because of the slow
evolution of high-performance, dependable, low cost and long-lived
membrane modules suitable for high capacity service.

Recent progress, as for example the preparations of UF
membranes with an high resistance in a large pH spectrum 1-13)able
to operate at high temperature (80°C),offers new possibilities to
the development of these separation techniques.

In this paper are reported some results obtained in a research
project started about three years ago in cooperation with small or
medium size industries, leaders in their fields. Scope of the
project was to demonstrate the potentialities of membrane separ
ation processes and particularly of ultrafiltration and reverse
osmosis in non-traditional areas or in situations where costs are
generally assumed prohibitive.

Another basic idea of the project was to operate in area where it was possible to interact with people having a detailed and deep knowledge of the overall industrial process. In fact membrane processes have not to be considered as a simple alternative to existing separation technique, but as a new approach able to solve also problems which where impossible to solve in the past. Therefore a detailed and complete knowledge of the process is necessary for offering alternatives to the standard procedures, often characterized by the creation of pollution problems, high energy cost, etc.

Three example will be discussed in this paper:
1) the introduction of ultrafiltration in an industrial field characterized by a very low technological content, and by huge pollution problems;
2) the use of combined UF and RO steps in the treatment of must before fermentation to wine;
3) the use of RO in the treatment of waste water from the dyehouse of textile industries.

We will also discuss shortly the preparation of new thermo-phylic UF and RO enzymatic membranes, able to operate at 70-80°C which appear a very promising possibility for future new industrial processes where selective mass transfer across the membranes will be combined to specific chemical reactions.

ULTRAFILTRATION IN TANNERIES

Membrane processes are still considered sophisticated and advanced technologies. Therefore their introduction in fields characterized by a low technological content is more difficult. However the conceptual semplicity of membrane separations and the semplicity of UF and also RO industrial plants, are factors which might facilitate their use in this kind of industrial applications. From February 1977 an ultrafiltration industrial plant is in operation in an italian tannery for protein and sulfide recovery from the dehairing waste water. This plant has been built up after one year of experiments on laboratories cells and on an UF pilot plant. All the previsions have been confirmed and today other plants are in operation or under construction in various italian tanneries.

The tanning industry is presently faced with the problem of controlling and reducing the pollution in the effluents from the various steps of the tanning process.

The new laws which are in force today require that tanning plants study and implement new technique to reduce pollution at costs which are not prohibitive to the viable continuation of basic tanning process. One of the possibilities for atacking this involves the study of processes, not directed toward the reduction of pollution in the various waste stream, but also to include in these process the possibility of recovering valuable chemical components presently eliminated in the waste effluents.

The research project have been focused initially upon two of the most polluting process streams which leave the tannery: the dehairing and the degreasing wastes. The objective of this research was to explore the possibility of using ultrafiltration to separate the organic components, present in these streams, from the inorganic chemical which are also present. This would permit, for example, the recycle of the permeate which is consequent to removal of the high molecular weight organic contaminants. This type of recycle is somewhat different than other reclamation techniques already considered in this field. Typically, alternative treatment schemes are limited to processes based upon the precipitation or sediment ation of a distinct solid phase. In contrast, ultrafiltration permits, particularly for the cases in question, a mean for controlling (and maintaining) the Na_2S-$NaHS$ concentration in the permeate while effecting a virtual elimination of the proteinaceous solutes and also of non-specific organic species which are colloi dally suspended[2].

It is well know that these effluents are among the most difficult to treat. It is typically convenient to treat these discharges separately from the other tanning effluents, since secondary and dangerous reactions, which liberate noxious gases, could result from the mixing of the various waste streams.

Table 1 contains the concentrations values of various species in the permeate and in the initial solution. The tests which the data of Table 1 refer to, were carried out at various pressures and at initial temperature of 40°C with HFM-180 tubular membranes.

Table 2, instead, includes some data achieved working on flat BM 500 membranes. As regards the rejection coefficient values, there are no significative difference between the two types of membranes even though the cut-off of the BM 500 membranes is higher (50,000) than the one of the HFM 180 (18,000). In practice, the membranes employed, as expected, exhibit an almost zero rejection

towards electrolytic species and in particular toward sulphide
(R = 0.02) and a high rejection (R > 0.85) towards proteinaceous
and colloidal substances.

Table 1. Membrane HFM 180, T = 40°C, P_{entr} = 3.2 atm

P_{exit} = 1.0 atm

	Na_2S	Protein-N	pH
Feed	3000 mg/l	4.96 g/l	12
Permeate	2866 mg/l	1.07 g/l	11
Concentrate	4120 mg/l	10.26 g/l	12

Table 2. Membrane BM 500, T = 40°C, P_{entr} = 3.0 atm

P_{exit} = 0.5 atm

	Na_2S	Protein-N	pH
Feed	6430 mg/l	4.27 g/l	12
Permeate	6590 mg/l	0.26 g/l	11
Concentrate	----	17.10 g/l	--

Fig.1 shows the trend of the permeation rate, after a
physical pretreatment of the effluent, in both membrane systems.
In practice, after an initial flux decay, the flow reaches an
asymptote of an acceptable values; then remaining constant during
the whole experiment; there were no remarkable variations in the
rejection values.

The tubular membranes experiments described were carried out
almost daily for a period of about seven months. At the end of
each experiment, the membranes were washed with hot water and then
with acidic or basic solutions. Some washings were even performed
with enzymatic detergents, when the flow was particularly low.
In all cases, at the end of the washes, permeate flow values equal
to those measured at the beginning of the tests were obtained.

The fact that the solutions permeated through the membranes
contained high sulphide concentrations, but were completely devoid
of colloidal species or proteinaceous residues, suggested their

re-employment in the hair removal baths. The recovery of the sulphides
present in the permeate was over 80% in practice. Consequently a
series of experiments was performed in which the permeated solutions
were re-utilized in the hair removal process. The results obtained
showed these baths were particularly suited for this purpose. The
presence of moderate amounts of amines and amino-acids derived from
protein degradation, seems to favor the hair removal process.

It must also be remarked that the characteristics of the baths
prepared with permeate are practically always the same; this is due
to the fact the various chemical species rejections remained
constant with time.

As concerns the concentrate we obtained a concentration increase
by a factor equal to about 4.5, without the permeation flow decreas
ing significantly.

Experiments have been also carried out on grease removal
wastes. The results indicate it is possible to drastically reduce
the concentration of the greases and surfactants in the waste water
and concentrate the emulsified greases retained by the membrane[3].
These emulsions have been employed in the hide fattening processes
that follow the tanning. Furthermore the operation of biological
plants that treat mixed wastes is made easier by the absence of
the surfactant recovered in the ultrafiltration process.

MEMBRANE PROCESSES IN MUST AND WINE TREATMENT

The potentiality of pressure driven membrane processes in
food applications is confirmed by an average annual growth rate
of about 37%. The major area for UF and RO however is today almost
totally whey purification. New promising applications appears in
other areas such as for example, wine technology. The fact that
these separation processes operate at molecular level, at room
temperature, without addition of other chemicals, are interesting
properties for a product so sensitive as the wine.

Today wine industries use already various separation techniques
in different steps of wine preparation which might be all, in
principle, substituted with membrane processes. And there are also
problems which have not been solved until now which might be solved
with membranes.

Addition of sulphur dioxide to the must is routinely carried
out for its antimicrobial activity and its antioxidative property.

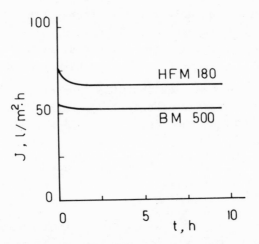

Fig. 1   Behaviour of permeate fluxes as function of time in the
          ultrafiltration of deharing waste water, after pretreatment.
          T = 40°C,   P = 2 atm

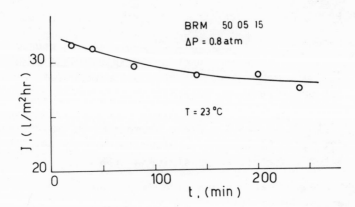

Fig. 2   A typical behaviour of permeate flux as function of time
          in white must ultrafiltration.

Practices which reduce some undesiderable effects in must
and in wines of polypheloxidase, present in the grapes are:
bentonite fining, thermal treatment (70°C for 3', for example).
With the today technology clouds or hazes caused by grape or yeast
proteins, tannins etc., are assisted in their separation from wine,
by the use of small amounts of fining agents which adsorb or combine
chemically and physically with the haze particles or colloids or
neutralize their electric charges causing them to agglomerate and
gravitate to the bottom. Tartrate instability, inhibition of malo-
lactic fermentation, wine sterilization are other examples of
problems where membranes might be used.

In cooperation with a medium size wine industry a research
project on must treatment with UF and RO has been started in 1977
at our Institute. The scope of the research was to analyze the
potentialities of these separation techniques in solving some of
the problems typical of the wine industry, and previously discussed.

Experiments have been carried out in particular for must
stabilization avoiding SO_2 addition, and for controlling poly-
phenols concentration in the must. This last problem is of significant industrial interest for white wine. Various polymeric
membranes in various configurations have been tested in this
research, most part of them already commercially used in other
food or biomedical applications[5].

In Table 3 and in Table 4 some experimental results are
reported which show it has been possible to change polyphenol concentrations in white must without changing sugar content. The must
was ultrafiltered on capillary membranes (Berghof, Tubingen,
Germany) with different cut-off (~50,000; 10,000; 2,000) and on
Abcor tubular membranes. The permeate from the BMR 500515 was used
as feed for the BMR10515, and the permeate from this one as feed
for the BMR 021006. The main problems in must ultrafiltration are
connected to flux decay and to membrane fouling. In Figure 2 a
tipical flux decay during UF of virgin white must with time is
reported. The steady-state fluxes generally obtained with the
various membranes tested are reported in Table 5.

The ultrafiltrate must has been also concentrate with RO
using commercial cellulose acetate and composite polyammide
membranes (PA 300). Interesting phenomena have been observed
concerning cations concentration in this step with PA300 membranes.
As shown in table 6 the rejection for the cations is generally

Table 3.

	BMR 50 05 15		BMR 10 05 15		BMR 02 10 06		MUST
	c_p (g/1t)	R%	c_p (g/1t)	R%	c_p (g/1t)	R%	c_i (g/1t)
Total nitrogen	0.129	8%	0.118	16%	0.118	16%	0.140
Total poliphen	0.241	9%	0.213	19%	0.114	57%	0.264
Sugars	162	0	162	0	162	0	162

Must ⟶ BMR 50 05 15 ⟶ Permeate ⟶ BMR 10 05 15 ⟶ Permeate ⟶ BMR 02 10 06 Permeate

Table 4. Membrane HFM 180, P = 2 atm, T = 25°C,
J = 20,2 1/m^2 h

	c_i	c_p	R%
TOTAL NITROGEN	741 mg/1	190 mg/1	74
SUGARS	18%	18%	0

Table 5. Asymptotic must ultrafiltrate fluxes.

Membrane type	Pressure atm	flux 1/^{2}h	T, °C
BMR 500515	0,5	30	20
BMR 100515	0,5	18	20
BMR 021006	0,5	2,5	20

Table 6. Membrane PA 300; P = 34 atm; T = 10°C

	c_i (mg/1)	c_p (mg/1)	R°(%)
N_2(total)	256	16	0.94
Polyphenols	576	12	0.98
Sugars	163	--	1.00
K^+	1255	18	0.98
Na^+	77	4	0.95
Ca^{++}	123	4	0.97
Mg^{++}	74	1	0.97
Fe^{+++}	11.5	1.6	0.86
Cu^{++}	0.9	0.9	0
Zn^{++}	2.3	1.9	0.17
Mn^{++}	0.9	--	1.00

higher than 98%, except for Cu^{++} and Zn^{++} which seem to permeate completely the membrane. This phenomenon, which is under investigation, might be attributed to an high specific interactions for these cations with the polymeric material, or to Donnan phenomena.

The results obtained increase the interest for the industrial application of membrane processes in the winemaking factories.

REVERSE OSMOSIS IN THE TEXTILE INDUSTRY

The reverse osmosis has been indicated as the only technique able, in principle, to meet the standards introduced by new regulations in many countries, for textile industries waste water discharge[6]. The other possible technology is carbon absorption which requires higher capital investiment, and also does not remove dissolved solids. Reverse osmosis not only removes the dye colors, but also removes 95% of the total dissolved solids, thereby meeting the requested standards.

The water permeated can be profitable recycled in the dye
house itself, decreasing the operating cost of the process.However
the reuse of the brine is in general a major problem, particularly
when the RO unit operates on the total discharge from the dyehouse.
Two possibilities can be considered in fact; the first is to reduce
the total volume concentrate by evaporating and finally destroying
it by usual technique; the second one is the concentrate reuse in
the process itself. The last one is possible however particularly
when the RO units operate on the discharge of a single line.

A research project has been carried out in a medium size
textile Company near Naples, Italy, to verify on a RO pilot plant
operating on the discharges of the dyehouse the real potentiality
of the method[7]. Experiments were made on the total discharge and
on the various operation lines. Waste water from the chlorite
bleaching, from the dye bath (using direct dyes), from the total
dyehouse discharge, have been particularly investigated. Spiral
wound cellulose acetate membranes (ROGA modulus) were used. A 90%
reduction of TSD and COD in the permeate was generally observed.
Only with chlorite bleaching waste water the reduction was of
about the 50%. The high magnesium and calcium rejections observed
permit the reuse of the permeate water in the process avoiding
the use of clustering agents. Asymptotic permeate fluxes were
generally obtained in less than two hours, with values of about
50% of the initial one.

An industrial RO plant with a capacity of 70 m^3/day has been
put in operation on the total waste water discharge at the end of
the experimental program. An interesting solution has been introduc
ed regarding the concentrate disposal. From an accurate analysis
of the energy consumption in the Company, it came out that the
building of the RO unit in combination with the steam generator,
already in operation, avoiding most part of the normal heat
dispersion, gave energy enough for evaporating the concentrate
to a sufficient level for traditional disposal. The cost of this
operation, in principle significant, was practically zero. In fact
only energy normally dispersed was used. The possibility of reusing
the permeate water as boiling water gives another economic benefit
to this solution. In Fig. 3. a schematic description of the plant in
operation is reported.

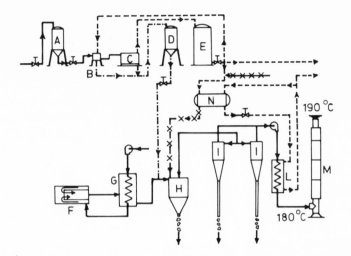

Fig. 3   Flow sheet of the dyehouse RO plant.
          A   Reservoir; B   Sand filter; C   RO unit; D   Concentrate
          reservoir; E   Permeate reservoir; F   Furnace; G   gas-gas
          exchanger; H   dryer; L   gas-liquid exchanger.

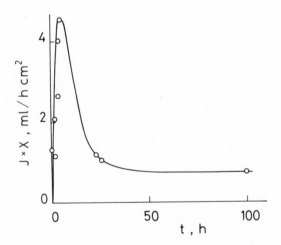

Fig. 4   Behaviour of the product JxX (flow rate per degree of
          conversion) as function of time. T = 70°C,   P = 2 atm;
          membrane surface = 12,56 cm^2; PS membranes filled with
          C.acidophila.

THERMOPHYLIC ENZYMATIC SEMIPERMEABLE MEMBRANES

Recent progress in enzyme engineering offers new possibilities to industrial applications of processes based on the use of immobilized enzyme systems and particularly of enzymatic membranes.

The main features of the more classical membrane preparation methods are the use of organic solvents and high temperature treatments in the annealing steps. These two requirements unfortunately are largely denaturing factors for most of the know enzymes. Therefore it has been impossible to prepare classical RO or UF membranes filled with enzymes or bacteria.

Interesting results have been obtained in a research project devoted to the preparation of enzymatic membranes, starting from a recently discovered extreme thermophylic bacterium, Caldariella acidophila, having enzymes showing high resistance to organic solvents, classical denaturing agents (e.g. urea, guanidine, etc.) and, most importantly, temperatures up to 100°C[8] The phase inversion technique has been used for preparing cellulose acetate (CA) membranes and polysulfone (PS) membranes filled with this bacterium[9]. The membranes have been already tested, at high temperature for their β/galactosidase activity.

All the membrane prepared, the CA and the PS ones, have shown an interesting enzymatic stability with time. The β/galactosidase activity of CA membranes stored at 4°C in buffer acetate after an initial transient shows a monotonic activity decay.
The half life time was of the order of 700 hours. In Fig. 4 a typical J·X curve is presented. After 100 hours the measured conversion (X) was of about 10%. The thermal dependence of the enzymatic activity, for the PS membranes is reported in Fig. 5. The data refer to the conversion measured in a series of parallel UF experiments performed at different temperature, when a steady state ultrafiltration flux was obtained.

The experimental results obtained show how is possible to prepare semipermeable membranes filled with bacteria, using standard phase invertion technique.

The performances of these membranes in ultrafiltration processes at high temperature are interesting and offer new potentialities to enzyme membrane industrial processes, where high temperature is required.

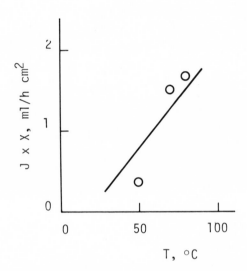

Fig. 5 Thermal dependence of β-galactosidase activity of PS
membranes filled with C.acidophila. The data refer to
the convertion measured in a series of parallel of UF
experiments at different temperature, when a steady-
state ultrafiltration flux was obtained. Membrane
surface = 12,56 cm^2; P = 2 atm.

REFERENCES.

1. R. Lacey, S.Loeb (Eds.) "Industrial Processing with membranes"
    Wiley-Interscience (1972)
2. B. Cortese, E. Drioli, C.P.M.C. 52(6), 511, (1976)
3. B. Cortese, E. Drioli, C.P.M.C. 54(2), 167, (1978)
4. A. W. Crull, "The evolving Membrane Market", P-041 Business
    Communications Co., Inc., Connecticut 06906
5. E. Drioli, G. Orlando, S. D'Ambra, A. Amati, "Membrane Processes
    in must and wine treatment", Proc.Food Proc.Eng., Helsinki
    1979, P.1.3.5. (ISBN 951-46-3997-9)
6. C. A. Brandon, J. J. Porter, "Hyperfiltration for Renovation
    of Textile Finishing Plant Wastewater", EPA 600/2-76-060,
    March 1976
7. E. Drioli, R. Napoli, "Il processo di osmosi inversa nel trat
    tamento delle acque di scarico dell'industria tintoria"
    Atti 2° Convegno sui Trattamenti Acque di Scarico Industria
    le, IRSA vol.45, 197, (1979)
8. M. De Rosa, A. Gambacorta, J. D. Bu'Lock, J.Gen.Microbiol.,
    86, 156-164 (1976)
9. E. Drioli, S. Gaeta, C. Carfagna, M. De Rosa, A. Gambacorta,
    B. Nicolaus, "Thermophylic enzymatic semipermeable membranes"
    J.of Membrane Science, submitted for publication

RECENT DEVELOPMENTS OF MEMBRANE ULTRAFILTRATION

IN THE DAIRY INDUSTRY

Jean-Louis Maubois

Dairy Research Laboratory - I.N.R.A.
65, rue de St-Brieuc
35042 RENNES CEDEX - France

INTRODUCTION

The word "ultrafiltration" is now familiar to the world dairy industry where this membrane technique is continuously expanding in such a rate that some are claiming ultrafilter will be so common, in a dairy plant, as a separator or as a plate heat exchanger (Roy, 1978). Proposed at the end of the sixties for the treatment of whey (Fallick, 1969) and for the manufacture of cheese (Maubois et al., 1969), membrane ultrafiltration was used, last year, to our knowledge, for the daily treatment of 8 000 tons of whey, roughly 3 % of the total world production and for the making of more than 60 000 tons of cheese. The membrane areas of UF units used in the dairy industry vary from 20 m^2 to more than 400 m^2.

GENERAL DESCRIPTION OF THE USE OF UF IN CHEESEMAKING

Traditionally, cheese is obtained by processing liquid milk into a gel. This is done by adding rennet to milk. The interstitial liquid of the gel (whey) is expelled progressively by syneresis. During syneresis, the main components of the gel (fat and proteins) gradually become more concentrated and the product acquire the characteristic shape, texture and composition of the particular cheese to be prepared. But, in practice, cheesemakers do not master all the factors which control whey drainage and the traditional cheeses are heterogenous in composition, quality and weight.

We found that the only way to minimize this heterogeneity was to keep the milk constituents which normally form the cheese in a homogenous liquid form. This idea implies that the whey drainage must be done before coagulation of the milk by the use of membrane ultrafiltration (Maubois and Mocquot, 1971). Indeed, thanks to this technique, it was possible to eliminate the necessary amount of water, lactose and minerals before coagulation and to obtain a product, the retentate or the liquid precheese, which has about the same composition as drained cheese.

Two stages are included in this process which has been developed by our team : the first is the preparation of a liquid precheese that means a liquid product obtained on the retentate side of the UF membrane and having a composition very closed or identical to that of the cheese to be prepared. The second step is the transformation of the liquid precheese into cheese by adding the coagulating enzyme (rennet), allowing the growth of lactic starters to take place and molding. The cheeses thus obtained are treated in the usual manner.

CRITERIA FOR IMPLEMENTATION OF ULTRAFILTRATION TO PREPARE LIQUID PRECHEESE

In addition to sanitary quality of the UF membrane and sanitary design of the UF equipment, implementation of ultrafiltration to prepare each type of cheese requires observance of general and particular criteria.

Bacteriological Criteria

Bacteriological growth may occur in the precheese just as it does in milk. For that reason, ultrafiltration of milk must be during a time and such a temperature that bacterial developments are kept under control.

Biochemical Criteria

Porosity of the membrane and hydrodynamic characteristics must be chosen so that no lactose in excess is retained during ultrafiltration. Indeed, any increase in lactose content in the retentate implies a decline in organoleptic quality of the cheese (sharp, acid or bitter flavor).

As shown by Brulé et al. (1974) (Fig. 1, 2, 3), mineral salts bound to the milk caseins are concentrated during ultrafiltration of normal milk at the same rate than proteins. That could lead to an increase of the buffering power of the precheese, a higher mineralization qualities (acid taste and/or sandy texture). Adjustment of mineralization of the fresh cheese 1 day after rennetting (calcium and phosphorus contents) is practically obtained by several ways :

- ultrafiltration of acidified milk generally by lactic starters,

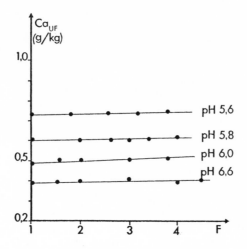

Fig. 1 - Calcium content of milk permeate at different pH
         in function of the concentration (Brulé et al., 1974)

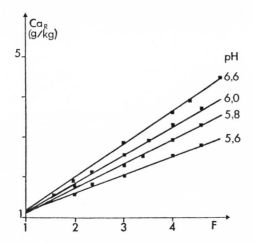

Fig. 2 - Calcium content of milk retentate at different pH
         in function of the concentration (Brulé et al., 1974)

        - addition of sodium chloride in the retentate during or
after ultrafiltration. Indeed, as shown in figure 4, sodium is
exchanged with calcium bounded to casein micelles,
        - diafiltration of the liquid precheese,
        - lowering of the rennetting pH.

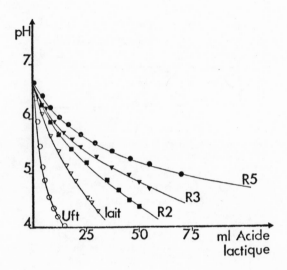

Fig. 3 - Buffering power of milk, permeate (UFt) and retentates
(Brulé et al., 1974)

Physico chemical Criteria

Viscosity of the retentate varies with the percentage of
protein in the retentate. This variation is influenced by the
temperature as shown by Culioli et al. (1974) (figure 5). Moreover,
the higher the protein content of the retentate, the higher is the
shear rate for which this product has a rheological Newtonian
behavior.

Consequently, the choice of UF parameters and especially UF
temperature to obtain the liquid precheese corresponding to the
cheese to be prepared will result from a compromise between an
optimum permeation rate, the effect of UF parameters on milk

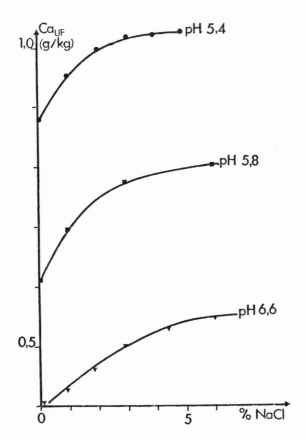

Fig. 4 - Effect of NaCl addition on calcium content of milk
permeate (Brulé et al., 1974)

components, the required mineral composition of the retentate and the control of microorganisms.

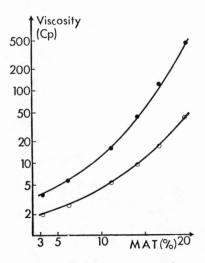

Fig. 5 - Variation of the viscosity of milk retentate at 2°C
and 30°C (shear rate : 437,4 s^{-1}) (Culioli et al., 1974)

RECENT DEVELOPMENTS IN UF CHEESEMAKING

Membrane ultrafiltration has been applied with success to the preparation of fresh soft cheese (30 000 tons in France), soft cheese of the Camembert type and Feta (Figures 6 and 7) and cheese made from goat's milk. But, recently progress in membrane technology and in dairy science have led to new applications.

Ricotta Cheese

The making of Ricotta and related cheeses is still largely an art because of the complexity of precipitation and the critical requirements for suitable texture and flavor. Proper flotation of curd particles resulting from the heat treatment applied to the acidified cheese milk is considered essential for the recovery and optimal curd texture of traditional Ricotta Cheese. Attempts to mechanize or otherwise improve the making of Ricotta Cheese have been diverse but none of the proposed processes is continuous or fully mechanized. Thanks to the utilization of UF, we (Maubois and Kosikowski, 1978) thought that it was possible to meet the challenge. Typical composition and yield at each step of the process that we

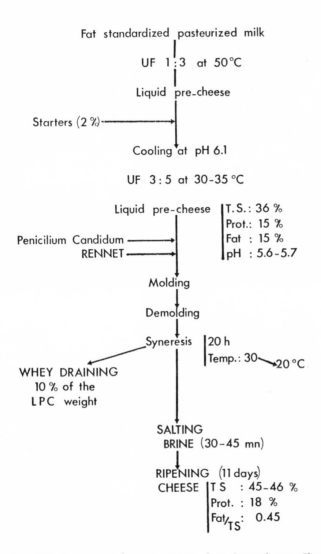

Fig. 6 – UF cheesemaking process for Camembert Cheese

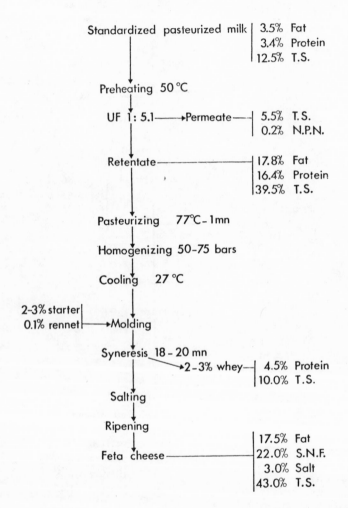

Fig. 7 - Processing of Feta Cheese 40 % (Hansen, 1977)

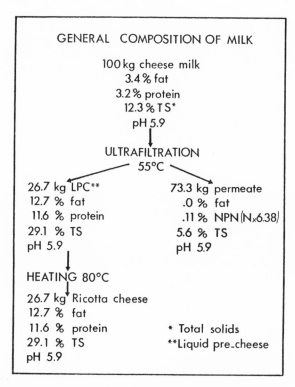

Fig. 8 - Typical composition and yield of Ricotta Cheese from
    ultrafiltration-high heat precipitation method (two-stage)
    (Maubois and Kosikowski, 1978)

are proposing are shown in figure 8. From this schema, a complete
mechanized process for continuous production is actually developed
with a dairy equipment manufacturer as : acidification of the cheese
milk to pH 5.9 with lactic starters, acid whey powder or food grade
acids, ultrafiltration at 55 to 60°C to approximatively 12 % protein
heating the acidified liquid precheese to 80°C in a scraped surface
heat exchanger and filling directly consumer containers. From the
organoleptic point of view, cheeses obtained by ultrafiltration were
preferred to fresh commercial Ricotta Cheeses by 70 % of the taste
panel.

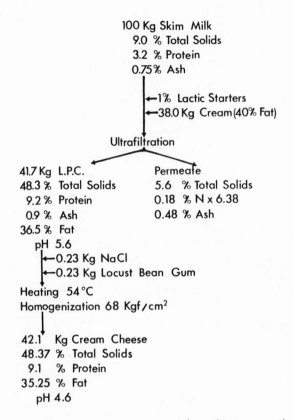

Fig. 9 – Cream Cheese making (Process 1)

Cream Cheese

    Utilization of membrane ultrafiltration for the making of Cream
Cheese was firstly proposed by Covacevitch and Kosikowski (1977)
but neither the process nor the final cheese were fully satisfactory.
Cheeses showed harder body and lower flavor and taste scores than
commercial samples. By taking in account the progress in membrane
equipment, especially the possibility to ultrafilter fatty dairy
products without any damage to the butterfat (Maubois et al., 1975)
and by adjusting mineralization of UF cheeses, we solved these
difficulties and the resulting cheeses were judged organoleptically
identical to the best commercial Cream Cheeses (Maubois and
Kosikowski, 1977). The schema of the proposed process is shown in
figure 9.

Saint-Paulin Cheese

    St-Paulin Cheese is a French variety of cheese related to more known types like Gouda or Edam. Application of UF to the making of this semi hard variety of cheese is becoming industrially feasible thanks to the development of a new membrane generation : the mineral one. Because of the very high temperature resistance (200 to 400°C) and the mechanical quality of this new type of UF membrane, liquid precheeses having a protein content higher than 21 % (that means

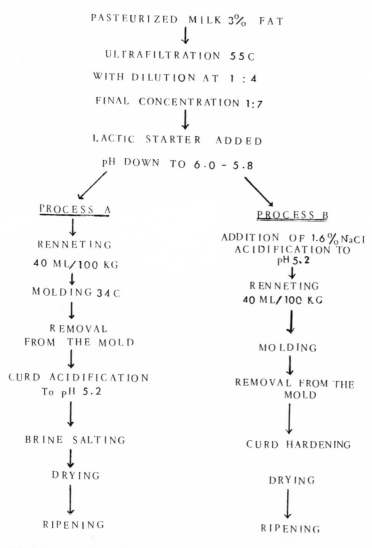

Fig. 10 - UF cheesemaking process for St-Paulin Cheese

a T.S. content higher than 45 %) can be obtained without any problem during subsequent cleaning and desinfection steps. The diagram of figure 10 shows the processes actually tested on a pilot scale by several French cheese firms.

OTHER DEVELOPMENTS OF UF IN DAIRYING

Spray drying the liquid precheese

The liquid precheese (or the retentate) can be dehydrated by usual spray-drying. The powder can be stored for several months before cheesemaking or/and sent economically to countries where milk production is scarce or even nonexistent. As shown by fig. 11 (Le Graet and Maubois, 1979), by adding the required small amount

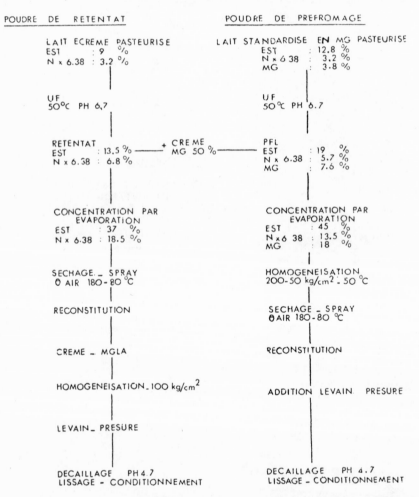

Fig. 11 - Cheesemaking from spray-dried liquid precheese

of water, lactic starters, fat and allowing it to clot and acidify, a cheese is obtained with organoleptic qualities notably superior to those of cheeses issued from the conventional process using normal milk powder. On the other hand, the spray drying of such a product represents an important savings of energy because of the possibility to use the UF permeate in an evaporated form.

Ultrafiltration and Thermisation of Milk at Farm Level

Because of the deep evolution of the milk production and in industrialization of dairying, all the dairy world is faced with a decrease of the bacteriological quality of collected raw milk. Particularly, expansion of the milking machine and the bulk tank has led to a contamination more and more frequent by psychrotrophic germs which produce very heat resistant enzymes. On the other hand, due to the continuous increase of petrol price, collecting milk every day or even every other day is becoming more and more expansive. From these considerations, we have proposed to realize an UF treatment coupled with an heat treatment in all the farms producing manufacturing milk. It is too early to present now some final conclusions but from our preliminary results, we can say that this treatment seems economically and technically feasible and leads to a possible collect of retentate (concentration 1 : 2 or slightly less) every four days.

Hydrolysates from Whey Protein Concentrates

Another very recent application of UF on dairy products was recently studied in our laboratory by Roger (1979). By using the enzymatic membrane reactor technique, a patented process for the production of small peptides (2 to 7 amino acids) from whey proteins concentrates was developed for satisfying the nutritional requirements of patients suffering of intestinal diseases or after surgery. Such an application is far from to be marginal, indeed, the needs were recently estimated to 6 000 tons per year of pure proteins.

REFERENCES

      Brulé, G., Maubois, J.-L. and Fauquant, J., 1974, Etude de la teneur en éléments minéraux des produits obtenus lors de l'ultrafiltration du lait sur membrane, Le Lait, 54 : 600.

      Covacevitch, H.R. and Kosikowski, F.V., 1977, Cream Cheese by ultrafiltration, J. Food Sci., 42 : 1362.

      Culioli, J., Bon, J.P. and Maubois, J.-L., 1974, Etude de la viscosité des rétentats et des préfromages obtenus après traitement du lait par UF sur membrane, Le Lait, 54, 538 : 481.

Fallick, G.F., 1969, Industrial ultrafiltration, Process. Biochem., sept. : 29.

Hansen, R., 1977, Feta Cheese production by ultrafiltration, Nordeuropaeisk Mejeri-Tidsskrift, 9.

Le Graet, Y. and Maubois, J.L., 1979, Fabrication de fromages à pâte fraîche à partir de poudres de rétentat ou de préfromage, Rev. Lait. Fr., 373 : 23.

Maubois, J.L., Mocquot, G. and Vassal, L., 1969, Procédé de traitement du lait et de ses sous-produits laitiers, Brevet Français 2 052 121.

Maubois, J.L. and Mocquot, G., 1971, Préparation de fromage à partir de "préfromage liquide" obtenu par ultrafiltration du lait, Le Lait, 51, 508 : 495.

Maubois, J.L., Brulé, G., Culioli, J. and Goudédranche, H., 1975, Applications de l'ultrafiltration sur membrane au traitement du lait en fromagerie. Symposium APRIA, A9-1.

Maubois, J.L. and Kosikowski, F.V., 1977, Making a cream cheese by ultrafiltration, Résultats non publiés.

Maubois, J.L. and Kosikowski, F.V., 1978, Making Ricotta cheese by ultrafiltration, J. Dairy Sci., 61 : 881.

Roger, L., 1979, Contributions à la recherche d'une meilleure utilisation en alimentation humaine des composants glucidiques et protéiques du lactosérum grace à l'emploi des techniques à membrane, Thèse de Docteur-Ingénieur, E.N.S.A. Rennes.

Roy, J., 1978, Table ronde sur l'ultrafiltration en fromagerie, La Technique Laitière, 921 : 9.

ULTRAFILTRATION OF WHOLE AND SKIM MILK

C. G. Hill, Jr., C. H. Amundson, and S. H. Yan[*]

Department of Chemical Engineering
Department of Food Science
University of Wisconsin
Madison, WI 53706

[*]Texas Instruments, Inc.
MS 947, Dallas, TX 75222

Tubular and hollow fiber ultrafiltration membranes have been used to concentrate and fractionate both whole and skim milk. The effects of the process variables (pressure, temperature, feed flow rate, and feed composition) on permeate rate and membrane rejection coefficients were determined. The data indicate the significance of concentration polarization in the ultrafiltration of milk. For the tubular membrane system used in the ultrafiltration of whole milk, a four-parameter model provides a good description of the observed variation of permeate flux with process variables. In these studies, the rejection coefficients for milk proteins, fat, lactose, and ash did not vary significantly with temperature, pressure or fluid velocity. The implications of preconcentration of skim and whole milk prior to cheese manufacture are discussed.

For an account of this work, see S. H. Yan, C. G. Hill, Jr., and C. H. Amundson, J. Dairy Sci. 62:23 (1979).

FACTORS AFFECTING THE APPLICATION OF ULTRAFILTRATION MEMBRANES IN THE DAIRY FOOD INDUSTRY

W. J. Harper

Department of Food Science & Nutrition
The Ohio State University, Columbus, Ohio

During the past decade, ultrafiltration and diafiltration have become accepted industrial processes in the dairy industry. (Bresleau and Kilcullen, 1977; Cotton and English, 1975; Delaney and Donnelly , 1977; Kosikowski, 1979; McKenna, 1978; Maubois, 1978 and Van et al., 1979.) Initially utilized for the processing of whey and skimmilk, more recent uses have included whole milk and churned buttermilk. Application in current pilot plant or commercial use includes:

1. Production of whey protein concentrates from both acid and sweet whey.

2. Preparation of concentrates for cheese manufacture including soft cheese, cottage cheese, hard cheeses and processed cheese.

3. Protein enrichment for fluid products and in the manufacture of cultured products.

4. Concentration of raw milk on the farm to reduce transport costs.

5. Preparation of specialty products for geriatric feeding.

6. Manufacture of calf replacers.

7. Enzyme recovery from lactose syrup preparation.

8. Recovery of cleaning compounds for reuse.

The successful application of ultrafiltration in the dairy industry depends upon the understanding of a number of diverse factors and the interrelationships between them. These factors include:

1. Variations in the characteristics of different raw materials, such as different wheys, and vairation in the behavior of a given feed stock at different times.

2. Variations in the characteristics of the desired end product with particular emphasis on desired functionality.

3. Differences in ultrafiltration membranes particularly in respect to cleanability and fouling.

4. Different characteristics of the available commercial equipment and differences between batch and continuous processing operations.

5. The effect of operating parameters and effect of operating parameter interactions on the characteristics on the finished product and the economics of operation.

6. Fouling of membranes.

7. Pre-treatment of feed stocks to minimize fouling and to maximize flux rate.

8. Water quality.

9. Cleaning procedures.

10. Scale up from pilot plant operation.

This presentation will touch on all of these factors with particular emphasis on those within the control of the dairy plant operating the membrane processing equipment.

Raw Material

Differences in fat, protein, carbohydrate and mineral composition of different feed stock have obvious effects on the operation of ultrafiltration plants. All other factors being equal, flux rates are generally in the order: whey> skimmilk> whole milk> churned buttermilk. This is related to difference in fat content. However, fouling rates are frequently higher with some whey and churned buttermilk than in skimmilk and whole milk.

In many instances, very small differences in composition can have a marked effect in flux rates. This is well illustrated by the work of Matthews (1979) in ultrafiltration of a lactic acid and sulfuric acid casein whey made from the skimmilk and processed in a four stage continuous system. Figure 1 shows a marked reduction in flux of the sulfuric acid whey as compared to the lactic acid whey. The exact reasons for this difference are not fully understood, but appear to be predominately related to differences in ionic constituents and resulting effects on protein complexes.

Marked day-to-day variations in flux rates in commercial whey ultra filtration plants results in varying efficiency. For sweet whey, this can frequently be traced to differences in fat content or in the treatment of the whey between cheese manufacture and ultrafiltration. Mishandling of the product and production of acid can have a significant effect on increasing fouling and decreasing flux rates.

Different types of whey vary markedly in behavior in UF plants. Sweet whey is generally easier to process than acid whey, and differences exist also between different wheys - such as Cheddar, Swiss and Italian wheys. These relate specifically to differences in heat treatment, mineral constituents and also in lipid content. As noted earlier, the introduction of lipid generally tends to increase fouling, decrease flux rates and also has a significant effect upon the functional properties of resulting whey protein concentrates.

End Product Use

Whey Protein Concentrate

The protein content of whey protein concentrates (WPC) commercially available ranges from 30 to 80 percent. For high protein concentrate: one or two diafiltration steps are employed in addition to the standard ultrafiltration. In addition, the ultrafiltration process may be varied according to the specific end use of the whey protein concentrate. Where minimal denaturation is desired for end products use, preheat treatment and shear during the UF process must be minimized. The functionality of the finished product affected by very subtle changes in processing conditions and most significant, includes: (a) preheat treatment, (b) pH adjustment prior to ultrafiltration, (c) the degree of concentration achieved and levels of lipid lactose and minerals in the final product, (d) pH treatment following membrane processing, (e) conditions of final concentration and drying. Each of these steps can have a significant effect upon the physical chemical properties of the protein system as illustrated in Table 1 which shows the effect of processing steps on the immunological activity of Bovine serum albumin (BSA) for a relatively undenatured WPC.

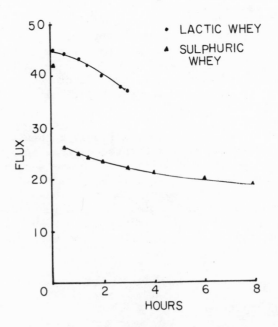

Figure 1.   Average system permeate system flux   rates
(liters hour^{-1}meter^{-2}) versus time (h)
during ultrafiltration of sulphuric (   )
and lactic (0) wheys in continuous mode.
Variation in flux rates of lactic and
sulfuric acid wheys from same skimmilk
in a 4-stage continuous UF module.

Matthews et al 1978

TABLE 1. EFFECT OF PROCESSING STEPS ON IMMUNOLOGICAL
          ACTIVITY OF BOVINE SERUM ALBUMIN

PROCESSING STEP	MG/ACTIVE BSA PER G WPC
RAW CONTROL	12.5
pH ADJUSTMENT TO pH 5.8	11.0
PASTEURIZATION (85C/FOR 15 SEC)	12.7
INITIAL HOUR OF ULTRAFILTRATION	11.8
COMPLETION OF ULTRAFILTRATION	10.7
pH ADJUSTMENT TO 6.5	11.0
EVAPORATION TO 40% SOLIDS	10.2
SPRAY DRYING	9.8

TABLE 2. CHARACTERISTICS OF SOME TYPICAL ULTRAFILTRATION MEMBRANES IN USE IN THE DAIRY INDUSTRY (DAIRY INDUSTRY, DELANEY AND DONNELLY 1977)

Manufacturer(a)	Chemical Nature of Membrane	Water(b) Permeability liters/m²/h	Rejection Characteristics Compound	Percent Rejection
Abcor	Cellulose acetate	500	mol.wt 24,000	98.2
Abcor	Copolymer	800	as for HFA 180	
DDS	Copolymer	148	NaCl	0
			Sucrose	3
			trypsin	100
DDS	Copolymer	40	NaCl	5
			Sucrose	10
			pepsin	100
DDS	Cellulose acetate	120	NaCl	0
			Sucrose	4
			trypsin	100
DDS	Cellulose acetate	80	NaCl	3
			Sucrose	10
			pepsin	100
PCI	Cellulose acetate	160	dextran	50
			trypsin	95
PCI	Copolymer	180	mol.wt 120,000	95

TABLE 2. (CONT.)

Manufac- Turer(a)	Max. Operating Temp., °C	Max. Operating Pressure kg/cm2	pH Operating Range
Abcor	50	4.0	2-14
Abcor	63	4.0	2-14
DDS	30	10	2-14
DDS	30	15	2-14
DDS	50	20	2-8
DDS	50	10	2-8
PCI	50	10	2-8
PCI	50	10	2-11

The changes in processing are frequently very subtle and are still not completely understood. Research is in progress in our laboratories to correlate effects of processing variables on physical chemical properties and functionality. Protein/protein complexing and protein/lipid complexing during ultrafiltration need to be more fully understood in order to optimize the process. As reported by Patel and Merson (1978) protein complexes appear to cause greater fouling than non-complexed proteins.

Ultrafiltration for Cheese Manufacture

Concentration of skimmilk and whole milk for cheese making ranges from approximately 14 percent total solids in the manufacture of cottage cheese, to a 2 to 1 concentration for Cheddar cheese manufacture and a 4 to 1 concentration for the manufacture of a number of soft cheeses. Most frequent application involves the utilization of skimmilk, although more recently whole milk is being concentrated for cheese making use. (Ernstrom et al, 1979) (Maubois and Kosikowski, 1978) added cream to skimmilk 0.5 hours before completion of UF to make Ricotta cheese. In whole milk applications, increased fouling can occur from free fat in the system and particular care needs to be taken to minimize excessive shear to insure that no partial churning has occurred prior to processing. (Homogenization appears to increase the flux rate and decrease fouling.) Yan et al., (1979) have presented a detailed study of processing parameter optimization for the UF processing of whole milk.

Membranes

Some types of membranes available for ultrafiltration are illustrated in Table 2 from the review article by Delaney and Donnelly (1977). Classically, the original membrane used in the dairy industry was made of cellulose acetate. A number of polymeric membranes have been available and new membranes are in continuous development. The major polymeric membranes are polysulfone, polyamide and polyimide membranes. The growth of commercial application of ultrafiltration in the dairy industry can be attributed to a major extent to a major extent to the development of noncellulose acetate membranes and their improved cleanability. The characteristics of the various membranes are illustrated in the table in terms of water fluxes, rejection characteristics, operating temperature, maximum operating pressures, pH range. Some of the newer membranes have higher maximum operating temperatures, pressures and improvement permeability and some are reportedly non-fouling in nature because of induced charges. Because of the need to use standard food plant cleaning practices to the greatest extent possible, an optimum ultrafiltration membrane should have a wide pH operating range (2-14), a high operating temperature, optimally up to 80C, and be resistant to chlorine up to 200 ppm.

The type of membrane, in addition to controlling its operating characteristics, also is significant in fouling. Each membrane has different chemical reactions and varies in its binding capability. This is illustrated in respect to phosphate binding for cellulose diacetate, polysufone and polyamide in Table 3.

The data shows that the binding of phosphate to the membrane can also be influenced by the presence of other materials.

Equipment

The basic configurations in principal manufacturers of filtration equipment currently being used in the dairy industry are:

Description/Type	Manufacturer
Open Tubular; Plate and frame;	Abcor, Patterson Candy, Int. Rhone-Polenc, DeDanske Sukkerfabrikker
Flat leaf	Dorr Oliver
Hollow fiber core	Romicon
Spiral Wound;	Tri-Clover, Abcor

Delaney and Donnelly (1977) summarized the advantage of various systems as follows:

Turbulent Flow Systems

Systems	Advantages	Disadvantages
1. Large tubes	1. Easily cleaned; no dead spaces,	1. High hold up vol. per unit area
a. 725mm		
b. 16–17mm		
c. 12–13mm		
d. 6mm		
	2. Well developed equipment	2. High pressure drops in tube connection
	3. Individual tube replacement possible	
2. Flat Plate	1. Well developed equipment	1. Difficult to design free of dead spaces

	1. No hold up volume	2. Entire module must be replaced on failure
3. Spiral Wound	1. Easy to change membranes	1. Large hold-up volume
	2. Improved sanitary design	2. Difficult to clean

Thin Channel Systems

1. Plate & Frame channel
  a. flat leaf, 0.7; 2.5mm
  b. flat retangular 0.7mm
	1. Economic for high viscous solution	1. Single tube replacement not possible

	2. Low hold-up volume	2. Difficult to design free of dead spaces
	3. Low pressure drop	3. Difficult to clean
	4. High conversion per pass	
	5. Equipment well developed	
2. Hollow fibers a. 1-2mm	1. Economic for highly viscous solutions	1. Small diameter tubes may be the results of some food application
	2. Low hold up volumes	2. Equipment less well developed
	3. Large area of membrane per unit volume-compact equipment	3. Single tube replacement not possible
	4. Potentially low cost	
	5. Low pressure drops	

As a general observation, there has been an increase in the trend towards the use of thin channel devices. Madsen (1978) has

TABLE 3.   EFFECT OF MEMBRANE TYPE AND COMPONENTS ON
PHOSPHATE BINDING TO MEMBRANES (6.5 μM po_4)

MEMBRANE	^{33}p BOUND (g/cm^2 x 10^{-2})			
	MILK BUFFER	MILK BUFFER + LACTOSE	MILK BUFFER + LACTOSE + WHEY PROTEIN	WHEY
POLY SULPHONE	20	14	69	6
POLYAMINDE	815	435	140	212
CELLULOSE DIACETATE	96	147	18	10

presented an extensive mathematical treatment of the mass transfer
processes in thin channel flows with the flat plate systems. The
recent review of Glover et al., (1978) provides a broader view of
the type of membrane processing equipment currently available.

A comparison of permeate fluxes with four different types of
equipment on the same feed is shown in Figure 2. The units with
the highest flux rate, gave the lowest protein rejection rate
(0.95).

One of the problems in research in this area has been the very
rapid development in equipment, so that very frequently the equipment being used for testing soon becomes obsolete. At the same
time, relationships between pilot scale performance and commercial
scale operation have been uncertain.

Physical Operational Parameters

Major operational parameters include pressure, flow rate temperature and pH, which are frequently interactive in nature
(Matthews 1979). Optimization of these parameters depends upon the
feed material, membrane type and equipment being used. No attempt
is made in this presentation to discuss these in detail, since
they have been covered in previous presentations and are covered
in reviews by Delaney and Donnelly (1977), Matthews (1977, 1979),
Glover et al., (1978) and Madsen (1978). In whole milk, Van et
al., (1979), found that the flux rate was independent of pressure
with their system at pressures above 100 kPa. They reported optimum operations at high flow rates, relatively low pressures and
relatively high temperatures (50-55C). These same relative conditions are also applicable to whey, skimmilk and churned buttermilk.

Increasingly, the industry is going to continuous operation
where the number of stages generally ranges from 4 to 6. The
effect of number of stages and flow rate on flux rates is illustrated in Figure 3.

Pretreatment and Fouling

These topics are interrelated, since most pretreatments used
for ultrafiltration have the minimization of fouling as a primary
objective. A secondary objective is the minimization of microflora. This later objective becomes less important as the industry tends to a greater reliance on continuous plants and minimization of total residence time. However, a pasteurization
pretreatment is essential to minimize undesirable outgrowth of
organisms that can cause either an increase in acidity (lactic
acid bacteria-Lactobacilli) or protein gelation (spore forming
Bacilli).

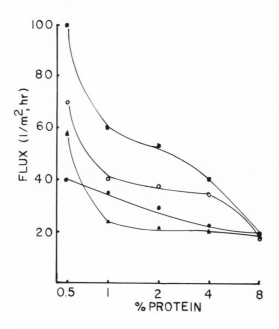

Figure 2.   Flux versus protein concentration for different
UF membrane plants on acid whey.

ABCOR: HFA-180

DDS: 600

Dorr Oliver: XP-24

Patterson Candy: T5/A

Matthews, 1977

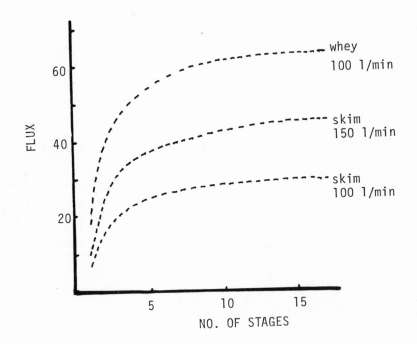

Figure 3.   Effect of number of stages on flux rate
            ($1/m^2/hr.$) for whey and skimmilk.

                        Delaney and Donnelly, 1977

Pretreatment

Considerable attention has been given to various heat treatments and pH adjustment in pretreatment of whey to maximize capital costs expenditures with skimmilk. Buttermilk pretreatment is limited to pasteurization to control microflora and centrifugation to reduce fat levels. In addition to pasteurization, the UF flux rate of whole milk is improved by homogenization.

One of the most notable methods of whey pretreatment has been that developed by CSIRO, Hayes et al., (1974) in treating whey to 80C for 15 seconds followed by a pH change to a value of between 5.2 and 5.9 (Figure 4). This has been confirmed in other countries and has caused a large increase in flux during processing of hydrochloric acid in whey. The system, however, is not equally applicable to all acid whey treatments and does not work nearly as well on sulfuric acid whey.

Dutch workers (Dewit et al., 1978) have reported a clarification technique for removing the lipid fraction whey prior to ultrafiltration. This process is demineralization, adjustment of pH to 4.6 floculation to remove 90 percent of the lipid mass, 99 percent of all bacteria and 10 percent of the protein orginally present. The protein derived from such pretreatment is claimed to have superior functionality because of lipid removal. A patented process (Attebury, U.S. Patent 56029) is also available for removal of calcium phosphate containing lipid and protein additional minerals. This constitutes a simple adjustment of pH of acid whey from 4.6 to 7 plus the addition of sodium or calcium hydroxide. The resulting sediment can be removed by gravity settling.

Membrane fouling is a complex phenomena that includes both the composition of the feed stock as well as the composition of the membrane. Muller and Harper (1979) reviewed chemical engineering aspects of preparing whey for ultrafilatration processes and have reviewed mechanisms involved in fouling. The major constituents in fouling are beta-lactoglobulin, bovine serum albumin, lipid and calcium phosphates. The degree of protein-protein interaction, ionic-protein interactions and lipid-protein interaction will be affected by pH, ionic strength, composition and preprocess treatment of the feed stock solution. The mechanism of fouling is still not completely understood and will require additional research, Patel and Merson (1979). However, protein interactions and calcium and phosphate interaction with membranes and proteins are definitely involved (Muller and Harper, 1979).

Another pretreatment that could be mentioned is that of preconcentration of whey for filtration (DeBoer et al., 1977) has shown the capacity of an ultrafiltration plant can be improved by

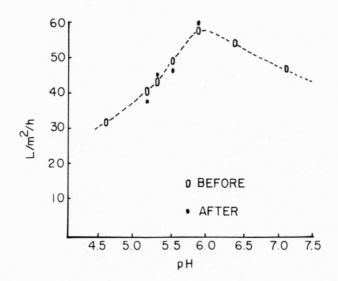

Figure 4.   The effect of pH adjustment before or after
            pasteurization at 80C for 15s on permeation
            rates for HCl casein whey in a Patterson
            Candy UF module.   The points represent the
            permeation rate after 30 minutes recirculation
            in a clean module for each trial.

                                        Hayes et al, 1974

feeding whey that has been preconcentrated to as high as 20 percent total solids. This is of benefit when whey has been concentrated for transportation to a central processing factory. Other pretreatments for whey are precentrifugation to remove casein fines that can accelerate fouling and pH adjustment to below pH 4 to decrease fouling and increase flux rates. This is thought to be primarily due to the permeation of undissociated lactic acid that permeates under these conditions. Also, a shift in the physical/chemical nature of calcium salts and calcium protein complexes has been observed.

Cleaning

Cleaning has two objectives; one is to eliminate microbial contamination and the second is to restore the flux rate of the equipment to its beginning level. Cleaning and sanitizing steps remain significant, although less attention is being given to them with the increased use of non cellulose acetate membranes. With cellulose acetate membranes, enzyme cleaning was essential and chlorine could not be used as a sanitizing agent. Cleaning was often incomplete and redeposition occurred as illustrated in Figures 5 and 6 that shows the cyclic nature of enzymatic cleaning. Three or four short cleaning cycles were more effective than one long cleaning cycle. This is also true of chemical cleaning used with the non-cellulose acetate membrane systems.

Cleaning/Sanitizing methods are relatively well established although they vary widely from company to company. A typical regime includes:

1. Preconditioned fresh water rinse to remove residual material. This may be performed at a temperature of 35-50°C depending on the membrane.

2. Recirculation of an alkaline detergent solution to remove protein and fat deposits. The complexing agent such as EDTA or hexametaphosphate may also be utilized to help remove mineral constituents. The temperature should be as high as possible in keeping with the stability of the membrane system. With some water supplies, an acid cleaning cycle may be desired.

3. A water rinse to remove detergent.

4. Sanitizing by circulation of 50-100 parts per million in free chlorine for 10 minutes at 20-30C and pH 7-9.0.

5. A final water flushed to remove the chlorine sanitizer, at 20-30°C.

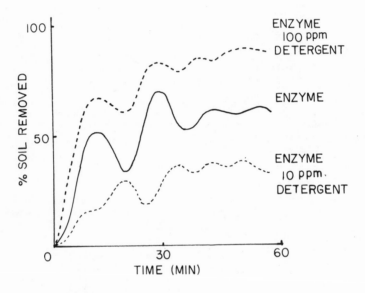

Figure 5. Effect of anionic detergent on soil removal
during enzymatic cleanin- of cellulose acetate
UF membrane (^{33}P soil)

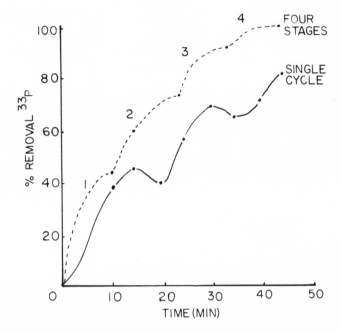

Figure 6.   Comparison of single cycle versus multiple
            cycle enzyme cleaning on soil removal in a
            tubular cellulose acetate UF system.

Iodine has been utilized as an alternative sanitizer to chlorine but is unsatisfactory with cellulose acetate systems and may react with some gasket materials. In sanitizing systems, it is necessary to be sure that the permeate side of the membrane is thoroughly flushed and sanitized. Some dairy plants use chlorine as a primary sanitizer and then store the unit overnight until starting up processing with 25 ppm of iodophore.

Water Quality

During the past several years a number of instances have been reported of contaminants in water supplies that can have irreversible adverse effects on ultrafiltration membranes. Inorganic deposits, which may be salts of iron, silicon, manganese or other elements have been observed in commercial ultrafilatration plants in New Zealand (Matthews 1979). There have been no major studies of water quality standards of ultrafiltration plants and this is an area of needed further research. Silicon in particular (at concentration as low as 50 ppm) appears to be deleterious to membrane flux and not readily removable by standard cleaning practices. To minimize the buildup of minerals in ultrafiltration membranes, an acid wash may also be included in the cleaning regime. In addition there has been some indication that at least one equipment manufacturer is considering the inclusion of a clause on minimum water quality standards in their guarantee.

Scale-up and Evaluation of Systems Performance

Matthews (1979) has summarized the problems involved in scaling up from pilot plant operations to full scale systems. He has indicated that questions that need to be answered prior to scale-up include:

1. The volume of product to be processed.

2. The end product specifications for compositions of physical properties of particular product.

3. Characteristics of the raw material.

4. Characteristics of the final product derived from the raw material meet those required by the customer.

5. Floor space availability.

6. Availability of services.

Pilot plant operations can be utilized to obtain a correlation between performance parameters (average flux and whey components retention coefficients, and operation variables for a particular supply). Reference can be made to various published performance studies that provide background information on particular systems such as those of Cotton and English (1975), Donnelly and Delaney (1974), Perri and Setti (1976) and Breslau et al., (1975). It is useful to use the same type of pilot plant study as planned commercial plants. Although batch mode studies can be useful, large scale ultrafiltration plants are rarely operated in batch mode. Generally, several stages of modules, each consisting of a membrane system and a recirculation pump to provide pressure and velocity, are linked in series to perform a continuous system. The mathematics of these systems has been previously described by Breslau (1976), Pepper and Smith (1977). Batch-mode pilot plants can be scaled up to give dimensional data for large continuous systems provided that:

1. The feed material is the same (this is extremely important)

2. The operation condition of the pilot plant, temperature, pH, recirculation flow rate, average transmembrane pressure are the same as employed in each stage of a multi stage plant.

Such information provides only first approximation. If the relationship between instantaneous flux and concentration factor is known in a number of stages, the total membrane area to achieve the required end product can be approximated. However, it is essential that the following characteristics of the raw material be known, Matthews et al., (1978a, b). When a DDS pilot plant was operated as a feed-and-bleed system for sulfuric acid whey, flux declined as a logarithmic function of the volume of whey processed and the unit area surface membrane. The extent of flux decline observed in this further study indicated that a large continuous plant would need at least 35 percent more membrane area than predicted from simple batch studies. A further consideration in scale-up is the end use of the permeate. If it is required that the permeate contain extrememly low concentration of protein, then selection of a plant with high protein retention characteristic is indicated. The component retention coefficient, however, will be influenced by the pH of the operation as well as the membrane type.

REFERENCES

Attebury, J. M., 1971, U.S. Patent 3 560 219.

Bakel, J. T., Morris, H. A. and Morr, V. C., 1974, J. Dairy Sci., 57:584.

Breslau, B. R., 1976, In Proceedings Whey Products Conference, p. 37, Ag. Res. Service, U.S. Dept. Ag., Philadelphia, Penn.

Breslau, B. R. and Kilcullen, B. M., 1977, J. Dairy Sci., 60(9)1379.

Coton, S. G. and English, A., 1975, In separation Processes by Membranes, Ion Exchange and Freeze Concentration in the Food Industry, p. D18.1, International Union of Food

Sci. and Technol, Paris, France.

de Boer, R., de Wit, J. N. and Hiddink, J., 1977, J. Soc.
    Dairy Technol., 30:112

de Wit, J. N., Klarenbeek, G. and de Boer, R., 1978, In Brief
    Commun., XX Int. Dairy Congr. p. 919.

Delaney, R. A. M. and Donnelly, J.K., 1977, In Reverse Osmosis
    and Synthetic Membranes, pp 417-443, Ottawa, Canada
    Nation Research Council, Canada.

Donnelly, J. K. and Delaney, R. A. M., 1974, Lebensmittel-
    Wissenschaft und Technologie, 7:162.

Ernstrom, C. A., Sutherland, B. J. and Jameson, G. W., 1978,
    J. Dairy Sci., (Suppl. 1)102.

Glover, F. A., Skidder, P. J., Stothart, P. H. and Evans, E.
    W., 1978, J. Dairy Research 45:45.

Hayes, J. F., Dunkerley, J. A., Muller, L. L. and Griffin, A.
    T., 1974, Austral. J. Dairy Technol. 29:132.

Hiddink, J., de Boer, R., Romijn, P. J., 1978, Netherlands
    Milk & Dairy J. 32(2)80.

Hiddink, J., de Boer, R. and Nooy, P. F. C., 1976, Zuivelzicht
    68:1126.

Kosikowski, F. V., 1979, J. Dairy Sc., 61(1)41.

McKenna, B., 1978, Irish J. Food Sci. and Tech 2(1)45.

Madsen, R. F., 1977, Hyperfiltration and Ultrafiltration in
    Plate-and-Frame Systems. Elsevier, Amsterdam,
    Netherlands.

Matthews, M. E., 1979, Proceeding Whey Research Workshop II,
    Palmerston, North, New Zealand.

Matthews, M. E., Doughty, R. K. and Hughes, I. R., 1978b,
    N. Z. J. Dairy Sci. and Technol., 13:37.

Matthews, M. E., Doughty, R. K. and Hughes, I. R., 1978b, N.
    Z. J. Dairy Sci. and Technol., 13:144.

Maubois, M., 1978. Technique Laitiere No. 927:1143.

Maubois, J. L. and Kosikowski, F. V., 1978, J. Dairy Sci.,
    61(7)881.

Muller, L. L. and Harper, W. J., 1979, J. Agr. and Food Chem.

Patel, P. C. and Merson, R. L., 1978, J. Food Sci. and Tech.,
    India, 15(2)56.

Peri, C. and Setti, D., 1976, Milchwissenschafft, 31:135.

Pepper, D. and Smith, G. A., 1977, In Advances in Enzyme and
    Membrane Technology. Symposium Series No. 51.
    Institution of Chemical Engineers, Rugby, England.

Smith, B. R. and MacBean, R. D., 1978, Austral. J. Dairy
    Technol. 33:57.

Yan, S. H., Hill, C. G. and Admundson, C. H., 1979, J. Dairy
    Sci., 62(1)23.

VEGETABLE PROTEIN ISOLATES AND CONCENTRATES BY ULTRAFILTRATION

Munir Cheryan

Department of Food Science
University of Illinois
Urbana, Illinois 61801

INTRODUCTION

Vegetable protein products have been consumed for centuries in the Orient. However, it was not until this century that the rest of the western world recognized them, and soybeans in particular, for their human food value. At present, soybeans are the major cash crop in the United States and are the major source of edible oil. Once the oil is removed, however, the major portion of the defatted soybean meal is used mostly as animal feed, with perhaps about 3-5% being used directly as human food. Although this may appear to be relatively small, it is apparent that a large number of food companies routinely incorporate vegetable proteins into some of their traditional food products and are developing new product formulations using vegetable proteins as a major ingredient or sole source of protein. The annual growth rate of this usage directly in human food is increasing at the rate of 10% per year [1].

Soybeans are by far the most important source of vegetable proteins. The desirable components of the soybeans are its protein and fat, but there are also some undesirable components that must be removed or reduced to increase the usefulness and functionality of soybean products. Some of the oligosaccharide components have been implicated as major flatus-causing factors [2]. Lipid-lipoxygenase interactions must be avoided to prevent painty off-flavors from developing. Phytic acid forms insoluble chelates with minerals or ternary phytate-mineral-protein complexes that reduce bioavailability of the complexed minerals [3]. Trypsin inhibitors are compounds that affect the efficiency of protein digestion.

Traditional processing methods of producing protein concen-

343

trates and isolates partially overcome these problems. Concentrates (> 70% protein, dry basis) are produced as a result of removal of soluble sugars from defatted soybean meal. This is done
by contacting the meal with either dilute alcohol, dilute acid or
moist heat. Filtration or centrifugation of the slurry results in
a residue containing essentially protein, insoluble polysaccharides
and ash. Protein isolates are a further stage of refinement with
essentially all of the nonprotein compounds removed. This is done
by extracting the meal in dilute alkali, centrifuging out the insolubles (polysaccharides), then bringing the pH down to the iso-
electric point of the proteins to insolubilize them, and centrifuging out the sugars. This type of product contains typically
more than 90% protein. In order to lower trypsin inhibitor activity, it is necessary to apply heat in some state of the process,
which, however, may result in irreversible denaturation of the
proteins. In addition, isolates prepared by traditional methods
may contain a substantial amount of the original phytic acid [3]
and some of the proteins that are soluble at the isoelectric point
is lost in the "whey", resulting in less than optimum yields.

Ultrafiltration (UF) has been suggested as a possible means
of overcoming these problems. Since oligosaccharides, phytic acid
and, to a lesser extent, trypsin inhibitors are smaller in molecular size than proteins and fat components, it should be possible,
by careful selection of membrane and operating parameters, to
selectively remove these undesirable components and produce a
purified protein isolate or lipid-protein concentrate (depending
on the raw material used) with superior functional properties.
Since Porter and Michaels [4] first suggested the use of UF for
processing soy extracts, a number of researchers have investigated
its use in various vegetable protein systems, among them alfalfa
[5], cottonseed [6], faba beans [7], lupinus albus [8], sunflower
seed [9] and soybean systems [6,7,10,11,12,13,14,15]. This paper
describes some of the studies done on the production of lipid-protein concentrates and protein isolates from water extracts of soybeans using ultrafiltration.

EXPERIMENTAL

Water extracts of soybeans. The procedure for the manufacture
of full-fat water extracts of soybeans has been described earlier
[2,14]. Essentially it consists of 4 steps: soaking of beans to
soften them and make grinding easier, inactivation of lipoxygenase
enzyme (which otherwise would cause oxidized off-flavors in the
final product) by blanching at 90°C for 3 minutes, grinding hot for
5 min., and removing the insolubles (primarily polysaccharides, fiber
and some protein) by filtration or centrifugation. The filter cake
or residue can be reextracted with water to increase yields. The
mixed filtrate is used as feed to the ultrafiltration unit.

Water extracts of defatted soy flour. This is prepared in a manner similar to the above [6,15]. Defatted soy flour of high nitrogen solubility is suspended in dilute alkali at room temperature for 30 min. The insolubles are removed by filtration or centrifugation and the residue re-extracted again. The pH is adjusted to 7 for UF, if necessary.

Ultrafiltration. All experiments were performed with a pilot-scale hollow fiber unit (Romicon, Inc.). Unless otherwise mentioned, all data presented here was obtained with the XM50(45) membrane cartridge of 1.39 sq. meters area. The retentate was recycled until a volume concentration ratio (VCR) of 5 was obtained, i.e., a reduction of initial feed volume by 80%. The retentate was then rediluted with water to the original volume and the feed reultrafiltered until the desired purity and concentration was achieved. Details of the analytical methods and operation of the system are available elsewhere [2,14,15,16,17].

RESULTS AND DISCUSSION

Table 1 shows the proximate composition of the extracts used as feed to the UF unit, in relation to the raw material. In general, it has been observed that the best UF performance in terms of flux and final product composition is obtained when the feed has a total solids of 2.5-4.5%, pH at least 2 units away from the isoelectric point of the proteins, suspended particle size less than 100μ and initial viscosity 1-3 cp.

The operating variables that have the greatest bearing on the two major performance characteristics (flux and rejection) are (1) transmembrane pressure (2) flow rate or velocity past the membrane surface (3) temperature (4) feed concentration and composition (5) pH and (6) membrane pore size. The optimum feed concentration was governed essentially by the extraction process. Yields were only marginally improved by doing more than 2 extractions; using more than 1:20 solids to water ratio diluted the extracts unnecessarily. The upper limit for feed concentration is about 12-14% total solids for full fat soy extracts and 15-17% for the defatted soy extract system. This is apparently due to the physicochemical properties of the rejected components; higher retentate concentrations resulted in incipient gelation, low pumping rates, low flux and severe fouling of the hollow fiber unit.

In general it was found that operating below the isoelectric point resulted in higher flux [2]. However, the resulting product had poor functional characteristics and off-flavors. Operating at high pH (8-10) had no beneficial effect on flux, or rates of removal of oligosaccharides [2] or trypsin inhibitors [14], but did retard the removal of phytic acid (see later). Hence all subse-

Table 1. Composition of water extracts of soybeans and
defatted soy flour (%w/w)

1 part soy solids   +   10 parts water ⟶ Extract I ⟍
        ↓   Combined
Residue I   +   10 parts water ⟶ Extract II ⟋ Extract
        ↓
Residue II

Component	Soybean	Full-fat Extract	Defatted Soy Flour	Defatted Soy Extract
Total solids	90.0	3.56	9.0	4.05
Protein(N x 6.25)	39.0	1.73	52.0	2.75
Fat	21.6	0.94	1.0	–
Ash	4.2	0.19	5.0	0.30
Oligosaccharides	11.9	0.58	13.0	1.00
Phytic acid	1.3	0.06	(b)	(b)
Other[a]	12.0	0.06	20.0	–
Protein yield	–	83%	–	94%

[a] By difference: includes fiber, insoluble carbohydrate, etc.
[b] Not measured

quent processing was done at the normal pH of soybean water extracts,
pH 6.6-7.0.

At the time these experiments were done, three membrane pore
sizes were available: PM-10(10,000 molecular weight cut-off),
XM-50(50,000) and GM-80(80,000). The XM-50 membrane was found to
give the best balance between flux and rejection [17]. In addition,
it was also observed that the larger diameter fibers (45 mil diameter) gave significantly higher flux and less fouling problems with
soybean systems than the 20 mil diameter fibers.   This is probably
due to the higher flow rates that were possible with the larger
fibers for the same pressure drop within the constraints of the
overall system [17] which in turn resulted in less mass transfer
resistance.

The effect of the other three variables on flux is shown in
Fig. 1. The data show classic effects due to concentration polarization for the full fat soy extract feed. Flux became independent
of pressure fairly rapidly and there was a marked hysteresis effect
on lowering the pressure from the highest value to the lowest. Also

There was a strong dependency on both temperature and flow rate in
the pressure-independent region. Attempts were made to correlate
the data in terms of classic laminar flow mass transfer models [16].
as shown in Fig. 2. As has been shown with a number of other cases
in the literature, experimental flux was much higher than that predicted from theoretical considerations such as the Graetz-Leveque
solution for the model. The correlation that best fit the data for
the full fat soy extract - XM50(45) was

$$Sh = 0.181 \ (Re)^{.47}(Sc)^{.33}$$

where Sherwood numbers (100<Sh<240) correspond to a flux of 20-50
1/sq.m./hr, Reynolds numbers (500<Re<3000) are typical of hollow
fiber systems and Schmidt numbers (16000<Sc<55000) are typical of
protein systems [16].

The behavior of some of the individual components are shown in
Figs. 3, 4, and 5. The less than ideal recovery of proteins can be
partly explained as due to the permeation of low molecular weight
nitrogenous compounds [2,17] and partly due to adsorption by the
membrane. Adsorption of feed components by the membrane can result
in a significant loss in yields if volumes processed are low [14].

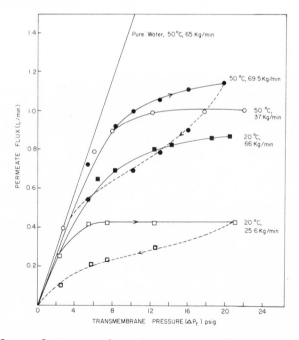

Fig. 1. Effect of transmembrane pressure, flow rate and temperature
on flux. (Arrows show direction of applied pressure.) [16]

Membrane adsorption is probably the reason for the apparent negative
rejections of sucrose and stachyose seen in Fig. 3. Ash components
exhibited a partial rejection indicating some degree of binding of
minerals by proteins.

Unlike the oligosaccharides which exhibited no apparent rejection by the membrane [2], neither trypsin inhibitors (Fig. 4) nor
phytic acid (Fig. 5) behaved as expected for a freely permeable,
non-interacting solute. With the former it is understandable since
trypsin inhibitors themselves are proteins of 8000–21000 in molecular
weight and hence would be substantially rejected by the XM50 membrane. Phytic acid and its known salts, however, have molecular
weights less then 1000, and its rejection by the membrane indicates
some kind of interaction between phytic acid and proteins, thus
retarding its removal by UF [3]. The existence of these complexes
is of great nutritional significance since it appears that these
complexed minerals are physiologically unavailable. The significance
of the pH effects shown in Fig. 5 has been explained in terms of the
nature of the phytate-protein complex under these environmental
conditions [3].

The production of soy isolates starting with extracts of de-

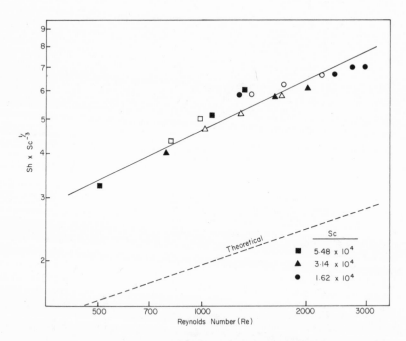

Fig. 2. Mass transfer correlation for XM50(45) hollow fiber-full
        fat soybean extract system. Theoretical line is based on
        Graetz-Leveque solution to mass transfer model. [16]

fatted soy flour has also been studied. It was observed [15,18]
that unlike the full fat system, at equivalent solids or protein
concentrations, concentration polarization effects were not flux-
controlling. Transmembrane pressure and temperature had significant
effects on flux, but flow rate had little or no effect. In addition,
little or no hysteresis effects were observed, unlike the full-fat
system. The combined effect of pressure and temperature could be
expressed in terms of a modified Poiseuille model for flow through
channels, $J = A (\Delta P_T/\mu)^n$, where μ is the viscosity and n is a
function of protein concentration, being lower at higher protein
concentrations, until the effects of concentration polarization
were obvious and mass transfer correlations such as those discussed
earlier were better representations of the data [15].

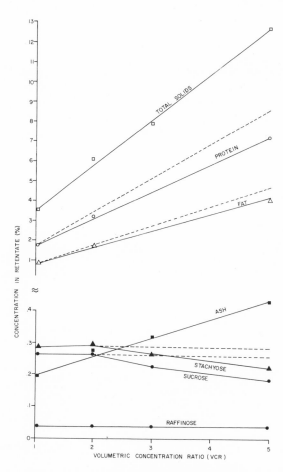

Fig. 3 Behavior of soybean components during ultrafiltration of full
       fat soybean extracts. (Broken lines indicate ideal behavior:
       100% rejection of proteins and fat, 0% rejection of sugars).

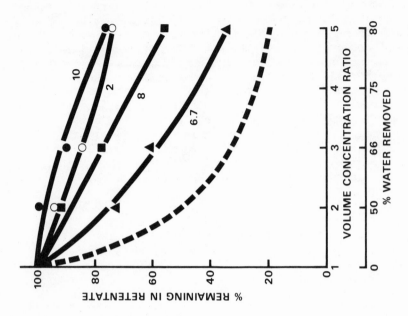

Fig. 5. Removal of phytic acid by ultrafiltration of full-fat soybean
water extracts. Variable is
processing pH [3].

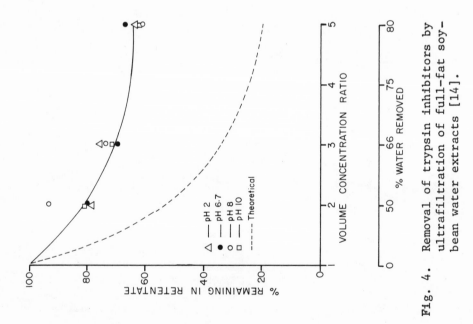

Fig. 4. Removal of trypsin inhibitors by
ultrafiltration of full-fat soybean water extracts [14].

Fig. 6 shows the behavior of the individual components during ultrafiltration of defatted extracts of soy flour. Removal of soluble carbohydrate resulted in an increase in protein content of the solids. Since severe fouling of the UF unit was experienced above 15-17% solids, it is necessary to wash the retentate solids with water by continuous or discontinuous diafiltration to improve protein content of the isolate.

CONCLUSIONS

The soybean products produced by ultrafiltration assayed typically 89% protein with defatted extracts and 60% protein, 34% fat and very low in undesirable oligosaccharides, phytic acid and trypsin inhibitor contents with full fat water extracts. Yields of protein were slightly higher than conventional isolation procedures due to the recovery of whey proteins. Due to the mild operating conditions, functional properties of these vegetable protein products are superior in some respects as compared to those produced by traditional means [19]. In addition, the economics may also be quite attractive [20].

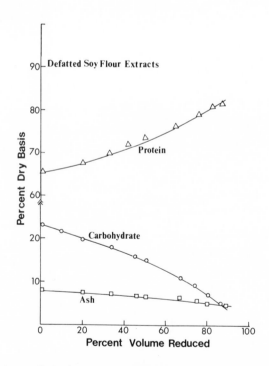

Fig. 6. Behavior of individual components during ultrafiltration of extracts of defatted soybean flour [15].

ACKNOWLEDGMENT

This research was partially supported by the Agricultural Experiment Station, University of Illinois, Urbana.

REFERENCES

1. G. N. Bookwalter, in "Nutritional Improvement of Food and Feed Proteins" M. Friedman, ed., Plenum Press, N.Y. (1978)
2. O. Omosaiye, M. Cheryan and M. E. Matthews, J. Food Sci., 43:354 (1978)
3. O. Omosaiye and M. Cheryan, Cereal Chem. 56:58 (1979)
4. M. C. Porter and A. S. Michaels, Chem. Tech. 1:633 (1971)
5. B. E. Knuckles, D. deFremery, E. M. Bickoff and G. O. Kohler J. Ag. Fd. Chem. 23:209 (1975)
6. J. T. Lawhon, D. Mulsow, C. M. Cater and K. F. Mattil, J. Food Sci. 42:389 (1977)
7. H. S. Olsen, Lebensm.-Wiss.u.-Technol. 11:57 (1978)
8. C. Pompei and M. Lucisano, Lebensm.-Wiss.u.-Technol. 9:338 (1978)
9. D. E. O'Connor, U. S. Patent 3,622,556 (1971)
10. D. R. Frazeur and R. B. Huston, U. S. Patent 3,728,327 (1971)
11. G. A. Iacobucci, D. V. Myers and K. Okubo, U. S. Patent 3,736,147 (1973)
12. C. Pompei, S. Maletto and M. Lucisano, Proc. IV Int. Congr. Fd. Sc. Technol. Madrid, Spain, Sept. 23-27 (1974)
13. K. C. Goodnight, H. H. Grant and R. F. Marquardt, U. S. Patent 3,995,071 (1976)
14. O. Omosaiye and M. Cheryan, J. Food Sci. 44:1027 (1979)
15. D. J. Nichols, M. S. Thesis, Univ. of Illinois, Urbana (1979)
16. M. Cheryan, J. Food Process Engineering, 1:269 (1977)
17. M. Cheryan and J. E. Schlesser, Lebensm.-Wiss.u.-Technol. 11:65 (1978)
18. M. Cheryan and D. J. Nichols in "Food Process Engineering 1979" Applied Sc. Pub., U. K. (1980) in press
19. C. L. Lah, M. S. Thesis, Univ. of Illinois, Urbana (1979)
20. D. W. Hensley and J. T. Lawhon, Food Technol. 33(5):46 (1979)

NEGATIVE REJECTIONS OF CATIONS IN THE

ULTRAFILTRATION OF GELATIN AND SALT SOLUTIONS

A. R. Akred,[+] A. G. Fane and J. P. Friend[+]

School of Chemical Engineering, University of
New South Wales, Australia and (+) Davis
Gelatine (Australia) Pty. Ltd., Botany, Australia

SUMMARY

Negative rejections of Calcium up to -400% and of Sodium up to
-700% have been measured in the ultrafiltration of gelatin solutions
containing these ions, using neutral membranes (DDS). Rejection
was found to be a function of gelatin concentration, pH, and transmembrane pressure with the cations exhibiting the strongest negative
rejection for high concentrations of gelatin, low pH and low values
of transmembrane pressure. However the effect was reduced by
operation in gel-polarised conditions (higher transmembrane pressures)
in which case the positively charged gel layer retarded the passage
of the cations, leading to positive rejection of the calcium ions.

A model, based on the convective and diffusive transport of
ions for normal ultrafiltration, coupled with Donnan membrane
theory, has been developed for the pre-gel polarised condition of
operation. This model successfully predicts the negative rejections
obtained experimentally.

INTRODUCTION

Gelatin solution can be readily concentrated by ultrafiltration
(1, 2). In an earlier feasibility study (2) it was shown that
dewatering of gelatin process liquors by ultrafiltration is cheaper
than by the more conventional multiple effect evaporation over the
concentration range 2 to 10 wt%. This study also showed that the
removal of ash (mainly calcium bisulphite) was greater than would
have been expected from applying ultrafiltration theory alone.

Since the removal of ash from the gelatin product is a requirement
(usually achieved by ion exchange) the 'super deashing' achieved in
ultrafiltration is an added bonus. This paper examines the conditions for negative rejection of ash and the mechanisms involved.

The phenomena of negative rejection have been reported in
several previous studies (Table 1). In addition several workers
have predicted its occurrence from theoretical considerations. In
general the studies may be divided into those with charged
membranes and those with neutral membranes.

Ion transport in reverse osmosis (hyperfiltration) has been
analysed theoretically by Dresner (9) who predicted the occurrence
of negative rejection of magnesium ions through a cation exchange
membrane, in chloride solution and at low pH. Hoffer and Kedem (10)
showed theoretically that negative rejection and anomalous osmosis
should operate together, and in a separate study (3) they observed
negative rejections of HCl and H_2SO_4 in the reverse osmosis of
copper salts at low pH. In contrast Yasuda et al. (11) measured
only positive rejections of NaCl in reverse osmosis with both
positively and negatively charged membranes. Similar effects have
been achieved with negatively charged ultrafiltration membranes,
which have shown positive calcium rejections up to 80% (12). Although
the results reported in the current work were obtained with neutral
primary membranes the situation is complicated by the probable formation of charged secondary membranes under some conditions.

Negative rejections have been obtained with neutral membranes
used for both reverse osmosis and ultrafiltration. Lonsdale et al.
(5) found negative rejections of -100% for Cl ions in reverse
osmosis in the presence of membrane-impermeable sodium citrate. They
successfully predicted their results using a model which coupled
Donnan membrane equilibrium with diffusive transport through the
membrane; convective solute transport was neglected because of the
relatively low solvent flux. Negative rejections of ions with
neutral ultrafiltration membranes have also involved the presence
of charged non-diffusible species, i.e. EDTA (7, 8), albumin (6),
or gelatin (2). In addition the Donnan membrane equilibrium theory
has been invoked to qualitatively explain the phenomena (2) (6).

EXPERIMENTAL

The equipment used was a Der Danske Sukker Fabrikker (DDS)
ultrafiltration module (20-0, 36-LAB, 20 cm diameter) with 20 membranes and a total membrane area of 0.36 m^2. This module was
incorporated in an ultrafiltration loop with a Rannie piston pump,
a 12 ℓ reservoir and a temperature controller capable of maintaining
conditions at ± 1°C. The membranes used were as follows:

Table 1.   Previously Reported Negative Rejections
in Pressure Driven Membrane Processes
(RO and UF) (N.R. = Negative Rejection)

Class	Process	Comments	Ref.
Charged Membranes	R.O.	N.R. (-5% to -15%) for HCl and H_2SO_4 through positive membrane.	3
	R.O.	N.R. (-10% to -27%) for phenol across charged cellulose acetate membranes.	4
	R.O.	N.R. for Cl^- ion in solution of NaCl and membrane-impermeable sodium citrate. -50% for weakly negative and strongly negative membranes; -100% for weakly positive membranes.	5
Un-charged Membranes	R.O.	N.R. (-100%) for Cl^- in solution of NaCl and sodium citrate.	5
	U.F.	N.R. (-20%) for Na^+ in solution of NaCl and albumin through cellophane and cuprophane membranes.	6
	U.F.	N.R. (-40%) for Cr^{6+} in solution of $CuCr_2O_7$ and EDTA through Amicon UM-2 membranes.	7,8
	U.F.	N.R. (-700% approx.) for 'ash' in solution with gelatin through Dorr Oliver Iopor membranes.	2

Membrane type A: DDS, GR8P, with cut-off 5,000 to 6,000
Membrane type B: DDS, GR6P, with cut-off 15,000 to 22,000

Both membranes were non-cellulosic co-polymer types and electrically neutral.

Experiments were made on gelatin production liquors and on artificial liquors. The production liquors had an initial composition of 3 to 4 wt % gelatin and ash (mainly calcium bisulphite) contents of 1 to 3% on gelatin (which is equivalent to 300 to 1000 mg/ℓ of solution). Adjustment of pH was made by addition of small quantities of HCl or NaOH. The artificial liquors were prepared by dissolving appropriate quantities of ash-free gelatin and calcium hydroxide in distilled water, with pH adjustment by HCl. One artificial liquor was prepared with equal quantities of calcium and sodium.

Variables adjusted during the series of experiments were pH (range 2.0 to 9.0) and transmembrane pressure (range 600 to 1800 kPa). Gelatin concentration was also varied over the range 2.0 to 23 wt % as the experiments were performed as batch-concentration runs. The operating temperature was 50°C. Following each experiment the membranes were cleaned with a proteolytic enzyme and rinsed with detergent solution. This procedure restored the membrane to its original water flux.

Concentrates and permeates were sampled and analysed frequently during each experiment. Determination of gelatin in the concentrates was by refractive index (at 35°C) and in the permeates by hydroxyproline analysis (13). Ash was determined on concentrates and permeates by evaporation to dryness followed by ashing in a muffle furnace at 550°C (14). Calcium and sodium concentrations were analysed by atomic absorption spectroscopy (15).

A separate series of measurements were made to evaluate the mass transfer coefficient for calcium through the GR6P membrane. Discs, 7 cm diameter, were cut from one of the membranes used in the ultrafiltration experiments and mounted in a stirred diffusion cell. This cell was located in a stirred water bath maintained at 50°C, and arranged so that calcium could diffuse from the cell chamber, through the membrane and into the (effectively) infinite volume of the bath. Calcium concentrations in the cell were measured as a function of time.

System Characteristics

Although the major aspect of the experimental work is concerned with ash/ion rejection, several other important characteristics of the system which were measured are summarised in Table 2. The gel concentration of 30 wt % is similar to that reported previously (1) (2), and the flux minima at about pH 5.0 is to be expected since this is close to the iso-electric point of gelatin proteins (16) (17).

Table 2. System Characteristics: Gelatin Solutions
and DDS   GR6P and GR8P Membranes

Gelatin rejection	GR6P 99.5% GR8P 99.7%
Gel concentration	$\sim$ 30 wt % gelatin (@ 50°C)
pH for minimum flux	$\sim$ 5.0 for 5 wt % gelatin $\sim$ 4.5 to 5.0 for 20 wt % gelatin

Pressure for gel-polarisation
(flux is pressure invariant)
	GR6P $\sim$ 700 kPa @ 4 wt %
	        $\sim$ 600 kPa @ 5 wt %
	GR8P $\sim$ 1400 kPa @ 3.5 wt %
	        $\sim$ 900 kPa @ 12 wt %

Mass transfer coefficient for Calcium and GR6P membrane at 50°C	2.8×10^{-3} m/hr (7.7×10^{-5} cm/sec)

RESULTS

The rejection of ash and separate ionic species may be reported
as the rejection coefficients,

$$\sigma_r = 1 - (C_2'/C_2) \tag{1a}$$

of $\sigma_\ell = 1 - (m_2'/m_2)$ (1b)

Where σ_r is based on molar (solution) concentrations and σ_ℓ based on
molal (solvent) concentrations. (Because the solvent is water the
molal concentration is defined as moles per litre of solvent.) In
most cases σ_r and σ_ℓ are identical or very similar, but when a
membrane-impermeable component is present in appreciable quantities
they will differ. The relationship between the two coefficients is,

$$\sigma_\ell = \sigma_r + C_V(1 - \sigma_r) \tag{2}$$

where C_V is the volume fraction of the membrane-impermeable species.
In this work ash and ion rejections are reported as σ_ℓ (solvent
basis, i.e. gelatin-free) so that zero rejection means equal ratios

of species to solvent in feed and permeate (the equivalent σ_r would be negative).

Ash rejections are also reported in terms of a batch performance ratio, R_p, which is defined by,

$$R_p = \frac{\text{ash removal factor}}{\text{concentration factor}} = \frac{x_{2,i}/x_{2,f}}{C_{1,f}/C_{1,i}} \qquad (3)$$

This ratio represents the performance over part or all of a batch concentration run, and is related to the 'batch averaged' rejection, $\bar{\sigma}_r$, by mass balance considerations. In general $\bar{\sigma}_r \propto (1-R_p)$, so that $\bar{\sigma}_r$ is zero for $R_p = 1.0$, is positive for $R_p < 1.0$ and is negative for $R_p > 1.0$ ($\bar{\sigma}_\ell$ varies in a similar fashion, modified by equation (2)).

The batch performance ratio is introduced because it requires ash analyses of the concentrate only; ash analyses of permeates are subject to considerable error. In addition, the ratio is of practical significance in the gelatin industry (21).

Figures 1 and 2 show the results for the production liquor in terms of the batch performance ratio, R_p, as a function of gelatin concentration, pH, transmembrane pressure drop and membrane type.

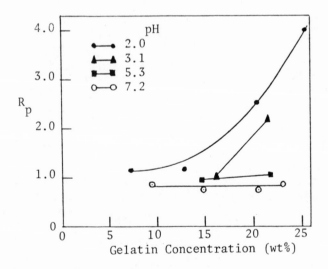

Fig. 1.   Effect of Gelatin concentration and pH on the batch
          performance ratio, R_p, for GR6P membranes and production
          liquor at ΔP of 940 kPa
          (R_p = (ash removal factor)/(concentration factor))

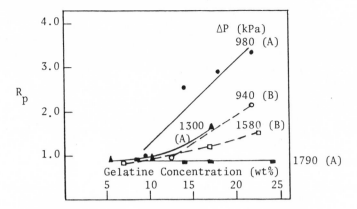

Fig. 2. Effect of Gelatin concentration and transmembrane pressure
drop (ΔP) on the batch performance ratio, R_p.
(A = GR8P membrane, B = GR6P membrane)

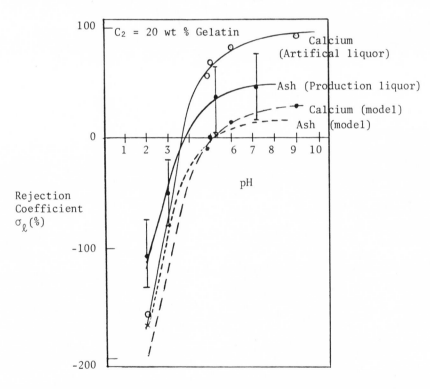

Fig. 3. Effect of pH on Rejection (σ_ℓ) for Calcium (artificial
liquor) and Ash (production liquor) with GR6P membranes
and ΔP of 950 kPa.

For pH values below about 5.0, $R_p > 1.0$ which indicates negative rejection of ash; the effect becomes more pronounced as gelatin concentration increases. Positive ash rejection is favoured by pH values above about 5.0. In addition, Figure 2 shows that rejection becomes less negative ($R_p \rightarrow 1.0$) and less sensitive to gelatin concentration as the applied transmembrane pressure is increased. Indeed at a pressure of almost 1800 kPa with the GR8P membrane the performance ratio is approximately 1.0 and independent of gelatin concentration in spite of the low pH. Figure 2 also shows that for similar operating conditions the GR6P membrane gives smaller values of R_p, which is equivalent to less negative ash rejection.

The effect of pH on ash rejection (equation (1b)) is shown in Figure 3; the error bars on the rejection values are a result of the imprecision of the ash analyses. However the Figure shows the dramatic influence of low pH, zero rejection at about pH 4.0, and significant positive rejections at high pH values.

Figure 3 also includes the more exact data obtained for calcium rejection with artificial liquors. The results follow a very similar trend to those observed for the ash rejections. It is of interest to note that zero rejections for both calcium and ash occur at about pH 4.0, and at the iso-electric point of the gelatin (pH $\sim$ 5.0) the rejections are significantly positive.

The effect of pH at different gelatin concentrations in artificial liquor is shown in Figure 4. These results fit the pattern already presented except for the apparently anomalous rejection value at the highest gelatin concentration and pH 2.0.

Figure 5 clearly demonstrates the interaction between trans- membrane pressure drop, gelatin concentration and calcium rejection at pH 2.0. The results are in qualitative agreement with Figure 2, and show negative rejections as great as - 380% at the lowest pressure and highest gelatin concentration.

Sodium rejections are considerably more negative than the calcium rejections (Figure 6) and they are rather less sensitive to changes in applied transmembrane pressure. The most negative rejections for sodium ions are about - 700%.

DISCUSSION

The results may be explained qualitatively in terms of the Donnan effect, a phenomena which has been invoked in several related studies to explain negative rejections, (2, 5, 6). In this study the most important agent is the gelatin molecule which is membrane- impermeable and significantly charged at the extremes of pH.

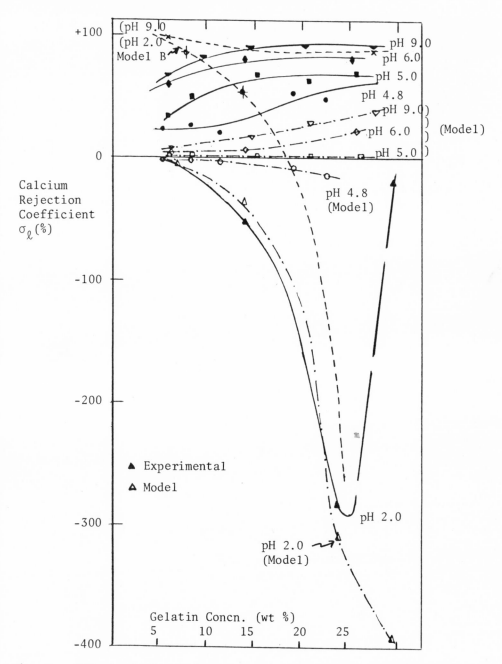

Fig. 4. Effect of pH and Gelatin Concentration on Rejection (σ_ℓ)
for calcium (artificial liquor) with GR6P membranes and
ΔP of 950 kPa.

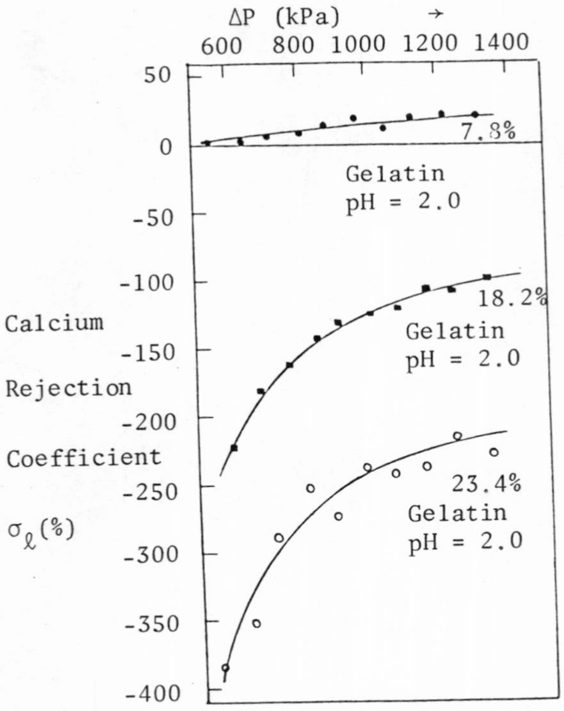

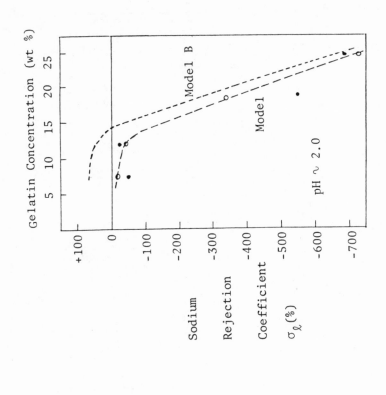

Fig. 6. Effect of Gelatin concentration on rejection (σ_ℓ) for sodium (artificial liquor) with GR6P membranes and P of 1000 kPa.
(● experimental data o model data)

Fig. 5. Effect of Gelatin concentration and transmembrane pressure drop (ΔP) on calcium rejection coefficient (GR6P membrane and artificial liquor)

Figure 7 shows the estimated variation of charge with pH for gelatin proteins.

However the Donnan effect is modified in this situation since there is significant solvent (water) flux (typically, 18 to 4 ℓ/m^2hr for the GR8P over the concentration range 3 to 20% gelatin and 60 to 10 ℓ/m^2hr for the GR6P over the same range). In addition the Donnan theory assumes the membrane-impermeable agent to be homogeneously dispersed throughout the upstream solution, whereas in ultrafiltration the phenomena of concentration polarisation can lead to the formation of a gel layer on the primary membrane which acts as a secondary membrane. This membrane will have its own solute rejection characteristics, and in the case of the gelatin it may be charged.

Viewing these experiments as ultrafiltration modified by Donnan effects provides the following explanation of the results. In the absence of primary and secondary membrane influences the ash and calcium would be expected to be carried by convective transport through the membrane. Rejection coefficients of zero and batch performance ratios of unity would be anticipated. However the presence of significant amounts of positively charged membrane-impermeable solute on the upstream side of the membrane would enhance the cation transport as the system attempts to establish a Donnan membrane equilibrium. This would result in permeates enriched in the

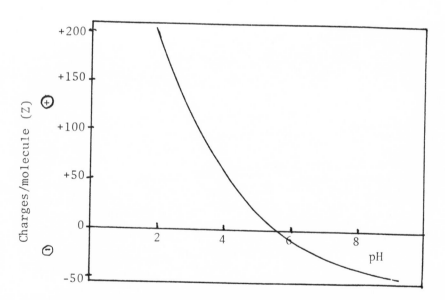

Fig. 7 Estimated variation of charge of gelatin molecule with pH

cations (and appropriate amounts of counter-ions to maintain electroneutrality). Thus negative rejections would be expected to be
strongest at high gelatin concentrations and when the gelatin carries
the most positive charges, namely at low pH (see Figures 1 to 6).

Since convective solute flux tends to dilute the Donnan effect
the negative rejections would be expected to be greater for the lower
flux membrane (all other factors being equal). Figure 2 supports
this reasoning with the GR8P membrane giving larger R_p values (more
negative rejection) at the same transmembrane pressure drop. However
the situation is complicated by the fact that the GR8P is less 'gel-
polarised' for a given transmembrane pressure drop (Table 2).

The way in which gel-polarisation tends to change the situation
is illustrated by the apparently anomalous point on Figure 4 at the
highest gelatin concentration and pH 2.0. Although at these
conditions the Donnan effect would be at its strongest and the
solvent flux would be at its lowest (bulk gelatin concentration is
approaching the gel concentration) the solute transport is enhanced
to a smaller degree than for less favourable conditions. However
it is probable that at the highest gelatin concentrations a gel-
polarised secondary membrane starts to exert a dominant role.
Assuming this membrane to be positively charged it will hinder
cation transport by Donnan exclusion. In other words the enhanced
transport resulting from a Donnan membrane equilibrium effect is
being counteracted by both convective flux dilution and Donnan
exclusion.

Similar reasoning accounts for the fact that negative rejections
are most pronounced at the lowest transmembrane pressures (Figures 2
and 5), where the degree of gel-polarisation is least and the
secondary membrane (if present) provides less of a barrier to cation
transport.

The results for sodium rejection are at first sight unexpected.
Donnan equilibrium theory would predict less enhancement of the
monovalent sodium ion compared with the divalent calcium ion.
However this effect is compensated for by two factors. Firstly, the
diffusion coefficient for sodium is approximately 30% greater than
that for calcium in aqueous solution (18), and a similar difference
could be expected for diffusion through the neutral primary membrane.
Secondly, the sodium has a lower charge than calcium but a similar
ionic radius (19), giving charge densities (Z/r^3) in the ratio of
2:1 for calcium and sodium. Consequently the calcium can expect to
be more positively rejected through Donnan exclusion by a
positively charged secondary (gel) membrane. The fact that sodium
rejection is less sensitive to transmembrane pressure supports the
view that the gel-polarised secondary membrane has a greater
influence on the calcium than on the sodium.

At pH values above about 4.0 the calcium rejections are positive
(up to 90%) and relatively insensitive to gelatin concentration. This
is explained by the strong tendency (20) of the calcium to bind to
the negatively charged carboxylic groups on the gelatin in solution.
In addition the secondary (gel) membrane will be negatively charged
and this can lead to positive rejection (by Donnan exclusion) of
the anions and corresponding retention of the cations to maintain
electroneutrality. Similar positive rejections (up to 90%) for
calcium have been found for the ultrafiltration of calcium chloride
solutions through negatively charged primary membranes (12). The
fact that positive rejections occur at the iso-electric point of
gelatin may also be attributed to calcium binding, since although
the net charge is zero there will still be a significant amount of
ionised carboxylic sites available.

Model for Donnan-modified Ultrafiltration

Previous models for Donnan-modified reverse osmosis (5) and
ultrafiltration (6) have neglected the effect of convective solute
transport on solute rejection. This assumption is reasonable for
the low (solvent) flux membranes considered ($\sim$ 5 ℓ/m^2hr (5),
$\sim$ 0.2 ℓ/m^2hr (6)), but is not acceptable for ultrafiltration with
high flux membranes (in the range 10 to 60 ℓ/m^2hr for this work).

In the simple model described below the transport of calcium
through the membrane is assumed to have two components, convective
transport due to solvent coupling and diffusive transport resulting
from a driving force caused by the Donnan effect; the two
components are assumed to be additive. This approach is consistent
with requirements that at very high solvent flux the solute transport
will be governed by convective transport and that at very low flux
it will be governed by the Donnan effect. Thus solute transport
through the membrane may be written,

$$J\ m_2 + \text{Donnan Diffusion} = J\ m_2' \qquad\qquad (4)$$

where,

$$\text{Donnan Diffusion} = B(m_2^* - m_2') \qquad\qquad (5)$$

In equation (5) the term B is the membrane mass transfer coefficient
for the ionic species considered. The driving force for diffusion
is assumed to be the difference between the permeate concentration
given by Donnan equilibrium theory, m_2^*, and the actual permeate
concentration, m_2'. In effect it assumes that at equilibrium (no
solvent flow) the ion concentration undergoes a step change at the
feed surface of the membrane.

From equation (1b)

$$m_2^{'} = (1 - \sigma_\ell)m_2 \tag{6}$$

Combining equations (4) (5) (6) and re-arranging gives,

$$\sigma_\ell = (\frac{B}{B + J}) (1 - \frac{m_2^*}{m_2}) \tag{7}$$

Noting that for dilute solutions without the presence of gelatin (i.e. within the membrane or in the permeate) the molal and molar concentrations are approximately equal,

$$m_2^* = C_2^* \tag{8}$$

Whereas in the feed solution allowance must be made for the gelatin fraction, so that

$$m_2 = C_2/(1-C_V) \tag{9}$$

where C_V is the volume fraction of the gelatin in solution.

Substituting (8) and (9) in (7) gives,

$$\sigma_\ell = (\frac{B}{B + J})(1 - (1-C_V) \frac{C_2^*}{C_2}) \tag{10}$$

The equilibrium concentration C_2^* may be obtained from Donnan equilibrium theory, assuming activity coefficients of unity. For divalent ions, such as calcium, (2) (20)

$$C_2^* = (C_2 (2C_2 + ZC_1)^2/4)^{1/3} \tag{11}$$

So that,

$$\sigma_\ell = (\frac{B}{B + J})(1 - (1-C_V) \{1 + \frac{ZC_1}{C_2} + \frac{Z^2C_1^2}{4C_2^2}\}^{1/3}) \tag{12}$$

Similarly for monovalent ions, such as sodium (2)

$$C_2 = (C_2^2 + C_2 ZC_1)^{\frac{1}{2}} \tag{13}$$

So that,

$$\sigma_\ell = (\frac{B}{B + J}) (1 - (1-C_V)\{1 + \frac{ZC_1}{C_2}\}^{\frac{1}{2}}) \tag{14}$$

Equations (12) and (14) satisfy the requirements that rejection (σ_ℓ) is zero for J >> B, and Z = 0. When J approaches zero the rejection tends to the value equivalent to the Donnan membrane equilibrium situation.

Rejection coefficients (σ_ℓ) may be calculated from equations (12) and (14), for specified concentrations of calcium (or sodium) and gelatin on the feed side of the membrnae, given the solvent flux (J), the mass transfer coefficient (B), and the charge (Z) and molecular weight of the gelatin.

The mass transfer coefficient for calcium was measured experimentally (Table 1) and the value for sodium was assumed to be 1.3 x the calcium value (18). Gelatin properties vary depending on the raw material and method of production. A molecular weight distribution was obtained by gel permeation chromatography followed by ultracentrifugation (21), from which the mean molecular weight was estimated to be 160,000. The charge on the gelatin molecule as a function of pH was estimated from a knowledge of the type and number of amino acids typically present (16) and the pKa values for the carboxylic and amine groups on the individual acids (22, 23). Figure 7 shows the estimated variation of charge with pH; details of the calculations may be found elsewhere (21). This method of calculation suggests an iso-electric point at pH 5.5 which is slightly higher than the reported value of pH 5.0 (16). Inspection of equations (12) and (14) indicates that the predicted rejection coefficients are relatively insensitive to uncertainties in the value of Z.

Predicted rejection coefficients for ash, calcium and sodium are included on Figures 3, 4 and 6. In order to calculate the coefficient for ash it is assumed that the ash is only in the form of calcium bisulphite $Ca(HSO_3)_2$. The model is seen to predict the negative rejections reasonably well, but with a tendency to overestimate the magnitude of the negative effect at high concentrations of gelatin. This is thought to be caused by the influence of the gel polarised 'secondary' membrane, the effect of which is not accounted for in the model.

As the pH is increased, and rejections become more positive, the discrepancy between the model and the experimental data increases. This is probably caused by the growing tendency of the calcium to bind to the carboxylic groups, a complicating factor not handled by the model.

Figures 4 and 6 also include predictions based on a model which excludes the effect of convective solute transport, i.e. the term J m_2 in equation (4) is taken to be negligible. This is equivalent to the model of Lonsdale et al. (5) for Donnan effects in reverse osmosis. The appropriate equations are,

(a) Model B, for calcium

$$\sigma_{\ell B} = (1 - (\frac{B}{B + J})(1 - C_V) \{1 + \frac{ZC_1}{C_2} + \frac{Z^2 C_1^2}{4C_2^2}\}^{1/3}) \qquad (15)$$

(b) Model B, for sodium

$$\sigma_{\ell B} = (1 - (\frac{B}{B + J})(1 - C_V)\{1 + \frac{ZC_1}{C_2}\}^{\frac{1}{2}}) \qquad (16)$$

As would be expected, these models are most satisfactory for the conditions of lowest flux, namely as gelatin concentration increases. The most serious discrepancy is that the model predicts positive rejection, approaching 100%, for pH 2.0 conditions at low gelatin concentration (high flux conditions). For this reason the original models (equations (12) and (14)) are judged to be more appropriate for Donnan effects using high flux ultrafiltration membranes.

CONCLUSIONS

Donnan effects can play an important role in the ultrafiltration of solutions containing membrane-impermeable charged proteins (such as gelatin) and membrane-permeable ionic species (such as calcium and sodium). For the case of gelatin and calcium ultrafiltered through neutral membranes negative rejections of calcium occur below pH 4.0, and values of at least -380% rejection are attainable.

In accordance with Donnan theory most negative rejections are obtained at the lowest pH and the highest concentrations of gelatin. However the Donnan-enhanced transport is increasingly counter-balanced as the membrane becomes more gel-polarised (with increase in transmembrane pressure drop and at very high gelatin concentrations).

The dependence of rejection of ash, calcium and sodium on pH and gelatin concentration can be predicted by a model which couples conventional ultrafiltration theory with Donnan equilibrium theory. This model is most useful for conditions which lead to negative rejections in pre-gel polarised conditions.

ACKNOWLEDGEMENTS

The authors would like to acknowledge helpful discussions with Associate Professor C. J. Fell and Dr. M. S. Lefebvre. They would also like to thank Davis Gelatine Pty. Ltd. for the opportunity to carry out this collaborative study and for permission to publish the findings.

NOMENCLATURE

B	=	membrane mass transfer coefficient (cm/sec)
C	=	molar concentration of solute (moles/ℓ of solution)
C_V	=	volume fraction of membrane-impermeable species in solution
J	=	solvent flux (ℓ/m^2.hr)
m	=	molal concentration of solute (moles/ of solvent)
R_p	=	batch performance ratio (equation (3))
x_2	=	concentration of ash per unit of gelatin solids
Z	=	charge per gelatin molecule
$\sigma_\ell \sigma_{\ell B}$	=	solute rejection coefficient based on molal concentrations (equation 1b), and predicted by Model B (equations (15) and (16))
σ_r	=	solute rejection coefficient based on molar concentrations (equation (1a))

Subscripts

1   Gelatin
2   Calcium (sodium)
f   final condition
i   initial condition

Superscripts

$'$ permeate
$*$ equilibrium value

REFERENCE

1   M. C. Porter and A. S. Michaels, "Membrane Ultrafiltration", Chem. Tech. July (1971) p445.

2   A. G. Fane and J. P. Friend, "Dewatering of Gelatine Liquor by Ultrafiltration" 5th Australian Conference on Chemical Engineering, Canberra, September (1977) pp 203-207.

3   E. Hoffer and O. Kedem, "Negative Rejection of Acids and Separation of Ions by Hyperfiltration", Desalination, 5:167-172 (1968)

4   H. Lonsdale, U. Merten and M. Togami, Journal of Applied Polymer Science, 11:1807 (1967).

5   H. K. Lonsdale, W. Pusch and A. Walca, "Donnan-membrane Effects in Hyperfiltration of Ternary Systems', J. Chem. Soc. Faraday Trans. 1, 71, 3:501-504 (1975)

6   K. D. Nolph, M. L. Stoltz and H. F. Maher, "Electrolyte Transport
     During Ultrafiltration of Protein Solutions", Nephron, 8:473-487
     (1971).

7   T. L. O'Neill, G. R. Fisette and E. E. Lindsey, "Separation of
     Metal Ions from Water by Chelation and Ultrafiltration : Part A"
     Water, A.I.Ch.E. Symp. Series, 71, 151:100-104 (1975).

8   A. Bulba, E. E. Lindsey, R. D. Archer and R. W. Lenz, "Separation
     of Metals from Water by Chelation and Ultrafiltration : Part B.
     Negative Rejections of Chromates", Water - A.I.Ch.E. Symposium
     Series, 71, 151:105-110 (1975).

9   L. Dresner, "Some remarks on the Integration of the Extended
     Nernst-Planck Equations in the Hyperfiltration of Multicomponent
     Solutions", Desalination, 10:27-46 (1972).

10   E. Hoffer and O. Kedem, "Hyperfiltration in Charged Membranes",
     Desalination, 2:25-39 (1967).

11   H. Yasuda, C. Lamaze and A. Schindler, "Salt Rejection by Polymer
     Membranes in Reverse Osmosis II. Ionic Polymers", Journal of
     Polymer Science Part A-2, 9:1579-1590 (1971).

12   D. Bhattacharya, J. M. McCarthy and R. B. Grieves, "Charged
     Membrane Ultrafiltration of Inorganic Ions in Single and Multi-
     Salt Systems", A.I.Ch.E. Journal, 20, 4:1206-1212 (1974).

13   Davis Gelatine Analytical Methods of Laboratory Testing - 1977
     as developed from Leaca, A.A., B.G.G.R.A. Res. Rep. B8 March,
     1957, R. M. Lollar, J.A.L.C.A., 53:2 (1958).

14   British Standard 757: 1974, Sampling and Testing of Gelatine.
     British Standards Institution.

15   P. Fodor, L. Polds and E. Pungor, "Determination of Calcium,
     Magnesium, Iron and Zinc, in Protein Concentrates by Atomic
     Absorption Spectrophotometry", Periodica Polytech. Chem. Engng.,
     18, 2:125-137 (1974).

16   A. G. Ward and A. Courts, "The Science and Technology of
     Gelatin", Academic Press Inc. London (1977).

17   A. G. Fane and C. J. D. Fell, "Recovery of Soluble Protein from
     Wheat Starch Factory Effluents", A.I.Ch.E. Symposium Series,
     73, 163:198-205 (1977).

18   R. C. Weast. ed., "Handbook of Chemistry and Physics", 54th
     edition 1973-1974, C.R.C. Press (1973).

19   G. H. Aylward and T. I. V. Findlay. eds., "Chemical Data Book", second ed. P.40, John Wiley & Sons, Sydney (1968).

20   S. Glasstone, "Textbook of Physical Chemistry", 2nd edition p 1252, Macmillian & Co. Ltd., London (1960).

21   A. R. Akred, "The Ultrafiltration of Mixed Solutes", M.App.Sci. Thesis, U.N.S.W. (1979).

22   H. R. Mahler and E. H. Cordes, "Biological Chemistry", pp 10-15, Harper and Row Publishers, New York (1966).

23   T. Matsuura, S. Sourirajan, "Reverse Osmosis Separation of Amino Acids in Aqueous Solutions using Porous Cellulose Acetate Membranes", Journal of Applied Polymer Science, 18:3593-3620 (1974).

THE APPLICATION OF ULTRAFILTRATION TO FERMENTATION PRODUCTS

N. C. Beaton

Dorr-Oliver, Inc.
77 Havemeyer Lane
Stamford, Conn.   06904

ABSTRACT

The industrial-scale application of ultrafiltration to the concentration and purification of products such as enzymes, amino-acids and antibiotics is reviewed. Processing techniques, such as diafiltration, for exercising control over product purity and product recovery are described. Special features in ultra-filtration plant and process design, which are necessary in the processing of certain enzyme liquors or fermentation broths, are considered. Finally, the economics of ultrafiltration are discussed in relation to the costs of frequently used competitive unit operations.

INTRODUCTION

Fermentation processes utilizing microorganisms such as yeasts, fungi and bacteria are the prime source of a wide variety of products for the food, beverage, chemical and pharmaceutical industries. The more common products include ethanol-based beverages (wines, beers and distilled spirits), organic acids (citric, lactic, glutamic and gluconic acids), antibiotics (penicillins, streptomycins and cephalosporins), enzymes (proteases, pectinases and amylases), steroids, and vitamins. Valuable products are also extracted from many industrial waste streams using fermentation technology. The stabilization of sewage and industrial effluents by means of the activated sludge process, is a further example of complex microbial action.

In many operations involving fermentation, the first

essential step in isolating the desired product is to separate
the microbial cells or cell debris from the broth. Operations
such as sedimentation, filtration or centrifugation would
typically be employed. The second step often involves the
purification of product in the clarified liquor, utilizing
processes such as crystallization (precipitation), ion-exchange
and carbon adsorption, with the product being recovered by
evaporation and drying. Finally, for certain pharmaceutical
products, a third or finishing step may be required, in which
the product is formulated into a sterile, pyrogen-free solution.

The separation process, ultrafiltration, can be used in
several such clarification, purification, recovery and finishing
operations, and it is the purpose of this article to review the
feasibility of applying this membrane process in a number of
industrial scale situations.

ULTRAFILTRATION BACKGROUND

Principles

Ultrafiltration has now been established for some ten years
as an industrial unit operation. The process utilizes
selectively - permeable membrane filters to separate the components of solutions and suspensions at the molecular size
level. Usually, the fluid streams are aqueous. Water, together
with solutes to which the porous membrane is permeable, flow
through the membrane when a trans-membrane pressure gradient is
applied. The liquid passing through the membrane is termed the
ultrafiltrate, or permeate, while the liquid retained on the
high-pressure side of the filter is called the retentate or
concentrate.

The dimensions of particles which can be retained by ultrafiltration membranes range from about 1.5 nm upwards. Separations
requiring the retention of species with diameters less than 1.5 nm,
which corresponds to a molecular weight of about 1,000, are considered to be applications of the related membrane process,
reverse osmosis. The upper limit in the size of particles to
which ultrafiltration is normally applied is generally taken to
be about 1-10 μm. Typical dimensions of small particles,
molecules and ions encountered in the fermentation industries
are detailed in Table 1. For practical purposes, membranes are
normally characterized according to their degree of retention
of a range of solutes of known molecular weight. A membrane
which, for example, retains over 95% of molecules of a particular
type (e.g., proteins) above 10,000 in molecular weight, would
be classified as having a 10,000 molecular weight "cut-off" or
exclusion limit.

TABLE 1. TYPICAL APPARENT DIMENSIONS OF SMALL
PARTICLES, MOLECULES AND IONS

SPECIES	RANGE OF DIMENSIONS (nm)
Suspended solids	10,000 - 1,000,000
Colloidal solids	100 - 1,000
Smallest visible particles	25,000 - 50,000
Oil emulsion globules	100 - 10,000
Bacterial cells	300 - 10,000
Yeasts and fungi	1,000 - 10,000
Viruses	30 - 300
Proteins/polysaccharides (m. wt. 10^4 - 10^6)	2 - 10
Enzymes (m. wt. 10^4 - 10^5)	2 - 5
Common antibiotics (m. wt. 300 - 1000)	0.6 - 1.2
Organic acids (m. wt. 100 - 500)	0.4 - 0.8
Mono- and di-saccharides (m. wt. 200 - 400)	0.8 - 1.0
Inorganic ions (m. wt. 10 - 100)	0.2 - 0.4
Water (m. wt. 18)	0.2

Most commercially available ultrafiltration membranes have
effective pore diameters within the range 2-10 nm, corresponding
to molecular weight exclusion limits in the 10^4-10^6 range. It
will be apparent that ultrafiltration membranes are permeable to
small water-soluble organic molecules (e.g., sugars, amino-
acids, carboxylic acids, alcohols, etc.) and to inorganic
ions (e.g., chloride sulphate, sodium, calcium, phosphate, etc.).
In general, three types of separation can be achieved:

(a) concentration of colloidal particles or soluble macro-
    molecules

(b) concentration of colloids or macromolecules, with simul-
    taneous separation of small molecules or ions

(c)   separation of colloids or macromolecules from small
      molecules or ions.

In conventional ultrafiltration, the membrane-retainable
components of the feed stream are concentrated, while freely-
permeable solids are transported, unchanged in concentration,
into the permeate stream. However, this transport rate, or
flux, decreases as the retentate concentration increases. The
degree of separation which is economically attainable with
direct ultrafiltration is therefore limited, unless the concentration process is terminated by the addition of water to the
retentate stream.

The ultrafiltration process whereby water (or another
fluid) is added to a retentate stream at a rate equal to or less
than the rate at which permeate is being withdrawn, is termed
diafiltration. The purpose of diafiltration, which is analogous
to cake washing in filtration operations, is either to improve
the recovery of a membrane-permeable product, or to increase the
purity of a membrane-retainable species. The higher separation
efficiencies resulting from diafiltration are, however, attained
at the expense of some dilution of the permeate stream.

The heart of all ultrafiltration systems is the selective
membrane used to effect the separation. The majority of ultrafiltration membranes are based on synthetic film-forming organic
polymers, such as polyamides, polysulphones, polyacrylonitrile
copolymers, or cellulose derivatives. The membranes themselves
consist of an integral, asymmetric two-layer structure. The side
of the membrane in contact with the pressurized feed solution
consists of a skin-layer, about 0.2 μm in thickness, which contains the fine pores at which separation takes place. The
remainder of the membrane thickness consists of a coarsely
porous sub-layer, which supports the skin, giving a total thickness of about 100 μm. In referring to membrane pore diameters,
it is the diameter of pores in the skin or active layer which is
implied. Apart from having suitable selectivity, industrial
membranes are also capable of extended operation over periods
ranging from 1-3 years. Under defined conditions of pressure
and flow specific to a particular design configuration, continuous operation within the pH range 2-12, at temperatures up
to 60°C, is generally permissible.

In addition to the requirement for adequate membrane life,
the filtration rate, or permeate flux, attained during ultrafiltration, must be sufficiently high to make the process
economically competitive. Special operating conditions are
required to maintain this filtration rate. When a static process
fluid is pressurized in contact with a membrane, a thick boundary
layer of retained substances will form, which reduces flux to a

very low level. To overcome this, the process fluid is forced
to flow across the membrane surface, frequently under conditions
of high turbulence. In this way boundary layer build-up is
limited, and flux, while not as high as for water, is sufficiently
high to make the separation process economically feasible. It
is in this requirement for flow to overcome so-called "concentration
polarization" effects, that the major energy requirement of the
process arises. Filtration capacity thus becomes dependent on
energy input. However, the capacity of a membrane system can be
increased both by increasing membrane area and by increasing
energy expenditure. Both approaches have associated costs, and
an economic study is required to determine the optimum flow
conditions in any equipment design.

The dependence of flux on operating pressure is related to
relative hydraulic resistances of boundary layers and membrane.
In the majority of applications, the boundary layer resistance
controls flux. In such cases, any increase in flux associated
with increased pressure is immediately counteracted by an increase in the boundary layer resistance. The result is that
flux often becomes virtually independent of the applied pressure
above about 100 kPa. Temperature increase, by comparison, is
much more effective in improving flux due to a decrease in fluid
viscosity, which leads to a reduction in boundary layer thickness.

Flux decreases with time in a manner which depends on the
extent to which surface formation of boundary layers is hydrodynamically reversible. The rate of flux decay is very variable,
depending on the nature of the species present. The deposition
of boundary layers which cannot be removed by turbulence, or
fouling, as the deposition phenomenon is normally termed, is
controllable by chemical cleaning processes. The frequency of
cleaning for regeneration of flux can range from daily to yearly
intervals. It depends to some extent on how much provision
is made for flux decay effects at the design stage.

Flux decreases as the concentration of membrane-retained
solids increases, due to the increasing viscosity of the fluid
and the concomitant increase in boundary layer thickness or
density, which creates increased resistance to permeate flow.
In each application, limits are therefore imposed on the maximum
attainable concentrations, for reasons which are both technically
and economically based. In applications where the required
product is the filtrate, it is apparent that the recovery which
can be realized without recourse to diafiltration depends directly
on the maximum allowable retained solids content.

When flow conditions, etc. are optimized, ultrafiltration
processes normally operate within the flux range 1-100

litres/m^2/hour, at pressures between 100 and 600 kPa, and
at temperatures up to 60°C. At a flux of less than about
1 litre/m^2/hour, the process will become prohibitively expensive
to operate. The maximum flux is determined by the intrinsic
membrane water flux which will lie in the range 100-500
litres/m^2/hour. As indicated above, process flux is generally
considerably less than water flux due to the formation at the
membrane surface of boundary layers which possess hydraulic
resistance.

Industrial Equipment and Process Design

The requirement for flow across the surface of ultrafiltration
membranes has already been identified. The various industrial-
scale ultrafiltration equipment design concepts differ mainly
in the size and shape of the channels in which such flow takes
place. The commoner designs include plate-and-frame stacks,
membrane-lined, pressure-sustaining porous tubes, and self-
supporting hollow-fibres (0.5-1.0 mm i.d.). In other designs,
flat-sheet membranes are incorporated either into spirally-
wrapped elements, or into flat-leaf cartridges. Such elements
or cartridges are contained inside separate external pressure-
sustaining housings. In most designs, the basic elements from
which complete ultrafiltration systems are constructed are
referred to as modules.

The required cross-flow velocity in ultrafiltration modules
can be attained independently of the processing capacity of the
system by means of recirculation pumping loops. Batch and
continuous ultrafiltration flow schemes incorporating such
recirculation loops are shown in Fig. 1. Process fluid is
introduced into the recirculation loops by means of a separate
feed pump.

Ultrafiltration systems can be operated either on a batch
or on a continuous basis. For a given membrane area, the
process mode selected will affect the capacity of the system.
This occurs because the membrane is required to process fluid
under different concentration conditions during the course of
the operation. Since filtration rate (flux) decreases with
increasing retentate concentration, it is advantageous to
operate under conditions such that the process fluid is
filtered at the lowest concentration for the longest possible
time.

In a batch system (Fig. 1), fluid is withdrawn continuously
from a tank, introduced into a recirculation loop, and recycled
back to the tank. The concentration of solids in the whole
system increases with time as the volume of retentate decreases.
The feed rate is maintained high in relation to the permeation

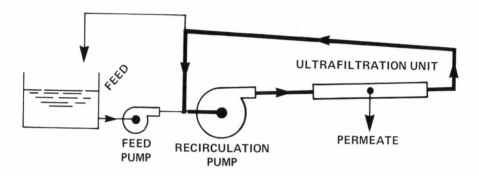

BATCH PROCESS MODE

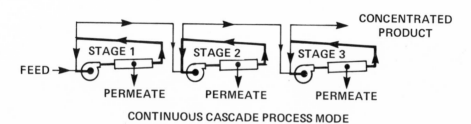

CONTINUOUS CASCADE PROCESS MODE

Fig. 1. Ultrafiltration process modes

rate to ensure that the concentration in the recirculation
loop is not significantly greater than that in the tank.
Typically the ratio of feed to permeate flow rates is in the
range of 10-20. Filtration continuously takes place at the
lowest possible retentate concentration, and the maximum time-
average process flux is attained.

Continuous processing may be accomplished in either single-
stage or multi-stage (cascade) systems. In a single-stage system,
the fluid to be concentrated (feed) is introduced to the recirculation loop by means of a feed pump. Concentrate is bled
from the system at such a rate that the ratio of feed to concentrate flows is equal to the volumetric concentration ratio
required for the process. When the equilibrium loop concentration
is established, the plant will contain fluid at the final concentration level, and will continue to operate at this concentration. Flux is therefore at its lowest level throughout the
duration of the process cycle.

If a steady stream of concentrated product is desired, it is
generally more economical to operate a multi-stage system in cascade
configuration. In a continuous multi-stage system (Fig. 1),
several recirculation loops are connected in series, with the
concentrate from one stage becoming the feed for the next. The
concentration of retentate within the loop increases from stage
to stage. In this way, it is possible to take advantage of the
higher fluxes obtained in the stages operating at lower concentration. The average process flux approaches that of a batch
operation, being typically greater than 80% of batch flux in a
4-stage system.

The importance of diafiltration procedures, either for
increasing the recovery of membrane-permeable solids or for
improving the purity of membrane-retainable solids, has been
referred to above. In batch operations, diafiltration is conveniently carried out by adding water to the feed tank to maintain a constant volume (Fig. 2). In continuous systems,
(Fig. 2) water is introduced into each stage at the same rate
as permeate is withdrawn. In each case, the retained solids
concentration remains constant, whereas the permeate stream
becomes more dilute with increasing recovery of permeable solids.
By selecting an optimum retentate concentration at which to
conduct the diafiltration, the process may be designed to minimize
either membrane area requirements or water usage (dilution of
the permeate).

Water volume additions and water flow rates associated with
diafiltration are normally expressed in terms of the turnover
ratio. In batch operations, this is the ratio of added water
volume to retentate volume, while in continuous systems, it is

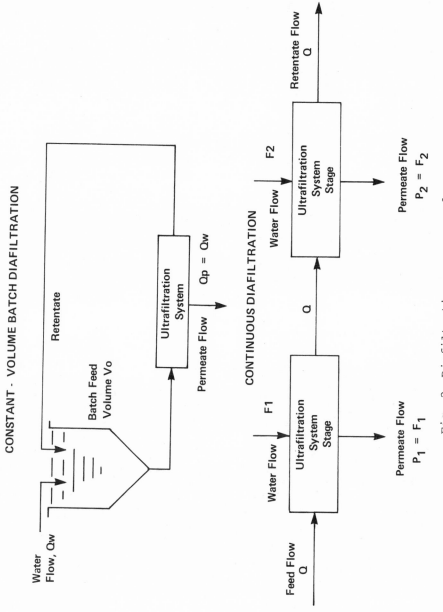

Fig. 2. Diafiltration process modes

the ratio of water flow to retentate flow. In the simplest case, water flow is equal to permeate flow. The dependence of membrane-permeable solids recovery in the ultrafiltrate on volume turnover ratio (constant-volume batch diafiltration) and on flow turnover ratio (continuous diafiltration with one and two stages) is illustrated in Fig. 3. It is apparent that in order to achieve a given product recovery, constant-volume batch diafiltration is the most efficient procedure as far as minimizing membrane area (turnover ratio) is concerned. The efficiency of the continuous diafiltration procedure approaches that of the constant-volume batch operation as the number stages increase.

Ultrafiltration in the Fermentation Industries

Currently, there are two principal industrial applications of ultrafiltration in the fermentation industries, viz. (i) enzyme concentration and purification, and (ii) fermentation broth clarification. These applications are discussed below in some detail. Various other potential or developing applications are briefly reviewed.

The advantages in applying ultrafiltration to fermentation products are as follows:

(a) low temperature operation is feasible, which enables heat-sensitive products to be processed.

(b) no change of state is involved, making the energy requirements modest in comparison to evaporation process.

(c) the separation takes place either with concentration, in the case of membrane-retainable species, or with minimal dilution, in the case of membrane-permeable products: with membrane-retained species, both concentration and purification functions can be combined in one unit operation.

(d) the filtrate is essentially free of suspended or colloidal solids and microorganisms, thereby facilitating the operation of downstream processes such as ion exchange, adsorption, etc.

(e) there is no addition of chemicals to the process stream, eliminating the need for either recovery or disposal operations.

Disadvantages of the ultrafiltration process include the following:

(a) filtration rates are relatively low, resulting in a requirement for large membrane areas, with substantial capital investments.

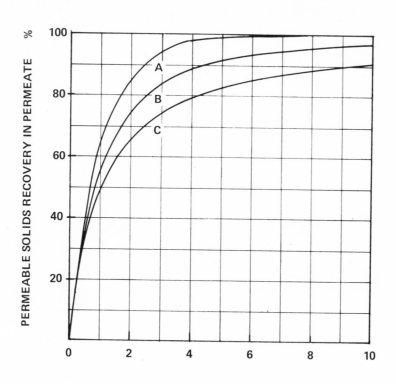

TURNOVER RATIO

A = Constant-Volume Batch Operation
B = Two-Stage Continuous Operation
C = Single Stage Continuous Operation

$$\text{Turnover Ratio} = \frac{\text{Water Volume}}{\text{Retentate Volume}} \text{ (Batch Diafiltration)}$$

$$\text{Turnover Ratio} = \frac{\text{Water Flow}}{\text{Retentate Flow}} \text{ (Continuous Diafiltration)}$$

Fig. 3. Membrane-permeable solids recovery by diafiltration

(b)   technical or economic restrictions exist with respect to the
      maximum retentate solids concentration which can be attained.
      Recourse to diafiltration (c.f. filter cake washing), with
      its associated cost, is thus often necessary either to
      increase the purity of a membrane-retainable product, or to
      improve the recovery of a membrane-permeable product.

ENZYME CONCENTRATION AND PURIFICATION

      The majority of microbial enzymes are produced in relatively
small batch fermentation operations. Ultrafiltration systems for
product concentration or purification may therefore also be conveniently operated in the batch process mode (Fig. 1). The
stability of the enzyme at the selected processing temperature
determines the duration of the batch process cycle. In general,
at temperatures in the 15°-25°C range, batch operation for 16-20
hours is acceptable. However, in some instances, primarily for
purposes of microbiological control, shorter batch process times
of 4-8 hours are essential. By way of comparison, in the dairy
industry, batch operations are normally restricted to 4-6 hours
in order to contain bacteriological contamination.

      Typical production schedules might require the processing of
25,000-50,000 litres of enzyme liquor in 20 hours. This would be
accomplished in plants containing, for example, 50-100 m^2 of
membrane area, with a proportionate increase in area for shorter
batch process times. A typical self-contained, 60 m^2 enzyme
ultrafiltration system based on Dorr-Oliver flat-leaf membrane
cartridges, is shown in Fig. 4. For purposes of flexibility,
multiple ultrafiltration units in this size range are generally
selected in preference to a single large system. It is also
common for several different enzymes to be processed on the
same plant, with variable batch fermentation schedules of 2-4
days being accommodated in rotation.

      Enzyme fermentation broths are normally clarified by con-
ventional filtration to remove cells and cell debris before being
subjected to ultrafiltration. The use of ultrafiltration or
cross-flow micro-filtration with highly porous membranes can in
principle be used to clarify enzyme solutions, but difficulties
in maintaining a high degree of enzyme transmission are invariably encountered, due to the development of dynamically-
formed, protein-retentive, surface films. In some instances,
concentration of unclarified or partially clarified enzyme broths
is practiced, with further clarification procedures being subsequently applied to the reduced-volume, high-solids concentrate
stream.

      Enzyme ultrafiltration plants incorporating non-cellulosic
ultrafiltration membranes are normally cleaned and sanitized at

Fig. 4. 60 m^2 flat leaf enzyme ultrafiltration plant

the end of each process cycle, or every 24 hours, whichever is
sooner. The principal cleaning solution is frequently an
alkaline detergent blend, with a pH in the range 10-13. This
solution is recirculated over the membrane surface for 30-60
minutes at a temperature of 40°-60°C. Supplementary acid cleaning
procedures are often beneficial for the removal of insoluble inorganic deposits. Plant sanitation is effected through the use
of dilute sodium hypochlorite solutions (150-300 mg/litre) at
40°- 50°C. At the end of the cleaning and sanitation cycles, the
cleanliness of the membranes is checked by water flux measurements.
The water flux is restored to its original value when the membranes
are clean.

In evaluating the feasibility of using ultrafiltration to
concentrate and/or purify enzyme streams, the following factors
must be considered:

Enzyme Activity Retention by Membranes

Ultrafiltration membranes with nominal molecular weight
exclusion limits in the range 10,000-20,000 are normally used
for enzyme processing. Such membranes are often characterized
with respect to retention capabilities by the use of model protein
or peptide solutes of defined molecular weight, e.g. α-
chymotrypsinogen (m.wt. 24,500), cytochrome C (m.wt. 12,400) and
bacitracin (m.wt. 1,400). With such membranes, enzyme activity
losses in the ultrafiltrate normally do not exceed 2-4% of the
feed activity. This conclusion is based on experiences in
operating commercial ultrafiltration systems on a range of enzymes
including amylases, amyloglucosidase, pectinases, microbial
rennins, cellulases, and various proteases. Since the reported
molecular weights of many enzymes lie within the range 20,000-
200,000, complete enzyme retention might be anticipated. However,
industrial enzymes are not pure well-defined, single species, nor
are the assay methods always specific for one particular enzyme.
It is possible that the permeate activity often observed is due
to low molecular weight contaminant fractions, which display
activity under the conditions of the enzyme assay procedure.

Low molecular weight metabolic products, substrate nutrients
(sugars and salts) and other low molecular weight constituents
of the fermentation broth will pass through the membrane into
the ultrafiltrate stream, essentially unchanged in concentration.
Partial retention of species in the species in the 1,000-
10,000 molecular weight range (peptides, oligosaccharides)
will be observed.

Recovery of Enzyme Activity

If an enzyme activity material balance is conducted between

initial and final batch processing conditions, allowing for any
residual hold-up losses in the ultrafiltration plant, then the
activity recovery in the combined concentrate and permeate streams
is typically 95-100% of that in the feed. By way of comparison,
activity recoveries in vacuum evaporation systems are reported
to lie in the 60-90% range. The small apparent activity losses
in ultrafiltration systems have been variously attributed to
such phenomena as (i) shear inactivation, associated with pumping, (ii) microbiological degradation, due to contamination of
either the fermentation liquor or the ultrafiltration equipment,
(iii) adsorption losses on the membrane surface, (iv) inactivation due to changes in the ionic environment, for enzymes
which are stabilized by membrane-permeable species. The extent
to which enzyme inactivation takes place by a particular
mechanism is highly dependent on enzyme type, and if the
mechanism can be identified, methods are often available for
reducing or even eliminating the loss of activity.

In industrial ultrafiltration plants, it is standard practice
to use conventional centrifugal pumps. Many enzyme streams can
be pumped in this way without loss of activity due to shear
degradation. With some products, recessed-impeller, torque-flow
centrifugal pumps, which exert lower shear forces on the fluid
than conventional centrifugal types, have been utilized. The
use of positive-displacement type pumps for handling very sensitive
enzyme streams has also been advocated.

Significant enzyme inactivation can occur if the fermentation
broth becomes contaminated with microorganisms. In such
circumstances, losses can be minimized by operation at lower
temperatures to control growth. To protect against contamination
introduced from the equipment, regular cleaning and sanitation
procedures are employed.

Degree of Concentration and Purification

The materials present in the feed to an ultrafiltration
system can be considered to consist of a mixture of membrane-
retainable species and membrane-permeable species. The concentration of membrane-retainable species in enzyme feed streams
varies typically between 0.5 w/o and 2.0 w/o, with membrane-
permeable solids ranging from 1 w/o to 10 w/o. During ultrafiltration, the membrane-retainable (enzyme containing) solids
concentration is increased in direct proportion to the volume
decrease attained. Volume reduction factors in the range 10-
fold to 50-fold are typical. The lower limit on retentate
volume, i.e., the maximum retentate total solids concentration,
which is economically attainable, is determined by the decrease
in process flux which accompanies the concentration process.

Since the enzyme is associated with the membrane-retainable solids fraction, and the membrane-permeable solids display very low enzyme activity, the enzyme purity or specific activity (the number of enzyme activity units per unit weight of dry solids) is increased during ultrafiltration. The purity of the retentate solids improves as ultrafiltrate is removed, and the proportion of membrane-retainable solids in the retentate total solids increases. The specific activity associated with the membrane-retainable solids fraction alone, represents the maximum enzyme purity which can be approached. The following example (Table 2) illustrates the various degrees of enzyme purification which can be achieved through direct volume reduction by ultrafiltration.

TABLE 2. THE EFFECT OF CONCENTRATION BY ULTRAFILTRATION ON ENZYME PURITY

Volume Reduction Factor	Membrane-Retainable Solids in Retentate (w/o)	Total Solids In Retentate (w/o)	$\dfrac{S \text{ (Retentate Total Solids)}}{S \text{ (Membrane-Retainable solids)}}$
1	1.5	4.5	0.33
5	7.5	10.5	0.71
8	12.0	15.0	0.80
10	15.0	18.0	0.83

NOTES:

(i)   S = Specific activity of enzyme (enzyme units per unit weight of solids)

(ii) Volume Reduction Factor = Volume of feed/volume of retentate

(iii) Membrane-permeable solids concentration = 3.0 w/o (assumed equal and constant in retentate and permeate)

The degree of enzyme purification achievable directly by means of ultrafiltration is restricted by the amount of volume reduction which can be economically attained. Further purification is, however, possible through use of the above described diafiltration procedures involving addition of water to the retentate stream. Taking the example considered in Table 2, the improvements in enzyme purity associated with constant-volume batch diafiltration of the 10-fold concentrate are illustrated in Table 3. It is apparent that diafiltration is an extremely valuable procedure for increasing the purity of enzyme streams, and the technique is widely used in conjunction with concentration by ultrafiltration. Depending on the amount of diafiltration which is necessary, additional membrane area will be required to accomplish the combined batch concentration-diafiltration process within the total batch operating time allowed. The value of the improved enzyme purity associated with diafiltration must, therefore, be balanced against increased ultrafiltration costs.

TABLE 3. THE EFFECT OF ENZYME PURITY OF CONSTANT-VOLUME BATCH DIAFILTRATION APPLIED TO AN ULTRAFILTRATION CONCENTRATE

Number of Volume Turnovers	Membrane-Permeable Solids in Retentate (w/o)	Total Solids in Retentate (w/o)	$\dfrac{S \text{ (Retentate Total Solids)}}{S \text{ (Membrane-Retainable Solids)}}$
0	3.00	18.00	0.83
1	1.10	16.10	0.93
2	0.41	15.41	0.97
3	0.14	15.14	0.99

NOTES:

(1)   S = Specific activity of enzyme (enzyme units per unit weight of solids)

(ii)   Volume Turnover = Volume of diafiltration water/volume of retentate

(iii)   Membrane retainable solids concentration = 15.0 w/o (assumed constant)

(iv)   Membrane-permeable solids in retentate obtained from Fig. 3.

Process Economics

The key parameter affecting both capital and operating costs is the average ultrafiltration process flux. This can vary from 5 to 50 litres/m^2/hour, depending on temperature, solids concentration and cross-flow velocity. Assuming a typical installed equipment cost of $800 per m^2 for systems in the 100 m^2 size range, and a flux of 20 litres/m /hour, the capital cost translates into $40,000 per m^3/hour filtrate capacity.

The components of the total cost associated with operating an ultrafiltration system are: (i) depreciation, (ii) membrane replacement, (iii) pumping energy, and (iv) labor, maintenance, and cleaning chemicals. Based on a 10 year plant life (straight line depreciation) with annual membrane replacement at $100 per m^2, and with an electrical energy consumption of 0.1 kw per m^2 (unit cost = $0.03 per kw h), the operating cost (16 h/day, 250 days/year) is $2.4 per m^3 of filtrate removed. An additional 10-20% of this cost will cover labor, cleaning chemical, and maintenance expenses.

Vacuum evaporation is the normal alternative to ultrafiltration for dewatering enzyme solutions. The evaporation process is generally considered to be more expensive to operate than ultrafiltration. The major benefits of ultrafiltration, however, arise from the purification capabilities of the process. In instances where the use of either salt or solvent precipitation processes has formerly proved to be necessary in order to reach the required enzyme purity, the membrane process can be used instead, with substantial cost reductions. In addition, the overall activity recovery will generally be higher with ultrafiltration than with a combined vacuum evaporation-chemical precipitation approach. Since enzyme product values are frequently of the order of $12 per kg, the overall economic balance is tipped strongly in favor of the ultrafiltration approach.

FERMENTATION BROTH CLARIFICATION

Since ultrafiltration membranes exhibit essentially zero retention of fermentation products such as organic acids, amino-acids or antibiotics, the ultrafiltration process can be used to separate suspended, microbiological and macromolecular solids from the principal product of fermentation. As in the case of microbial enzymes, fermentation to produce various low molecular weight products is generally conducted on a batch basis, but the scale of operation is often much larger. Either batch or continuous ultrafiltration systems may, therefore, be preferred, depending on particular circumstances. If the stability of the product is low, which would restrict batch operation to, say,

4 hours, then continuous ultrafiltration would be used. Continuous processing would, in all probability, also be used if
the filtration capacity of the system were required to exceed
about 10 m^3/hour, unless suitable storage tanks were already
available. Operating temperatures range typically from 15°C to
50°C, depending primarily on product stability.

The size of ultrafiltration plants required to match the
levels of production of fermentation products can vary from, say,
10 m^2, in the case of a high cost, low-volume amino-acid, to
1,000 m^2 for a lower cost organic acid or bulk antibiotic. Because fermentation broths vary quite widely in their physical
and chemical properties, ultrafiltration plants would normally
be custom-designed for one application. Multi-application use,
while possible in principle, is, therefore, much less convenient
than in the case of enzyme streams.

Ultrafiltration plants for fermentation broth clarification
are normally cleaned either after every batch operation (e.g.,
every 10 hours), or every 20 hours, if the operation is continuous.
The flux decay experienced during the process cycle is fully
reversed by cleaning, with measurements of water flux being used
to determine when the membranes have been fully cleaned.

The cleaning and sanitation procedures involved are essentially
similar to those described for enzyme ultrafiltration systems.
The total duration of cleaning and sanitation procedures is 2-4
hours. Particularly close attention must be paid to plant
rinsing before cleaning, since the suspended solids level in
concentrated broths is very high. The bulk of the loosely-
adhering solids must first be removed from the system in order
that the principal cleaner operate with maximum effectiveness.
In the Dorr-Oliver flat-leaf system described below, such solids
removal is facilitated by reducing the trans-membrane pressure
to zero while maintaining a high flow velocity across the
surface of the membrane.

Product Purity

The ultrafiltration membrane retention of organic acids,
amino-acids and antibiotics with molecular weights below about
600, is essentially zero, i.e., the solute concentration is the
same in both retentate and permeate streams. The product, therefore, constitutes the major part of the membrane-permeable
solids fraction, c.f. enzymes, where the product is contained
in the membrane-retainable solids fraction. During ultrafiltration,
suspended solids, cell debris, and soluble proteinaceous materials
remain in the retentate stream, giving a clear, essentially micro-
organism-free filtrate. Gross suspended or fibrous material is
usually removed upstream of the ultrafiltration system by coarse

screening processes.

The purity of products clarified by ultrafiltration is
potentially superior to those clarified using coarser separation
processes, such as pre-coat rotary vacuum filtration, or centrifugation. This is because soluble macromolecular material such
as proteins, nucleic acids, and polysaccharides, with molecular
weights above, say, 20,000 will be removed along with the
membrane-retained cellular material. Substances in the 1,000-
20,000 molecular weight range will be partially retained. These
properties of ultrafiltration membranes, whereby macromolecules
are removed, are of substantial benefit. One example is in the
case of antibiotics, where downstream purification processes
may be required to remove antigenic high molecular weight substances. Further benefits may arise from the fact that the
clarity of the filtrate is also excellent, which will assist
in the operation of any downstream adsorption processes, such
as ion-exchange.

Product Recovery

Typical fermentation broths contain 5 w/o-25 w/o total
solids, with the principal product being present at concentrations
within the range 2.0 w/o to 20 w/o. Direct reduction in volume
(or flow, in the case of continuous systems) by means of ultrafiltration is normally limited to a maximum within the range
5-10 fold. This is due to the increase in retentate viscosity
which accompanies concentration of the membrane-retainable
solids. However, a degree of volume reduction in the 5-10 fold
range permits 80-90% recovery of low molecular weight fermentation
products to which the membrane is freely permeable. Further
product recovery is attainable through the use of batch or
continuous diafiltration procedures, as described above. This
is achieved at the expense of some increase in the capital and
operating costs of the ultrafiltration system.

In batch diafiltration, water addition is preferentially
commenced after a period of concentration during which the
volume of the batch has been appreciably reduced. The optimum
concentration point for minimizing diafiltration time can be
calculated by balancing a decreasing volume to be filtered
against a decreasing filtration rate. To minimize dilution of
the filtrate, diafiltration is frequently carried out at the
final retentate concentration. The dependence of product
(permeable-solids) recovery on batch volume turnover ratio is
illustrated in Table 4 for an ultrafiltration process involving
5-fold volume reduction, followed by diafiltration of the
concentrate.

TABLE 4. EFFECT OF CONSTANT VOLUME BATCH DIAFIL-
TRATION ON MEMBRANE-PERMEABLE SOLIDS
RECOVERY FROM A 5-FOLD CONCENTRATE

5-Fold Concentrate Batch Volume Turnover Ratio	Recovery of Membrane-Permeable Solids in Ultrafiltrate (%)		
	Concen- tration Stage	Dia - filtration Stage	Combined Stages
0.5	80	7.9	87.9
1.0	80	12.6	91.6
2.0	80	17.3	97.3
3.0	80	19.0	99.0
4.0	80	19.6	99.6

In continuous ultrafiltration, several concentration stages
would typically be followed by either combined concentration-
diafiltration stages or by "pure" diafiltration stages, in which
water injection flow is equal to permeate flow. In conventional
continuous diafiltration (no permeate re-injection), the permeate
becomes more dilute with increasing recovery of product. However, if required, this situation can be altered by the use of
counter-current diafiltration procedures, in which the permeate
from one or more stages is injected into upstream stages. The
overall effect is that the same recovery as in conventional
diafiltration can be achieved with substantially less dilution
of the product stream. A greater membrane area is, however,
required, and a balance between the costs associated with product
dilution and membrane system operating costs must be made.
Continuous countercurrent diafiltration finds application
primarily in cases where the product must be recovered by
evaporation downstream of the ultrafiltration system.

Equipment Design Configuration

The suspended solids loading in most fermentation broths
is high (1 w/o-15 w/o). There is, therefore, a potential danger
of flow-channel blockage with compacted microbiological solids,
if the channel dimensions are small. On the other hand, the
process fluxes in this application are relatively low, i.e., the
membrane area required per unit of capacity is high. This
would indicate preference for an ultrafilter system with a

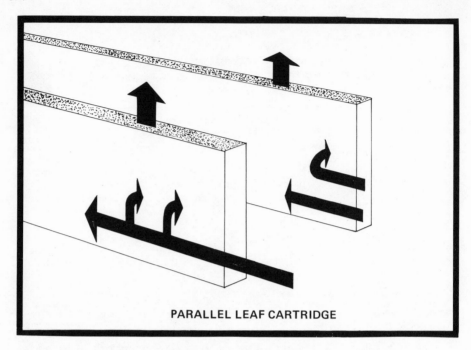

PARALLEL LEAF CARTRIDGE

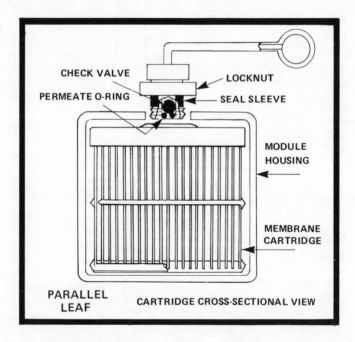

CHECK VALVE   LOCKNUT

PERMEATE O-RING   SEAL SLEEVE

MODULE
HOUSING

MEMBRANE
CARTRIDGE

PARALLEL
LEAF   CARTRIDGE CROSS-SECTIONAL VIEW

Fig. 5a. Dorr-Oliver parallel-leaf ultrafiltration system
        cartridge cross-section.

Fig. 5b. Dorr-Oliver parallel-leaf ultrafiltration system 18 ft^2 cartridge.

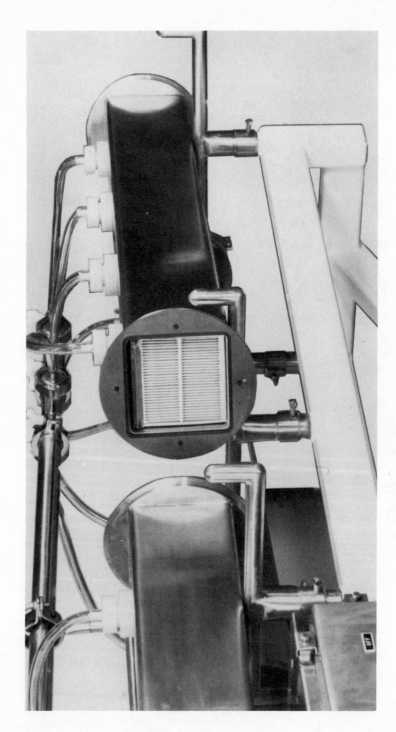

Fig. 5c. Dorr-Oliver parallel-leaf ultrafiltration system cartridge installed in housing.

moderately high membrane packing density, in order to keep the
physical dimensions of the plant at an acceptable level. While
these two requirements are somewhat conflicting, they can be
satisfactorily met in the Dorr-Oliver parallel-leaf membrane
cartridge system illustrated in Fig. 5.

In the Dorr-Oliver design, the flow channel inter-leaf
spacing is 0.25-0.30 cm, and the equipment is operated at linear
surface velocities of 2.5-3.0 m/sec. Under these conditions,
no channel blockage has been observed in continuous long-term
operation on fermentation broths. The equipment is also
relatively compact. For a 300 m^2 system, the overall ultrafiltration plant packing density is approximately 10 m^2 of membrane
area per m^3 of total system volume, including pumps and piping.
Floor area requirements amount to 4 m^2 per 100 m^2 of installed
membrane area. A 320 m^2 installation containing equipment of
this type is shown in Fig. 6.

Hydraulic Systems Design

The viscosity of fermentation broths normally lies within
the range 2-100 cp at 25°c. The fluids are often non-Newtonian,
with viscosity being dependent on shear rate. The increase in
viscosity during concentration by ultrafiltration can be quite
dramatic, with values as high as 500 cp being attained. This
viscosity increase imposes certain requirements in the area of
hydraulic systems design. As the viscosity increases, the
pressure difference across the membrane modules in a recirculation
loop will increase.

The recirculation flow will also decrease in accordance with
the flow-head characteristics of the recirculation pump.
Eventually, if the viscosity becomes sufficiently high, the
maximum head which can be generated by the pump will be
approached. Problems of pump cavitation are likely to be encountered if over-concentration takes place, and limits on
the degree of concentration must be imposed.

In a system where the maximum and minimum allowable transmembrane pressures are defined, the viscosity-related increase
in pressure difference across the ultrafiltration modules in a
recirculation loop, governs the maximum number of series-connected
modules which the loop can contain. Shorter recirculation flow
paths than are used with low viscosity fluids must be selected
in order to account for the increasing parasitic pressure drops
resulting from concentration. Thus, within a recirculation
pumping station, greater usage of parallel-connected modules
is warranted, in order to maintain ratios of module costs to
pump/ancillary equipment costs, which are as favorable as those
for low viscosity fluids.

Fig. 6. 320 m^2 CIP ultrafiltration plant

Ultrafiltration plants incorporating the Dorr-Oliver flat-leaf designs have been operated successfully on fluids with nominal viscosities as high as 600 cp, using feed and recirculation pumps of the conventional centrifugal type. Operation on broths with viscosities in the 200-400 cp range at final concentration is being undertaken on a commercial basis.

TABLE 5. ASSUMPTIONS FOR COST COMPARISON BETWEEN ULTRAFIL-TRATION AND PRE-COAT ROTARY VACUUM FILTRATION

Parameter	Unit Operation	
	Pre-coat Rotary Vacuum Filtration	Ultrafiltration
Throughput (flux)	400 1/m^2/h	17 1/m2/h
Membrane Life	--	1 year
Membrane Replacement Cost	--	$100/m^2
Filter-Aid Consumption	24 kg/m^3 filtrate	--
Filter-Aid Cost	$0.22/kg	--
Power Consumption	1.0 kw/m^2	0.1 kw/m^2
Energy Cost	$0.03 /kw-h	$0.03/kw-h
Annual Process Duty	250 days, 16 h/day	250 days, 16 h/day
Labour Requirements	16 h/day/100 m^2	1 h/day/100 m^2
Labour Cost	$10/h	$10/h
Annual Maintenance Cost	$90/m^2	$7.5/m^2
Cleaning Fluids Cost	--	$5/day/100 m^2
Capital Cost	$3000/m^2	$750/m^2
Depreciation Period (SL)	10 years	10 years

Process Economics

As in all filtration systems, the filtration rate, or flux, is the critical parameter of establishing operating and capital costs. The permeate flux levels attained in the ultrafiltration of various fermentation broths, differing widely in origin, lie typically in the range 30-50 litres/m^2/hour at the initial feed concentration, and decrease to 5-10 litres/m^2/hour at the final retentate concentration. Flux decreases with increasing retentate concentration, and the inverse linear flux-log (retainable-solids concentration) relationship, frequently observed with macromolecular systems, is sometimes obeyed over part of the concentration range. There is no sound theoretical basis for this, since boundary-layer erosion-type effects are likely to occur with the large particulate and colloidal materials present. In addition, as described below, viscosity changes during concentration can be highly significant, with the result that cross-flow velocity will decrease in the centrifugal pumping systems normally employed. Variable flux-time decay rates at different concentration levels also complicate the theoretical modeling of the flux-concentration relationship.

During a diafiltration stage, the membrane-retainable solid concentration remains constant. Flux, which is dominated by the membrane-retainable solids concentration in the retentate, therefore, tends to remain relatively constant. A small flux decrease may be observed if there are significant flux-time decay effects. While such effects are specific to individual streams, a 10-20% flux decrease in 20 hours is typical. On the other hand, if the permeate viscosity is appreciably reduced as a result of diafiltration, a small flux increment may be observed.

One of the more common traditional operations for clarifying fermentation broths is pre-coat rotary vacuum filtration. Making a series of assumptions as to operating parameter costs (Table 5), the economics of operating an ultrafiltration-diafiltration system and a vacuum filtration-cake washing system are compared in Table 6. Equal product recoveries are assumed in the two approaches. There will generally also be a significant cost penalty in disposing of the spent filter-aid, over and above those costs involved in disposing of an ultrafiltration concentrate stream. This cost disadvantage for pre-coat rotary vacuum filtration is not brought into the economic comparison because of widely different local situations.

It is apparent from Table 6 that while the capital investment required for pre-coat filtration is some six times lower than that for ultrafiltration, the operating cost is some 40% higher. The additional capital associated with the pre-coat filter operation would be recovered in about three years, and if the filter aid disposal cost penalty approaches $1.0 per m^2, the pay-back time is reduced to a period of the order of two years.

TABLE 6. OPERATING COST COMPARISON BETWEEN ULTRA-
         FILTRATION AND PRE-COAT ROTARY VACUUM
         FILTRATION

Cost Item	Operating Cost ($/m^3 filtrate)	
	Pre-Coat Rotary Vacuum Filtration	Ultra-Filtration
Filter Aid	5.30	—
Membrane Replacement	—	1.50
Energy	0.08	0.18
Labour	0.26	0.37
Maintenance	0.06	0.11
Cleaning Chemicals	—	0.18
Depreciation	0.19	1.10
TOTAL OPERATING COST	5.89	3.44

NOTE:
    Capital Cost = $470 per m^3/day (pre-coat filtration)
    Capital Cost = $2760 per m^3/day (ultrafiltration)

MISCELLANEOUS APPLICATIONS

     There are numerous potential applications of ultrafiltration in the fermentation and related industries, some of which have already achieved limited commercial-scale acceptance. The more important examples are as follows:

Wine Clarification: Many wines contain proteinaceous materials which, unless removed, will sediment on storage. The conventional approach is to add finings, e.g. bentonite, to the wine in order to adsorb the proteins, and then filter off the residue (lees),

often with a pre-coat rotary vacuum filter. For some wines, it
may be possible to remove the proteins directly, and at lower
cost, by means of ultrafiltration.

Activated Sludge Mixed-Liquors Clarification: The stabilization
of sewage and industrial effluents is frequently achieved by
the complicated microbial activated sludge process. The use of
ultrafiltration to separate the bio-mass from the mixed liquor
enables a high quality effluent, free of microorganisms and
suspended solids, to be produced. The effluent can either
be recycled as of low-grade water source, or be further purified
for potable use. Both these approaches are being practiced
commercially on a limited scale.

Pyrogen and Microorganism Removal from Fermentation Products:
Ultrafiltration membranes are almost totally retentive of microorganisms, and fully retentive of pyrogens. Pyrogens are predominantly high molecular weight metabolic products of certain
bacteria. Several applications of ultrafiltration relating to
sterilization and pyrogen removal have been found in finishing
operations for fermentation products, and in the preparation of
water for use in finishing operations.

Enzyme Retention in Membrane Reactors: Continuous enzyme reactors,
utilizing an ultrafiltration membrane system may be used to
separate low-molecular weight reaction products from the enzyme
and high-molecular weight substrate species, e.g., amino-acid/
protein hydrolysate isolation from protein-protease systems, or
sugar isolation from starch-amylase/amyloglucosidase systems.
Ultrafiltration has also been used to retain or isolate β-
galactosidase in batch and continuous operations involving
hydrolysis of lactose from milk and cheese whey ultrafiltration
permeates.

Bio-mass Retention in Membrane Fermentors: Continuous membrane
fermentors, utilizing ultrafiltration to separate low molecular
weight products from the bio-mass, have been evaluated, e.g., in
the production of ethanol from sugars using yeast. The above
mentioned continuous clarification of sewage effluent in an
activated-sludge reactor is a good example of this approach.

BIBLIOGRAPHY

Beaton, N.C., 1977, Applications and Economics of Ultrafiltration,
Inst. Chem. Eng. Symp., 51:59.

Beaton, N.C., and P.R. Klinkowski, Industrial Ultrafiltration-
Design and Application of Diafiltration Processes, J. Sepn.
Process Tech., in press.

Beaton, N.C. and Steadly, H., Industrial Ultrafiltration, Recent Developments in Separation Science, VI, in press.

Charm, S.E., and Wong, B.L., 1970, Enzyme Inactivation by Shearing, Biotech. Bioeng., 12:1103.

Flinn, J.E., ed., "Membrane Science and Technology", Plenum Press, New York (1970).

Gebbie, P., Fane, A.G., and Fell, C.J.D., 1976, Ultrafiltration and Its Role in Activated Sludge Waste Water Treatment, Amer. Water Works Assoc., 3:17.

Jeffries, J.W., Omstead, D.R., Cardenas, R.R., and Gregor, H.P., 1978, Membrane-Controlled Digestion: Effect of Ultrafiltration on Anaerobic Digestion of Glucose, Biotech. Bioeng. Symp. No. 8:37.

Klinkowski, P.R., 1978, Ultrafiltration, Chem. Eng., May 8:165.

Kowalewska, J., Pozanski, S., Bendnarski, W., and Sulima, K., 1978, The Application of Membrane Techniques in Enzymatic Hydrolysis of Lactose, and Repeated Use of β - Galactosidase, Nordeuropaeisk Mejeri-Tidsskrift, 44:20.

Madsen, R.F., "Hyperfiltration and Ultrafiltration in Plate-and-Frame Systems, Elsevier, Amsterdam (1977).

Madsen, R.F., Olsen, O.J., Nielsen, I.K., and Nielsen, W.K., 1972, Experiences with Concentration and Separation of Industrial Solutions with Reverse Osmosis and Ultrafiltration, Filtration and Separation, 9:567.

Nielsen, W.K., and Wagner, J., May 1976, Novel Developments in Ultra- and Hyperfiltration as Applied to the Biochemical Industries, Proc. Symp. "Separation Processes in the Biochemical Industries", London.

Payne, R.E., and Hill, C.G. Jr., 1978, Enzymatic Solubilization of Leaf Protein Concentrate in Membrane Reactors, J. Food Sci., 43:385.

Roger, L., Maubois, J.L., Thapon, J.L., and Brule, G., 1978, Hydrolyse du lactose contenu dans l' ultrafiltrat de lait ou de lactosérum en reacteur enzymatique à membrane, Annales de la Nutrition et de l' Alimentation, 32:657.

Roozen, J.P. and Pilnik, W., 1973, Ultrafiltration Controlled Enzymatic Degradation of Soy Protein, Process Biochem., 8:24.

Tachauer, E., Cobb, J.T., and Shah, Y.T., 1974, Hydrolysis of

Starch by a Mixture of Enzymes in a Membrane Reactor, Biotech.
Bioeng., 16:545.

Wysocki, G., 1977, Stofftrennung: Erfahrungen mit KBM - Ultra-
filtrations-anlagen, Ernahrungswirtschaft, 6:278.

CONCENTRATING FRUIT JUICES BY REVERSE OSMOSIS

R. L. Merson, G. Paredes, and D. B. Hosaka

Department of Food Science and Technology
University of California
Davis, CA 95616

INTRODUCTION

The difference between reverse osmosis and ultrafiltration
lies principally in the retention characteristics of the membranes.
With respect to food processing, ultrafiltration membranes are
able to retain protein-sized molecules, whereas reverse osmosis
membranes retain lower molecular weight solutes such as sodium
chloride or sucrose. In order to concentrate fruit juice, it is
necessary to retain sugars, e.g., disaccharides such as sucrose
(molecular weight 342, typically present in fruit juices as 5% of
the total weight), and monosaccharides such as glucose and fructose
(MW 180, also present in fruit juices at about 5% total weight).
It is also necessary to retain organic acids such as citric acid
(MW 192) or malic acid (134), typically present in fruit juices
at the 1 to 2% level, and various salts present at less than half
a percent. In addition, it is desirable to retain trace amounts
of low molecular weight (<150) aroma compounds such as alcohols,
esters, aldehydes, etc.; typically there might be 100 such compounds
in a fruit juice with a total concentration of less than 400
parts per million.

There is a large industry associated with water removal from
fruit juices. In the past, satisfactory products have been
obtained by multiple-effect vacuum evaporation coupled with
either aroma recovery or the addition of small amounts of fresh
juice or essential oils. For fifteen years, it has seemed attractive to use reverse osmosis to concentrate fruit juice in order
to retain "fresh juice" quality and reduce water removal costs.

The goals are to minimize any thermal damage that may occur
during evaporation, to maximize aroma retention, to minimize
energy use, and to minimize capital costs of water removal equipment.

There are several concerns that have limited the use of
reverse osmosis for fruit juice concentration in the past. The
first is the high osmotic pressure of the juice. Commercial
concentrates such as orange or apple juice are typically 40 to
45% solids with corresponding osmotic pressures in the range of
1200 to 1300 psi. Such high osmotic pressures require high
operating pressures; and, therefore, membranes and equipment are
needed that can withstand these high operating pressures. Both
membranes and pumps have been a problem in the past, limiting the
final concentration of the fruit juices to about 30-35% solids.

The second problem has been the selectivity of membranes
with respect to aroma compounds. It is principally the aroma
which differentiates the character of different juices, but these
molecules are usually hydrophilic and have been difficult to
retain with membranes available heretofore. Other concerns have
been the fouling, cleaning, and sanitation of membranes, modules,
and ancillary equipment. Also low fluxes and corresponding high
equipment costs have been obtained with fruit juice applications
in the past.

Some characteristics of traditional (Loeb-type) cellulose
acetate membranes are given in Table 1. Due to the relatively
poor compaction stability of these membranes, the maximum operating
pressure is on the order of a 6.9 megapascals (1000 psi), limiting
the maximum concentration of solids to 30-35%. The maximum
temperature for processing or cleaning is on the order of 40°C.
The optimum pH range is near 4.5 which is ideal for fruit juice
processing, but it does preclude caustic cleaning which is common
in the food industry. Solute retention with cellulose acetate
membranes has been satisfactory for most dissolved solids (sugars,
salts, etc.). However, aroma retention has been marginal (Merson
and Morgan, 1968) except for very low flux membranes; rejection
coefficients with cellulose acetate membranes are usually low
(typically zero), but at least they are not negative as would be
the case for stripping of volatiles from the juice as occurs
during evaporation.

In 1976, Riley et al. described a poly(ether/amide) membrane
which has since become commercially available as PA 300 (Universal
Oil Products, Inc., Fluid Systems Division, San Diego, California).
This membrane has a thin film composite construction; that is, it
is made by interfacial polymerization [cross-linking of a poly(ether/
amine)] on a polysulfone ultrafiltration backing. With this
construction, it should have superior compaction stability and be
able to operate at relatively high pressures. It is able to

Table 1. Characteristics of Reverse Osmosis Membranes

	Cellulose Acetate	Poly(ether/amide) thin film composite
Maximum operating temperature	40°C	55°C
pH range allowable	3-7	3-12
Pure water permeation rate at 1000 psi	20-30 gfd	50-60 gfd
NaCl rejection	97%	98-99%
Compaction stability	poor	good
Chlorine stability	fair	poor

withstand higher temperatures and higher pH values than cellulose acetate membranes and also gives high permeation rates with superior rejection of solutes. In particular, solute rejections reported by Riley et al. (1976) for low-molecular-weight aldehydes, alcohols, esters, and ketones (which might be considered model compounds for fruit juice aroma) are remarkably high from solutions in the 100 to 1000 ppm range (see Table 2). The principal drawback to the polyamide membrane appears to be a very low tolerance for chlorine.

EXPERIMENTAL

Tests were made at a pilot scale level with two fruit juices, orange juice (Paredes, 1980) and tomato juice (Hosaka, 1978). A typical batch reversis osmosis loop was used, consisting of a temperature-controlled feed tank, a pump (Manton Gaulin homogenizer with the homogenizing valve removed), the membrane module, and a back pressure regulator. Pressure pulsations from the triplex feed pump were minimized using an in-line, flow-through pulsation dampener described by Kavanagh (1975). This device consisted of a length of rubber tubing of the same diameter as the stainless steel feed line. The sanitary rubber tubing passed through a cylindrical housing of a slightly larger diameter steel pressure pipe. The annular space between the tubing and the housing was filled with nitrogen at a pressure slightly lower than that of the feed inside the tube. Thus, the tubing could expand with each pulse of the pump piston, reducing the pressure fluctuations in the fluid.

For the orange juice experiments, a spiral-wound module was used containing a poly(ether/amide) membrane (PA 300, Universal Oil Products, Inc., Fluid Systems Division, San Diego, California). Membrane area was 1.86 m^2 (20 ft^2). The module, designed to maximize membrane surface area and minimize concentration polariza-

Table 2.   Solute Rejection of PA Membrane at 6.9×10^6 Pa
(1000 psi) and 25°C (Riley et al., 1976)

Solute	MW	Concentration (ppm)	pH	Rejection (%)
Acetaldehyde	44.05	660	5.8	70-75
Ethyl Alcohol	46.07	700	4.7	90
Acetic Acid	60.05	190	3.8	65-70
Urea	60.06	1,250	4.9	80-85
Methyl ethyl ketone	72.10	465	5.2	94
Ethyl acetate	88.10	366	6.0	95.3
Phenol	94.11	100	4.9	93
Phenol	94.11	100	12.0	>99
Butyl Benzoate	178.2	220	5.8	99.3
Citric Acid	192.12	10,000	2.6	99.9

tion, required careful preparation of the feed to remove suspended
matter. The California Valancia orange juice was extracted,
finished, and pasteurized at 74-79°C for 10 seconds commercially.
It was then chilled to 40°F and shipped by refrigerated truck to
Davis in two 55 gallon drums, stored overnight at -4°C, and
centrifuged at 15,000 rpm in a refrigerated centrifuge (Sharples
Model No. AS-14) to remove suspended solids. The juice was then
frozen in 5 gallon lots to be thawed as needed for reverse osmosis
trials by submerging the 5 gallon plastic containers in water at
30°C. Reverse osmosis trials were conducted at 5, 15, and 25°C
and 3.45, 5.17, 6.90 MPa (500, 750, 1000 psi). The feed rate was
8 l/min (2.1 gal/min).

For the tomato juice experiments, cellulose acetate membranes
were used mounted on ceramic cores (Rev-O-Pak, Inc., Newbury
Park, California). The feed juice flowed in an annular space
between the 1.6 cm diameter membrane-coated core and 2.3 cm
diameter stainless steel housing. Part of the appeal of this
module design was that the membrane cores could be removed for
cleaning and observation of any membrane fouling which may have
occurred. Plastic wires were wrapped spirally around the cores
to promote turbulence. Membrane area (7 cores in series) was
0.317 m^2 (3.41 ft^2).

The tomatoes (var. GS-12) were crushed (broken) at three
different temperatures: for hot break juice, steam was injected
to raise the temperature in less than 5 sec to 104°C and held for
45 sec; for medium break juice, the holding temperature was 80°C.
These juices were finished hot through a Brown Citrus extractor
(Model #3600) with a 0.114 cm screen size and stored frozen in No.

10 cans until reverse osmosis runs could be made. Some tomatoes
were crushed at room temperature and centrifuged batchwise at 2000
rpm for 20 min to remove suspended solids and produce a tomato
serum. Reverse osmosis experiments were carried out at 3.45 MPa
(500 psi) at 25°C, at a feed flow rate of 7.7 1/min.

Solute concentrations were measured with a refractometer.
Preliminary evaluation of the aroma retention was obtained by a
chromatographic analysis similar to that of Wyllie et al. (1978);
aroma was stripped from the juice by bubbling nitrogen through
the sample. The aroma was adsorbed onto a porous packing material,
subsequently desorbed into chloroform, and analyzed on a chromato-
graph consisting of an open column glass capillary tube coated
with Carbowax-20-M and a flame ionization detector.

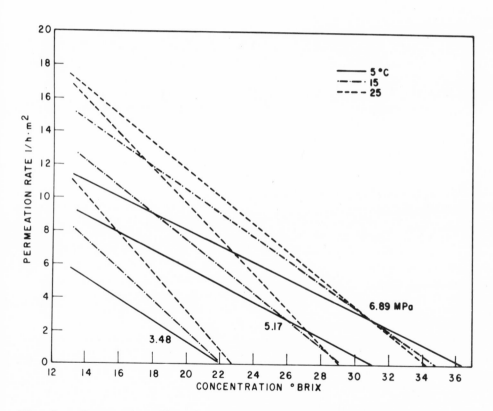

Fig. 1. Permeation flux for orange juice serum with a PA-300
         spiral wound poly(ether/amide) membrane at three
         pressures and three temperatures as a function of feed
         tank concentration in the batch-type reverse osmosis
         system.

RESULTS AND DISCUSSION

Experimental permeation rates for orange juice serum in the spiral wound poly(ether/amide) membrane are presented in Figure 1. The concentration scale, common in the fruit juice industry, is the concentration of pure sucrose which would give the same refractive index as the fruit juice; it is very close to the percent soluble solids in the juice. As expected, the permeation rate increased with pressure and also with temperature. The extrapolation of the curves to zero permeation rate gives the maximum concentration that could be achieved at each condition. These values correspond roughly to the osmotic concentration of fruit juice at those temperatures and pressures. Obviously, maximum achievable concentrations would be somewhat less than these under commercially acceptable permeation rates.

The permeation rate at 6.89 MPa (1000 psi) at 25°C is compared with literature data in Table 3. It can be seen that the permeation flux with the present pilot scale poly(ether/amide) membrane is substantially improved over the previously reported data for cellulose acetate membranes which were obtained on an experimental scale.

Figure 2 shows chromatographic aroma analysis. These data, which must be considered very preliminary, indicate that ethyl alcohol may not be well retained in orange juice but that higher molecular weight aroma constituents appear to be well retained. In comparing the peak sizes in Figure 2, note that the permeate sample was five times as large as the feed sample. Further chromatographic analysis, and analysis for other constituents, is continuing.

Table 3.   Concentration of Orange Juice by Reverse Osmosis. Average permeation rates at 6.89 X 10^6 Pa (1000 psi) and 25°C

Membrane	Support	Conc. Range °Brix	Avg. Permeation Rate l/h·m^2	gal/ft^2 day
Cellulose acetate	Experimental cell	10.5-32 10.4-22.5	0.85 10.0	0.5* 5.9**
Poly(ether/ amide)	Spiral wound module	13.4-34 13.4-22.5	8.5 13.6	5.0 8.0

*Merson and Morgan (1968).

**Matsuura et al. (1972).

The permeation rate for tomato juice is presented in Figure 3. These results are surprising in that the hot break tomato juice and the tomato serum gave similar permeation rates, whereas the medium break tomato juice gave higher permeation rates. Rapid steam injection to 104°C (hot break) inactivates the pectic enzymes in the juice and maintains a high viscosity feed juice. Heating only to 80°C (medium break) allows the enzymes to remain active long enough to reduce the viscosity somewhat and an increase in permeation rate was expected. Leaving the juice without heat treatment results in maximum enzyme activity causing floculation and settling of all of the insoluble solids and much of the pectin material; this results in a clear serum of low viscosity which was further clarified by centrifugation. It was expected that this serum would result in the least amount of fouling of the membrane surface and give the highest permeation rate. It was surprising that this preparation gave the same permeation rate as the hot break juice and also appeared to show the same

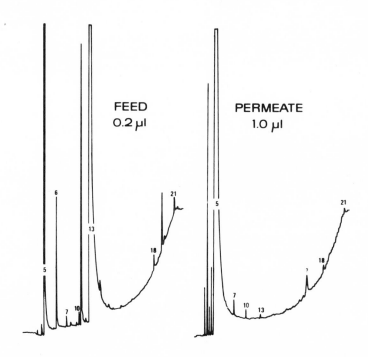

Fig. 2. Chromatographic analysis of orange aroma in headspace. Ordinate is recorder response, abscissa is elution time. Permeate sample (right hand side) was 5 times as large as the feed sample. Peak 5 is ethyl alcohol; peak 13 is limonene.

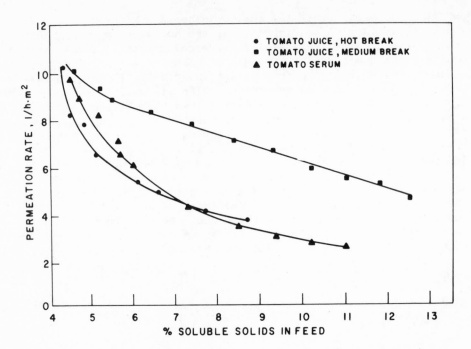

Fig. 3.   Permeation flux for tomato juice as a function of concen-
           tration in the feed tank.   Cellulose acetate membranes
           mounted on porous ceramic cores.   Pressure was 3.45 MPa
           (500 psi); temperature was 25°C.

degree of fouling as evidenced by the rapid drop in permeation
rate in the concentration range from 4.5 to 5% solids as indicated
in Figure 3.   These results should be repeated and verified
before they are used for further design.   The average permeation
rate for these three trials at 3.45 MPa (500 psi) for the actual
concentration range of the experiment is given in Table 4.   These
values are somewhat lower than the value of 23.4 $1/h \cdot m^2$ (13.8 gal/ft^2
day) reported by Matsuura et al. (1973) for tomato juice in about
the same concentration range (5.9 to 12.9% solids) but under
conditions of no concentration polarization and at a much higher
pressure (6.9 MPa; 1000 psi).

    For tomato juice, the concentration obtained in these experi-
ments may be sufficiently high for practical purposes.   Tomato
juice processors are not particularly worried about heat damage
to the flavor since consumers prefer a slightly cooked tomato
flavor; and, therefore, there is less quality incentive to replace
evaporation by reverse osmosis, although heat damage to color and
vitamin C need to be optimized.   With tomato juice, the primary
concern is to reduce energy consumption during water removal and
to increase evaporator capacity.   Therefore, the goal of reverse

Table 4. Concentration of Tomato Juice by Reverse Osmosis.
Cellulose acetate membranes on ceramic cores at
3.45 MPa (500 psi) and 25°C

Juice	Concentration Range % solids	Average Permeation Rate $1/h \cdot m^2$	$gal/ft^2 day$
Hot break (104°C)	5.1- 9.1	11.7	6.9
Medium break (80°C)	4.5-13.1	16.5	9.7
Serum (centrifuged)	4.3-10.6	10.4	6.1

osmosis is to concentrate the solids two- to three-fold, that is,
to remove 50 to 75% of the water and then adjust the final concentration of the product by evaporation. Therefore, a reverse osmosis
concentrate of 12 to 18% solids might be reasonable for tomato juice,
whereas that would be near the starting concentration for orange
juice.

CONCLUSION

It must be emphasized that these results are preliminary and
the conditions that were used are not necessarily the optimum
conditions. The modules in these experiments were not pushed to
their limits; that is, to the highest pressures, the highest
recirculation rates, or the highest temperatures that might be
used. Nevertheless, it appears that sufficient progress has been
made in membrane and module development to continue to be optimistic
about the potential success of reverse osmosis for concentrating
fruit juices.

REFERENCES

D. B. Hosaka, 1978, "Concentration of Tomato Juice and Serum by
     Reverse Osmosis," M.S. Thesis, University of California, Davis.
J. A. Kavanagh, 1975, New Zealand J. of Dairy Sci. and Technol.,
     10:58-59.
T. Matsuura, A. G. Baxter, and S. Sourirajan, 1973, Acta Alimentaria
     Academiae Scientarium Hungaricae, 2(2):109-150.
R. L. Merson, and A. I. Morgan, 1968, Food Technol., 22:631-634.
G. Paredes, 1980, "Orange Juice Concentration by Reverse Osmosis,"
M.S. Thesis, University of California, Davis.
R. L. Riley, R. L. Fox, C. R. Lyons, C. E. Milstead, M. W. Serdy,
     and M. Tagami, 1976, Desalination, 19:113-126.
S. G. Wyllie, S. Alves, M. Filsoof, and W. G. Jennings, 1978,
     Headspace sampling: use and abuse, pp. 1-15 in "Analysis of
     Foods and Beverages: Headspace Techniques," G. Charalambous,
     ed., Academic Press, New York.

SURFACTANT MICELLE ENHANCED ULTRAFILTRATION

Pak S. Leung

Union Carbide Corporation
Tarrytown Technical Center
Tarrytown, New York   10591

INTRODUCTION

A large pore size ultrafiltration membrane usually cannot be used to separate small molecules. Even some tight and low-flux reverse osmosis membranes cannot separate small molecules, eg. phenol molecules. The high phenol solubility in the membrane phase is the difficulty. If the small molecules are held by larger particles, separation can be easily done by ultrafiltration and in high flux. We are reporting the ultrafiltration of small molecules, eg. phenol, bromophenol and metallic-ion chelating agents by the addition of micelle-forming surfactants. The affinity of these small molecules to the large micelle particles contributes to a high efficiency ultrafiltration separation.

EXPERIMENTAL

Ultrafiltration was performed using zirconia-precoat membranes at 60 psi and 40°C. The flux was usually about 100 gallons per square foot per day. All surfactants used were commercial products without further purification.

RESULTS AND DISCUSSION

Surface tension was used as a measure of the concentration of surfactant in the filtrate. The surfactant critical micelle concentration (cmc) was determined by surface tension measurement. Figure 1 shows the cmc of the surfactant, dimethyl dicocoammonium chloride (DDAC) solution. Measurements of filtered solutions with concentrations near the cmc (point A) were carried out at a four-fold dilution to render the surface tension method more sensitive.

It is seen that 50 ppm surfactant monomers leak through the ultrafiltration (point A). Table I lists the retention percentage of
some other surfactants. The surfactant DDAC is used to enhance the
separation of phenol. The percent rejection of phenol as a function
of pH at 1% surfactant concentration is depicted in Figure 2. It
is of interest to see that the rejection of phenol is higher at
higher pH. Although phenol is more soluble at high pH, the pheno-
lates probably interact more strongly with the positive micelle
particles. Figure 3 shows ultraviolet spectra of phenol in DDAC
solution as compared to that in aprotic solvents and in aqueous
solution at different pH. The spectral similarity between the
phenol molecules are "bound" to the micellar surface rather than
in micellar interior. At pH = 12, 87% of a 0.1% phenol solution is
retained. At the same surfactant concentration, 87% of a 0.2%
8-hydroxyquinoline, a chelating agent, and 98% of a 0.1% o-bromo-
phenol are retained.

It was found that certain micelles can enhance the retention
of certain chelating agents. Differences in retention of two chelating agents can be used to separate two metallic ions in a mixture. The principle will work as long as one of the two chelating
agents complexes one ion more strongly than the other and vice versa.
These principles are demonstrated in a preliminary way using dimethyl
dicoco ammonium nitrite as surfactant, and thiourea and catechol as
respective chelating agents for Ag^+ and Cu^{++} ions.

The percent retention by the surfactant micelle can be expressed in a simple relationship with a partition coefficient.
Figure 4 shows a mathematical derivation of ultrafiltration retention efficiency based on the partition principle. The theoretical
calculation compares favorably with the experimental data. The
results are depicted in Figure 5.

The surfactant retained in the ultrafiltration can be recovered by salting out or by solvation processes. Figure 6 shows the
decrease of phenol retention in the presence of methanol. Salt
can be used to enhance the separation of the surfactant from the
bulk phase. Both the surfactant and the methanol can thus be recycled in a continuous process.

TABLE I

RETENTION OF SURFACTANT MICELLES

SURFACTANT	ELECTROLYTE ADDED	% RETENTION
0.25% Na dodecyl sulfate	–	26%
0.25% Na dodecyl sulfate in 0.5% Tergital 12-P-6	–	69%
1% Tergital 12-P-6 (ethoxylated dodecyl phenol)	–	95%
1% dimethyl didecyl ammonium chloride 50% active	–	50%
1% dimethyl didecyl ammonium chloride 50% active	0.5% NaCl	97%
1% dimethyl dicoco ammonium chloride 50% active	–	99.5%

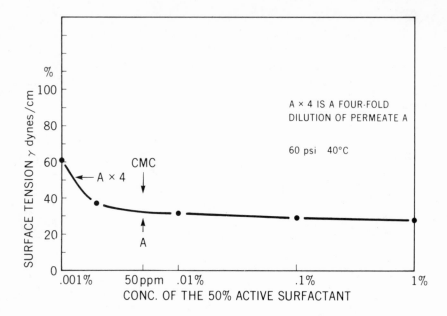

Figure 1. The retention of surfactant micelle from 1% dimethyl dicoco ammonium chloride 50% active as measured by surface tension.

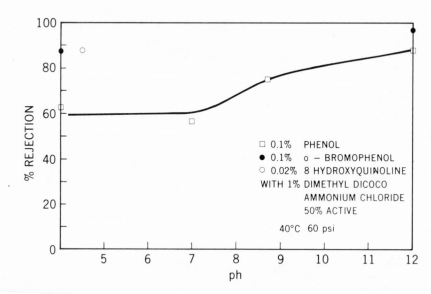

Figure 2. Phenol rejection as a function of pH at 1% surfactant concentration.

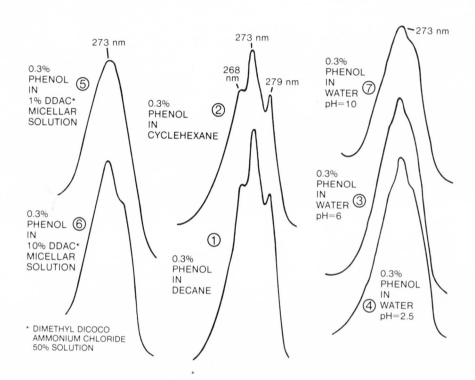

Figure 3.   The UV spectra of phenol.

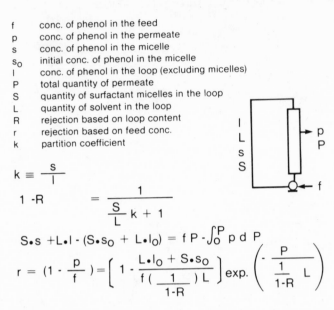

f   conc. of phenol in the feed
p   conc. of phenol in the permeate
s   conc. of phenol in the micelle
s_0 initial conc. of phenol in the micelle
l   conc. of phenol in the loop (excluding micelles)
P   total quantity of permeate
S   quantity of surfactant micelles in the loop
L   quantity of solvent in the loop
R   rejection based on loop content
r   rejection based on feed conc.
k   partition coefficient

$$k \equiv \frac{s}{l}$$

$$1 - R = \frac{1}{\dfrac{S}{L} k + 1}$$

$$S \cdot s + L \cdot l - (S \cdot s_0 + L \cdot l_0) = f P - \int_0^P p \, d P$$

$$r = \left(1 - \frac{p}{f}\right) = \left[1 - \frac{L \cdot l_0 + S \cdot s_0}{f\left(\dfrac{1}{1-R}\right)L}\right] \exp\left(-\frac{P}{\dfrac{1}{1-R} L}\right)$$

Figure 5. Comparison of theoretical calculation and experimental data.

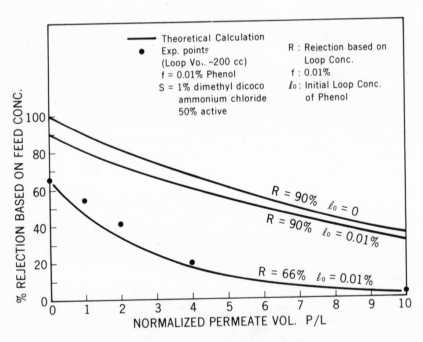

Figure 4. Derivation of ultrafiltration retention efficiency based on the partition principle.

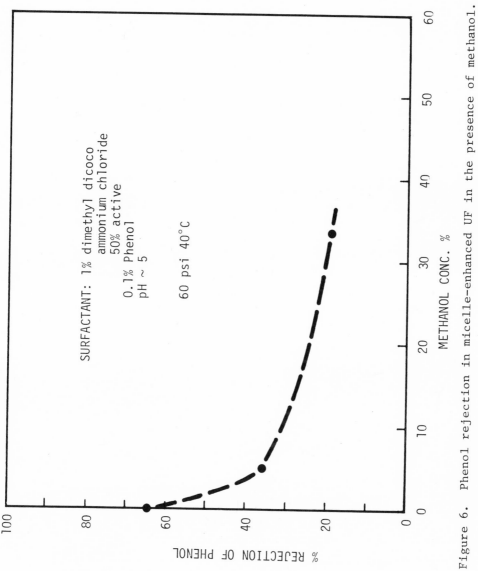

Figure 6. Phenol rejection in micelle-enhanced UF in the presence of methanol.

THIN-CHANNEL ULTRAFILTRATION, THEORETICAL AND EXPERIMENTAL APPROACHES

R.F. Madsen and W. Kofod Nielsen

A/S De Danske Sukkerfabrikker
Driftteknisk Laboratorium
DK-4900 Nakskov, Denmark

INTRODUCTION

During its approximately 20 years of life, as a unit operation ultrafiltration has had quite a turbulent history. During the first 10 to 14 years, the real fact was that very much work was done, showing that almost everything was possible, but very few applications were really successful. Then came a period in which it was a general opinion that the process could in reality be used successfully for very few applications only, and now we are in a period where, during the last few years, a great number of large plants have been built and are in successful operation.

Apart from a few early experiments and results, the history of ultrafiltration is closely connected to the history of reverse osmosis, beginning with the membranes made in the late fifties and the early sixties by Loeb and Sourirajan[1] and Manjikian et al.[2] at the UCLA. In the following years, different types of systems and membranes were developed. For some years, much enthusiasm was put into these processes, but with a few exceptions the practical results were very limited.

During the first years of development, almost all work was made in the U.S.A., and very little seemed to go on in other countries. A very large part of the total membrane and membrane-system developments was controlled by the US Office of Saline Water.

At that time, within the field of membranes, Europe did not seem to be active at all. This is probably true, but approximately in 1965, the membrane development was started in some European research centres.

In many respects, the development in Europe was different from the development in the U.S.A. This was possibly mainly due to the fact that in Europe the development was made primarily by a few private companies, looking at the membrane process not only as a method of producing drinking water from sea water or brackish water, but primarily as a new unit operation to be used for a number of different applications.

In the Danish Sugar Corporation, we were a few people who in 1965 began to look into the membrane processes. Being very far away from the main development centres within this field, we started the development of the membranes and membrane processes in some directions other than those in the U.S.A.

In our paper, I shall try to describe the status we have achieved so far, especially in the ultrafiltration field.

THE SYSTEM

The ultrafiltration system we use today (Fig. 1) is a system with flat sheet membranes. The liquid to be filtered is pumped between membranes situated at a distance of 0.8 mm or 30 mils. The units are built from moulded, perforated plastic plates which give a porous substructure for the membranes and form the channels where the liquid to be filtered flows between two membranes.

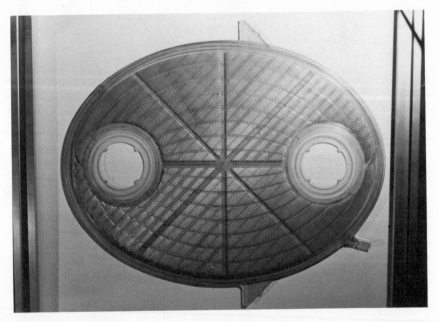

Fig. 1. Plate for ultrafiltration system, DDS-module 35.

Fig. 2. Plate with packing rings for DDS-module 35.

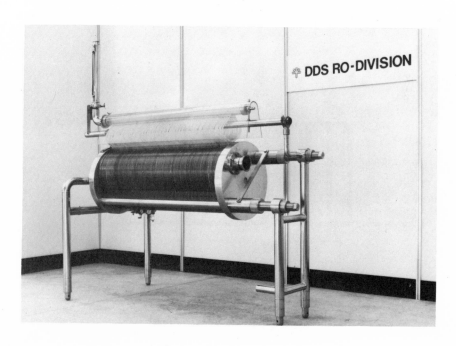

Fig. 3. DDS-module 35.

The two circular holes form the tubes from where the liquid is supplied to the flow channels and runs out. A membrane is placed on each side of these plates and is sealed around the holes by special packing rings (Fig. 2). The permeate leaves the plates through a tube in the edge of the plate. These plates are assembled in a module (Fig. 3) with up to 42 m^2 filtering surface (420 sq.ft.) The module is kept together by two bolts. This means that an exchange of one pair of membranes is possible just by opening the module at the place of the wanted exchange.

Figure 4 shows the internal flow principle of the module. The liquid to be filtered is flowing through a number of parallel membrane pairs. The module can be assembled to give as many steps in series as you want (uneven numbers only). This basic module can then be used in building large systems with many steps in series or parallel, each stage normally having its own recycling pump.

A large number of plants are now in operation. Figure 5 shows a typical plant for ultrafiltration of 500 tons whey per day, consisting of 30 modules.

In our case, the development of this module was one of the main reasons why ultrafiltration from being a difficult process suddenly became a unit operation which could be used successfully.

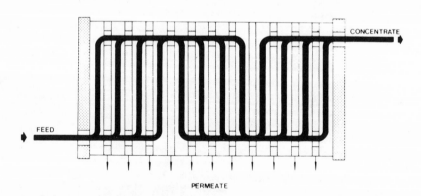

Fig. 4.   Internal flow principle of DDS-module 35.

Fig. 5.   DDS-module 35 plant for whey ultrafiltration.

Membranes

For a number of years, the main trouble for us as well as for other manufacturers was the bad thermal and chemical resistance of ultrafiltration membranes. However, during the latest years, a number of very resistant ultrafiltration membranes has been developed.

Table 1 gives a brief survey of the membrane types we produce today. The three ranges of membranes cover a very large number of potential applications. The GR types are, except for some applications on organic solutions and emulsions, the most all-round membrane. The FS types are specially well suited when organic solvents are involved, because they are attacked by very few organic solvents only. This has been of importance for ultrafiltration of penetrant oil and paints.

For high-alkaline resistance, the GR and FS membranes are produced on polypropylene backing paper; for other applications, they are often produced on polyester paper.

In most respects, the cellulose-acetate membranes have less chemical resistance than the two other membrane types. They are, however, still in operation in a number of plants, and in a few cases they have advantages over the two other membrane types.

Table 1. Properties of DDS-membrane types

Membrane type	GR	FS	CA (cellulose acetate)
Max. temperature $^{o}C(^{o}F)$	80 (180)	80 (180)	50 (122)
Acid resistance	Dilute mineral acids		pH 2
Alkaline resistance	>10% NaOH in operation	0.5% NaOH for cleaning	pH 8
Chlorine resistance (as NaClO)	Some 2% Some 0.01%	Good resistance Long-time experiments not finished	0.002%
Hydrogen peroxide	10%	do.	1%
Aliphatic alcohols	No limits	No limits	20%
Esters	No limits	No limits	Rather poor
Formaldehyde	5%		1%
Phenol	No harmful effect		Destroyed
Aliphatic hydrocarbons	No harmful effect		Destroyed
Other organic solvents, ketones, etc.	Limited resistance	Good resistance to most. Detailed information available	Very limited resistance
Water flux g/ft.^{2} d at 20°C			
Cut-off value ∿ 60,000 ∿ 20,000 ∿ 6,000	∿ 200 ∿ 130 ∿ 60	∿ 200	∿ 180 ∿ 100 ∿ 50

Design of ultrafiltration systems

Our experience in ultrafiltration has given some more general
rules for the design of ultrafiltration systems, which has been part
of the success achieved during the last few years.

This experience, I think, is of a more general nature and of
importance not only for our system. Below is given a short survey
of the important factors.

Permeate side: Liquid filled
                      Low volume
                      Liquid flow everywhere

Membrane exchange: Easy permeate control necessary
                      Easy exchange

Product side: Good liquid flow everywhere (no dead spots)
                      Minimum volume
                      Correct pressure and differential
                        pressure control
                      Minimum membrane stress
                      Energy economy

The permeate side

In some early constructions, very little care was taken to the
permeate side. In principle, the permeate side should have a clean,
sterile solution, and you might think that very little happens here.
However, a plant is not always in operation, and when a plant is
stopped, a risk of contamination exists. Important points are:
(A) The whole permeate volume must be filled with liquid, and
    no volumes must exist in which the liquid is not changed
    systematically.
(B) The permeate volume should be as small as possible.

The reason for this is that if the whole permeate volume is
not flushed completely during the cleaning operation, a great risk
of growth of bacteria and mould on the permeate side will exist.

For example in dairy operation, it is important that all lactose
is washed out of the system prior to a chemical sterilization. In
case of a large permeate volume, this will be almost impossible
without a very high time loss and the production of large amounts
of washing water. This means also that a reasonable flow of liquid
must exist everywhere at the permeate side.

Membrane exchange and product control

It is important to have an easy membrane exchange on site. In our modules, we have solved this problem in such a way that one pair of membranes can be easily exchanged everywhere in the plant just by opening the modules. The permeate from each pair of membranes can be visually controlled.

Prevention of "dead spots"

In each type of ultrafiltration equipment, the flushing of various parts of the membrane is somewhat different. Most important is that no place is flushed really badly, giving growth of scaling. With the cellulose-acetate membranes, it was a must that such places did not exist, because very often the membranes were destroyed behind such scaling. This does not occur with the modern, chemical-resistant membranes, but bacteriological growth may still occur, and a decreased product quality may be found because a little material remains in the plant and gives chemical reactions with the cleaning solvent.

Pressure and differential-pressure control

In the childhood of ultrafiltration, the knowledge of how to control the flow on the concentrate side and the pressure was limited and a number of plants were built, in which either the flushing had to be too small, or part of the membranes had to operate at too high a pressure.

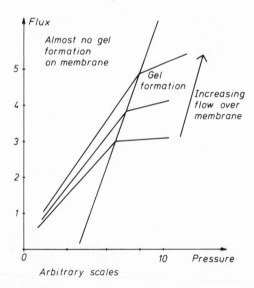

Fig. 6.   Variation of flux with flow and pressure.

Figure 6 gives a typical example of how, with arbitrary scales, the flux of a system varies with pressure and flow.

In a large number of experiments, operating with a transparent module and a coloured liquid (iron-dextran solution), we have compared this curve with the visual impression of gel formation on a membrane. By comparing these experiments, we can get a figure as shown in Fig. 7, showing the parameters we can allow for ultrafiltration and the parameters we cannot allow. The curve is only an indication, but it has the typical characteristics:

(1) A certain flow through the plant is necessary, even at low pressures and fluxes.

(2) With increasing pressures, increasing flows are necessary, and above a certain pressure (or flux), unrealistic, high flows are necessary.

It is extremely important that all the membranes in a plant are kept under conditions corresponding to the usable area on Fig. 7. On the other hand, an exact definition of the usable area for a specific plant, fluid, membrane, and concentration is not a very easy task.

As long as we keep ourselves within the usable area, unfortunately the economy of a plant increases the closer we get to the membrane-clogging area. This problem brings us into the main question in ultrafiltration: How to design the optimum flow channel.

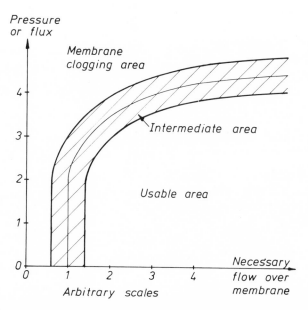

Fig. 7.   Parameters usable for ultrafiltration.

Flow-channel design

Ultrafiltration plants exist with different types of flow channels. Among these are tubular with a large diameter, thin tubes, spiral wound, and plate-and-frame. In this paper, we shall not go into a detailed discussion on the best design, but we shall try to give a description of the parameters important in designing. The main parameters are:

Energy consumption/volume unit permeate (operating costs)
Capacity/membrane-area unit (investment and membrane costs)
Capacity/volume unit (retention time, start-up time, losses during stops)

The best equations used for calculations of the most reasonable dimensions are those experimentally proved to be reasonably correct.

Goldsmith[3] showed that for turbulent cases, the equation

$$\frac{W \times D_H}{D} = K_1 \times Re^{0.913} \times Sc^{0.346} \tag{1}$$

where W is limiting mass-transfer coefficient, D_H hydraulic diameter, D diffusion coefficient, Sc Schmidt number (ν/D), ν kinematic viscosity, and K_1 a constant, derived from analogy to thermal transfer coefficient, agrees with experimental results.

For low Reynolds number laminar flow, Goldsmith has found a reasonable agreement with the equation

$$\frac{W \times D_H}{D} = K_2 \left(Re\frac{D_H}{X} \right)^{1/2} Sc^{1/3} \tag{2}$$

where X is the channel length.

In the range Re 100 - 3000 for flat channels, we have found the empirical relation[4],

$$W = \frac{K_3 \times D_H^0}{K_4 + K_5 \times \Delta P^{-0.6}} \simeq K \times \Delta P^{0.6} \times D_H^0 \tag{3}$$

with constant liquid and temperature. Channel heights 0.04 to 0.1 cm.

We can transform these three equations as functions of D_H, X, and either ΔP (pressure loss/cm channel), τ (shear force), or ΔE (energy loss/cm channel). In the turbulent case, we use a Fanning friction factor between $K_f \times Re^{-1/4}$ and $K_f \times D_H^{-0.3}$, corresponding to smooth and rough channels, and we get:

Goldsmith turbulent case (4)

$$W_\ell = K_A \times \Delta P^{0.52-0.45} \times D_H^{0.46-0.50} = K_A \times \tau^{0.5} \simeq K_B \times \Delta E^{0.33}$$

Goldsmith laminar case

$$W_\ell = K_C \times \Delta P^{0.5} \times D_H^{0.5} \times X^{-0.5} = K_C \times X^{-0.5} \times \tau^{0.5}$$

$$\simeq K_D \frac{\Delta E^{0.25}}{X^{0.5} \times D_H^{0.25}} \qquad (5)$$

Our empirical equation for slit flow

$$W_\ell = K_E \times \Delta P^{0.6} \times D_H^0 = K_E \frac{\tau^{0.6}}{D_H^{0.6}} = K_F \frac{\Delta E^{0.3}}{D_H^{0.9}} \qquad (6)$$

W_ℓ is the maximum obtainable fluxes.

The equations (4), (5), and (6) have almost the same exponents for ΔP and τ, but they are in complete disagreement in the dependence on dimensions.

We know that our equation is not valid for flat channels with $D_H > 0.1-0.2$ cm and have pointed out[4] that equations (5) and (6) are almost equal if the channel length X is defined to be proportional with D_H. This will be the case if secondary flows exist as vortices with diameters proportional to the channel height.

We do not know of published experiments where the dependence on D_H for really turbulent flows has been experimentally proved.

In our experiments, it has been proved[4] that the Fanning friction factor, even at low Reynolds numbers (100), is 1.5 to 3 times the expected for purely laminar flow for channel heights of 0.06 to 0.1 cm, but the average velocity is nearly proportional to the pressure loss.

Then, what are the reasons why the theoretical considerations as known today do not give us a reasonable tool for dimensioning flow channels for ultrafiltration?

I think, in our theories we have overlooked some factors. From a large number of experiments, we know that a flat membrane, operating in our systems, does not get a gel layer as an even layer but as a striping almost parallel to the flow direction and with a wavelength which increases with increasing channel height. This type of gel layer is easily removed, and it disappears when the operating pressure is removed, or the system is changed to water. An even layer is formed only if the operating pressure is increased above the point where the flux is almost independent of pressure.

I shall point out three known phenomena which may give some explanation of our observations. The phenomena are

    (a) Normal pressure formation in gels,
    (b) Structural turbulence,
    (c) Slip velocities.

Weissenberg[5] described the effect of the development in gels and other viscoelastic fluids of pressures normal to the direction of stress. These pressures[6] have been shown to produce very easily waves in gel films under stress or strain very similar to the waves we have seen as striping on the membrane surface.

Furthermore, a gel has the property[7] that the viscosity decreases with flow, this property being time dependent. It is also known that a so-called "structural turbulence"[7] exists where the critical Reynolds number is proportional to D_H^2.

The solutions interesting for ultrafiltration are almost all solutions where the gels and at least the concentrated solutions have this behaviour. Consequently, let us look at a hypothetical case (Fig. 8).

We suppose that the mechanism of ultrafiltration is the following:

The gel tries to start as an even layer on the membrane, but due to stresses in the membrane, normal pressures are developed, giving the gel structure a cross section as shown in Fig. 8, 2). When the gel "tops" increase, the stresses per cm^2 gel will decrease, and the flow into the gel tops decreases. Then, there will be space for new tops until a certain equilibrium has been obtained.

The existence of slip velocities in polymeric solutions has been predicted by several authors, and it is a well-known fact that viscoelastic fluids flow preferably where the highest shear stress exists and give often almost plug flow. The phenomena are described by Oldroyd[8], among others.

A mathematical treatment of all these phenomena is very difficult and without much meaning, because the measurements of the rheological behaviour of the ultrafiltration gels do not exist.

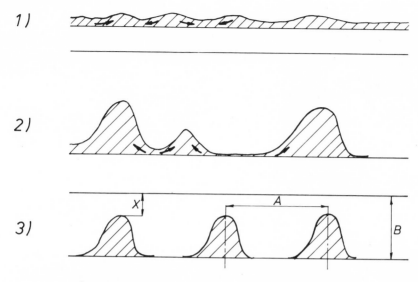

Fig. 8.   Cross section of gel formation.

Our observations and measurements give, however, some general rules, important for the dimensioning of ultrafiltration plants.

Limiting conditions for pressure/pressure difference are never to be exceeded. Gel formation is reversible for a given liquid, membrane, and system as long as the flux is kept below the limiting flux obtained at a high filtration pressure. If for a short time the flux and operation conditions are changed to conditions for strong gel formation, this gel formation can often be removed only by a cleaning procedure.

First of all, this gives construction limitations for normal operating conditions, preventing especially gel formation on the membrane with the highest filtration pressure. It also gives limitations on start-stop procedures, preventing pressure without flow on the membranes, even for short periods.

Temperature control is very important, especially on plants operating at an increased temperature and a high pressure. It is important to note that it can have a very bad effect to cool down a plant containing high-concentration protein. For example, mistakes have been made in sending cold water into such a plant or by letting it stand cooling down. This can clog a plant completely.

At last, the operating conditions during cleaning must be satisfactory. The problem is that during cleaning with thin solutions, the fluxes are very high. At an increased temperature, the fluxes

might be so high that special care must be taken to ensure that the
flow in the last flow channels in a plant is sufficient.

STATUS OF ULTRAFILTRATION AND FUTURE FOR THE PROCESS

Today, ultrafiltration is a process widely used in the dairy
industry, the enzyme and pharmaceutical industry, for electrophoretic
paint and oil emulsions, and in the pulp industry applications have
been initiated.

In the dairy industry, the major types of application are:

Ultrafiltration of whey. Fig. 5 shows one of our large plants
in operation, making products with protein contents from 35% to 90%.
Skimmilk and whole milk are ultrafiltered as a pretreatment before
the production of various acidified milk products and some cheese
products (the MMV process developed by Maubois[9]). Our main plants
today produce acidified milk products and Feta cheese.

For a number of other cheeses, the process is now in development, but some difficulties have still to be overcome. The main
difficulty is calcium having to be removed from the concentrate
in order to produce a good cheese. This means that the milk to be
treated must be acidified to a pH of approx. 6-6.2 at ultrafiltration. If the milk is somewhat acidified before ultrafiltration,
this is possible, but it also makes ultrafiltration more difficult
to perform.

The problems are solved if the first acidification can be made
by lactic acid and not by direct culture.

The capacity of a plant depends on four main factors:
(1)   Type of whey or milk
(2)   Pretreatment
(3)   % protein in final product
(4)   Operating temperature.

The capacity is very dependent on the protein concentration
and the pH. All dairy products give higher fluxes if they have been
exposed to a strong heat treatment before ultrafiltration. In many
cases, however, this is not possible, because some of the properties
of the final products are lost.

In all cases, the flux depends on $\ln(c/c_0)$, c being actual
protein concentration and c_0 a limiting concentration.

From a capacity point of view, the operating temperature is
very important. In most cases, a temperature of 50-55°C will be the
ideal temperature. At higher temperatures, calcium-phosphate complexes easily clog the membranes, at lower temperatures, the flux

decreases by 2-3% per degree centigrade. However, most plants in operation today have cold operation, from 4 to 12°C. This is partly a bacteriological problem, but with our present membranes where cleaning and sterilization are made by NaOH, HNO_3, H_2O_2, or hypochlorite, this problem is not a serious one.

A main problem is the functional properties, and in the plants for production of cheese and high-quality protein, the operating conditions are always chosen according to quality criteria and not to capacity criteria.

During the last few years, in the pulp and paper industry, the development has resulted in some large-scale pilot plants being in operation for production of lignins or lignosulfonates. A number of research laboratories are developing on the use of lignosulfonates, and it may be expected that within the next few years, the process will be developed into large plants.

For decolorization of bleaching liquor, a large pilot plant has been in successful operation for 2 years in Sweden.

In the pulp industry, ultrafiltration at 60-70°C has been a great success, because fluxes of 100 to 300 $1/m^2/h$ (70-200 g/sq.ft.d) have been obtained, giving a much better economy than at low temperatures.

The high fluxes obtained at high temperatures make it economically reasonable to use the process where it was previously impossible.

Ultrafiltration of organic solutions is another coming field. So far, it has been really possible only for a short time. It is difficult to guess what kind of applications will succeed.

Very much has been said about membranes more specific than those produced today, but in this respect the success has been limited. Personally, I am not too optimistic. For economical reasons, the membranes must have very good characteristics, with rejection of one component of more than 90% and rejection of another component of less than 10%.

In oil-emulsion ultrafiltration, a major problem has been some penetrant oils destroying the early membranes. The new FS membranes have solved this problem.

In the pharmaceutical and enzyme industries today, a large number of applications are in operation, including concentration and purification of enzymes, iron dextran, blood plasma, insulin, gums, gelatine, etc.

In the near future, ultrafiltration of low-concentration streams at high temperatures and high fluxes will make most thought processes economically possible.

Personally, I think that the use of ultrafiltration will grow rapidly but that we have reached a point of development of equipment and membranes where, of course, future developments will still give improvements with steps forward, but the steps will be smaller than seen in the years behind.

In applications, a lot has yet to be done, many processes given up a few years ago have to be reconsidered, because things impossible three years ago are now being possible or even easy to make.

REFERENCES

1. S. Loeb and S. Sourirajan, UCLA Department of Engineering Report 60-60 (1960).
2. S. Manjikian, S. Loeb and J.W. McCutchan, Proc. of the First Int. Symp. on Water Desalination, Washington D.C., Oct. 3-9, 1965, Vol. 2, pp. 159-173, USDI, Office of Saline Water.
3. R.L. Goldsmith, Ind. Eng. Chem. Fund., Vol. 10 (1971) 113.
4. R.F. Madsen, "Hyperfiltration and Ultrafiltration in Plate-and-Frame Systems", Elsevier, Amsterdam (1977).
5. K. Weissenberg, Proc. 1st Intern. Rheol. Congr., Scheveningen, p. I-29 (1948).
6. A. Jobling and J.E. Roberts, Goniometry of Flow and Rupture, in: "Rheology", F.R. Eirich, ed., Academic Press Inc., New York (1958), Vol. II, pp. 503-535.
7. A.G. Ward and P.R. Saunders, The Rheology of Gelatin, in: "Rheology", F.R. Eirich, ed., Academic Press Inc., New York (1958), Vol. II, pp. 313-362.
8. J.G. Oldroyd, Non-Newtonian Flow of Liquids and Solids, in: "Rheology", F.R. Eirich, ed., Academic Press Inc., New York (1956), Vol. I, pp. 653-682.
9. J.-L. Maubois, Séparations par membranes, échanges d'ions et cryo-concentration dans l'industrie alimentaire, A.P.R.I.A., Paris (1975).

AUTOMATED HOLLOW FIBER ULTRAFILTRATION:

PYROGEN REMOVAL AND PHAGE RECOVERY FROM WATER

Georges Belfort,[1] Thomas F. Baltutis[2] and William F. Blatt[3]

[1]Department of Chemical and Environmental Engineering
Rensselaer Polytechnic Institute, Troy, NY 12181; [2]Harza
Engineering Company, Chicago, IL 60606; [3]Amicon Corp.
Lexington, MA 02173

ABSTRACT

The operational aspects of automated hollow fiber ultrafiltration systems are reviewed both with respect to filtrate utilization
as well as a recovery system for membrane-retained species. Typify-
ing the former is the removal of pyrogens, while an example of the
latter is the removal and subsequent recovery of f2 bacteriophage
from both tap and lake water. While preparation of colloid-free filtrate can be easily explained by sieving governed by the pore size of
the membrane, the recovery of retained species necessitates a more
complex model. With respect to phage, logarithmically decreasing
concentrations of phage are recovered in the concentrate solution by
successive backflushes of the membrane with filtrate. A model is
proposed to explain the logarithmic decrease in concentration as a
function of the backflush flux and volumetric throughput. The model
can be used to predict the total number of original phage in a water
sample filtered through the unit on the basis of an assay of the
phage in a _single_ backflush. Phage was selected as a model system to
establish operational parameters for extending these studies to
encompass enteroviruses in water.

SYMBOLS

A - Permeation surface area of membrane, L^2
C - Ratio of pore to approach velocity, -
C_R - Group of constants defined in Eq. 9, L^{-3}
D - Drag force, MLT^{-2}
d_p - Particle diameter, L

439

F_A – Attachment force, MLT^{-2}
F_R^A – Net removal force, MLT^{-2}
G_i^R – Pressure gauges for inlet ($i = 1$) and the accumulate
reservoir ($i = 2$), MLT^{-2}
I – Backflush impulse, MLT^{-1}
i – Backflush number, –
k – Proportionality constant defined in Eq. (6), $TM^{-1}L^{-1}$
L_i – Limit settings on the pressure gauges, MLT^{-2}
N^i – Number of attached particles, –
N_i – N after the ith backflush, –
N^i – N before first backflush, –
N_R^o – Number of particles removed in a backflush, $N_R = N_o - N_i$, –
n^R – Integer number of backflush, –
Q – Backflush volumetric flow rate, L^3T^{-1}
r – Correlation coefficient, –
t – Duration of the experiment, T
t_i – t after the ith backflush, T
t' – Summation of the duration of n backflushes, T
V^n – Accumulated backflush volume, L^3
V_i – V after the ith backflush, L^3
V'_n – Summation of the accumulated volumetric backflushes for n
backflushes, L^3

Greek

μ – Dynamic viscosity, $ML^{-1}T^{-1}$

INTRODUCTION

The use of flat sheet permselective membranes to specifically remove pyrogens from parental solutions has recently been reviewed by Nelsen[1], while the general efficacy of ultrafilters in virus retention has been the subject of several papers (2-7). Although the problems of pyrogen removal and virus concentration would at first glance appear highly disparate, basically both present the same challenge, i.e., the capture of material at extremely low concentration from large process volumes.

Conventional flat membrane systems are characterized by relatively low membrane area to process volume ratios. For high volumetric throughput with reasonably small systems this limitation could be serious. Accordingly, a possible alternate approach would be to use fine hollow fibers which considerably augment the effective membrane area. In this system, however, one limitation in concentrating and recovering materials from very dilute solutions has been solute loss during a recirculation regimen. The seiving of macromolecular solutes during the filtration process results in a build-up of these solutes on or near the membrane surface. This process is termed concentration polarization. To maintain high permeation fluxes, these solutes must be removed using adequate hydrodynamic forces resulting from cross-flow operation and/or backflushing the membranes by reversing the permeation direction. Chemical and physical cleaning techniques are also used for effective flux restoration [8].

By using periodic backflushing for pyrogen removal and virus capture, it became apparent that one could use total entrapment (no recirculation) with highly dilute solutions, and minimal flux reduction would be observed. In the pyrogen application, this would mean sustained performance with short down times for backflush and cartridge restoration. With virus capture, this could afford a means of processing substantial quantities of fluid volumes using relatively small modules with recovery of the entrained virus in a highly reduced volume.

Details of the method were first reported by Belfort et al., (4-7) who used hollow fiber ultrafiltration to concentrate poliovirus 1 from a 100 l of tap water. After processing, several backflushing steps (with and without solutes) were used to recover the polarized solutes including the poliovirus 1, which was then assayed-for using tissue culture techniques. Excellent virus recoveries were attained in a few hours.

Carrying the operational logic further, it became apparent that adsorption or entrainment via "dead-ended" systems would result in pore occlusion, either reducing flow at constant driving pressure, or alternately, leading to an increase in fluid inlet pressure at constant flow. It was this latter consideration that led us to a simple method for cycle automation.

In the sections that follow we will prove out the efficacy of a hollow fiber system subjected to varying levels of pyrogen challenge, further develop the rationale and operational control in the automated system and discuss our preliminary results with the system for solute (bacteriophage f2) capture. In addition, a theoretical model is presented to describe the recovery of macromolecular solute during the backflush steps.

THEORETICAL

Backflush Recovery Model

During ultrafiltration, both solutes and solvent are carried by convective forces toward the membrane surface. Those solutes that are partially or totally rejected concentrate in the viscous boundary layer adjacent to the membrane-solution interface. The concentration profile reaches its maximum at the membrane surface interface where solutes are exposed directly to various surface forces. These include non-covalent (physical) interactions such as ionic and hydrogen bonding which are positive attractive forces. Hydrophobic

interactions, on the other hand, originate from a net repulsion
between the water and the non-polar regions of the membrane surface
allowing non-polar regions on the solute to associate with similar
regions on the membrane surface. Clearly, the lower the interaction
energy between membrane and solute the easier the solute can be
recovered by backflushing.

Assume that a number, N_0, of macromolecular solute particles are
attached in a monolayer on a membrane with surface area, A, by a
force, F_A. During a backflush, the hydrodynamic drag force from
Stoke's Equation is,

$$D = 3 \pi \mu d_p C [Q/A] \tag{1}$$

where C is the ratio of the pore velocity to the approach velocity,
$\frac{Q}{A}$, and d_p is the particle diameter, (see Fig. 1). The net removal
force F_R acting on a particle is

$$F_R = D - F_A \tag{2}$$

The removal of particles from the surface is not assumed to be
instantaneous when $F_R > 0$. Rather, it is assumed that the particles
must be subjected to the removal force for some length of time to
overcome the membrane-particle interaction force. The removal
impulse, I, is defined as

$$I = \int F_R \, dt \tag{3}$$

The incremental impulse with respect to time is

$$dI = F_R \, dt \tag{4}$$

The incremental decrease in the number of particles attached to
the surface for an incremental impulse is assumed to be proportional
to the number of particles attached to the surface,

$$\frac{-dN}{DI} \propto N \tag{5}$$

A proportion constant k may be introduced so that Eq. (5) may be
written as

$$\frac{-dN}{dI} = k N \tag{6}$$

Substituting for dI from Eq. (4) gives

$$\frac{-dN}{F_R dt} = kN \tag{7}$$

Rearranging Eq. (7) gives

$$\frac{dN}{N} = k F_R \, dt \tag{8}$$

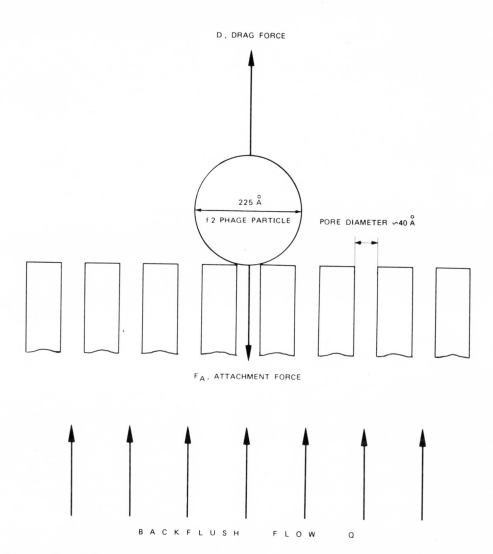

Figure 1.   Schematic of Bacteriophage f2 on Membrane Surface During Backflush

Substituting for F_R from Eq. (2) gives

$$\frac{dN}{N} = \left(\frac{-3 \, k \, \mu \, \pi \, d_p}{A} Q + k \, F_A\right) dt \tag{9}$$

The group of constants in the first term on the right side are specific for a defined particle-membrane system. This group of constants will be designated C_R, the particle volumetric removal rate.*

During a backflush, the number of particles on the membrane surface decreases from N_o to N_i as the time goes from t_o to t_i. Eq. (9) may be written as

$$\int_{N_o}^{N_i} \frac{dN}{N} = \int_{t_o}^{t_i} C_R \, Q \, dt + \int_{t_o}^{t_i} k \, F_A \, dt \tag{10}$$

Noting that the integral of flow Q over time is the backflush volume, V_i, Eq. (10) may be integrated to obtain

$$\ell n \, \frac{N_i}{N_o} = C_R \, V_i + k \, F_A \, t_i \tag{11}$$

Dividing by the time of backflush t_i gives

$$\frac{1}{t_i} \, \ell n \, \frac{N_i}{N_o} = C_R \, \frac{V_i}{t_i} + k \, F_A \tag{12}$$

Equation (11) states that the change in the number of particles attached to the membrane surface is a function both of the volume of backflush water and the duration of the backflush. Eq. (12) enables one to evaluate the constants C_R and $k \, F_A$ by graphing

$$\frac{1}{t_i} \, \ell n \frac{N_i}{N_o} \quad \text{against} \quad \frac{V_i}{t_i}$$

The number of particles remaining on the surface after one backflush may be removed by backflushing several more times. Eq. (10) may be extended as follows:

$$\int_{N_o}^{N_1} \frac{dN}{N} + \int_{N_1}^{N_2} + \cdots \int_{N_{n-1}}^{N_n} \frac{dN}{N} = C_R \left[\int_{t_o}^{t_i} Q \, dt + \int_{t_1}^{t_2} Q \, dt \cdots \int_{t_{n-1}}^{t_n} Q \, dt \right]$$

$$+ k \left[\int_{t_o}^{t_1} F_A \, dt + \int_{t_1}^{t_2} F_A \, dt + \cdots \int_{t_{n-1}}^{t_n} F_A \, dt \right] \tag{13}$$

---

*C_R has dimensions L^{-3} because k must have dimensions $TM^{-1}L^{-1}$ to satisfy Eq. (6).

Integration gives

$$\ell n \frac{N_1}{N_o} + \ell n \frac{N_2}{N} + \cdots \ell n \frac{N_n}{N_{n-1}} = C_R[V_1 + V_2 + \cdots V_n]$$

$$+ k F_A[t_1 + t_2 + \cdots t_n] \tag{14}$$

Thus for n backflushes,

$$\ell n \frac{N_n}{N} = C_R V_n' + kAt_n' \tag{15}$$

where N_n designates the number of particles remaining after the nth backflush and V_n' and t_n' are the cumulative backflush volume and duration respectively.

Use of the Predictive Model

Equation (11) can be used to predict the total number of particles initially attached to the membrane surface on the basis of number of particles recovered in the backflush volume. Noting that the number of particles recovered N_R in the backflush volume is the difference between the number of particles attached before and after backflush, i.e.,

$$N_R = N_o - N_i \tag{16}$$

This can be used to predict N_o based on N_R. Substituting Eq. (16) into Eq. (11) gives

$$\ell n (\frac{N_o - N_R}{N_o}) = C_R V_i + kF_A t_i \tag{17}$$

Rearranging gives

$$N_o = \frac{N_R}{1 - exp\ (C_R V_i + kF_A t_i)} \tag{18}$$

Having evaluated the constants C_R and kF_A for a particular macromolecular solute particle-membrane system using Eq. (15) for which recovery data on each backflush must be obtained, N_o may be calculated based on N_R, V_i and t_i for a single backflush.

MATERIALS AND ANALYTICAL METHODS

Testing for Pyrogen

Lexington tap water (fortuitously, an ideal pyrogenic test solution), served as our basic process stream. Evaluation of pyrogen content was by the Limulus Amebocyte Lysate Assay using Pyrotell Kit #1 (associates of Cape Cod., Inc., Woods Hole, Mass.), while Pyrotell Positive Control Lot #3 served as the endotoxin standard. Dilutions were performed with non-pyrogenic, sterile water from Abbott Labs

(Chicago, Illinois) using sterile disposable plastic vials (also demonstrated free of pyrogenic response in our test system).

Test on the Lexington water was as follows: approximately 4mls of tap water was collected into a sterile tube after allowing the water to run for one full minute. Dilutions of this sample were performed in sterile water at 1/3 and 1/10. Each of these basic samples was then serially diluted, as above, and the titers recorded for both samples. Values are expressed as half log dilutions. Table I summarizes the results. Lexington tap water had a positive titer of 1/100 and a $\pm$ reaction at 1/300. The control (100 ng/ml) is positive at 1/100 and negative at 1/300 which corresponds to the advertised claim of 0.125 ng/ml sensitivity.

For system evaluation, three basic challenges were offered to check retentivity of the cartridges under varying "load conditions".

Cell Cultures and Methods

E. Coli (Strain K13) cells were used for the growth and plaque assay of the f2 phage throughout this study. A seed culture of these cells were initially obtained through the courtesy of Dr. Vincent D. Olivieri, John Hopkins University, Maryland. The cells were grown on TYE-Agar medium in plastic tissue culture plates. The TYE-Agar (19 gms of Bacto-Tryptone Yeast Extract and 15 gms of Agar in 1 l H_2O), was autoclaved at 120°C for 20 minutes. Incubation of E. Coli cells was 37°C in 90 to 95 percent relative humidity and 5% CO_2 atmosphere.

Bacteriophage f2

The bacteriophage f2 stock solution was prepared from a sample kindly supplied by Dr. V. Olivieri. The f2 was allowed to infect E. Coli Strain K13 for a lytic period of about five to six hours at 37°C. In a TYE solution (19 gm TYE/1), the resultant bacteria by centrifugation at 7000 rpm for a half an hour. The resultant stock solution was stored at 4-5°C and usually contained from 10^9 to 10^{12} PFU/ml.

Virus Assay

The viral containing sample (i.e., stock dilutions) was added to Tryptone - Agar containing plates via a soft agar overlay lawn (19gms of Tryptone Yeast Extract and 7gms of Agar in 1 liter H_2O) and four drops of E. Coli Strain K13/$CaCl_2$ solution. After an eight hour incubation period, the plaques were counted.

EXPERIMENTAL

Two sets of experiments were conducted. The objective of the

Table I

PYROGEN TESTS USING LEXINGTON TAP WATER

Sample	Sterile water ml	Tap water ml	Dilution factor	Reaction[1] of 0.2 ml sample
starting material	0	starting material	no dilution	+
A	2	1	1/3	+
B	9	1	1/10	+
C	9	1 (of A)	1/30	+
D	9	1 (of B)	1/100	+
E	9	1 (of C)	1/300	(±)[2]
F	9	1 (of D)	1/1,000	–
G	9	1 (of E)	1/3,000	–
H	9	1 (of F)	1/10,000	–
I	9	1 (of G)	1/30,000	–
J	9	1 (of H)	1/100,000	–

Sterile Tube Check

K	4	—	—	negative
L	4	—	—	negative
M	4	—	—	negative

[1] Positive control (100 ng/ml) positive at 1/100; negative at 1/300 (manufacturer claims .125 ng/ml sensitivity).

[2] Thick cloudy fluid not a typical gel, but not completely fluid.

first set was to determine the efficiency of pyrogen removal for
different pyrogen challenges. Experiments with the preliminary and
the automated system are discussed. The objective of the second set
was to determine if a model virus, bacteriophage f2, could be
concentrated from a tap and lake water feed using the automated
system. Virus recovery during successive backflushes was also
studied. The Backflush Recovery Model is also tested and then used
to predict the concentration of phage in the original feed solution.

Pyrogen Studies

1. Preliminary System

In all studies, an Amicon H10P10 (nominal retentivity, 10,000
M.W.; area $\equiv$ 10 ft^2) was used. Insofar as pyrogen testing involves
scrupulous preparation of materials in contact with the filtrate, a
relatively rigorous cleaning of the cartridge is necessary. A
scheme depicting this is shown in Fig. 2. Initially some 10-15 l
of distilled deionized water are put through in forward flow
operation to remove all glycerine from the fibers. This is then
followed by 10 l of 0.1 N NaOH in backflush mode and 10 l of 0.1
N HCL in forward flow. To remove all traces of acid and alkali,
sufficient distilled, deionized water in forward flow is used until
the pH of the filtrate tests near neutrality (between pH 6-8). The
initial cartridge preparation involves all the steps utilized above.
For cleaning between runs, one goes directly to the NaOH backflush,
or steps 2-5 shown on Fig. 2. The basic system employed for the
initial studies on pyrogen retention and retrieval is shown on
Fig. 3. A pressure gauge atop the cartridge and a lower elbow
fitting with exit valve for backflush comprise the primary aspects
of the system; a peristaltic pump, process stream and filtrate
reservoirs complete the unit. The left portion of Fig. 3 depicts
the unit in filtrate production. Process fluid is introduced into
the fiber lumen (egress is blocked by the gauge) while the filtrate
expressed through the fiber walls is collected in a reservoir
connected to the upper filtrate port. The right hand side of the
figure illustrates the backwash mode whereby solvent (generally
ultrafiltrate or cleaning solution) is reverse-flowed through the
fibers carrying membrane-retained species down the lumen and out
the lower elbow valve into a collection vessel.

The larger part of our detailed studies on pyrogen removal from
low and high challenge solutions was done on this basic system.
Practical considerations for unattended operation led us to develop
the automated system described below.

2. The Automated System

As noted earlier, pressure actuation provided the logic to
control the system, i.e., filtrate production at constant flow is

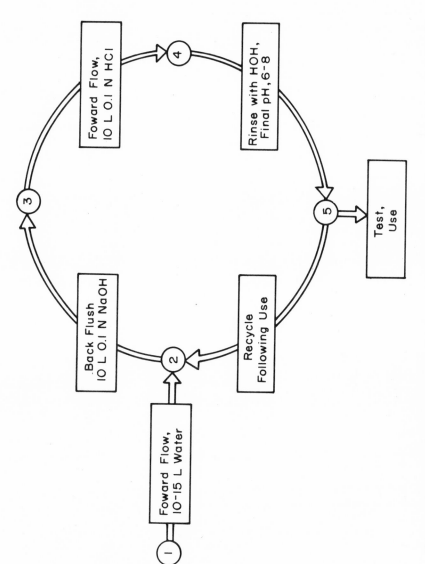

Figure 2. Hollow Fiber Cleansing and Restoration Cycles for Pyrogen Use.

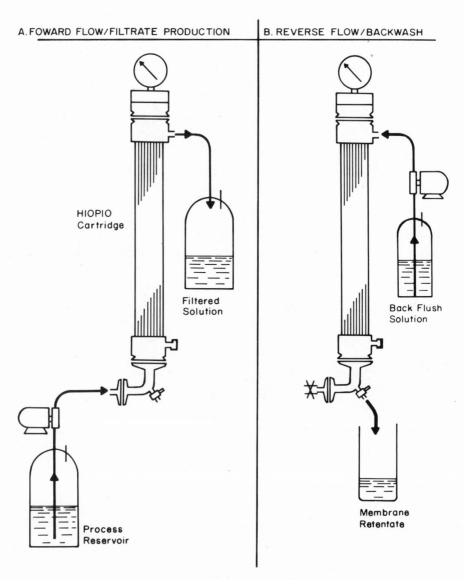

Figure 3.   Basic Hollow Fiber Test System for Ultrafiltrate Production and Retentate Collection

sustained until sufficient back pressure is engendered to trigger an automated backwash followed by restoration of fiber permeating flow.

How this is accomplished can be visualized by the rather simple rendition of the single fiber unit with shell casing shown in Fig. 4. In the filtration phase, shown atop the figure, process fluid at constant rate is directed down the fiber "dead ended" by a valve at the bottom. Solvent and permeating species are expressed through the fiber while retained species are concentrated within the fiber lumen and/or deposited on the fiber wall. Eventually, by virtue of a progressively increasing accumulate layer of membrane retained species, inlet pressure will be raised to a level that triggers the second phase, backwash accumulate. This occurs when the valve in the filtrate line is closed, but process flow is maintained thus causing fluid buildup in an accumulator reservoir attached to the upper arm of the casing. When pressure in this reservoir reaches a pre-set level, inlet flow is stopped and the valve leading to the fiber lumen opens; reverse flow commences and accumulated species are expressed into the collecting vessel (backflush phase). This would be waste in the case of pyrogen removal, but represents the content of interest in the application involving phage or virus entrapment.

A detailed schematic of the operational aspects of the actual system is shown in Fig. 5. Two pressure gauges, each with two pressure limit settings serve to monitor inlet pressure and backwash reservoir pressures (G_1 and G_2 respectively). Three shut-off valves (v_1, v_2, v_3) serve to control process flow, filtrate flow and fiber content. A peristaltic pump and flow meter in the process flow line complete the system. With no slippage or backflow through the pump, process flow rate is equivalent to the filtrate production rate.

The lower part of the figure summarizes the operational phases and changes that occur. During filtrate production, valves v_1 and v_2 remain open while v_3 is closed. The inlet pressure (G_1) is slowly rising as the result of solute deposition. When G_1 reaches the first limit setting (L_1), valve v_2 closes and backwash accumulation begins. When G_1 reaches the second limit setting (L_2) backwash commences. The pump is stopped, v_1 closes, and v_3 opens, permitting reverse flow under the pressure engendered within the accumulate reservoir. The lower limit setting of G_2 (L_3) is the mechanism for restarting the initial phase, i.e., restoration of filtrate production by pump restart, closure of v_3, and opening of v_1 and v_2.

An overview showing the operational changes and "triggering" aspects as reflected by the pressure levels changes in gauges G_1 and G_2 is shown in Fig. 6. While curve shapes representing the three phases remain essentially the same, one can induce temporal effects by varying the different limit settings. For example, setting L_2 substantially above L_3 will lead to a longer accumulate cycle and a

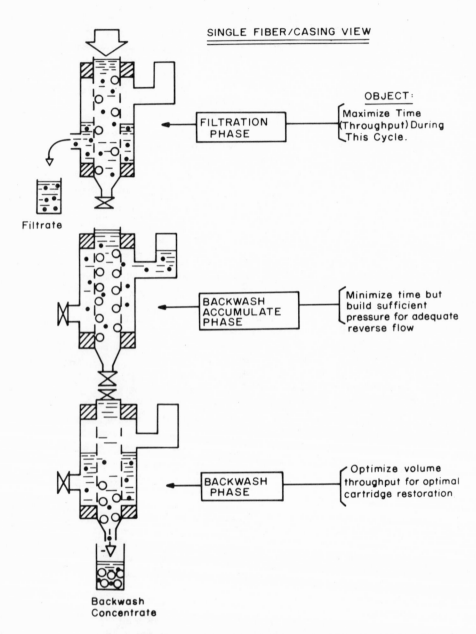

SINGLE FIBER/CASING VIEW

OBJECT:
Maximize Time
(Throughput) During
This Cycle.

FILTRATION PHASE

Filtrate

BACKWASH ACCUMULATE PHASE

Minimize time but
build sufficient
pressure for adequate
reverse flow

BACKWASH PHASE

Optimize volume
throughput for optimal
cartridge restoration.

Backwash Concentrate

Figure 4. Rationale for Automated Control (Single Fiber View)

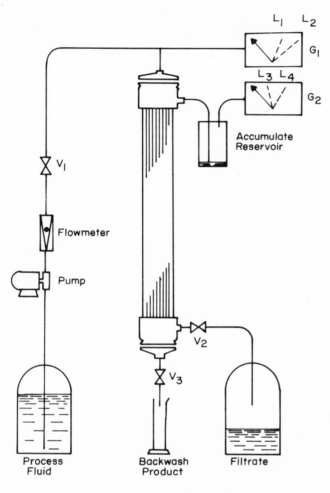

OPERATIONAL PHASES

	Filtrate	Accumulate	Backwash
Pump	Running	Running	Off
G_1	Rising	Rising	Falling
G_2	Stable	Rising	Falling
V_1	Open	Open	Closed
V_2	Open	Closed	Closed
V_3	Closed	Closed	Open

Figure 5. Operational Aspects of an Automated Hollow Fiber System for Solute Entrapment.

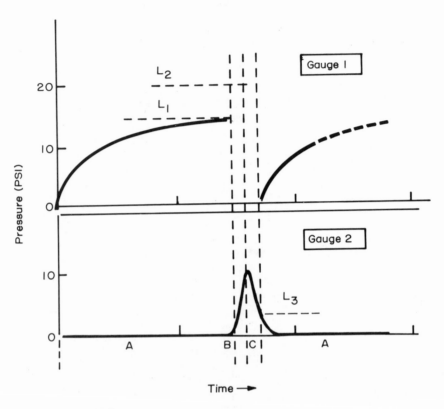

Figure 6.   Pressure-Activated Changes in the Automated Hollow Fiber
System.

greater backwash pressure. Lowering L_3 can provide far more sustained backwash time (and volume). The entire pattern, of course, can also be altered by a change in the process flow rate.

3. Pyrogen Removal Procedure

a. Processing of 100 liters of tap water.
Process fluid was introduced at a flow rate of approximately 420 ml/min. Samples of ultrafiltrate were collected for pyrogen testing every 4 liters of filtrate; inlet pressure was continuously recorded and the flow rate adjusted to the start level if minor corrections were necessary. Following this study, the backwash procedure (using ultrafiltrate) was used to express the entrapped material within the hollow fibers. Roughly 100 ml sequential aliquots were taken and checked for "titer" and the data used to estimate recovery.
b. Processing 500 liters of tap water.
At a flow rate of 500 ml/min the general procedure outlined above was followed. In this case it was necessary to continuously add tap water to replenish our stock 100 l reservoir. The system was shut down overnight but continued for several hours on additional working days until a total volume of 500 l was processed. This study was two-fold in purpose. By pooling of the retentates, one objective was to collect sufficient pyrogenic material for the high challenge system (the 20 l of "heavy load" solution, described below). The other was to evaluate the capacity of the system in sustained operation (by monitoring the inlet pressure as a function of volume processed). The effectiveness of pyrogen removal was examined by taking samples of ultrafiltrate every 50 l. In the backflush mode, initial samples were taken every 500 ml and later every liter with recovery estimated as described above.
c. Processing of 20 liters of "heavy load" solution.
The total backwash material recovered in the above experiment (from 500 l of tap water) was collected in a reservoir and brought up to a final volume of 20 l with additional tap water. This sample was considered the "heavy load" starting material. Prior to this final study, the system was thoroughly cleaned as outlined in the standard procedure, then finally rinsed with 4 l of "pure" ultrafiltrates. The sample was processed as above with aliqouts of filtrate taken every four liters for pyrogen test. Backflush sampling and analysis were the same as in the second study.

Bacteriophage f2 Concentration Studies

In these concentration experiments, only the automated system was used to filter from 20 to 100 l of seeded tap water containing bacteriophage f2. After the entire sample was filtered, the filtrate was substituted for the sample water on the feed side of the pump, and the unit was switched from the "filtering" mode to the "accumulating" mode. The unit would automatically go into a backflush when the preset pressure set point was reached (L_2 on G_1 in Fig. 5). The set point L_3 was set such that the unit would stop backflushing when the desired volume of backflush water was obtained. The unit was then allowed to go through an "accumulating" and "backflushing" cycle several times. Phage concentration was measured individually for each backflush.

RESULTS AND DISCUSSION

Pyrogen Challenge

1. Preliminary System.

    a. 100 liters tap water – The processing of 100 l of tap water required 267 minutes and at the flow utilized attained a final inlet pressure of only 3.5 psi; evidently we were operating well below the capacity of the system and the process flow probably could have been set much higher. All evaluation tests on the filtrate proved negative for pyrogen. The larger emphasis in this initial study of the sensitivity level employed (approximately 0.125 ng/ml) was entrapment of pyrogen and validation of same. Table II summarizes the results obtained by titering the sequential backwash aliquots and calculating the percent recovery. The cumulative total of 52% recovered would appear to be extremely reasonable considering the low content and high membrane area that might afford some measure of irreversible adsorption under the conditions of this study.

    Fig. 7, which plots the reciprocal of the last positive titer taken from the aliquot data against backwash volume, demonstrates the backwash "peak", i.e., the volume at which the highest content of pyrogen positive material is released. In this study then a backwash volume of approximately 500 ml would prove effective in removal of a substantial portion of the entrapped material. Also note (from Table II) that the total pyrogen positive material is estimated at 1.25 mg, and consequently a recovery value of 50% from the 100 l starting volume is indeed encouraging.

    b. 500 liters tap water – Processing 500 l of tap water required a total running time of 17.7 hours; a final inlet pressure of 8.6 psi was noted at the conclusion. Once again, this indicates that a higher flow rate could have been employed. The pyrogen content of the starting material was determined by pooling 1 ml aliquots from each of the five 100 l refills of the reservoir. Upon evaluation, half log dilution was titered out and found to be 1/300. (With respect to the total content: sensitivity x titer = 0.125 ng/ml x 300 = 37.5 ng/ml x 500 l = 18.75 mg.)

We have noted variable content in the tap water, no doubt relating to daily sampling time and general usage flow. The sequential analysis on ultrafiltration samples taken during the study are shown in the upper portion on Table III. All samples

Table II

BACKWASH ANALYSIS FROM THE COLLECTION OF 100 LITERS OF TAP WATER

Sample	Backwash volume	Last positive titer	mg recovered (Vol. x Titer x Sensitivity) Individual	Total	% Recovery Individual	Total
1	90.0 ml	1/3,000*	0.034	0.034	2.7	2.7
2	97.5	1/30,000	0.365	0.399	29.2	31.9
3	98.0	1/10,000	0.123	0.522	9.8	41.7
4	99.0	1/3,000	0.037	0.559	3.0	44.7
5	99.0	1/1,000*	0.012	0.571	1.0	45.7
6	104.0	1/1,000*	0.013	0.584	1.1	46.8
7	102.0	1/1,000*	0.013	0.597	1.0	47.8
8	103.0	1/1,000*	0.013	0.610	1.0	48.8
9	99.0	1/1,000	0.012	0.622	1.0	49.8
10	95.5	1/300*	0.004	0.626	0.3	50.1
11	80.0	1/300	0.030	0.656	2.4	52.5
12	UF samples	all negative	—	—	—	—

Starting Material

100 Liters: clots at 1/100 dilution x .125 ng/ml sensitivity

= 12.5 ng/ml in tap water

= 12,500 ng/liter

= 12.5 μg/liter x 100 l

= 1.25 mg in 100 liters

* Result ± at next ½ log dilution.

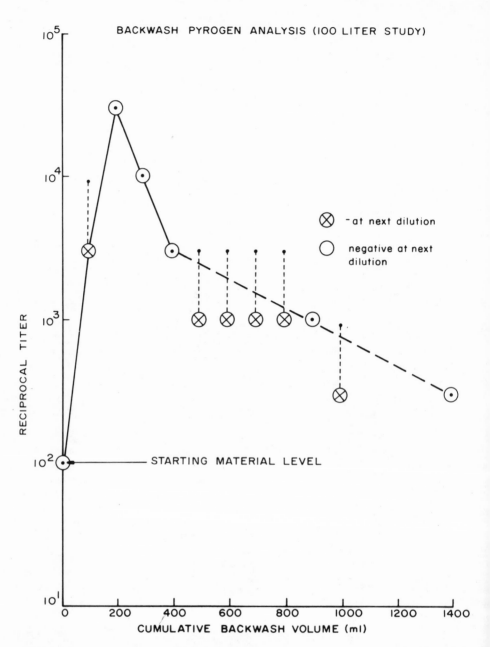

Figure 7.   Sequential Backwash Analysis for Pyrogen Content in a
            100 1 Tap Water Preparation

Table III

FILTRATE PYROGEN TESTS

500 Liter Pyrogen Test. The tap water used here titered out
1/300. At 50-liter intervals, the UF
stream was sampled and checked for
pyrogens.

UF SAMPLE @	PYROGEN TEST
50 liters	negative
100 liters	negative
150 liters	negative
200 liters	negative
250 liters	negative
300 liters	negative
350 liters	negative
400 liters	negative
450 liters	negative
500 liters	negative

20 Liter Pyrogen Test. The water used here titered out 1/10,000.
At 4-liter intervals, the UF stream was
sampled for pyrogens.

UF SAMPLE @	PYROGEN TEST
4 liters	negative
8 liters	negative
12 liters	negative
16 liters	negative
20 liters	negative

were negative indicating complete retention within the limits of the
assay and sustained integrity of the cartridge and fiber unit in the
more protracted study.

Table IV summarizes the titers obtained on backwash evaluation.
In this study a rather low total recover (14%) was obtained. Partial
explanation for this may be interference by other colloids found in
the tap water. This study does involve a substantial water volume
and our backflush concentrates are considerably enriched in other
membrane-retained species as evidenced by the rather dark and murky
appearance of the expressed retentates.

Fig. 8 (taken from the backwash data) indicates that a more
definitive "peak" should have emerged but probably was somewhat less
than the next half log dilution. The final samples have titers less
than the starting material because of the dilution with UF filtrate.

Insofar as this was a protracted study, inlet pressures were
monitored throughout. Fig. 9 illustrates the overall increase
resulting from progressive solute deposition but there are some
variations as the result of overnight shutdown.
    c. 20 liters "heavy load" – The 20 l heavy load required 40
minutes to process at a flow rate of 500 ml/min with a final inlet
pressure of 5.8 psi attained at the conclusion. The lower portion
of Table III demonstrates the effectiveness of the treatment – a
starting solution with 4 log content of pyrogen has been cleared
with no evidence of leakage. This is an extremely high challenge
and far above the level that would be normally presented. With
respect to backflush and recovery (see Table V), 25% was obtained
in the cumulative backflush.

While the three experiments demonstrated reduced recovery
levels, our water source undoubtedly was not the ideal choice for
studies of this type; endotoxin standard diluted in sterile water
will certainly be considered in further studies. It would have
been interesting to examine the first aliquots of NaOH backwash, but
unfortunately, alkali destroys the pyrogen. Peak evaluation of the
final studies is shown in Fig. 10 and once again the characteristic
early appearance is noted.

2. Automated System

While this system was validated for production of pyrogen-free
solutions, the majority of our initial studies were directed at
examining the varying settings to provide optimal cycle time for
sustained operation. With the presumption that the system would be
best suited for the final "polishing" of parenterals to insure
absence of pyrogenicity, the following operational estimates might
be drawn.

Table IV

BACKWASH ANALYSIS FROM THE COLLECTION OF 500 LITERS OF TAP WATER

Sample	Backwash volume	Last positive titer	mg recovered (vol. x titer x sensitivity)		% Recovery	
			Individual	Total	Individual	Total
1	0.5 l	1/10,000	0.625	0.625	3.33	3.3
2	0.5 l	1/10,000	0.625	1.250	3.33	6.6
3	0.5 l	1/10,000	0.625	1.875	3.33	10.0
4	0.5 l	1/1,000	0.625	2.500	3.33	13.3
5	2.0 l	1/100	0.025	2.525	0.13	13.4
6	2.0 l	1/100	0.025	2.550	0.13	13.6
7	2.0 l	1/100	0.025	2.575	0.13	13.7
8	2.0 l	1/100	0.025	2.600	0.13	13.8
SM	—	1/300				

18.75 mg available
(500 l x 1/300 x .125 ng/ml)

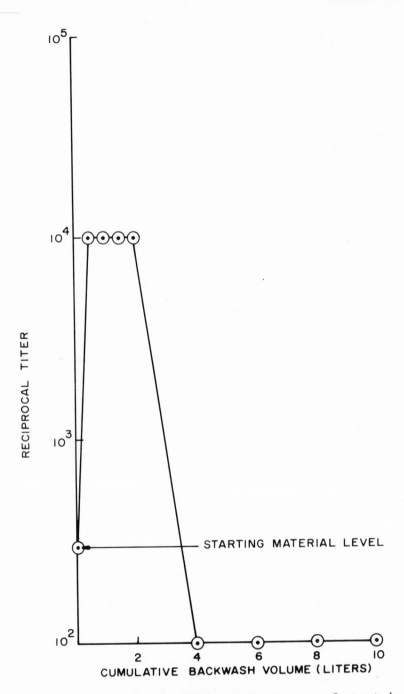

Figure 8.   Sequential Backwash Analysis for Pyrogen Content in a
            500 1 Tap Water Preparation

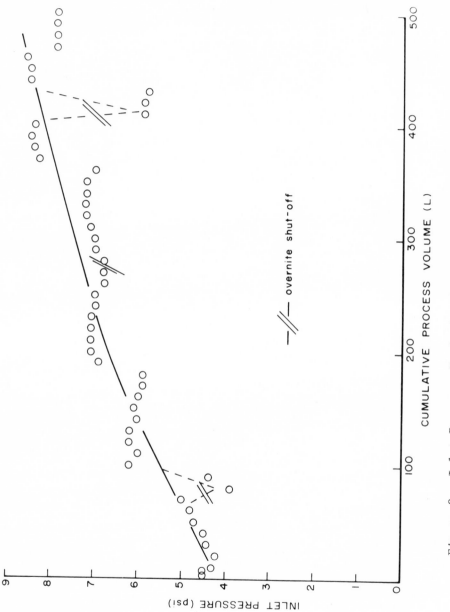

Figure 9. Inlet Pressure Variation in Processing 500 1 of Tap Water.

Table V

BACKWASH ANALYSIS FROM THE COLLECTION OF 20 LITERS OF "HEAVY LOAD" WATER

Sample	Backwash volume	Last positive titer	mg recovered (vol. x titer x sensitivity) Individual	Total	% Recovery Individual	Total
1	0.5 l	1/100,000	6.250	6.25	25.0	25.0
2	0.5 l	1/1,000	.063	6.31	0.3	25.3
3	1.0 l	1/100	.013	6.33	0.05	25.3
4	1.0 l	—	—	—	—	
5	1.0 l	1/10	.001	6.33	0.004	25.3
SM	—	1/10,000	25 mg available (20 l x 1/10,000 x .125 ng/ml)			

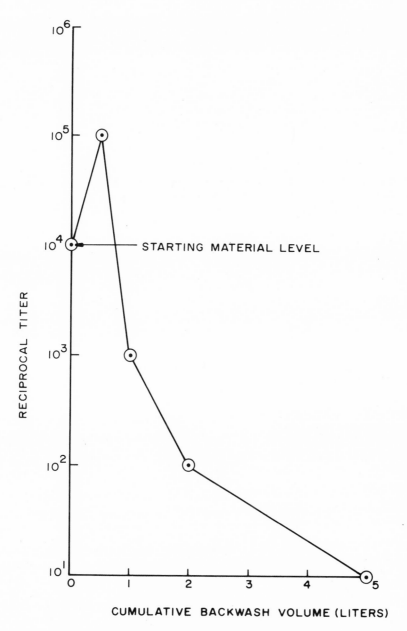

Figure 10.   Sequential Backflush Analysis for Pyrogen Content in a
20 1 "Heavy Load" Solution

Flows of approximately 1500–2000 ml/min per 10 ft^2 of membrane area would undoubtedly be sustained for several hours at a minimum, with approximately 2–3 minute down times for backflush and cartridge restoration. At times, we have observed an early performance decrement, i.e., a fall in the second filtrate production time, but subsequent cycles at similar values which may relate to some irreversible solute entrapment in larger pores, or an operational membrane change, as yet unexplained. When later decrements are of such magnitude so as to prove operationally impractical, then the chemical restoration cycle is recommended.

Currently, we are beginning evaluation on a more advanced system permitting utilization of considerably higher membrane area and operational controls that facilitate cycle adjustment (repeated backflushing, etc.). What has been documented in the first aspect of the study is the capacity of the fiber cartridge to withstand a high pyrogen challenge, to provide sustained operation, and backflushing for recovery and/or cartridge restoration. The more extensive validation of the automated system lies in the blue dextran studies (not reported here) and phage studies (reported below). These latter results are presented in Table VI and are discussed below.

Bacteriophage f2 Concentration

A summary of the phage experiments is shown on Table VI. The concentration ratios C_i/C_o and recovery R of bacteriophage f2 for each experiment are presented in Table VII. Concentration ratios are of the order of 10^2, and recovery varies from 39 to 350%.

The quantities $\frac{1}{t_i} \ln \frac{N_i}{N_0}$ and $\frac{V_i}{t_i}$ are calculated and plotted in Fig. 11. With the exception of experiments 22 and 24, the graph shows reasonably good correlation (r = 0.904) between the two quantities.* A low count of the initial numbers of bacteriophage may be responsible for the poor correlation experiments 22 and 24. The slope and intercept (-0.0102 ml^{-1} and 0.0483 sec^{-1}) are the quantities C_R and kF_A respectively.

The data from the experiments with bacteriophage f2 appear to support the particle removal model expressed in Eq. (15). Fig. 12 shows the correspondence between $\ell \frac{N_i}{N_0}$ and the sum of $C_R V_n$ and $kF_A t_n$ for phage f2.

The correlation coefficient for all experiments is 0.838. Although this is not as close to the absolute as would be desired, it must be pointed out that virological experiments rarely give ideal results because of the nature of the viruses themselves and

---

*Replacing f2 with a polysacharide (Blue Dextran 2000), a correlation of r = 0.996 was obtained.

Table VI

SUMMARY OF BACTERIOPHAGE f2 EXPERIMENTS

Experiment No.	Macromal Solute	Solvent	Initial Concentration PFU/ml	Initial Process Volume l	Average[a] Backflush Volume ml
16	f2	tap water[b]	40	19.4	480
17	f2	tap water[b]	320	19.4	440
18	f2	tap water[b]	250	100.	115
19	f2	lake water[c]	4900	19.4	360
21	f2	tap water[b]	47	19.5	180
22	f2	tap water[b]	90	19.5	93
23	f2	tap water[b]	570	19.5	560
24	f2	tap water[b]	850	19.5	145
25	f2	tap water[b]	790	19.5	143

[a] Excludes last backflush, which was significantly longer, in order to maximize total recovery.

[b] Dechlorinated with 50 mg/l sodium thiosulfate.

[c] Lake Michigan water; 1.5 JTU turbidity.

Table VII

SUMMARY OF RESULTS OF EXPERIMENTS WITH BACTERIOPHAGE f2

Experiment No.	CONCENTRATION c_i/c_o						RECOVERY	
	Backflush Number						Based on Stock %	Based on Feed %
	1	2	3	4	5	6		
16	40	3.0	0.50	0.25			—	115
17	50	12	0.97	1.6	0.19		103	95
18	450	110	209	47	12		98	252
19	50	8.0	3.0	1.7	0.53	2.5	105	105
21	75	21	12	3.6			—	99
22	210	120	98	8.7	23		240	190
23	14	5.3	2.4	1.2	0.70		66	39
24	48	7.4	87	5.1	0.51		352	127
25	140	14	7.0	3.4	1.7		310	125

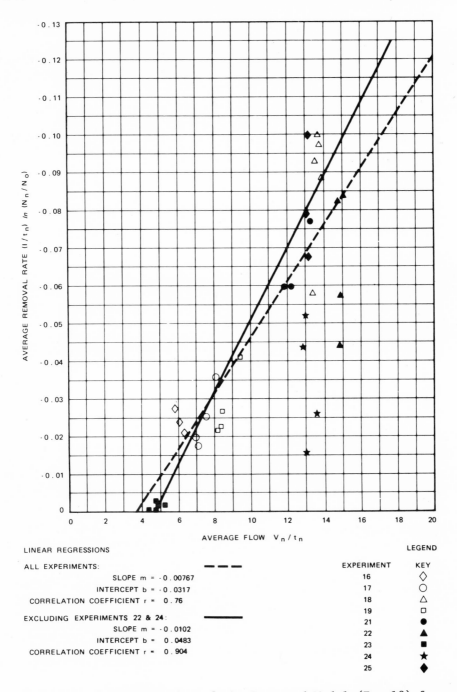

Figure 11.   An Evaluation of the Proposed Model (Eq. 12) for
            Bacteriophage f2

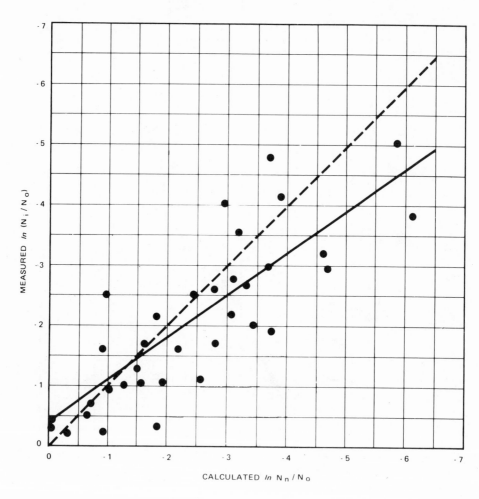

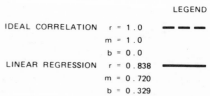

Figure 12.   Correlation of Measured and Calculated Values of
$\ell n\ N_i/N_o$ for Bacteriophage f2 Experiments

the problems associated with virus assay procedures. These
problems are discussed below.

One of the major requirements for counting viruses is the need
to obtain a relatively concentrated sample. As shown previously,
the automated system is capable of producing in the first backflush
a concentration of over 400 times that of the original sample (see
Table VII, Exp. 18). The concentration ratio decreases for
subsequent backflushes. However, using Eq. (18) the total number
of virus particles on the membrane may be estimated knowing the
number recovered in the first backflush, the volume of the backflush
water, and the time of the backflush. The predicted and actual
number of viruses initially held on the membrane surface is
presented in Table VIII. The predicted number is generally of the
same order of magnitude as the actual number, and in half the
experiments it is within 15% of the actual number. The average
error for the experiments listed in Table VIII is about $\pm 25\%$.

In the experiments performed for phage concentration, a
chemical cleaning agent was never used in the backflush water. This
was done so as to preserve the purity of the concentrated
bacteriophage f2 as may be desired in a commercial application.
However, cleaning agents (detergents, sodium hydroxide, etc.), are
commonly used to clean badly fouled membranes. Note that Eq. (11)
predicts that if the net attractive force F_A is reduced, the volume
V_i which results in the removal and consequent recovery of a number
of particles N_i will be reduced. Note that C_R is negative. This
confirms the intuitive suggestion mentioned in the introduction to
the theoretical section.

Experimental Errors

The errors associated with the experiments presented here may
be classified as either systematic or procedural errors. Systematic
errors are those attributable to the automated system. These errors
include the hold-up volume of filtrate in the tubing between the
membrane module and filtrate measuring reservoir, the diluted
residual module volume in the tubing between the module and the
outlet tube, and the volume of unfiltered water within the fibers at
the time of the nth backflush.

The systematic errors are for the most part minor with the
exception of the diluted residual volume in the tubing. This volume
was about 25 ml. It is considered significant when the volume of
backflush water was less than 100 ml per backflush, as in experiment
23. After the first backflush, the volume of water in the tubing
was the residual volume of the previous backflush. This volume was
collected with the subsequent backflushes. Because of the linearity
of the model with respect to volume, this error was not considered
to be significant in the overall experiment.

Table VIII

PREDICTED AND MEASURED N_o FOR BACTERIOPHAGE
CONCENTRATION EXPERIMENTS

Experiment No.	Predicted N_o PFU ml^{-1}	Actual N_o PFU ml^{-1}	Error % Actual N_o
16	1.4×10^6	9.0×10^5	+55
17	5.8×10^6	6.7×10^6	-13
18	6.1×10^7	6.3×10^7	-3.2
19	9.4×10^7	1.0×10^8	-6.0
21	8.6×10^5	9.1×10^5	-5.5
22	3.6×10^6	4.2×10^6	-14
23	1.4×10^7	2.2×10^7	-36
24	9.5×10^6	2.1×10^7	-55
25	2.6×10^7	1.9×10^7	+37

Procedural errors include errors in the measurements of volumes of water, errors in procedures for assaying viruses, and errors in the measurement of the phage concentration.

The most important procedural errors were in the virus assay procedures. Clumping of several virus particles in concentrated solutions - especially in the laboratory stock - resulted in the net recovery of more viruses than thought to have been added. Break-up of the clumps probably occurred during the filtering and backflushing steps.

A second procedural error was in the dispersion of the stock virus solution into the sample volume. Although the sample was agitated for over one-half hour prior to being filtered, in several experiments the number of viruses counted in the sample were less than that added. Incomplete mixing and adsorption losses onto the reservoir walls are believed to be responsible for this error.

CONCLUSIONS

The experiments and results presented and discussed here have shown that the fine hollow fiber concept can be automated into a useful filtration (for pyrogens) and concentration (for bacteriophage f2) device. The major results and conclusions obtained during this study are listed below.

- Hollow fiber ultrafilters successfully removed pyrogens from Lexington tap water (100 and 500 liters) and from 20 liters of "heavy load" solution. In no case were pyrogens detected in the permeate.

- The automated system was successfully used to concentrate bacteriophage f2 from 100 liters Evanston, Illinois, tap water and 20 liters Michigan Lake water.

- For the bacteriophage f2 concentration studies, initial solute concentrations ranged from 40 to 4900 PFU/ml, with concentration ratios from 14 to 450, and recovery efficiencies from 39 to 252% (mean value for 9 experiments was 127%).

- A two parameter model, The Backflush Recovery Model, based on a time-dependent net removal force, was developed to describe the recovery of solutes. The model can be used to estimate the number of phage initially held onto the membrane with an average error of about 25%.

Based on the above study, it is our contention that the automated fiber system is an important new development in process and analytical removal and concentration technology. The operating procedures and methods should be further developed. One obvious application of this automatic unit is for the concentration of enteroviruses and other pathogenic organisms from water. There is an urgent need for a reliable, simple to operate, lightweight pathogenic concentrator.

ACKNOWLEDGEMENT

The authors thank Mr. Jim McGuire, Research Engineer, Scientific Systems Division, and Mr. Peter N. Rigopulos, Vice President, both of Amicon Corporation, for their contribution and support during this study. The authors also thank Dr. Jimmy Quon, Chairman of Civil Engineering and Dr. Herman Cember, Group Leader of Environmental Engineering, both at Northwestern University, for their encouragement during this project.

REFERENCES

1. L.L. Nelsen, "Removal of Pyrogens from Parental Solutions by Ultrafiltration," Pharmaceutical Technology, 47-50 (May 1978).
2. T. Tamura and T. Takano, "A New Rapid Procedure for the Concentration of C-type Viruses from Large Quantities of Culture Media: Ultrafiltration by Diaflo Membrane and Purification by Ficoll Gradient Centrifugation," J. Gen. Virol. 41, 135-141 (1978).

3. A.W. Morrow, C.J. Whittle and W.A. Eales, "A Comparison of Methods for the Concentration of Foot-and-Mouth Disease Virus for Vaccine Preparation," Bull. Off. int. Apiz, 81, (11-12) 1155-1167 (1974).

4. Belfort, G., Rotem, T. and Katzenelson, E. (1975), "Virus concentration using hollow fiber membranes," Water Research, 9, 79-85.

5. Belfort, G., Rotem, Y. and Katzenelson, E. (1976), "Virus Concentration using Hollow Fiber Membranes," II, Water Research, 10, 279-284.

6. Belfort, G., Rotem, Y. and Katzenelson, E. (1978), "Virus Concentration by Hollow Fibers: Where to Now?" Progress in Water Technology, Vol. 10 NO 1/2, 359, 364.

7. Belfort, G., Rotem, Y. and Katzenelson, E., "Concentration of Viruses from Water Using Capillary Ultrafiltration."

8. Belfort, G. (1976), Cleaning of reverse osmosis membranes in wastewater renovation. In Water 75, published by AIChE, N.Y., p. 76-81.

DEPYROGENATION OF HUMAN CHORIONIC GONADOTROPIN

David S. Johnson, Karl Lin, Edward Fitzgerald,
and Bernard LePage

Millipore Corporation
Bedford, MA 01730

ABSTRACT

Using ultrafiltration, the successful removal of endogenous pyrogens from a complex glycoprotein (molecular wieght = 59,000) has been achieved. Several lots of lyophilized human chorionic gonadotropin were pyrogenic, according to the official USP Rabbit Pyrogen Test. Subsequent filtration using a 100,000 nominal molecular weight limit membrane removed a significant amount of the endogenous pyrogenic material. Molecular filtrates showed an average rise of $0.27°C$ in rabbits, whereas original samples showed an average rise of $1.01°C$. A weighed amount of the raw lyophilized hormone was dissolved in sterile pyrogen-free 0.9% saline to yield an estimated biological activity of 500 USP units per milliliter. The activity of the hormone was determined by the official USP Test, using female weanling (22-23 days old) Wistar rats. Nearly 70% of the original activity was recovered in the filtrate. Appropriate experiments were run before the human chorionic gonadotropin solution was processed. For negative controls, the entire filtration system was initially flushed with six liters of hot ($180°C$) Water for Injection and then subsequently rinsed with same volume of pyrogen-free 0.9% saline. Solution modifications in ionic strength and pH, using a constant volume molecular wash of the retentate volumes, may improve recoveries of the original hormone activity.

ULTRAFILTRATION OF PROTHROMBIN COMPLEX

Gautam Mitra and Kathryn J. Fillmore

Cutter Laboratories, Inc.
Biochemical Development
Berkeley, CA 94710

SUMMARY

'Prothrombin Complex' is a mixture of a partially purified
group of proteins (Factors II, VII, IX, and X) dependent upon
vitamin K for biological activity. Treatment of congenital Factor
IX deficiency is the best documented use for this mixture of proteins generally isolated from plasma or its subsequent fractions
by adsorption/elution with calcium phosphate, aluminum hydroxide
gel, or ion exchange resin. We have concentrated Prothrombin Complex
by membrane ultrafiltration with no loss of biological activity.
Adjustment of salts, etc. by diafiltration has also been demonstrated.
A statistical model is presented relating the flux with bulk protein
concentration and bulk liquid velocity. With successful membrane
ultrafiltration/diafiltration, more efficient purification schemes
for isolation of Prothrombin Complex becomes available. Utilizing
10^6 daltons cut-off membranes, Factor IX activity is demonstrated
both in the retentate as well as in the filtrate.

INTRODUCTION

Prothrombin complex (PTC) is a mixture of partially purified
group of proteins isolated from human plasma and dependent upon
vitamin K for biological activity. These proteins are Factors II,
VII, IX, and X. The most prevalent clinical use of this complex
has been for congenital Factor IX deficiency or hemophilia B patients.
Several methods for purification are known starting from various
source materials during Cohn's cold ethanol process (1), and these
are: plasma, cryosupernatant, effluent I, Fraction III, and Fraction
IV-1. In the purification processes, vitamin K dependent factors are
adsorbed on aluminum hydroxide, calcium phosphate, or ion exchange

resin (2-6) followed by elution at high ionic strength buffers.

Concentration of eluate and salts adjustment has generally been achieved by any of the following: use of volatile buffer, desalting using dialysis sacs, desalting using Sephadex gel filtration, and reprecipitation with either ethanol or polyethylene glycol. We have evaluated ultrafiltration/diafiltration as an effective method for this purpose. Earlier, we had reported the suitability of this approach for human serum albumin (HSA) and immune serum globulins (ISG) (7-9). Availability of large-scale ultrafilters within the last couple of years has made ultrafiltration/diafiltration a viable unit process for production level operations of human plasma proteins.

Materials and Methods

Laboratory scale ultrafiltration experiments were run utilizing Amicon TCF 10 thin channel unit with PM10 membrane (10,000 daltons cut-off). Pilot scale runs for salts exchange/concentration were carried out with Amicon DC 30 hollow fiber unit with PM10·membrane. For fractionation purposes, Millipore Corporation's Pellicon cassette with PSVP membrane (10^6 daltons cut-off) was used. Air pressure operated diaphragm pump was used for circulation of protein solutions in all pilot scale operations.

Factors II and VII were assayed according to the method of Quick (10) using deficient substrates. Factor IX was assayed by modified partial thromboplastin time using deficient substrate according to the method of Langdell et al (11). Factor X was assayed by the method of Bachman et al (12) using Russell's Viper/Venom Cephalin. Non-activated partial thromboplastin time (NAPTT) tests were performed according to Kingdon et al (13). Protein assays were done by Kjeldahl. Antibody was raised in rabbits against non-ultrafiltered PTC and utilized for Ouchterlong plates (14). SDS-PAGE electrophoresis was run according to Weber and Osborn (15).

Results and Discussion

Freeze-dried PTC was utilized for initial feasibility experiments. Reconstituted PTC was diluted to approximately 4.5 mg/ml protein with 0.088M sodium chloride-0.05M sodium citrate, pH 6.9 buffer. This diluted protein solution was ultrafiltered back to its original concentration with the Amicon TCF 10 thin channel system utilizing PM10 membrane. Operating pressure was 25 psig of nitrogen at a temperature of +5 degrees C. Table 1   shows the experimental determination of flux (J) at various bulk fluid velocities (U) and bulk protein concentrations (C_b). For concentration polarization model, the data are fitted to a non-linear equation of the form:

$$J = AU^B \ln (C_w/C_b)$$

Table 1

Ultrafiltration Flux as a Function of Protein
Concentration and Bulk Fluid Velocity

Observation No.	Fluid Velocity (cm/sec.)	Protein Concentration (g/100 ml)	Ultrafiltration Flux (cc/cm^2 sec.)
1	62.79	0.64	5.21×10^{-4}
2	62.79	0.97	4.58×10^{-4}
3	62.79	1.55	3.67×10^{-4}
4	62.79	2.34	2.96×10^{-4}
5	84.26	0.43	7.29×10^{-4}
6	84.26	0.53	6.17×10^{-4}
7	84.26	0.78	5.50×10^{-4}
8	84.26	0.91	5.21×10^{-4}
9	84.26	1.77	3.63×10^{-4}
10	150.05	0.45	7.71×10^{-4}
11	150.05	0.55	6.42×10^{-4}
12	150.05	0.88	5.42×10^{-4}
13	150.05	1.07	4.88×10^{-4}
14	150.05	2.40	3.33×10^{-4}
15	187.35	0.45	10.29×10^{-4}
16	187.35	0.56	8.75×10^{-4}
17	187.35	0.76	6.67×10^{-4}
18	187.35	1.30	5.46×10^{-4}
19	234.30	0.45	10.92×10^{-4}
20	234.30	0.57	8.75×10^{-4}
21	234.30	0.89	7.38×10^{-4}
22	234.30	1.51	5.46×10^{-4}
23	234.30	2.24	5.00×10^{-4}

Table 2

Non-Linear Regression Analysis of Concentration Polarization Model

Parameter	Value	Asymptotic Standard Error	Std. Error as % Parameter Value	Asymptotic 95% Confidence Limits
A	0.3252×10^{-4}	0.1239×10^{-4}	38.1	0.0854×10^{-4} $- 0.567 \times 10^{-4}$
B	0.4384	0.07126	16.3	$0.300 - 0.576$
Cw	7.3245	1.4574	19.9	$4.487 - 10.153$

Mean Absolute Error (MAE): 0.5461×10^{-4}
MAE as % of the Mean of J: 8.55
Std. Dev. of MAE: 0.3319×10^{-4}

Where Cw = protein concentration at the membrane surface, A and B = constants. The model on page 2 does not consider the correction factor for the variation of diffusivity and viscosity with protein concentration. Kozinski and Lightfoot had shown (16) the correction factor to be quite close to unity.

From the preliminary computer run (BMDP3R - non-linear regression program), the model does not have residuals with constant variance with respect to 5 different values of U. Consequently, the weighted least square (weights inversely proportional to the variance of the individuals) program was used and the results are shown in Table 2 . Residuals are normally distributed according to the Kolmogorov-Smirnov test at 5% significance level (maximal deviation = 0.1315, 5% critical level = 0.278). Since the parameter values are all significant, the MAE is small, and the residuals are independently and identically normally distributed, we conclude that the model is acceptable.

The Reynolds number for all the bulk fluid velocities were < 500 depicting laminar flow situations. The exponent on the bulk fluid velocity is 0.43807, compared to the Lévêque solution (17) which predicts an exponent of 0.33 for laminar flow. Table 3 shows the protein species distribution by molecular weight as determined by SDS-PAGE electrophoresis. The relative migrations of the protein peaks were compared with HSA and its polymers as standard. Four major protein peaks are noted in the molecular weight range of 82,000-317,000 daltons. The predicted protein concentration at the membrane surface (Cw) is 7.32 gm/100 ml.

Table 3

SDS-PAGE Electrophoresis of PTC Before and After Ultrafiltration

| Molecular Weight | SDS-PAGE Scan (% Total Protein) | |
	Before Ultrafiltration	After Ultrafiltration
82,000	30.05	32.20
136,000	11.13	11.59
163,000	13.64	12.70
223,000	45.18	43.52
317,000	Trace	Trace
> 317,000	Trace	Trace

Table 4

Biological Activities of PTC Before and After Ultrafiltration

Sample	A_{280}	II (u/ml)	VII (u/ml)	IX (u/ml)	X (u/ml)	NAPTT (secs)
Before ultrafiltration	22.0	24.2	9.9	22.8	13.8	181 (1:100 dilution)
After Ultrafiltration	25.0	26.2	10.2	25.5	16.0	184 (1:100 dilution)

Recovery of biological activities following ultrafiltration is depicted in Table 4 . These assays were done for the maximum bulk fluid velocity tested (234.30 cm/sec). The specific activities of the Factors II, VII, IX, and X remain unchanged, so does the NAPTT value. No loss of biological activities and no activation of the product is seen by these assays. PTC thus appears to be immune to shear, at least in the bulk fluid velocity range investigated in this study. SDS-PAGE electrophoresis results shown in Table 3 also show no discernible change in protein species distribution before and after ultrafiltration. Antibody raised against the non-ultrafiltered PTC in rabbit was used to run Ouchterlony plates in which the non-ultrafiltered and ultrafiltered PTC showed identity.

Fractionation of the PTC was attempted by using Millipore's 10^6 daltons cut-off membrane (PSVP). The PTC (composition as in Tables 3 and 4) was centrifuged, filtered through 0.45 u filter, and diluted to an A_{280} of 0.815 with 0.09M solium chloride, 0.05M sodium citrate, pH 6.9 buffer. Diafiltration against the same buffer was carried out for 0.86 volume replacements using the Millipore PSVP membrane ($1/2$ ft.^{2} area). Sixty-two percent of the initial protein was collected in the permeate. The permeate was concentrated by Amicon Corporation's 10,000 daltons cut-off membrane to an A_{280} of 16.1. The pertinent biological activities were: II = 25.2 u/ml, VII = 4.3 u/ml, IX = 15.9 u/ml, X - 15.9 u/ml, NAPTT = 247 secs. (dilution 1:100). Comparison of the specific activities with that of the starting material is shown in Table 5 . There seems to be no significant change in Factor IX purity in the permeate. Factors II and X are purified whereas Factor VII is depleted. Although the bulk protein recovery in the permeate is according to a first-order constant volume molecular washing, the individual factors show different amounts of rejection at the membrane interface. This rejection seems to be a function of the bulk protein concentration, since the total protein recovery in the permeate decreases significantly at increasing bulk protein concentrations. No activation is noted in the permeate.

Based on the preliminary work, large-scale pilot lots have been run. The PTC is eluted off the DEAE-Sephadex gel in a volatile buffer (Ammonium Bicarbonate) and divided in 2 equal parts. One half is lyophilized and the powder is dissolved in 0.088M sodium chloride-0.05M sodium citrate at pH 6.9. The other half of the eluate is diafiltered to 5 volume replacements against the same buffer using 10,000 daltons cut-off hollow fiber membrane (Amicon) followed by ultrafiltration for concentration. Five different starting lots were investigated, and the results are summarized in Table 6 . In each case, the experimental material has longer NAPTT time compared to respective control. The specific activities are comparable. The dia/ultrafiltered material appears to be less activated. Freeze drying as a unit process is perhaps contributing to the deleterious characteristics of the controls.

Table 5

Comparison of Specific Activities Before and After Membrane Fractionation

Sample	Units Factor II mg Total Protein	Units Factor VII mg Total Protein	Units Factor IX mg Total Protein	Units Factor X mg Total Protein
Starting Material	1.22	0.50	1.15	0.69
Fractionated Material	1.74	0.30	1.10	1.10

Table 6

Pilot Scale Diafiltration/Ultrafiltration of PTC vs. Appropriate Controls

Run No.	Dia/Ultrafiltration			CONTROL		
	IX (u/ml)	Specific Activity (IX/A_{280})	NAPTT (secs) (1/100)	IX (u/ml)	Specific Activity (IX/A_{280})	NAPTT (secs) (1/100)
1	27.6	1.08	360	24.6	0.94	201
2	24.9	1.07	290	14.4	0.94	240
3	24.1	1.11	370	21.6	0.92	208
4	30.8	1.10	340	27.6	0.70	152
5	34.8	1.01	200	26.8	0.65	142

Similar to HSA and ISG (7-9) membrane dia/ultrafiltration appears quite suitable for salts exchange/concentration of PTC. Fractionation in dilute solutions using 10^6 daltons cut-off membrane tend to deplete the permeate with respect to Factor VII, and there is no appreciable change of Factor IX specific activity.

REFERENCES

1. E. J. Cohn, L. E. Strong, W. L. Hughes, D. L. Mulford, J. N. Ashworth, M. Melvin, and H. L. Taylor, J. Amer. Chem. Soc., 68, 459 (1946).
2. E. Bidwell, J. M. Booth, G. Dike, and K. Denson, Brit. J. Haemat 13, 568 (1967).
3. S. Hoag, F. Johnson, J. Robinson, and P. Aggeler, New England J. Med., 280, 581 (1969).
4. S. Middleton, I. Bennet, and J. Smith, Vox Sang., 24, 441 (1973)
5. J. Heystek, H. Brummelhuis, and H. Krijnen, Vox Sang., 25, 1113 (1973).
6. H. Suomela, G. Myllyla, and E. Raaska, Vox Sang., 32, 1 (1977).
7. P. K. Ng, G. Mitra, and J. L. Lundblad, Sep. Sci., 11 (5), 499 (1976).
8. G. Mitra and J. L. Lundblad, Sep. Sci., 13(1), 89 (1978).
9. P. K. Ng, G. Mitra, J. L. Lundblad, J. Pharm. Sci., 67(3), 431 (1978).
10. A. J. Quick, J. Biol. Chem., 109, LXXIII, (1935).
11. R. D. Langdell, R. H. Wagner, and K. M. Brinkhous, J. Lab. Clin. Med., 41, 637 (1953).
12. F. Bachman, R. Duckert, and F. Koller, Thromb. Diath. Haemorrh., 2, 24 (1958).
13. H. S. Kingdon, R. L. Lundblad, J.J. Veltkamp, and D. L. Aronson, Thromb. Diath. Haemorrh., 33, 17 (1975).
14. O. Ouchterlony, Prog. Allergy, 5, 1 (1978).
15. K. Weber and M. Osborn, J. Biol. Chem., 244, 4406 (1969).
16. A. A. Kozinski and E. N. Lightfoot, AIChE J., 18, 1030 (1972).
17. M. A. Leveque, Ann. Mines., 13 (1928).

Acknowledgments:

We wish to thank Dr. M. Coan for coagulation assays. P. Chen provided advice and assistance with the statistical analysis. We are grateful to J. L. Lundblad for his interest and encouragement of this project.

PYROGEN REMOVAL BY ULTRAFILTRATION - APPLICATIONS IN

THE MANUFACTURE OF DRUGS AND U.S.P. PURIFIED WATER

L. L. Nelson and A. R. Reti

Millipore Corporation
Ashby Road
Bedford, MA   01730

ABSTRACT

In almost all practical cases, the nature of pyrogenic material
is the lipopolysaccharide outer coat of gram negative organisms.
The pyrogen retention efficiency of an ultrafiltration membrane is
strongly influenced by the size and state of aggregation of such
lipopolysaccharide molecules. Data is presented, showing the retention characteristics of a number of nominal molecular weight retention membranes when challenged with solutions of LPS in different
states of aggregation. Practical examples of pyrogen removal from
parenteral drugs, including dextrose solutions, antibiotic preparations and U.S.P. purified water, are presented, with special emphasis
on system design approaches and overall process economics.

485

HIGH FLUX CELLULOSIC MEMBRANES

AND FIBERS FOR HEMOFILTRATION

E. Klein, F. F. Holland, R. P. Wendt and K. Eberle

Gulf South Research Institute
P. O. Box 26518
New Orleans, LA   70186

ABSTRACT

The diffusive and convective permeabilities of a series of new cellulosic membranes and hollow fibers have been measured. Increased hydraulic permeability is accompanied by a shift in the response spectrum of molecular weight versus limiting rejection, with very little change in the diffusive permeability of small solutes. A pore theory model is used to compare experimental and theoretical permeabilities. Reasonable fits are obtained for diffusive transport, with poorer correlations found for convective transport. The rejection of the hemofilter barriers is compared to the response of the glomerular capillary as a function of solute size.

MICROPOROUS MEMBRANE FILTRATION FOR CONTINUOUS-FLOW PLASMAPHERESIS

Barry A. Solomon[1], Clark K. Colton[2], Leonard I. Friedman[3], Franco Castino[3], Tom B. Wiltbank[4] and Duane M. Martin[5]

[1]Research Division, Amicon Corporation, Lexington, MA 02173; [2]Department of Chemical Engineering, Massachusetts Institute of Technology, Cambridge, MA 02139; [3]American National Red Cross, Bethesda, MD 20014; [4]Piedmont Carolinas Red Cross, Charlotte, NC 28203; [5]Biotek Research, Inc., Shawnee, KS 66216

INTRODUCTION

The increased use of specific plasma components for therapeutic purposes as compared to whole blood transfusions, the greater frequency with which plasma can be donated compared to whole blood, and the growing use of therapeutic plasmapheresis has created a need for an improved plasma collection technique. Whereas whole blood donations require about 30 min, the $1\frac{1}{2}$ to 2 hr typically required with current centrifugal procedures has restricted plasma collection from volunteer donors. We have developed and clinically evaluated a new microporous membrane filtration technique for use in continuous-flow donor plasmapheresis. The procedure allows for the efficient, safe, and relatively simple acquisition of source plasma without requiring the more cumbersome and often time-consuming currently-available centrifugal methods. The results from our in vitro experiments suggested that this approach would be practicable for human donor plasmapheresis (1,2). In this paper we describe the initial clinical trials of continuous donor membrane plasmapheresis using a device with a 400 cm^2 membrane surface area. Our goal was to provide a procedure which would be attractive not only to the plasma collection industry but to the donors themselves in terms of time, safety, and ease of donation. Our results confirm that the technique is safe for the donor, provides plasma of equal or improved quality, and can be carried out within a time period required to attract volunteer donors.

In addition to the results of human donor plasmapheresis, we also report on an investigation of the use of continuous micro-porous membrane filtration for the acquisition of antibody-rich plasma from animals immunized to produce specific antibodies. Current procedures involve whole blood collection (350 to 700 ml) from animals followed by centrifugal separation to yield 200 to 400 ml of plasma approximately every five to six weeks. The use of currently-available centrifugal plasmapheresis systems has not proved practical with animals, and the red cells are consequently discarded. This routine bleeding results in low-yield, high-cost products with concomitant animal health problems exacerbated by the chronic large-scale blood-cell losses. We have modified the human donor membrane plasmapheresis system for use in animals. Our results show that it provides a rapid, highly-efficient, safe, and simple continuous plasmapheresis technique which will be useful in the production of animal blood products.

PREVIOUS WORK

Continuous-flow membrane plasmapheresis is carried out by pumping anticoagulated whole blood through a thin-channel filtration device in which the blood flow path is parallel to a microporous membrane, as depicted schematically in Figure 1. The use of a microporous membrane insures that the formed elements of the blood are retained in the perfusate, whereas the plasma proteins freely pass across the membrane. Successful separation of plasma from the formed elements requires that the red cells do not hemolyze on contact with the membrane surface or as a result of severe deformation within the membrane pores.

The feasibility of membrane plasmapheresis was confirmed in our initial studies with a small test cell having a total filtration area of 6.5 cm^2 (Figure 2). The results established that hemo-globin-free plasma of a quality comparable to that obtained from centrifugal techniques could be separated from whole blood by this process. There was no retention of plasma proteins including clotting factors V, VIII, and IX by membranes having 0.4 to 0.6 μm pore diameters, as assessed by antisera agglutination, bidimensional immunoelectrophoresis, plasma prothrombin time, and specific clotting factor assays. The effects of operating conditions were also examined, including wall shear rate, average transmembrane pressure difference, hematocrit, and the chemical nature of the membrane. The results demonstrated a relative insensitivity of the plasma filtration rate to the transmembrane pressure over a range of 50 to 500 mm Hg and to the nature of the microporous membrane.

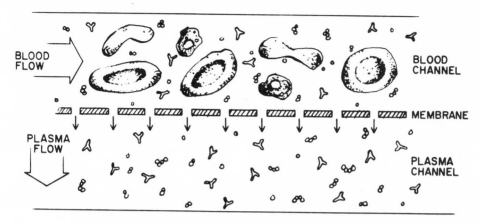

Fig. 1.   Schematic illustration showing the principle of cross-
flow microporous membrane filtration of plasma from whole
blood in which the blood flow is parallel to the membrane
while the plasma filtration is perpendicular to the
membrane.

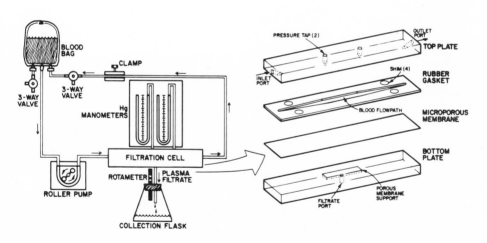

Fig. 2.   Schematic diagram of the experimental apparatus used for
preliminary in vitro feasibility studies illustrating in
detail the 6.5 cm^2 test filtration module (2).

The plasma filtration rate increased with the wall shear rate and decreased with increasing hematocrit. Red cell hemolysis occurred when the transmembrane pressure reached a critical value, the magnitude of which increased with increasing wall shear rate.

Based on results with the small cell, a larger cell was fabricated having a filtration area of 400 cm^2. It consisted of four polymethylmethacrylate plates, three membranes, and associated gaskets which provided six parallel channels, each 40 cm long, as shown schematically in Figure 3. Each channel had a divergent blood path with a smaller width, and thus higher wall shear rate, at the inlet than at the outlet. This channel design served to suppress hemolysis in the inlet region where the transmembrane pressure was the highest. A polycarbonate membrane with 0.60 μm pore diameter (Nuclepore Corp., CA) was chosen for large-scale experiments. Medical-grade silicone gaskets (Dow-Corning Co., Midland, MI) were used to establish the various channel heights. The blood entered each channel from an angular port into an entry region in which no filtration took place and in which the channel width gradually expanded. A similarly shaped exit region eliminated any disturbance caused by the outlet port. In vitro experiments with this larger, full-sized cell corroborated the design approach. About 500 ml of hemolysate-free and cell-free plasma could be obtained from anticoagulated, whole blood in approximately 30 min when the inlet wall shear rate was about 1600 sec^{-1}. As with the smaller cell, biochemical and hematological assays of the perfused blood and filtered plasma indicated that the process caused no significant blood damage to the formed elements or clotting mechanisms and that the plasma collected was quite similar to that obtained by centrifugal techniques.

CLINICAL EVALUATION OF DONOR PLASMAPHERESIS

Apparatus

The experimental filtration module was similar to that used in the in vitro studies. In order to achieve a target of 500 ml human-source plasma in approximately 30 to 40 min, it was necessary to attain a wall shear rate comparable to that used in vitro. This corresponds to a blood flowrate through the device greatly in excess of that generally available from human donors. To meet this requirement, a recirculating pump was incorporated into the blood flow loop in order that the flowrate entering the device could be increased independently of the blood flowrate to and from the donor. The complete flow circuit and control system is shown schematically in Figure 4.

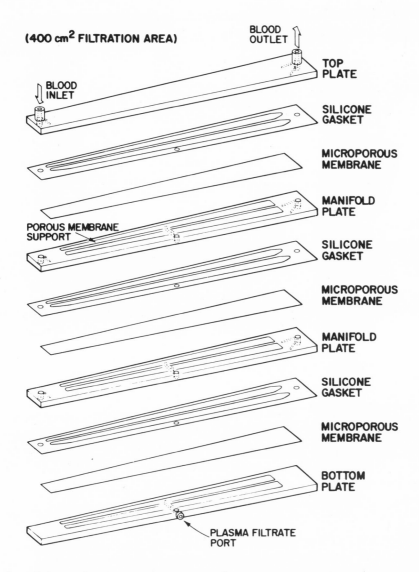

Fig. 3. Schematic diagram of 400 cm² filtration module illustrating
         the 4 plates, 3 membranes, 3 double-channel gaskets, and
         blood and plasma porting. The module is 54 cm long, 5 cm
         wide at the inlet, 8 cm wide at the outlet, and
         approximately 5 cm high when fully assembled.

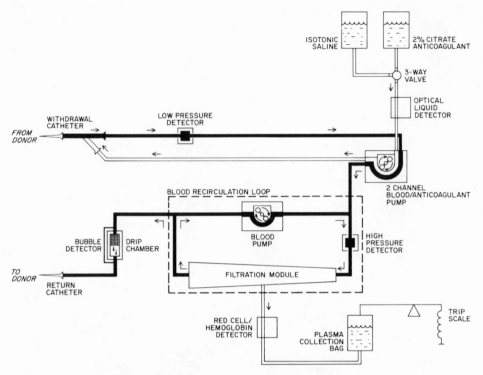

Fig. 4.   Schematic diagram showing complete flow circuit and
          control system for in vivo clinical studies of donor
          plasmapheresis.

The prototype control system was designed to perform the
following required functions: (1) withdrawal of blood through a
needle from a donor's antecubital vein at a fixed rate; (2) anti-
coagulation of blood at the hub of the needle at a predetermined
ratio of blood to anticoagulant (2% sodium citrate in isotonic
saline); (3) increase of the blood flowrate entering the device to
achieve the desired wall shear rate; (4) return of the blood
continuously to the donor; and (5) collection of the plasma into an
appropriate container.

Several safety devices were incorporated into the control
system. Each had the capability to terminate the procedure and
shut down all pumps in the event of malfunction. The individual
safety devices included:

   1. A detector of low donor blood flowrate which used a
       compliant, polyvinyl chloride "pillow" resting on a

spring. Upon deflation, the spring would move forward and
trigger an electromechanical contact to stop the pumps.

2. A pressure monitor at the filtration module inlet which
    used the same pillow as described above. An abnormal
    increase in the inlet pressure (e.g., above 200 mm Hg)
    would expand the pillow and activate a microswitch which
    shuts off the pumps.

3. A bubble catcher and ultrasonic bubble detector in the
    return blood line to eliminate the potential of air
    embolization.

4. An optical detection system to insure adequate anti-
    coagulant flow.

5. An optical detector which monitored red cell and/or
    hemoglobin in the filtrate line to indicate membrane
    rupture and/or hemolysis.

6. A mechanical trip scale which interrupted the procedure
    upon collection of a predetermined amount of filtrate.

Experimental Procedure

An experimental clinical protocol was designed to insure safety
by proper pre-screening of the donors and by detailed start-up, run,
and shutdown instructions for the operator. The entire system was
primed with sterile isotonic saline prior to use. Two 15-gauge
needles were used on the withdrawal and infusion lines. The system
is also compatible with use of a double lumen needle for withdrawal
and reinfusion of whole blood to and from the donor. Blood was
withdrawn from the donors at flowrates ranging from 45 to 60 ml/
min. The recirculating pump was adjusted to give the maximum blood
flowrate through the filtration module without triggering the over-
pressure alarm system.

Samples of blood from the donor at the inlet and outlet of the
module and samples of plasma filtrate were taken at various times
during the procedure. Biochemical and hematological assays carried
out included hematocrit, white blood cell with differential plate-
lets, plasma hemoglobin, ionized calcium, clotting factors VIII and
IX, and clotting mechanisms. Bidimensional immunoelectrophoretic
analysis of 30 plasma proteins was carried out on all samples. In
several experiments analysis for the activation of complement was
carried out in samples taken every two minutes and ^{51}Cr-labeled red
blood cell survival was studied three weeks after the procedure.

Results and Discussion

Of the 61 in vivo procedures initiated, 40 resulted in successful collection of 570 ml of plasma filtrate (500 ml plasma, 70 ml
saline prime and anticoagulant) before termination of the procedure.
The primary failure mode was membrane rupture, either during the
priming stage or early during the actual run. Generally, small
holes in the membrane gave rise to red cell leakage into the plasma
filtrate which lead to system shutdown by the plasma filtrate
hemoglobin/red cell detector. These holes were thought to occur as
a consequence of manufacturing defects or during handling and
sterilization of the reusable module. In three cases the complex
tubing harness leaked, and in another three cases insufficient
blood flowrate from the donor caused premature termination.

The total blood flowrate at the filtration module inlet was
designed to be 270 ml/min. However, it was seldom possible to
achieve this flowrate without triggering the inlet high pressure
monitor switch. The problem was believed to be caused by unaccount-
ably high pressures downstream of the filtration module. By
minimizing the pressure drop in the return lines by removing all
noncritical constrictions, plasma filtration flowrates of about 12
ml/min were reproducibly achieved. This allowed 570 ml of plasma
filtrate to be collected in an average of 48 min. While this
collection rate is slightly slower than that achieved in vitro with
essentially the same device, it nevertheless provides a significant
advantage over currently-available centrifugal techniques. Further
attention to reducing extraneous pressure drops within the system
can further decrease the collection time by allowing for higher
blood flowrates (and therefore higher wall shear rates) through the
filtration module.

Red cell survival studies were carried out according to the
technique of Hollison (3). Donor blood was sampled periodically
over a three-week period. No significant departure from normality
associated with the plasmapheresis procedure was observed. Plasma
protein analysis fully corroborated our earlier findings that no
significant rejection took place, although a reduction of protein
concentration occurred in the plasma filtrate because of dilution
with the anticoagulant. However, in several cases with donors
having a high lipoprotein level prior to plasmapheresis, the plasma
filtrate lipoprotein concentration was significantly lower, after
correction was made for anticoagulant dilution, than that found in
the perfusing blood plasma.

In Table I a summary of the hematological analysis is
presented. The hematocrit of the donor's blood remained essentially
constant. There was a very small amount of red blood cells found
in the plasma filtrate because of the pinholes described above.

Table I. Summary of Hematological Assays

| | Donor Blood | | Plasma Filtrate | Number of Donors |
	Before Procedure	After Procedure		
Hematocrit (%)	44±3	42±3	$(\frac{200\ RBC}{mm^3})$	40
WBC ($\times 10^{-3}/mm^3$)	7±3	6±2	--	40
Platelets ($\times 10^{-5}/mm^3$)	2.5±0.8	2.4±0.8	--	21
Plasma Hemoglobin (mg/dl)	5±3	6±4	24±16	34
Prothrombin Time (sec)	12±5	13±2	15±2	34
Partial Thromboplastin Time (sec)	37±9	41±11	38±9	34
Factor VIII (sec)	37±3	39±4	40±4	34
Factor IX (sec)	40±4	40±4	42±5	23

The concentration of red blood cells was, nonetheless, well below the 400 cells/mm^3 that is normally found in centrifugally prepared plasma, and it represents a rejection of intact red cells greater than 99.999%.

Of particular interest was the differential white blood cell count since alteration of the granulocyte count reportedly accompanies activation of the complement system. This phenomenon, which has been observed with hemodialysis (4) and leukopheresis (5) procedures, involves a sudden granulocytopenia followed by an increase of the granulocyte count to values much higher than those obtained prior to the procedure. In the membrane plasmapheresis procedures, a significant change in differential white blood cell count was not observed.

In an effort to determine whether complement activation occurred, a series of studies was undertaken with frequent sampling (two-minute intervals) so that transient variations in titre could be observed. These samples were analyzed for granulocytes and for CH$_{50}$ and C$_3$ hemolytic titres. The only effect observed was a slight elevation of the hemolytic titre of C$_3$ in two of the four procedures in which it was measured. In view of the limited sample size and known variation in reactivity amongst individuals,

no conclusions can be drawn about possible complement activation
as a result of the procedure. Additional studies are necessary
to clarify this matter.

Analysis of the clotting mechanisms (prothrombin time and
partial thromboplastin time) as well as specific clotting factors
(factors VIII and IX) did not reveal abnormalities. The slight
increase in times in the post-procedure sample and in the plasma
filtrate may be due to the small amount of citrate present in these
samples.

Plasma hemoglobin concentrations were essentially unchanged in
the blood flow circuit but were slightly elevated in the plasma
filtrate. The level of 24 mg/dl is comparable to that found in
source plasma prepared by centrifugal techniques. The presence of
a higher hemoglobin concentration in the filtrate may result from
red cell damage during passage through the pores or the subsequent
hemolysis of red cells that passed intact through a pinhole in the
membrane.

ANIMAL PLASMAPHERESIS

Prior to its use for plasma collection from goats, which are
widely utilized for antibody production, the human donor plasmapheresis system was modified and tested in vitro in order to
eliminate the necessity of a recirculation perfusion loop. This
was accomplished by using thinner (230 to 430 μm versus 500 μm with
human plasmapheresis) medical-grade, precision, silicone gaskets
(Dow Corning, Midland, MI) and higher blood flowrates from the
animal. Whereas the flowrate from the human antecubital vein is
about 60 ml/min, blood can be withdrawn from the jugular vein of a
goat at rates up to about 180 ml/min. As a result, plasma filtration rates of 17 to 19 ml/min were achieved in vitro with a blood
flowrate of 110 ml/min using a 380 μm thick gasket, thereby demon-
strating the feasibility of operating the system without a recirculating pump.

Experimental Apparatus

A schematic diagram of the experimental apparatus for the goat
plasmapheresis studies is shown in Figure 5. In addition to containers for the isotonic saline priming solution and the citrate
anticoagulant, provision was made for flowing saline into the blood
withdrawal catheter by means of a dual injection port. If the
catheter became occluded, the valve on the safety saline bag could
be opened to clear the catheter, thus allowing continuation of
plasmapheresis without full shutdown of the system. A single

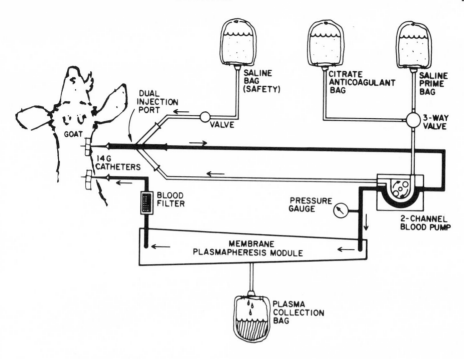

Fig. 5. Schematic diagram of the experimental setup used for goat
         plasmapheresis

two-channel pump was used for the coupled pumping of blood and
anticoagulant. A pressure gauge was utilized to monitor pressure
at the filtration module inlet, and a blood filter/bubble drip
chamber was utilized in the return line to the animal. The same
$400~cm^2$ module was used as in the human donor system.

Goats (95 to 150 pounds) were placed on a padded, squeeze-chute
bleeding table and laid on their sides. The jugular vein region
was shaved, washed with an antiseptic soap, and thoroughly wiped
with an iodine solution, Betadine. Two 14-gauge by 2-inch catheters
were inserted into the jugular vein approximately 7 to 12 cm apart.
The catheter proximal to the jaw was connected to the inlet line and
the distal catheter served as the return line. The proximal
catheter was inserted first, the inlet line to the filtration module
was connected, and the blood pump was started immediately. The
return catheter was inserted after blood flow reached the blood
filter, and the return line was immediately connected. The plasma
collection line was then opened, and the filtrate was directed to a
collection bag after the presence of undiluted plasma was detected.
The pump speed was then slowly increased until the desired maximum
inlet pressure was reached. When a predetermined plasma volume

(600 ml) was obtained, saline flow was initiated and the feed catheter removed. The return catheter was removed when the blood flow cleared the filtration unit.

Results and Discussion

Initial experiments were carried out to evaluate the effect of operating parameters (gasket thickness and blood flowrate) on plasma filtration rate. Blood flowrate was adjusted so as not to exceed an inlet pressure of 250 mm Hg or higher. Although greater than the maximum value permitted during human donor plasmapheresis at comparable shear rates, no hemolysis occurred because the critical transmembrane pressure for the onset of hemolysis with goat red blood cells was much greater ($\geq$ 360 mm Hg) than the value for human red cells, presumably because the goat cells are more rigid (6).

The results of these exploratory experiments are summarized in Table II. In all cases where operating conditions were otherwise

Table II. Effect of Operating Parameters on Plasma Filtration Rate During Goat Plasmapheresis

Gasket Thickness (μm)	Anesthetized	Number of Tests	Inlet Blood Flowrate (ml/min)	Inlet Pressure (mm Hg)	Inlet Shear Rate (sec^{-1})	Plasma Filtration Rate (ml/min)
430	No	3	135	240	1100	12
	Yes	4	170	245	1380	16
380	No	1	90	300	940	17
	Yes	4	135	290	1400	23
280	No	2	125	305	2420	27
	No	6	145	360	2814	31
230	Yes	2	60	300	1740	23*
	Yes	1	75	335	2175	26*
	Yes	1	90	370	2610	29*

* Plasma hemoglobin visibly detectable in filtrate.

fixed, the blood flowrate achieved with the anesthetized animals
was greater than with the awake animals. This led to higher wall
shear rates and correspondingly higher plasma filtration rates. We
believe that the venous pressure in the awake animals was greater
than in the anesthetized animals, perhaps because of fear of the
procedure, thereby reducing the driving force for blood flow through
the module. In the course of the experimentation, it was found that
inlet pressures in excess of 360 mm Hg could be utilized with no
significant hemolysis of the goat red blood cells. In one experiment with an inlet pressure of 360 mm Hg, a gasket thickness of
280 μm, and an inlet blood flowrate of 146 ml/min, a plasma filtration rate of 31 ml/min was attained, which corresponds to the
collection of 600 ml of plasma in less than 20 minutes. Since the
volume of the anticoagulant-diluted plasma represented about 65% of
the total blood volume, the inlet plasma flowrate was about 95 ml
in that experiment, and the plasma flowrate of 31 ml/min represented
a fractional yield of about 32%. In an attempt to increase the
fractional yield even further, a filtration module utilizing a 230
μm gasket was assembled and evaluated in vivo at much lower blood
flowrates (62 to 90 ml/min). Hemolysis was observed in all cases as
was to be expected at the low shear rates and high inlet pressures
employed and due to the increased hematocrit (60% v/v) at the outlet
region of the module due to the increased fractional yield (59%).
As the blood flowrate increased, there was a corresponding increase
in the plasma filtration rate and decrease in the amount of
hemolysis due in part to the decreased fractional yield (49%). As
a result of this evaluation, all subsequent experiments were carried
out with a 280 μm gasket and a blood flowrate of 146 ml/min.

   Membrane plasmapheresis was carried out on 40 goats and the
analysis is summarized in Table III. A sample of blood was drawn
from the animal prior to the procedure and anticoagulated with
ethylenediaminetetraacetic acid. The concentration of its plasma
constituents (total protein, albumin, globulin) are compared to
that of the samples of plasma filtrate and blood plasma in the
return line. Samples taken from the plasma filtrate and blood
return line were collected after 250 ml and 500 ml of plasma had
been collected. No significant difference between the returning
blood plasma and plasma filtrate was observed. The reduced concentrations of these samples compared to the levels in the starting
samples result from the 1:8 dilution ratio of anticoagulant to whole
blood. Furthermore, the decrease of plasma protein concentration
grows with time over the course of the plasmapheresis procedure.
At a fractional yield of about 32%, nearly 68% of the anticoagulant solution is infused back into the animal's circulatory
system. After the collection of approximately 500 ml of plasma
with a blood flowrate of 146 ml/min, more than 200 ml of anticoagulant has been returned to the animal. This alone represents
a dilution effect of approximately 6%. This explains why the plasma

Table III.   Analysis of Plasma Concentrations of
Measurements with 40 Goats

	Starting Plasma	After 250 ml Plasma Collected		After 500 ml Plasma Collected	
		Returned Plasma	Plasma Filtrate	Returned Plasma	Plasma Filtrate
Total Protein, g/dl	7.6	6.4	6.2	6.2	6.0
Albumin (A), g/dl	2.6	2.0	2.1	2.1	2.1
Globulin (G), g/dl	5.0	4.5	4.5	4.4	4.4
A/G Ratio	0.50	0.45	0.45	0.45	0.45

protein concentrations in samples taken after 500 ml of filtrate
were collected are lower than comparable samples taken after
collection of 250 ml.

The use of membrane filtration as an alternative to the existing
process for antibody collection is particularly exciting. More
frequent plasmapheresis may be carried out by continuously returning
the red blood cells to the animal. The effect of plasmapheresis on
circulating and filtered plasma protein constituents was studied in
goats that underwent plasmapheresis twice within a 48-hour period.
In Table IV the results of a typical case are presented. The total
protein, albumin, and globulin concentrations did not recover to
their starting concentrations within the 48 hr. The albumin-to-
globulin ratio, however, was constant throughout the entire run.
Most surprising was the titre of anti-rabbit IgG. Whereas the
concentration of the entire globulin fraction was reduced by nearly
8% after 48 hr, the antibody titre did not decrease at all. This
may be due to a rebound phenomena associated with rapid depletion of
circulating antibodies (7). Nonetheless, in view of the decrease of
total protein concentration, a 48-hour rest period between plasmapheresis procedures does not appear sufficient. More recent data
indicates that a 96-hour rest period is sufficient to allow total
protein concentrations to recover to pre-plasmapheresis levels.
Experiments are currently underway to evaluate the long-term effect
of frequency (every 96 hr) plasmapheresis on a herd of hyperimmune
goats with respect to the antibody titre and frequency of and need
for re-immunization.

Several goats immunized against second antibody (anti-IgG) or
primary antibody (anti-specific biochemical) were evaluated in order
to assess the effect of plasmapheresis on the yield of a wide
spectrum of antibodies. The results are presented in Table V.

Table IV.   Results of Sequential Goat Plasmapheresis
            Within 48 Hours

	First Procedure			Second Procedure		
	Starting Plasma	Returned Plasma	Plasma Filtrate	Starting Plasma	Returned Plasma	Plasma Filtrate
Total Protein, g/dl	7.9	6.7	6.6	7.5	6.1	6.0
Albumin (A), g/dl	2.6	2.2	2.2	2.6	2.1	2.1
Globulin (G), g/dl	5.3	4.5	4.4	4.9	4.0	3.9
A/G Ratio	0.49	0.49	0.50	0.53	0.53	0.54
Anti-rabbit IgG (Titre)	1:25.0	--	1:22.0	1:25.3	--	1:22.1

Table V.   Goat Antisera Titre Evaluation

Goat Antisera Description	Number of Goats Tested	Average Working Titre Range	% Dilution Due to Plasmapheresis
Second Antibody			
Anti-rabbit IgG	10	1:18.5	13.4
Anti-guinea Pig IgG	6	1:17.5	13.9
Anti-human IgG	3	1:25	13.0
Primary Antibody			
Anti-digoxin	3	1:20,000	12.5
Anti-porcine insulin	2	1:35,000	12.9

The average working titre range for the various antibodies collected by membrane plasmapheresis were all consistent with the dilution expected because of anticoagulant infusion.

CONCLUSIONS

These results demonstrate that continuous membrane filtration plasmapheresis is a viable alternative to conventional centrifugal techniques for the collection of source plasma from human donors and antibody-rich plasma from immunized goats. The prospect of attracting a pool of volunteer donors for plasmapheresis is more realistic now that a relatively rapid and safe procedure for the large-scale collection of plasma has evolved. Although further evaluation of the reliability of the process and the suitability of the source plasma for subsequent component fractionation is needed, our research has shown the practicability of this new concept in continuous plasma separation from whole blood. Application of this technique to therapeutic plasmapheresis also appears likely in view of the recent medical interest in plasma exchange procedures (8,9) and extracorporeal modalities requiring access to plasma (10,11).

Plasmapheresis of hyperimmune goats every four days to collect 600 to 800 ml of plasma could increase productivity by a factor of 15 to 40 times over current procedures. This increase could easily justify the moderate increase in handling time. In addition, increased productivity could lead to smaller herds of animals, further reducing the overall cost of antisera. Through use of this technique, less expensive antisera could become available, which would greatly increase the use of immunoassays and might open up new applications for immunochemicals.

REFERENCES

1. Castino, F., Friedman, L.I., Solomon, B.A., Colton, C.K., and Lysaght, M.J., The filtration of plasma from whole blood: A novel approach to clinical detoxification, in: "Artificial Kidney, Artificial Liver, and Artificial Cells," T.M.S. Chang, Ed., Plenum Press, New York (1978).
2. Solomon, B.A., Castino, F., Lysaght, M.J., Colton, C.K., and Friedman, L.I., Continuous flow membrane filtration of plasma from whole blood, Trans. Am. Soc. Artif. Intern. Organs, Vol. XXIV, 21 (1978).
3. International Committee for Standardization in Hematology, Recommended methods for radioisotope red cell survival studies, Brit. J. Hematol. 21:241 (1971).

4. Jenson, D.P., Brubaker, L.H., Nolph, K.D., Johnson, C.A., and
    Nothum, R.J., Hemodialysis coil-induced transient neutropenia
    and overshoot neutrophilia in normal man, Blood 41:399 (1973).
5. Rubins, J.M., MacPherson, J.L., Nusbacher, J., and Wiltbank,
    T., Granulocyte kinetics in donors undergoing filtration
    leukapheresis, Transfusion 16:56 (1976).
6. Zweifach, B.W., Microcirculation, Ann. Rev. Physiol. 35:117
    (1973).
7. Terman, D.S., Tavel, T., Petty, D., Tavel, A., Harbeck, R.,
    Buffaloe, G., and Carr, R., Specific removal of bovine serum
    albumin (BSA) antibodies by extracorporeal circulation over
    BSA immobilized in nylon microcapsules, J. Immunol. 116:1337
    (1976a).
8. Israel, L., Edelstein, R., Mannoni, P., and Radot, E.,
    Plasmapheresis and immunological control of cancer, Lancet 2:
    7986 (1976).
9. Bier, M., Zukoski, C.F., Merriman, W.G., and Beavers, C.D.,
    Rapid extracorporeal complement inactivation, Trans. Am. Soc.
    Artif. Intern. Organs 19:130 (1973).
10. Terman, D.S. and Buffaloe, G., Extracorporeal immunoadsorbents
    for specific extraction of circulating immune reactants, in
    "Artificial Kidney, Artificial Liver, and Artificial Cells,"
    T.M.S. Chang, Ed., Plenum Press, New York (1978).
11. Bansal, S.C., Bansal, B.R., Rhoads, J.E., Jr., Cooper, D.R.,
    Boland, J.P., and Mark, R., Ex-vivo removal of mammalian
    immunoglobin G: Method and immunological alterations, Int. J.
    Artif. Organs 1:94 (1978).

ACKNOWLEDGEMENTS

This work was supported in part by NHLBI Contract N01-HB-6-2928. We thank Dr. John Sanderson and Mr. Charles Oberhauser of Dynatech R&D, Cambridge, MA, for their assistance in the fabrication of the prototype hardware for the donor plasmapheresis system. We also thank Mr. Aldo Pitt, Ms. Beth Grapka, Ms. Susan L. Martin, and Dr. Duane N. Tinkler for their invaluable assistance on this project.

DETERMINATION OF GRAETZ SOLUTION CONSTANTS IN THE IN-VITRO HEMO-

FILTRATION OF ALBUMIN, PLASMA, AND BLOOD

Kenneth Isaacson, Patricia Duenas, Cheryl Ford, and
Michael Lysaght

Biomedical Products Division
Amicon Corporation
Lexington, Mass. 02173

INTRODUCTION

The clinical ultrafiltration of uremic blood followed by re-
plenishment of water and vital solutes is termed hemofiltration.
The process is an alternative to dialysis for the treatment of
chronic uremia and was designed to offer higher rates of clearance
for intermediate molecular weight solutes (1-5). Several clinical
investigators have reported both acute and chronic medical benefits
for patients treated with hemofiltration rather than dialysis.
These include decreased adverse symptomatology during treatment
sessions (6-9) and amelioration of hypertension (10,11) and lipid
mishandling (12-15) when these are secondary to uremia. The clinical evaluation of hemofiltration is continuing particularly through
an NIH sponsored multicenter crossover study (16-18).

Hemofiltration employs special biocompatible filters whose
membranes are highly permeable to water, electrolytes, and uremic
toxins but are retentive to blood protein and formed elements.
Under clinical conditions, transport in hemofilters is governed by
complex boundary layer effects termed concentration polarization.
In 1975, Colton, Henderson, Ford, and Lysaght (1,2) published the
basic quantitative description of this phenomena. Their analysis
was based upon in-vitro studies with plasma and whole blood and
confirmed with in-vivo experience with one patient.

Since those early publications, investigators have noted that
certain patients filter "faster" than others, even after allowance
of protein composition and hematocrit. We have observed in laboratory studies that filtration rate on the same or identical devices
varies substantially with differently sourced aliquots of plasma or

blood.

In this report, we have repeated the earlier process studies of Colton, et al (1,2) using well characterized bovine serum as the test fluid in place of plasma or blood. This was motivated by (1) a desire to overcome the ambiguity associated with the intrinsic variability of naturally occurring biologic fluids and (2) the possibility that increased quantitation of transport of individual constituents of plasma would further the understanding of the relationship between transport rate and blood or plasma composition.

Cohn Fraction V bovine serum albumin, BSA, is economical, is readily obtained and can be stored almost indefinitely. Complete characterization of the material can be accomplished by such standard procedures as column chromatography, immunoelectropheresis and biuret reaction.

METHODS

Experimental – Filtration studies were performed with Cohn Fraction V bovine serum albumin (United States Biochemical, Cleveland, OH) in 0.1M phosphate buffer, pH = 7.4, preserved with 0.1 percent sodium azide and stabilized with 0.04M sodium caprylate. BSA solution concentration was determined by the biuret reaction (19) and monitored with a hand held refractometer throughout testing. Each lot of BSA was checked by gel chromatography with Sephadex G-150; albumin monomer concentration always exceeded 98%. BSA solutions were discarded after 40 determinations or after 2 weeks storage at 4°C whichever occurred first.

Whole blood and plasma were obtained from the Children's Hospital (Boston, MA) after storage of 21 days or longer, post-collection, and were mixed with isotonic saline in order to provide solutions with protein concentration and hematocrit corresponding to the earlier work of Colton, et al. Plasma protein was determined by the biuret reaction and hematocrit by standard clinical micro-centrifugation.

Membranes of XP-50 formulation (Amicon Corp., Lexington, MA) were fabricated into cartridges with exposed lengths ranging from 6 to 25cm, containing from 25 to 6500 fibers. Inside diameters were measured by photomicroscopy and ranged from 190 to 1100 μm. Area varied from 45 to 10000cm^2.

Solutions were pumped from a well mixed reservoir through the test circuit by roller pump (Masterflex, Cole-Parmer, Chicago, IL) selected to produce a relatively pulseless flow. A heat exchanger (500cm^2, 316 stainless steel) surrounded by a water bath maintained the temperature of the solution measured at the ultrafilter inlet ± 0.5°C in the range of 22 to 37°C. Inlet and outlet pressures

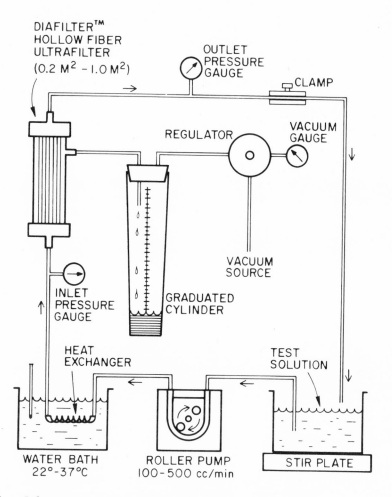

Fig. I.   Schematic representation of the test circuit used for characterization of clinical ultrafilters.

were monitored with gauges and could be increased as desired with a clamp on the ultrafilter outlet tubing. Flow rates of solutions into the ultrafilter were determined with a tachometer on the roller pump previously calibrated ± 2 percent. Ultrafiltrate was collected in a stoppered graduate cylinder connected to a regulated vacuum source equipped with a vacuum gauge. Ultrafiltration rate was determined by timing the ascent of the fluid into the graduate cylinder. Mean transmembrane pressure (ΔP_{TM}) was defined as the average of the inlet and outlet fluid pressures plus the negative pressure external to the fibers. ΔP_{TM} was varied from 150 to 1500mmHg of which approximately one half of the total was positive pressure and one half negative pressure. Wall shear rate per unit length $\gamma w/L$ ranged from 20 $(cm \cdot sec)^{-1}$ to 400 $(cm \cdot sec)^{-1}$.

The test circuit is shown in Figure I.

Calculations - Flux measurements are reported as the length averaged ultrafiltrate flux, J_F:

$$J_F = \frac{Q_F}{A} \qquad (1)$$

which is the total flow rate of ultrafiltrate per unit membrane area.

In concentration polarization, the ultrafiltration flux, J_F is approximated by the simple film theory of mass transfer (20):

$$J_F (X) = k \ln \left(\frac{C_g}{C_{Pb}}\right) \qquad (2)$$

where C_g is the so-called "gel concentration" of the protein and C_{Pb} is the bulk protein concentration averaged over the length of the fiber. k is the local mass transfer coefficient given as (21):

$$k = 32 \left(\frac{\gamma w \ D^2}{L}\right)^{1/3} \left(\frac{X}{L}\right)^{1/3} \qquad (3)$$

We express the length averaged ultrafiltrate flux J_F as a simple function of two empirically determined constants:

$$J_F = a \left(\frac{\gamma w}{L}\right)^b \ln \left(\frac{C_g}{C_{Pb}}\right) \qquad (4)$$

Constants a & b are known as the Graetz-solution constants (22). Note that pressure does not appear in equation 4. The average wall shear rate per unit length ($\frac{\gamma w}{L}$) conveniently represents all parameters relating to flow conditions and channel geometry which for tubular channels, is well approximated by the expression:

$$\frac{\gamma w}{L} = \frac{8\overline{V}}{d} \tag{5}$$

where $\overline{V}$ is the average linear velocity of the fluid, and d equals the inner diameter of the hollow fiber. Alternatively:

$$\frac{\gamma w}{L} = \frac{32 \ Q_B}{\pi d^3 NL} \tag{6}$$

where Q_B = the average of the inlet and outlet flow rate of fluid through the hollow fibers; and N equals the number of hollow fibers in the bundle.

RESULTS AND DISCUSSIONS

The measured ultrafiltrate flux, J_F, for 10% BSA solution at various values of wall shear rate per length is plotted as a function of mean transmembrane pressure in Figure II. As predicted, ultrafiltrate flux reached a stable, pressure-independent plateau indicating that concentration polarization was indeed the controlling factor of filtration rate. However, the threshold transmembrane pressure necessary to produce polarization controlled filtration was approximately 1200mmHg for BSA as compared to the value of 500-600mmHg reported by Colton for solutions of human blood or plasma (1,2). This difference is not significant so long as all of the filtration rate data is collected at transmembrane pressure differences above the threshold value, a condition that was satisfied throughout this study. The difference did suggest, however, that there might be a significant variation in dependence of transport on protein concentration between plasma proteins and bovine serum albumin.

The effect of protein concentration on ultrafiltrate flux is shown in Figure III for bovine serum albumin at 22 and 37°C. Colton's data for plasma proteins is also shown. (It will be seen below that the difference in fiber formulation is insignificant.) The data fits nicely to a straight line on a semi-log plot confirming the proper choice of concentration dependence in equation 3. The extrapolation of the data to the extinction of J_F on the x-axis gives the so-called gel-concentration, C_g. As expected the value of C_g is significantly higher for BSA than for a solution plasma proteins, since the latter contains many, larger, slower diffusing species of lipoprotein. The value of C_g for BSA clearly does not correspond with the solubility limit of nearly 60 gms/100cc reported by Keller (23), and Kozinski (24). Nor is there a significant difference of C_g for BSA as temperature was increased from 22 to 37°C despite the large differences in ultrafiltrate flux at lower protein concentrations. These facts support the theory that osmotic pressure forces as well as hydrodynamic resistance are the limiting factors governing ultrafiltration rates of protein solu-

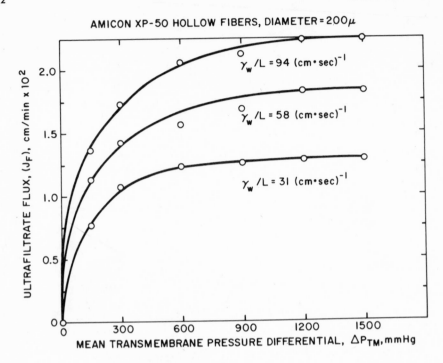

AMICON XP-50 HOLLOW FIBERS, DIAMETER = 200μ

Fig. II. Dependence of ultrafiltrate flux on mean transmembrane pressure difference with 10% BSA.

tions. The term gel-concentration therefore contains no real physical significance, but it is highly useful as a tool for representing the contribution of total protein concentration on measured ultrafiltrate flux. Subsequent filtration studies with BSA were conducted at a concentration 10% BSA, selected arbitrarily.

Figure IV shows the temperature dependence of ultrafiltration with 10% BSA. Ultrafiltrate flux, J_F, was found to increase linearly with temperature over a range of 22 to 37°C, which encompasses the temperature variations likely to be seen in in-vivo as well as in-vitro testing. The overall increase was 34% over this range.

The dependence of ultrafiltrate flux, J_F, on wall shear rate per length, $\gamma w/L$, is shown in Figures V, VI, and VII for 10% BSA, diluted plasma and diluted blood, respectively, at 22 and 37°C plotted on a log-log scale. The protein and red cell concentrations of the diluted plasma and blood were adjusted to those of Colton and correspond to hemofiltration in the pre-dilution mode. Colton's data taken with the acrylic copolymer XM-membrane is shown in the figures with a dashed line. There is excellent correlation of our

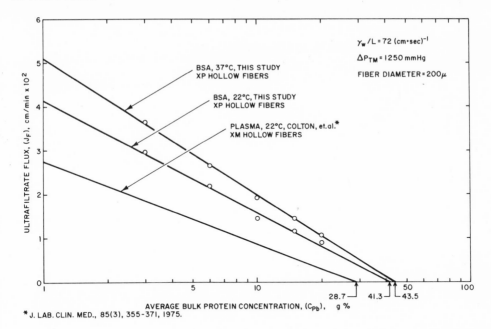

Fig. III. Effect of average bulk protein concentration on ultrafiltrate flux for bovine serum albumin at 22 and 37°C,
and for plasma proteins at 22°C as reported by Colton.

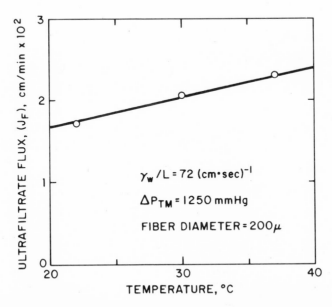

Fig. IV. Effect of temperature on ultrafiltrate flux for 10% BSA.

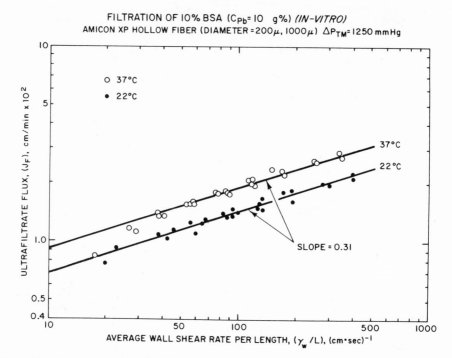

Fig. V.   Dependence of ultrafiltrate flux on average wall shear rate
per unit length for 10% BSA at 22 and 37°C.

data, taken with the XP, polysulfone membrane to the previous work
of Colton which confirms that protein ultrafiltration is boundary
layer controlled and suggests that the predictive equations developed here can be considered valid for other protein retentive membrane of sufficiently high water permeability.

The satisfactory correlation of the data with a straight line
on the log-log plot indicates that the form of the mass transfer
equations chosen here is correct.

The data were fitted to an equation of the form:

$$J_F = a \, (\frac{\gamma w}{L})^b$$

by linear regression analysis, and the results are presented in
Table I.

The exponential dependence of ultrafiltrate flux on $\gamma w/L$, given
by the exponent b in Table I, which may be seen in the plots as the
slope of the lines is exactly as predicted by equation (3) for

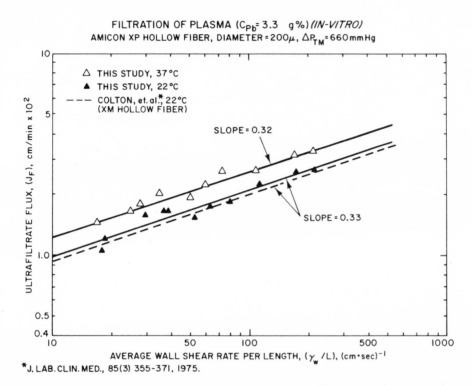

FILTRATION OF PLASMA (C_{Pb}= 3.3 g%) *(IN-VITRO)*
AMICON XP HOLLOW FIBER, DIAMETER = 200μ, ΔP_{TM} = 660 mmHg

△ THIS STUDY, 37 °C
▲ THIS STUDY, 22 °C
--- COLTON, et. al.,* 22°C
(XM HOLLOW FIBER)

SLOPE = 0.32

SLOPE = 0.33

ULTRAFILTRATE FLUX, (J_F), cm/min x 10^2

AVERAGE WALL SHEAR RATE PER LENGTH, (γ_w /L), (cm·sec)$^{-1}$

*J. LAB. CLIN. MED., 85(3) 355-371, 1975.

Fig. VI.   Dependence of ultrafiltrate flux on average wall shear
rate per unit length for diluted plasma (C_{Pb} = 1.8g%) at
22 and 37°C for this study, and for Colton et al, at 22°C.

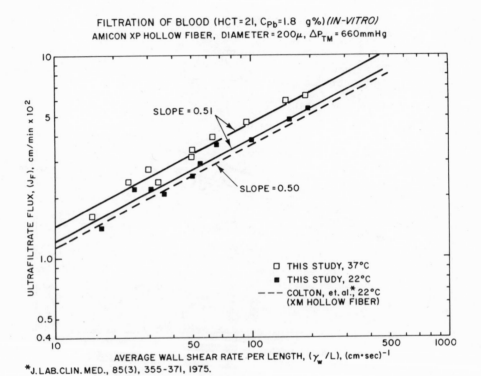

FILTRATION OF BLOOD (HCT = 21, C_{Pb} = 1.8 g%) *(IN-VITRO)*
AMICON XP HOLLOW FIBER, DIAMETER = 200μ, ΔP_{TM} = 660mmHg

*J. LAB. CLIN. MED., 85(3), 355-371, 1975.

Fig. VII.   Dependence of ultrafiltrate flux on average wall shear
            rate per unit length for diluted blood (HCT = 21, C_{Pb}
            3.3g%) at 22 and 37°C for this study and for Colton e
            al, at 22°C.

bovine serum albumin and for plasma proteins. The significant increase in shear dependence for diluted blood has been attributed to
the augmentation of protein diffusion away from the boundary layer
caused by rotation and translation of red blood cells in shear flow
(25).

Table I.   Summary of empirical constants for the ultrafiltration
           of BSA, plasma, and blood.

$(J_F$ = cm/min; $\gamma w/L$ = $(cm\ sec)^{-1}$; C_{Pb} = gm %)

$$J_F = A \left(\frac{\gamma_w}{L} \right)^B$$

	HCT	C_{Pb}	$A \times 10^2$		B	
			22°C	37°C	22°C	37°C
BSA	--	10.0	0.47	0.63	0.31	0.31
PLASMA, THIS STUDY, XP-FIBERS	--	3.3	0.49	0.61	0.33	0.32
PLASMA, COLTON, et. al.[*], XM-FIBERS	--	3.3[**]	0.44	–	0.33	–
BLOOD, THIS STUDY, XP-FIBERS	21	1.8	0.37	0.44	0.51	0.51
BLOOD, COLTON, et. al.[*], XM-FIBERS	21	1.8[**]	0.36	–	0.50	–

[*] J. LAB. CLIN. MED., 85(3), 355-371, 1975.   [**] EXTRAPOLATED TO THIS VALUE.

It is interesting to note that the effect of increased temperature is simply to increase the relative height of the lines
which is reflected as an increase in the coefficient a. The magnitude of the increase in diffusivity was substantially larger than
predicted by the Stokes-Einstein relation (26). The overall increase in flux as temperature was increased from 22 to 37°C was
34%, 24% and 19% for 10% BSA, diluted plasma and diluted blood
respectively.

The dependence of ultrafiltrate flux, J_F, with wall shear rate
per length, $\gamma w/L$ can be combined with the dependence on temperature
and protein concentration to yield an equation of the form of equa-

tion (4) with the addition of a factor for temperature dependence, yielding a predictive equation for the dependence ultrafiltrate flux on all controllable parameters which is summarized in Table II.

These equations have been validated for protein ultrafiltration with XM and XP membranes and is safely considered valid, over the ranges of variables studied, for other similar membranes.

Table II. Predictive equation summarizing the effects of controllable parameters on ultrafiltrate flux for BSA, blood and plasma.

$$ J_F = A \left(\frac{\gamma_w}{L} \right)^B \left[1 + \left(\frac{T-22}{15} \right) (D) \right] \ln \frac{C_g}{C_{Pb}} $$

where: J_F = ULTRAFILTRATE FLUX, cm/min

γ_w / L = WALL SHEAR RATE PER UNIT LENGTH, $cm \cdot sec^{-1}$

C_{Pb} = BULK PROTEIN CONCENTRATION, g %

T = TEMPERATURE, °C

	A	B	C_g	D
BSA	3.25×10^3	0.31	42.4	0.34
PLASMA	2.27×10^3	0.33	28.7	0.24
BLOOD (HCT-21)	1.32×10^3	0.51	28.7	0.19

CONCLUSIONS

1. The polarization limited flux of bovine serum albumin is qualitatively similar to that of plasma or whole blood. The major quantitative differences are that, relative to plasma or blood, albumin exhibits a higher threshold transmembrane pressure for the onset of fully developed polarization AND a higher flux at a given concentration.

2. For plasma, blood, or albumin, flux rate increases sharply with increasing temperature. Care must be taken to distinguish data taken at ambient temperature from that at 37°C.

3. The results of our studies with plasma and blood indicate good
   agreement with those of Colton although we used a different
   membrane material and operated over a wider range of fiber
   diameters and device areas.

REFERENCES

1. Colton, C.K., Henderson, L.W., Ford, C.A., and Lysaght, M.J.,
   Kinetics of Hemodiafiltration -- In-Vitro Transport Characteristics of a Hollow Fiber Blood Ultrafilter, J. Lab.
   Clin. Med., 85 (3), 355-371 (1975).
2. Henderson, L.W., Colton, C.K., and Ford, C.A., Kinetics of
   Hemodiafiltration, II -- Clinical Characterization of a New
   Blood Cleansing Modality, J. Lab. Clin. Med., 85 (3), 372-
   391 (1975).
3. Lysaght, M.J., Ford, C.A., Colton, C.K., Stone, R.A., and
   Henderson, L.W., Mass Transfer in Clinical Blood Ultrafiltration Devices -- A Review; Chapter in "Technical Aspects
   of Renal Dialysis" (T.H. Frost, Ed.), Pitman Medical Publishing Co., Ltd., United Kingdom, pp. 81-94 (1978).
4. Quellhorst, E., Fernandez, E., and Scheler, F., Treatment of
   Uremia Using an Ultrafiltration -- Filtration System, Proc.
   Eur. Dial. & Transplant Assn., 9, 584 (1972).
5. Quellhorst, E., Schuenemann, B., and Doht, B., Haemofiltration --
   A New Method for the Treatment of Chronic Renal Insufficiency,
   Chapter 10 in Technical Aspects of Renal Dialysis (T.H.
   Frost, Ed.), Pitman Medical (1978).
6. Bosch, J.P., Kahn, T., Glabman, S., von Albertini, B., Geronemus,
   R., Moutousis, G., Acid-base Homeostasis in Patients on Hemo-
   dialysis and Hemofiltration, Abstracts American Society of
   Nephrology, 12, 112 (1979).
7. Koch, K.M., Ernst, W., Baldamus, A., Brecht, M.H., George, J.,
   Fassbinder, W., Sympathetic Activity and Hemodynamics in
   Hemodialysis, Ultrafiltration, and Hemofiltration, Abstracts
   American Society of Nephrology, 12, 121 (1979).
8. Shaldon, S., Deschodt, G., Bead, M.C., Claret, G., Mion, H., and
   Mion, E., Vascular Resistance and Stability During High Flux
   Hemofiltration Compared to Dialysis, Abstracts American
   Society of Nephrology, 12, 129 (1979).
9. Pierides, A.M., Schnepp, B., Johnson, W.J., Hemofiltration in
   the Treatment of Acute and Chronic Renal Failure, Abstracts
   of the 9th Annual Clinical Dialysis and Transplant Forum,
   47 (1979).
10. Quellhorst, E., Schuenemann, B., and Doht, B., Treatment of
    Severe Hypertension in Chronic Renal Insufficiency by Hemofil-
    tration, Proc. Eur. Dial. & Transplant Assn., 14 (1977).
11. Henderson, L.W., Ford, C.A., Lysaght, M.J., Grossman, R.A., and
    Silverman, M.E., Preliminary Observations on Blood Pressure
    Response with Maintenance Diafiltration, Kidney Int., 7,
    413 (1975).

12. San Felipo, M., Grundy, S., Henderson, L., Transport of Very
    Low Density Lipoprotein Triglyceride: Comparison of Hemo-
    dialysis and Hemofiltration. Abstracts 1979 American
    Society of Nephrology, 12, 98 (1979).
13. San Felippo, M., Barg, A., Henderson, L.W., Altered Lipid and
    Insulin Response in Hemodialysis Patients After Hemofiltra-
    tion, Abstracts 1979 American Society of Nephrology, 12,
    128 (1979).
14. Geronemus, R., von Albertini, B., Glabman, S., Kahn, J.,
    Rayfield, I., Bosch, J.P., Glucose Tolerance in Patients on
    Hemodialysis, Sequential Ultrafiltration and Dialysis, and
    Hemofiltration, Proceedings 1979 EDTA (in press).
15. Henning, H.V., and Balusek, E., Lipid Metabolism in Uremia:
    Effect of Regular Hemofiltration and Hemodialysis Treat-
    ment, J. Dial., 1 (6), 595-605 (1977).
16. Quellhorst, E., Schuenemann, B., "Controlled Study to Compar-
    ing Hemodialysis and Hemofiltration. Treatment in Patients
    with Chronic Renal Insufficiency, 1st Annual Report on NIH
    NO1-AM-8-2229, 31 August 1979. Report #AK-1-8-2229-1.
17. Koch, K., and Baldamus, C., Hemodialysis/Hemofiltration: A
    Comparison of Medical Technical, and Cost Factors, 1st
    Annual Report on NIH Contract NO1-AM-8-2227. Report #AK-
    1-8-2227-1.
18. Bosch, J.P., Long Term Hemofiltration in Comparison with Hemo-
    dialysis, 1st Annual Report on NIH Contract NO1-AM-7-2228.
    Report #AK-1-7-2228-1.
19. Kingsley, G.R., The Determination of Serum Total Protein by the
    Biuret Reaction, J. Biol. Chem., 131, 197-200 (1939).
20. Blatt, W.F., David, A., Michaels, A.S., et al: Solute Polar-
    ization and Cake Formation in Membrane Ultrafiltration:
    Causes, Consequences, and Control Techniques, Membrane
    Science and Technology, Flinn, J.E., Editor. New York,
    Plenum Corporation, 47-97 (1970).
21. Knudsen, J.G., Katz, D.L., Fluid Dynamics and Heat Transfer.
    New York, McGraw-Hill Book Company, 365-367 (1958).
22. Graetz, L., Ann. d., Physik, 337-357 (1885).
23. Keller, K.H., Camales, E.R., Yum, S.I., J. Phys. Chem., 75,
    379 (1971).
24. Kozinski, A.A., Ultrafiltration of Protein Solutions, PhD
    Thesis, Univ. of Wisconsin (1971).
25. Keller, K.H., Effect of Fluid Shear on Mass Transport in Flow-
    ing Blood. Fed. Proc. 30, 1591-1599 (1971).
26. Einstein A., Über die von Molekular Kinetischen Theorie der
    Wärme Geforderte Bewegung von in Ruhenden Flüssigkeiten
    Suspendierten Teilchen. Ann. Phys., 17, 549-560 (1905).

APPENDIX I: NOMENCLATURE

A	=	Hemofilter area, cm^2
C_g	=	"gel" concentration, $gm/100cm^3$
C_{Pb}	=	Average bulk protein concentration, $gm/100cm^3$
d	=	Fiber inside diameter, μ
D	=	Molecular diffusion coefficient, cm^2/sec
J_F	=	Ultrafiltrate flux, cm^3 per (min x cm^2)
k	=	Mass transfer coefficient cm/min
L	=	Fiber length, cm
N	=	Number of fibers in hemofilter
P_{TM}	=	Transmembrane pressure, mmHg
Q_B	=	Average volumetric flow rate of fluid through hemofilter, cm^3/min
Q_F	=	Volumetric flow rate of filtrate, cm^3/min
$\overline{V}$	=	Average linear velocity, cm/sec
γw	=	Wall shear rate, sec^{-1}

ULTRAFILTRATION IN PATIENTS WITH ENDSTAGE RENAL DISEASE

Juan P. Bosch, Robert Geronemus, Sheldon Glabman,
George Moutoussis, Thomas Kahn and Beat von Albertini

From the Renal Division of the Department of Medicine
Mount Sinai School of Medicine of the City University
of New York, 1 Gustave L. Levy Place, N.Y., N.Y. 10029

INTRODUCTION

For the past 15 years maintenance hemodialysis has been the main
technique utilized in the treatment of end-stage kidney disease. It
has been successful in maintaining life in patients whose outlook was
completely hopeless. Even if transplantation offers an attractive
alternative, hemodialysis has remained the single most important
treatment modality. More widespread use is limited only the technical complexity and the considerable economical impact of the procedure. Hemodialysis maintains patients with end-stage renal disease
by replacing, to a certain extent at least, the excretory function of
the failing kidneys. Considerable technical progress has been made over
the years to improve this function. Despite the improved efficiency and
design of the dialysis apparatus, body composition remains deranged. It
has been demonstrated that the size of the various fluid compartments are
often abnormal, and plasma levels of a number of small molecular weight
compounds remain above normal values (1-4). Moreover, there is evidence
that large molecular weight substances normally excreted by the kidneys
are incompletely cleared from the body, even by the most efficient dialysis apparatus (5-7). Furthermore, all of the clinical manifestations of
renal failure are not caused by the retention in the body of substances
which are normally excreted by the kidneys (8). The kidney also has important metabolic functions. For example, polypeptide hormone degradation occurs in the kidney, vitamin D metabolism requires normal kidney
tissue, and important hormones and enzymes (renin erythropoietein,
prostaglandins) are produced in the kidney. It is therefore not surprising that the functionally anephric well-dialyzed patient is only
partially restored to good health since all of the metabolic and biochemical abnormalities associated with renal failure are not corrected.

The well being of patients is not only impaired by the incomplete correction of the manifestations of the uremic state by maintenance hemodialysis, but also may be secondary to the treatment per
se. Dialysis is not always well tolerated. Muscle cramps, angina,
hypotension and postdialysis fatigue are common, especially when large
amounts of fluid have to be removed (9,10). Rapid changes in plasma
osmolality, induced by dialysis may result in a host of neurological
symptoms (11-13). Hypoxemia is frequently noted during dialysis, and
may be related to ventilatory changes induced by rapid alterations in
the acid-base status (14-16). Psychosocial adjustment by patients
and their families and rehabilitation of patients depend, to a considerable extent, on correction of most of the manifestations of
uremia and tolerance to the procedure.

The finding that conventional hemodialysis does not correct many
of the manifestations of the uremic state, has led investigators to
explore other techniques. It has been observed that patients do
better on maintenance hemodialysis when there is some residual renal
function (17). Solute transport in the human glomerulus occurs not
by diffusion but by convective mass transfer in the bulk flow of
water. This process is largely size independent for solutes as large
as myoglobin (17,000 daltons) (18). The amount of solutes of large
molecular size removed even at very low glomerular filtration rates
may be a significant factor in the well being of these patients (17).

Although both ultrafiltration of the plasma water and diffusion
of solute occurs with conventional hemodialysis, this procedure depends primarily on the diffusion process. Passive diffusion of
solute occurs down a concentration gradient from plasma water, across
a semipermeable membrane, into the dialysate solution. The diffusion
process in inherently size-discriminatory with mass transfer rapidly
decreasing as solute size increases. The removal of small size
solutes which are rapidly diffusible, depends mainly on blood and
dialysate flow rates, while the clearance of larger sized solutes
is predominantly a function of the permeability and the surface area
of the dialysis membrane. It has become apparent that the clinical
adequacy of dialysis, as measured by the improvement of nerve con-
duction time, does not necessarily correlate with the removal rate
of the readily measured small solutes but depends rather on the
duration of dialysis. This finding has led to the formulation of the
"middle molecule hypothesis" (6). This proposal argues that some of
the manifestations of uremia, in particular neuropathy, are caused
by larger molecular weight solutes in the range of 500 to 5000 daltor

In an attempt to remove these large-size solutes, dialyzers with
a larger surface area and more permeable membranes have been designed
While duration of dialysis has been reduced using such dialyzers (19)
reports of clinical improvement are conflicting, suggesting that the

proposed toxic "middle molecules" are not adequately cleared or
that these solutes are not the only cause of the uremic state (17).

A technique based solely on the principle of convective solute
transport has been proposed by Henderson and Bluemle (20,21). In
this procedure, termed hemofiltration, an ultrafiltrate of the plasma
is created by hydrostatic pressure exerted across a semi-permeable
membrane. Simultaneously, the blood volume is reconstituted by the
administration of a simulated normal plasma ultrafiltrate.

New thinner and more porous synthetic membranes have been
developed which allow the passage of solutes up to a molecular weight
of 50,000 daltons (22). Driven by the hydrostatic pressure gradient
such solutes cross the membrane with the bulk of the ultrafiltrate
in concentrations similar to those in the plasma water. As clearance
equals ultrafiltration rate for all solutes not rejected by the
membrane, the removal rate is proportional to the respective plasma
concentrations and independent of the size of such a solute. At
comparable blood flow rates the removal of small solutes by ultrafiltration is less effective than by diffusion but for larger solutes
of possible pathophysiological importance ultrafiltration is far
more effective (23).

Considerable volumes of ultrafiltrate have to be removed to
lower the plasma concentrations of smaller solutes to that achieved
with hemodialysis. Removal of about 20 liters of ultrafiltrate
three times weekly is necessary to obtain pretreatment blood urea
concentrations similar to these obtained by conventional hemodialysis
(24). New hollow fiber filters are available which are capable of
ultrafiltration rates up to 120 ml/min. At such rates replacement
of the removed ultrafiltrate in equal volumes is critical. The
replacement fluid, usually a physiological electrolyte solution
with lactate or acetate as base, is added to the blood either
proximal (pre-dilution) or distal to the filter (post-dilution)
(Fig. 1). The advantage of predilution is that it permits higher
ultrafiltration rates without leading to excessive hemoconcentration
in the filter. The disadvantage of the pre-dilution method is that
it requires twice the volume of replacement fluid as compared to the
post dilution technique. Accurate and fail-safe matching of the
replacement fluid with the ultrafiltrate is required by both methods.
Automatic systems for volume replacement based on volumetric or
gravimetric control have been developed (25).

Henderson has developed hemofiltration to clinical practibility
using a pre-dilution technique. Long term use of hemofiltration
in place of hemodialysis has been well tolerated and has resulted
in an improvement of the patients, (18). Quellhorst and others
have successfully used hemofiltration with post-dilution in recent
years (26,27).

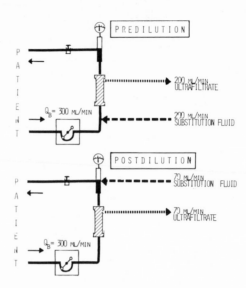

Figure I. Substitution models for hemofiltration

During hemodialysis the extent to which ultrafiltration can be effectively performed is limited by two factors: 1) the low hydrauli permeability of the cellulosic membranes which necessitates high transmembrane pressure gradients and 2) the poor tolerance of the patients. Effective fluid removal in volume overloaded patients is often limited by the occurrence of hypotension and other symptoms related to intravascular volume depletion.

Bergström and other investigators have recently found that flui removal is better tolerated if ultrafiltration is done separately from regular dialysis (28,29). Ultrafiltration has been accomplishe before regular dialysis by the use of conventional flat plate dialyzers in which large transmembrane pressure gradients have been obtained by generating a negative pressure inside the empty dialysat compartment. Although large amounts of fluid have been removed, no signs and symptoms of volume depletion have developed. In addition, the blood pressure has been better controlled in hypertensive patien with ultrafiltration followed by dialysis than by regular dialysis alone (28,29).

We have performed a controlled study: 1) to determine if conven tional dialysis can be made more efficient by modifying prevailing techniques using ultrafiltration followed by conventional dialysis. 2) to evaluate an alternate technique, hemofiltration, for the treat ment of end stage kidney disease and to compare its effectiveness to conventional dialysis as well as to sequential ultrafiltration and dialysis.

COMPARISON OF HEMOFILTRATION (HF) AND SEQUENTIAL ULTRAFILTRATION
AND DIALYSIS TO CONVENTIONAL HEMODIALYSIS (HD)

METHODS

14 patients on maintenance hemodialysis for at least 6 months
prior to study were selected. Their age averaged 40 years with a
range between 28-50 years. Their residual kidney function averaged
1.6ml/min with a range between 0 and 3 ml/min. The patients were
studied for a 5 month control period on hemodialysis and then randomly selected in two groups: 7 patients underwent 4 months of
sequential ultrafiltration and dialysis and 7 patients 4 months of
hemofiltration. Both groups returned to hemodialysis for 4 months.
11 patients completed the above protocol. All three treatment
modalities were adjusted to obtain the same pre-treatment BUN of
90 ± 10mg/dl.

Hemodialysis was performed with a $1.5m^2$ Travenol hollow fiber
dialyzer.

Sequential ultrafiltration and dialysis was done with the same
$1.5m^2$ dialyzer and consisted of 1½ hours of ultrafiltration followed
by standard hemodialysis. The mean amount of fluid removed during
ultrafiltration was 4.5 liters. Body weight was maintained constant
during the dialysis part of sequential ultrafiltration and dialysis
by replacing volume losses.

Hemofiltration was performed with a $1.3m^2$ Amicon Diafilter 40
in post dilution and a combined pre and post dilution substitition
mode. In each hemofiltration an average of 26.2 liters ultrafiltrate
was removed and replaced with an acetate containing substitution
fluid by an automatic gravimetric system.

The mean treatment time for all 3 treatment modalities is
shown in Table 1.

In sequential ultrafiltration and
dialysis treatment time was only 16
more minutes than hemodialysis, this
was due to higher blood flows used in
this modality. In hemofiltration treatment time averaged 250 minutes despite
the large quantities of fluid exchange.
This was possible due to the use of new
hemofilters and use of pre and post
dilution substitution mode which has
been previously reported (30).

TABLE 1
Treatment Time

HD	219 ± 6 min
HF	250 ± 8
HD	230 ± 11
UFD	246 ± 12

In Table II are shown the body weights, urea generation and the volume of ultrafiltrate needed per treatment in order to maintain a pre treatment BUN at 90 ± 10mg/dl in the 6 patients on hemofiltration. It is shown, that in order to maintain the same pretreatment BUN, the amount of fluid exchange per hemofiltration is independent of body weight but directly related to urea generation. Further studies are necessary to determine if BUN is the right parameter to use in prescribing the amount of fluid to exchange per treatment.

TABLE II

Patient	Weight kg	Urea Generation gms/24 hrs	Fluid exchange L/treatment
1	81.7	8.9	25.4
2	47.2	9.7	27.7
3	44.6	7.6	21.7
4	93.9	10.9	31.4
5	54.9	7.1	20.3
6	57.0	8.4	24.0

RESULTS

1. SYMPTOM ANALYSIS: To measure the patients well being under the various treatment modalities, a survey was undertaken for both symptoms occurring during the treatment and during the interval between the treatments. Patients were requested to fill out a questionnaire comprising a list of several common symptoms for both the time during the treatment and the inter-treatment interval. Additionally, all treatment records were analyzed for occurrence of symptoms. Results are shown in Table III. In both treatment modalities an improvement of symptomatology was noted for both the treatment and the period between treatments.

TABLE III
Symptom Analysis
a. Symptoms during the treatment

	Headache	Weakness	Tiredness	Dizziness
HD	19.1	15.6	12.1	13.1
HF	4.4	9.0	7.4	5.1
% Change HF/HD	-76%	-42%	-39%	-61%
HD	12.3	39.6	36.5	22.4
UFD	6.7	18.8	18.5	11.2
% Change UFD/HD	-46%	-53%	-49%	-50%

b. Symptoms between the treatments

	Headache	Weakness	Nausea	Vomiting	Cramps
HD	18.2	12.8	15.5	2.7	22.2
HF	4.2	3.2	4.5	.4	11.8
% Change HF/HD	−77%	−75%	−71%	−85%	−47%
HD	7.0	24.1	12.2	3.9	16.8
UFD	1.9	8.3	2.5	2.1	31.9
% Change UFD/HD	−73%	−60%	−80%	−46%	+100%

In both experimental treatment modalities an improvement of symptomatology was noted for both the treatments and the period between treatments. A greater sense of well being and absence of fatigue was expressed in hemofiltration (Table III). Despite the removal of intertreatment weight gains equal or greater than in dialysis there was a marked reduction of symptoms during HF and UFD. A marked difference was found in the occurrence of cramps: while cramps were reduced in hemofiltration, they markedly increased during UFD.

2. MEAN PRETREATMENT ELECTROLYTES: In Table IV are shown the mean pretreatment electrolytes during the control and experimental period. The low initial plasma bicarbonate reflects the chronic metabolic acidosis of these patients, whole depletion of body buffers is only partially corrected by dialysis. Pretreatment plasma bicarbonate increased significantly during the hemofiltration period and subsequently fell during the second control period on dialysis.

HEMOFILTRATION

	Na meq/l	K meq/l	Cl meq/l	CO_2 meq/l	Anion Gap meq/l
HD	138	5.3	101	18.5	23.9
HF	136	5.6	102	20.3*	21.0*
HD	137	5.2	100	20.3	21.2

ULTRAFILTRATION − DIALYSIS

	Na meq/l	K meq/l	Cl meq/l	CO_2 meq/l	Anion Gap meq/l
HD	139	5.5	103	17.1	24.3
UFD	139	5.3	105	16.0*	23.2
HD	139	5.2	103	18.7*	22.4

(*=$p<.05$ by paired t compared to previous period)

In sequential ultrafiltration and dialysis a further deterioration of the metabolic acidosis was seen: plasma bicarbonate fell significantly during this treatment modality and rose again in the second control period on hemodialysis. This phenomenon may be partially explained by an unreplaced loss of bicarbonate during the ultrafiltration part of this form of therapy.

3. CALCIUM, PHOSPHASE METABOLISM: In Fig. 2, Fig. 4 are shown the monthly changes for serum calcium, phosphorous and alkaline phosphatase. Parathyroid hormone (PTH) levels for each patient are shown in Figure 3 and Figure 5.

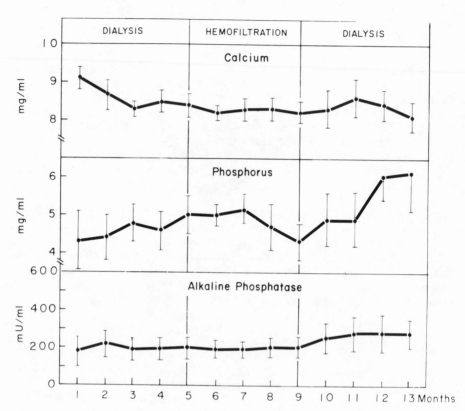

Fig. 2: Monthly values for serum calcium, phosphorus and alkaline phosphatase for 6 patients on hemofiltration. (Mean ± S.E.M.)

There were no significant changes in calcium, phosphorus and alkaline phosphatase for either group where the total period averages are considered. However, as seen in Figure 2 there is a clear trend toward decreased phosphorus in the hemofiltration group during the experimental period. In hemofiltration, PTH fell during the experimental period and rose again in the second control period in spite of unchanged serum calcium. This lowered PTH may result from 1) the lowered plasma phosphorus, 2) loss of PTH via ultrafiltration 3) or more likely, a greater positive calcium balance in hemofiltration than in hemodialysis.

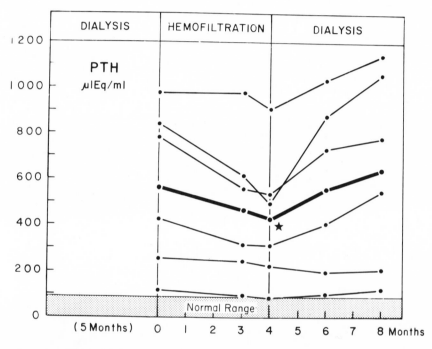

Fig. 3: Parathyroid hormone levels by radioimmunoassay for 6 patients on hemofiltration. First value (0) is at end of 5 months control hemodialysis. Mean values are shown with heavy line. (* = p<.05 by paired t test, compared to previous period.)

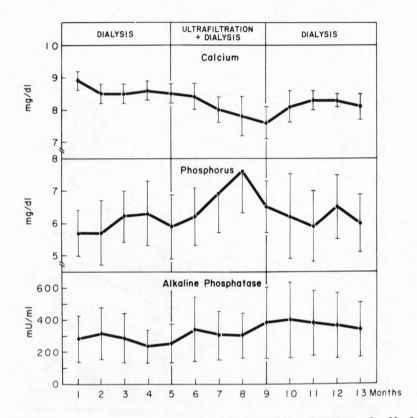

Fig. 4: Monthly values for serum calcium, phosphorus and alkaline phoshpatase for 5 patients on sequential ultrafiltration and dialysis. (Mean ± S.E.M.)

        In the sequential ultrafiltration there was a trend toward increased phosphorus and decreased calcium during the experimental period.

        PTH rose gradually possibly due to the rise in phosphorus and decrease in calcium during the experimental period.

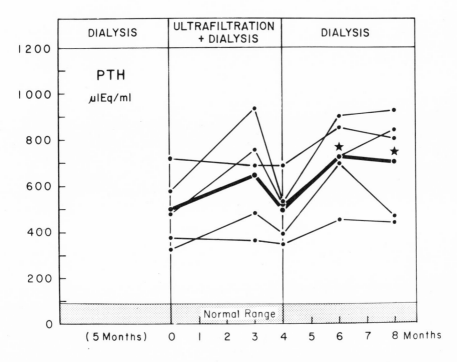

Fig. 5: Parathyroid hormone levels by radioimmunoassay for 5 patients on sequential ultrafiltration and dialysis. First value is at end of 5 months control hemodialysis. Mean values are shown with heavy line.

4. NUTRITIONAL EVALUATION: During the course of the study we used
urea kinetics as described by Gotch et al not only to prescribe
the treatment necessary to maintain a constant pretreatment BUN
but also as a method to evaluate the nutritional status of these
patients. These data have been published recently and demonstrated
that in sequential ultrafiltration and dialysis patients lowered
their protein intake significantly while on hemofiltration protein
intake remained unchanged (31).

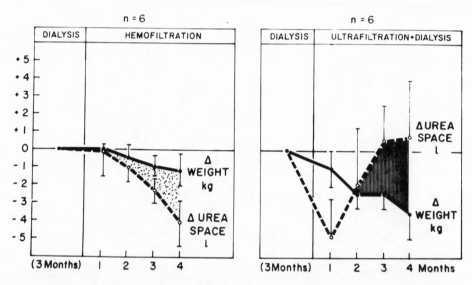

Figure 6: Changes in body weight and urea space during the experimental periods. The dotted area in hemofiltration represents the
net gain in total body solids, the shaded area in sequential ultrafiltration and dialysis the net loss in total body solids (Mean ±
S.E.M.).

In Figure 6 are summarized the monthly changes in body weight
and urea space measured by urea kinetics in hemofiltration and
sequential ultrafiltration and dialysis. It has been suggested that
urea space approximates total body water. Accordingly it may be
possible to estimate total body solids as the body weight minus the
urea space. During HF body weight decreased 1.3kg while urea space
decreased significantly 4.5 1. Since urea space decreased more than
body weight these findings suggest that during hemofiltration body
solids increased despite the decrease in body weight. It was not
possible to demonstrate a significantly higher dietary protein
intake or a lower urea generation in these patients. More precise
techniques may be necessary to detect small positive nitrogen
balance.

In sequential ultrafiltration, body weight decreased significantly 2.9kg while urea space increased minimally suggesting a net loss of total body solids.

The lower protein intake demonstrated in this group may explain the loss of total body solids in these patients. Several investigators have shown that changes in protein intake are rapidly reflected in alterations in plasma amino acids concentrations and that amino acids measurements provide a valuable index of nutritional status.

Plasma amino acids were determined 2 times in each patient during control, experimental and second control period. Samples were obtained in the fasting state and measurements were performed using column chromatography. All determinations included the measurement of the following amino acids: aspartate, serine, glutamine, glutamate, glycine, alanine, threonine, valine, methionine, isoleucine, leucine, phenylalanine, cysteine, tyrosine, citrulline, taurine, alpha-aminobutyric acid, cystathioine.

TABLE V

HEMOFILTRATION

	HD	HF	HF
Total amino-acids (μmol/dl)	230	200	213
Non-essential amino-acids (μmol/dl)	169	141*	156
Essential amino-acids (μmol/dl)	61	59	56
Valine/Glycine Ratio	.45	.52	.44
Tyrosine/Phenylalanine Ratio	.74	.54	.87
Essential/Non essential Ratio	.36	.42	.36*

SEQUENTIAL ULTRAFILTRATION AND DIALYSIS	HD	UFD	HD
Total amino-acids (μmol/dl)	275	239*	249
Non essential amino acids (μmol/dl)	202	178	185
Valine/Glycine Ratio	.50	.46	.48
Tyrosine/Phenylalanine Ratio	.72	.71	.74
Essential/Non-essential Ratio	.38	.34	.35

$* = p < .05$ when compared to preceeding period
These findings are summarized in Fig. 7.

For these calculations essential amino-acids included are: threonine, valine, methionine, isoleucine, leucine, phenylalanine. Non-essential amino-acids are: aspartate, serine, glutamine, glutamate, glycine, alanine. Total amino-acids includes essential plus non-essential amino-acids. Mean values are shown in Table V. Normal values for our laboratory are: Total amino-acids 187.4 μmol/dl. Essential amino-acids 54.4 μmol/dl and non-essential amino-acids 133.7 μmol/dl.

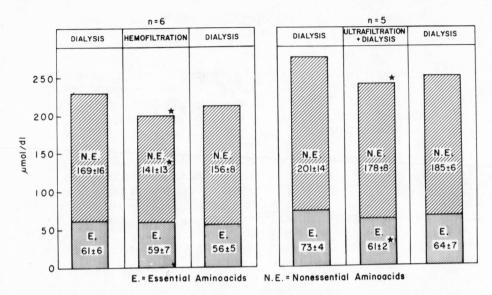

Figure 7: Mean profile for essential amino acids (threonine, valine, methionine, isoleucine, leucine, phenylalanine) and non-essential amino acids (asparate, serine, glutamine, glutamate, glycine, alanine) for the two experimental groups. (* = p<.05 by paired t test, compared to previous period).

In the control period studies in these patients show evidence of malnutrition: increased non essential amino-acids, increased glycine, valine, and decreased E/NE ratio. Other abnormalities in amino-acids apparently not related to malnutrition were also demonstrated in the control period such as increased citrulline, cystine 1- and 3-methylhistidine. A decrease in tyrosine/phenylalanine ratio was also demonstrated.

During hemofiltration non essential amino-acids decreased significantly from 168.6 to 140.8 μmoles/dl and increased to 155.9 in the second dialysis period. (Fig. 7, left panel). Essential amino-acids remained unchanged during the 3 periods. Consequently E/NE amino-acids increased from .36 to .42 with hemofiltration and returned in the second hemodialysis period to .36, a value which is significantly lower than .42.

In UFD's significant decrease in total amino-acids was demonstrated (Fig. 7, right panel). This was due to a fall in non-essential and essential amino-acids. A decrease in essential to non-essential amino acids ratio from .38 to .34 occurred during this period. A significant decrease in valine and leucine was seen. Valine to glycine ratios decreased from .46 to .37 during UFD. Significantly lower values for citrulline and cystine were also found during UFD but this finding is difficult to interpret because these values were lower or unchanged during the second dialysis period.

In summary the amino-acid changes demonstrated during HF suggest an improvement in the nutritional status of these patients. Other demonstrated changes in amino-acids may not be related to nutritional status, but may reflect better clearance of these substances. However, the lowering of the tyrosine/phenylaline ratio due to lower levels of tyrosine may suggest a depletion syndrome. This finding may be due to losses of tyrosine by filtration or increased conversion of tyrosine to phenylalanine. During UFD the amino-acid changes suggested a deterioration in the patients nutritional status. It is possible that the lower levels of some amino acids were due to loss by clearance in the dialyzer. However, because levels remained low in the second hemodialysis period, it would appear that there was a lower generation rate of these amino acids.

CONCLUSIONS

HEMOFILTRATION: was well tolerated and clinical improvements were observed:
- Nutritional status: a greater loss of total body water (urea space) than weight, suggesting an increase in body solids and improvement in plasma amino acids
- Decrease in PTH levels with a tendency of serum phosphate to decrease
-Marked increase in pretreatment plasma bicarbonate
-Improved symptomatology during and between treatments

SEQUENTIAL ULTRAFILTRATION AND DIALYSIS: no clinical improvements were noted except for better treatment tolerance. Clinical outcomes included:
Weight loss without change in the urea space/weight ratio suggesting a loss of serum calcium to decrease and of phosphate to increase
-a tendency of serum calcium to decrease and of phosphate to increase
-a decrease in pretreatment plasma bicarbonate and worsening of metabolic acidosis.

Hemofiltration appears to be superior to hemodialysis in several areas, while sequential ultrafiltration and dialysis as a maintenance therapy appears to offer no clinical benefits over conventional hemodialysis and may result in the deterioration of several clinical parameters.

REFERENCES

1. Coles, G.A.: Body composition in chronic renal failure. QJ Med, 41, 161: 25, 1972.

2. Giovanetti, S., and Maggiore, Q.: A low-nitrogen diet with proteins of high biological value for severe chronic uremia. Lancet 1: 1000, 1964.

3. Franklin, S.S., Gordon, A., Kleeman, C.R. and Maxwell, M.H.: Use of a balanced low-protein diet in chronic renal failure: JAMA 202: 477, 1967.

4. Kerr, D., Robson, A., Eliott, R.W. and Asheroft, R.: Diet in chronic renal failure. Proc. Roy. Soc. Med. 60: 115, 1967.

5. Gordon, A., Berström, J., Fürst, P., and Zimmerman, L: Separation and characterization of uremic metabolites in biologic fluids. A screening approach to the definition of uremic toxins. Kidney Int. 7: S45, 1975.

6. Babb, A.L., Popvitch, R.P., Christopher, T.G. and Scribner, B.H.: The genesis of the square-meter hour hypothesis. Trans. Am. Soc. Artif. Intern Organs 17: 81, 1971.

7. Scribner, B.H., Babb, A.L.: Evidence for toxins of "middle" molecular weight. Kidney Int. 7 (suppl 2): 349, 1975.

8. Teschan, P.E.: On the pathogenesis of uremia. Am. J. Med. 48: 671, 1970.

9. Kim, K.E., Neff, M., Cohen, B., Somerstein, Chinitz, J., Onesti, G., and Swartz: Blood volume changes and hypotension during hemodialysis. Trans. Am. Soc. Artif. Intern. Organs, 16: 508, 1970.

10. Stewart, W.K., Fleming, L.W., and Mannel, M.A.: Muscle cramps during maintenance hemodialysis. Lancet 1: 1049, 1972.

11. Wakim, K.G.: The pathophysiology of the dialysis disequilibrium syndrome. Mayo Clin. Proc. 44: 406, 1969.

12. Cerra, F.B., Anthone, R., and Anthone, S.: Colloid osmotic pressure fluctuations and the disequilibrium syndrome during hemodialysis. Nephron 13: 245, 1974.

13. Arieff, A.L., Massry, S.G., Barrientos, A. and Kleeman, C.R.: Brain water and electrolyte metabolism in uremia: effects of slow and rapid hemodialysis. Kidney Int. 4: 177, 1973.

14. Bischel, M.D., Scoles R.G. and Mohler, M.D.: Evidence for pulmonary microemboli during hemodialysis. Trans. Am. Soc. Artif. Intern. Organs 19: 492, 1973.

15. Hurwitz, S., Milne, J., Goldman, H.L.: Blood gas abnormalities in patients on chronic hemodialysis. Abstract Nephron 13: 266, 1974.

16. von Albertini, B., Kirpalani, A., Goldstein, M., Glabman, S., and Bosch, J.: Changes in pCO_2 during and after hemodialysis. Proc. Clin. Dialysis Transpl. Forum 6, 1977.

17. Nolph, K.D.: Short Dialysis, Middle Molecules and Uremia. Ann. Intern. Med. 86: 93, 1977.

18. Henderson, L.W., Silverstein, M.E., Ford, C.A. and Lysaght, J.L.: Clinical response to maintenance hemofiltration. Kidney Int. 7: 6-58, 1975.

19. Rosenzweig, J., Babb, A.L., Vizzo, J.E., Scribner, B.H., and Ginn, H.E.: Larger Surface area hemodialysis. Proc. Dialysis Transpl. Forum. 1: 56, 1971.

20. Henderson, L.W.: Peritoneal Ultrafiltration Dialysis: Enhanced urea transport using hypertonic peritoneal dialysis fluid. J. Clin. Invest. 45: 960, 1966.

21. Henderson, L.W., Besarab, A., Michaelis, A., Bluemle, Jr., L.W.: Blood purification by ultrafiltration and fluid replacement (diafiltration) Trans. Amer. Soc. Artif. Organs. 13: 2116, 1967.

22. Henderson, L.W., Colton, C.K. and Ford, C.A.: Kinetics of hemofiltration. II. Clinical characterization of a new blood cleansing modality. J. Lab. Clin. Med. 85: 372, 1975.

23. Colton, C.K., Henderson, L.W. and Ford, C.A., et al.: Kinetics of hemodiafiltration I. In vitro transport characteristics of a hollow-fiber blood ultrafilter. J. Lab. Clin. Med. 85: 355, 1975.

24. Quellhorst, E., Scheler, F., et al.: Arbeitstagung ueber Haemofiltration, Braunlage/Harz. Wissenschaftl. Inform. Fresenius Stiftung 4: 85, 1976.

25. Quellhorst, E., Schnenemann, B. and Borghardt, J.: Clinical and technical aspects of hemofiltration. Artif. Organs. II. 4: 334, 1978.

26. Schaefer, K., V. Herrath, D., Gullberg, C. et al.: Chronic
    Hemofiltration. Artif. Organs. II, 4: 386, 1978.

27. Kopp. K.F.: Hemofiltration (Editorial). Nephron 20: 65, 1978.

28. Bergström, J., Asaba, H., Fürst, P. and Oueles: Dialysis,
    ultrafiltration and blood pressure. Proc. Europ. Dial.
    Transpl. Assoc. 13: 293, 1976.

29. Asaba, H., Bergström, J., Fürst, P., Lindh, K., Mion, R.,
    Oueles, R., and Shaldon, S.: Sequential ultrafiltration and
    diffusion as an alternative to conventional dialysis. Clin.
    Dial. Transpl. Forum. VI: 129, 1976.

30. Geronemus, R., von Albertini, B., Glabman, S. et al: Enhanced
    small molecular clearance in hemofiltration. Proc. Clin. Dial.
    Transpl. Forum. VIII: 147, 1978.

31. Bosch, J., Geronemus, R., von Albertini, B. et al.: Urea kinetics
    in hemodialysis, hemofiltration and sequential ultrafiltration and
    dialysis. Proc. Clin. Dial. Transpl. Forum. VIII: 142, 1978.

DEVELOPMENT OF NOVEL SEMIPERMEABLE TUBULAR

MEMBRANES FOR A HYBRID ARTIFICIAL PANCREAS

Clark K. Colton[1], Barry A. Solomon[2], Pierre M. Galletti[3], Peter D. Richardson[3], Chieko Takahashi[4], Stephen P. Naber[4], and William L. Chick[4]

[1]Department of Chemical Engineering, Massachusetts Institute of Technology, Cambridge, MA 01239; [2]Research Division, Amicon Corporation, Lexington, MA 02173; [3]Division of Biological and Medical Sciences, Brown University Providence, RI 02912; [4]Joslin Research Laboratory, Harvard Medical School, Boston, MA 02215

Lack of normal insulin release by the pancreas may contribute to the long-term complications of diabetes. As an approach to restoring physiological insulin delivery, we are developing an implantable hybrid artificial pancreas consisting of pancreatic cells from islets of Langerhans cultured on the outside surface of semipermeable tubular membranes which are permeable to glucose and insulin but retain antibodies and lymphocytes. The device is implanted in the cardiovascular system as an arteriovenous shunt so that the cells have access to oxygen from the blood stream. The insulin-producing tissue is protected from the immune rejection process by the semipermeable membrane. This protection of the islet cells eliminates the problems associated with immunosupression which is required in ordinary transplantation procedures. Our approach offers the possibility of utilizing islet cells from animals in place of human tissue, and it relies upon natural biological feedback for insulin release and control of blood glucose concentration.

Our initial in vitro[1,2,3] and in vivo[4,5] feasibility studies were carried out with bundles of capillary hollow fibers having an internal diameter of 200 μm. These devices were implanted in rats and functioned successfully for periods up to half a day. Clotting problems, which may preclude lonter experiments, are reduced by using tubes with larger internal diameter. Therefore, for long-term implantation we have fabricated semipermeable tubular membranes

with diameters from 1 to 2.5 mm which have properties similar to
those of the hollow fibers and which provide a markedly less throm-
bogenic environment. In this paper we briefly review our previous
results with hollow-fiber devices, and we summarize our current in
vitro and in vivo studies with larger diameter tubes.

All hollow-fiber and tubular membranes were prepared from
Amicon XM-50 acrylic copolymer with a nominal molecular weight
cutoff of 50,000. The cross section of the membrane wall is not
uniform. At the lumen surface there is a thin skin which determines
the hydraulic permeability and is the solute discriminating barrier.
The remainder of the wall is a spongy matrix which provides mechanical support and is believed to determine the solute diffusive transport properties. It provides an excellent substrate for cell
attachment. The interstices in the spongy layer are sufficiently
large to permit intrusion of some single cells and small cell
clusters into the wall of the fiber.

CAPILLARY HOLLOW FIBERS

In our early studies we routinely employed bundles of 100
parallel capillary fibers potted in a cylindrical jacket, as shown
in Figure 1. These hollow-fiber membrane devices were similar to
those originally used to grow cancer cells in high density[6].
The jacket had one or more ports through which insulin-producing

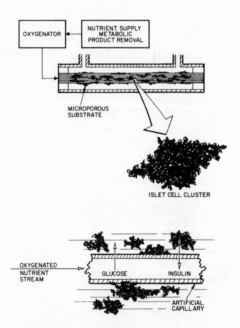

Fig. 1. Islet cell culture on hollow fibers[2].

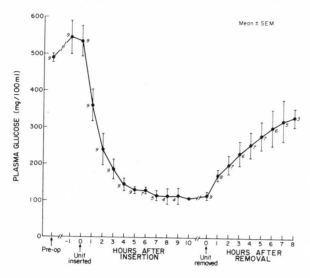

Fig. 2. Effect of the artificial pancreas on plasma glucose concentration in rats as a function of time after insertion[4]. Data are mean values for the number of measurements indicated by numerals.

cells were seeded onto the outside surfaces of the fibers. The lumen of each fiber was perfused with an oxygenated stream of nutrient fluid such as tissue culture medium or blood. An increase in the glucose concentration in the perfusate was sensed by the beta cells (which comprise the major component of the islet cells) on the outside of the fiber. The cells released appropriate amounts of insulin which permeated across the wall of the fiber into the perfusate medium. Since the wall of the fiber separated the transplanted cells from the perfusion fluid, contact with antibodies and lymphcytes was blocked.

Our initial in vitro experiments with units seeded with cultured neonatal rat islet cells showed that insulin output varied in direct relation to the glucose concentration of the medium perfusing the device. In subsequent in vivo experiments, devices were implanted as external iliac artery to vein shunts in rats with alloxan-induced diabetes. The effect of device implementation on the mean plasma glucose levels in a group of these diabetic animals is shown in Figure 2, which is a plot of plasma glucose concentration versus time. Within one hour following insertion of the pancreatic device in the external shunt, the mean plasma glucose level begins to fall. Approximately five hours after insertion, the mean plasma glucose concentration reaches its normal level and is maintained at approximately 100 mg/dl. Following removal of the device, the plasma glucose concentration rises back to the diabetic range. These findings indicate that such devices are capable of restoring normal

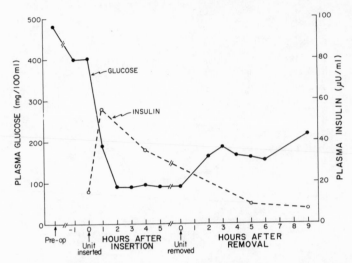

Fig. 3. Effect of the artificial pancreas on plasma glucose and insulin concentrations in a single representative alloxan-diabetic rat as a function of time after insertion[4].

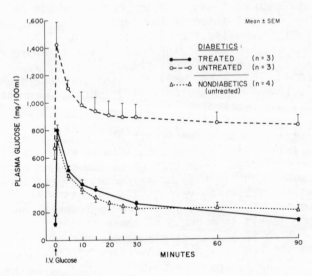

Fig. 4. Intravenous glucose tolerance tests in alloxan-diabetic rats with the artificial pancreas inserted ex vivo in an arterio-venous shunt[4].

blood glucose concentration in diabetic recipients.

Figure 3 shows a typical result for plasma glucose and insulin concentrations in one of the rats. Following insertion of the unit in the shunt, circulating insulin levels rise significantly while glucose levels decline. As the glucose concentration approaches the normal range, insulin concentration falls, suggesting feedback regulation between glucose and insulin. Following removal of the unit from the shunt, plasma glucose concentration again rises back to the diabetic range while circulating insulin concentration further declines.

Intravenous glucose tolerance tests were also conducted in device recipients, and the results are shown in Figure 4. Three groups of rats were used in this study, including non-diabetic control animals, untreated diabetic animals, and diabetic animals treated with device implantation. As anticipated, plasma glucose concentrations in untreated diabetic rats receiving intravenous glucose were grossly elevated. In contrast, plasma glucose concentrations in normal non-diabetic and in treated diabetic rats were similar. These data indicated that, when implanted as an arteriovenous shunt in the vascular system of diabetic rats, the device restores glucose homeostasis essentially to normal.

SEMIPERMEABLE TUBULAR MEMBRANES

More recently we elected to fabricate larger-bore semipermeable tubular membranes in order to increase device biocompatability with respect to blood clotting. Figure 5 is a schematic drawing which shows one device prototype in which these new semipermeable tubes were employed. The circular device is approximately 5 cm in diameter and is fabricated from a single length of semipermeable tubing with an internal diameter of approximately 1 mm. The tube is wound in a spiral configuration and is encased in a plastic jacket designed so as to minimize the dead volume in the outer chamber compartment. Ports are provided for blood flow in the lumen and for seeding of cells into the outer chamber. We have fabricated similar devices utilizing semipermeable tubes with an internal diameter of 2.5 mm. Devices incorporating a single large-bore tube have proven to be more biocompatible when tested in non-anticoagulated sheep and heparinized and non-heparinized dogs than devices fabricated from a multiplicity of parallel tubes or hollow fibers. Although parallel devices tend to clot within hours in heparinized animals, newer prototypes fabricated with a single larger-bore tubular membrane have remained patent for considerably longer. For example, in one of our most recent experiments in which a device was implanted but not seeded with cells, the unit remained patent for 25 days in a non-heparinized normal dog (after which it was electively removed for inspection), despite the fact that the dog has a very active clotting system.

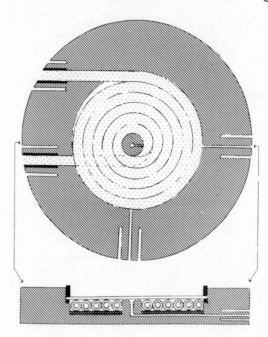

Fig. 5.   Schematic diagram of spiral-wound unit fabricated from a
single length of large-bore semipermeable tubular membrane. Cells
are seeded into outer chamber through one port, and a second port
is provided to extract fluid simultaneously from the chamber.

      Experiments were conducted in diabetic dogs. Vascular access
was provided by implanting external iliac artery to vein shunts in
normal dogs. Diabetes was then induced by chemically destroying
the animal's own beta cells with streptozotocin or a combination of
streptozotocin and alloxan. Results from one experiment using a
unit seeded with neonatal rat islet cells are shown in Figure 6
which is a plot of plasma glucose concentration versus time.
Although the time required to normalize plasma glucose concentration
is greater than that observed with hollow fiber devices in rats,
these initial data show that a device with one large-bore tube is
capable of restoring blood glucose concentration to normal in a
diabetic dog and that glucose concentration rises again when the
device is removed. Taken together, these in vivo data from the rat
and dog suggest that the hybrid artificial pancreas consisting of
living islet cells cultured on the outside surfaces of semipermeable
fibers and tubes constitutes an effective means for restoring carbohydrate tolerance in experimental diabetes.

      Obviously, many problems must be overcome before such a hybrid
pancreas device can be used for treatment of diabetic patients.
Among them are biocompatability with respect to blood clotting. At

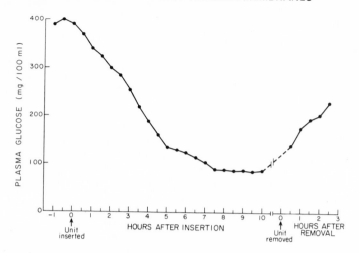

Fig. 6. Effect of the artificial pancreas with the design shown in Figure 5 on plasma glucose concentration with a streptozotocin-diabetic dog as a function of time after insertion.

present, this problem appears to be greatly reduced by using a larger diameter semipermeable tube in the device instead of a parallel array of capillary fibers. However, the performance of insulin-secreting cells in these new single-tube device prototypes must now be investigated in greater depth. The larger diameter tubes used in these devices must have thicker walls than those of capillary fibers in order to ensure satisfactory mechanical strength. The greater wall thickness increases the diffusional resistance and the lag time associated with glucose and insulin diffusion. These phenomena may account for the slower response seen in Figure 6.

In order to provide a rational basis for the design of devices suitable for long-term implantation, we are examining in detail the transport properties of the tubular membranes, the relationship between wall thickness and mechanical properties, and the effect of wall thickness on insulin secretory dynamics in these devices. Other design-related questions include the membrane surface area required to support a given number of islet cells and the possible influences of cell distribution within the device on secretory performance. Another area undergoing further study is the relationship between internal diameter and the propensity for thrombosis. In the remaining portion of this paper, we sketch some of our current efforts in these areas, specifically membrane mechanical properties, transport properties, and in vitro insulin secretory dynamics.

Mechanical Properties

When the units are seeded with beta cells, the outer chamber

is at a higher pressure than the lumen, and fluid is ultrafiltered
from the outer chamber to the lumen so as to deposit islet cells on
the membrane surface. If the pressure difference between the outer
compartment and the lumen is high enough, the membrane wall will
collapse. A similar pressure differential can occur in a unit after
implantation or in any situation where flow occurs through the lumen.
Since there must be a drop in lumen pressure between the inlet and
the outlet, and the volume of the outer chamber is fixed, fluid
ultrafilters out of the membrane at the device inlet and is absorbed
at the outlet in a manner similar to Starling's flow in the microcirculation. As a futher consequence, the pressure in the outer
chamber is higher than in the lumen at the outlet. This differentia`
is transiently increased if the outlet lumen pressure is suddenly
reduced, for example, by a change in an animal's hemodynamic status.

     To investigate this phenomenon we developed an apparatus to
study, under carefully controlled conditions, the collapse of a
single tubular membrane potted in a cartridge, as shown in Figure 7.
Water is pumped through the lumen while the air pressure in the outer
chamber is increased with time by use of a syringe pump. The lumen
inlet and outlet pressures and the outer chamber pressure are all
monitored. At some point, the pressure differential between the
outer chamber and lumen becomes large enough to collapse the fiber,
and the water is shunted toward the parallel branch which has a flow
resistance ten times higher than that of the fiber.

     The plots of pressure versus time in Figure 8 qualitatively
illustrate typical results. The outer chamber pressure increases
linearly with time, and the lumen inlet pressure is initially constant. When a previously undamaged tube suddenly collapses, the
lumen inlet pressure rises rapidly at the same time that extensive

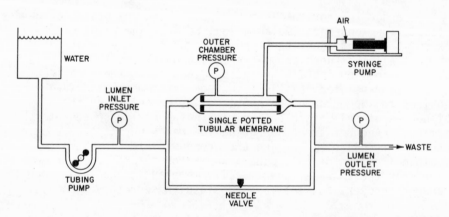

Fig. 7. Schematic diagram of apparatus for measurement of critical
collapse pressure of a tubular membrane.

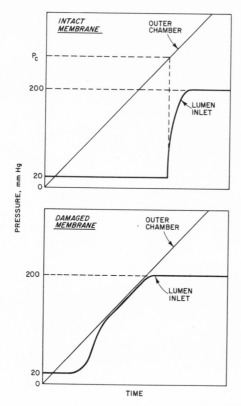

Fig. 8. Qualitative behavior of results from membrane collapse
experiments.

deformation is observed visually. The difference between the chamber
and the mean lumen pressures at this point is denoted as the critical
collapse pressure (P_c). These damaged membranes do not subsequently
leak red cells or pass blue dextran (MW 2,000,000) except after
repeated collapse. We infer that the initial structural damage
occurs only to the spongy matrix and not to the skin. When the experiment is repeated with a damaged membrane which has previously
undergone collapse, deformation begins at a much lower pressure difference. Such damaged membranes are consequently unsuitable for in
vivo use because of the possibility of tube deformation and subsequent thrombus formation.

Figure 9 shows how the critical collapse pressure varies with
the ratio of wall thickness to mean tube radius. If one assumes
that (1) the critical radial stress (σ) for fiber collapse is independent of wall thickness (h) and depends only on the material properties of the wall, and (2) that the wall thickness is small compared to the mean tube radius (R), then the cylindrical shell

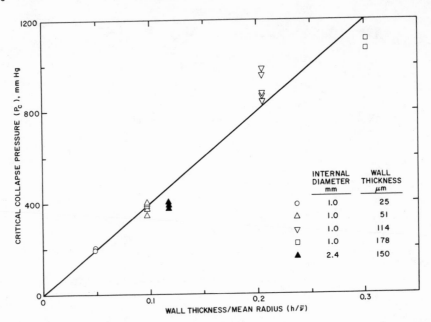

Fig. 9. Critical collapse pressure as a function of ratio between wall thickness and mean tube radius.

approximation of stress distribution under transmural loading leads to an expression for the critical collapse pressure: $P_c = \sigma(h/R)$. The validity of this expression is corroborated by the fact that all the data in Figure 9 lie on a straight line. By keeping the outer chamber pressure below the critical value during beta cell seeding, damage to the fragile tubes is prevented.

Transport Properties

Measurements of the diffusional mass transfer resistance of semipermeable tubes are underway with a variety of solutes. The tubes are potted in a cartridge and operated in cocurrent dialysis mode. These measurements pose a difficult challenge for two reasons. First, the large diameter of the tubes causes a large fraction of the total resistance to reside in the adjacent mass transfer boundary layers. Even with correction for these liquid-phase resistances, very high accuracy in concentration measurement is required for meaningful results. Second, the high membrane hydraulic permeability coupled with different pressure drops in the lumen and dialysate sides precludes complete elimination of simultaneous convective transport.

Figure 10 is a plot of the true membrane mass transfer resistance for permeation of sodium chloride as a function of wall

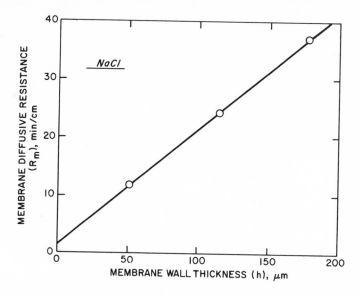

Fig. 10. Membrane diffusive mass transfer resistance to sodium chloride as a function of membrane wall thickness.

thickness. The resistance is linearly proportional to wall thickness with an intercept close to zero which indicates that the skin makes little or no contribution and that the controlling resistance is in the spongy matrix. The effective diffusion coefficient for sodium chloride is about one-half of its value in water, a finding which is compatible with the highly open structure of the spongy matrix. Our preliminary results with inulin which has a molecular weight of 5200 (nearly the same as insulin) also suggest a linear increase in resistance with wall thickness. In view of these results, and if one neglects the possible influence of transmural convection, then one expects that the transient equilibration time for diffusion of glucose and insulin across the membrane wall will increase with the square of the wall thickness.

In Vitro Insulin Secretory Dynamics

To investigate whether the wall thickness of the tubular membrane affects the insulin secretory dynamics, a simple model unit was constructed with a single short length of semipermeable tubular membrane. Figure 11 shows a schematic view of the model device. It is approximately 5 cm in length. The semipermeable tube lies in a narrow trough into which the islets can be conveniently placed with a micropipette. In experiments with these devices, we utilized adult rat islets of Langerhans freshly isolated from pancreases by collagenase digestion and Ficoll density gradient centrifugation[7]. In a typical experiment, approximately 300 to 400 islets are placed

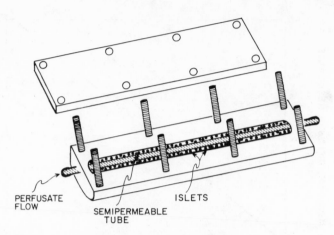

PERFUSATE
FLOW

SEMIPERMEABLE
TUBE

ISLETS

Fig. 11. Schematic diagram of model hybrid artificial pancreas unit for in vitro study of insulin secretory dynamics.

within the small trough and adjacent to the semipermeable tube. After seeding the unit, it is sealed by fastening the top of the chamber in place.

The experimental apparatus is illustrated in Figure 12. In a typical experiment, the lumen of the semipermeable tube is perfused with a Krebs-Ringer bicarbonate solution containing either 100 or 300 mg glucose/dl. After an initial control period of perfusion at the lower glucose concentration, the islets are stimulated for one hour with a step change to a higher glucose concentration, after which the glucose level is returned to the lower concentration of 100 mg/dl. When the glucose concentration in the perfusate is increased, glucose passes across the wall of the semipermeable tube and stimulates the islets in the trough. Released insulin then passes back across the membrane and is taken up by the perfusate. This perfusate is collected at one minute intervals by means of a fraction collector and is analyzed both for glucose and insulin concentrations.

In order to permit comparison between results of different experiments, appropriate controls are included to allow for variation in the performance of batches of islets prepared at different times. To accomplish this, each batch of islets is tested by perfusing them in a small Millipore filter chamber in which the islets are placed on the surface of a microporous membrane[7]. The islets come in direct contact with the perfusate without intervening membranes so that the insulin secretory dynamics represent intrinsic insulin release kinetics of the islets without the extraneous influence of adjacent stagnant fluid layers. A Millipore filter chamber is run with the same batch of islets simultaneously with the model

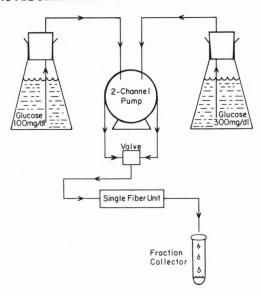

Fig. 12.   Schematic diagram of perfusion apparatus for studying
insulin secretory dynamics.

pancreas device in each experiment.

    For our initial studies we chose to examine the effect of mem-
brane wall thickness on insulin secretory dynamics. Figure 13 shows
results of a typical experiment with a unit constructed with a semipermeable tube of 1 mm internal diameter and 170 μm wall thickness.
The insulin secretory response is plotted as microunits of insulin
released per islet per min versus time. The insulin secretory
response of the hybrid pancreas unit is shown with the solid line and
that of the Millipore chamber with the dashed line. The glucose
concentration profile is shown in the upper portion of the figure.
As anticipated, the initial rise in insulin output in the model
device is delayed, and the time required to achieve maximal release
rate is increased, when compared to the results with the Millipore
chamber. Similarly, the drop in secretion rate when the glucose
concentration in the perfusate is returned to its initial level is
delayed. Despite this delay compared to the Millipore chamber, the
insulin output begins to rise significantly within 10 min following
onset of glucose stimulation. We are presently studying whether use
of thinner membranes will further accelerate this response.

    In summary, in vivo data obtained in our hybrid artificial
pancreas project over the past several years suggest that a hybrid
artificial organ consisting of living islet cells cultured on the
outside surfaces of semipermeable fibers and tubes constitutes an
effective means for restoring carbohydrate tolerance in experimental

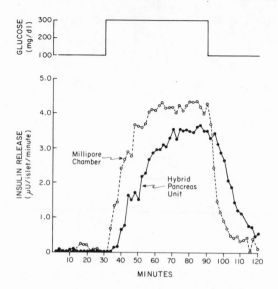

Fig. 13.   In vitro insulin secretory response following step increase and decrease in perfusate glucose concentration.

diabetes. Additional experiments, however, are required to optimize further the biocompatibility with respect to blood clotting and the dynamics of the insulin secretory response.

Acknowledgement

        This work was supported in part by NIH grants AM-15398 and AM-21222, by grants from the Kroc Foundation, the Wolfson Family Foundation, the Kenyon Family Foundation, and by the Harvard-MIT Program in Health Sciences and Technology. W.L.C. is an Established Investigator of the American Diabetes Association. S.P.N. is a NIH Postdoctoral Fellow (No. AM05398). C.T. is supported by the Foundation for In-Service Training and Welfare of the Private School Personnel, Japan. We thank Mr. Steve Kramer and Mr. Peter Seissler for technical assistance with the transport and mechanical property measurements.

REFERENCES

1.   W.L. Chick, A.A. Like, and V. Lauris, Science 187:847 (1975).
2.   W.L. Chick, P.M. Galletti, P.D. Richardson, G. Panol, T.W. Mix, and C.K. Colton, Trans. Amer. Soc. Artif. Intern. Organs 21:8 (1975).
3.   W.L. Chick, D.L. King, and V. Lauris in "Diabetes Research Today," E. Lindenbloom, ed., Schattauer Verlag, Stuttgart (1976).

4. W.L. Chick, J.J. Perna, V. Lauris, D. Low, P.M. Galletti, G. Panol, A.D. Whittemore, A.A. Like, C.K. Colton, and M.J. Lysaght, Science 197:780 (1977).
5. A.D. Whittemore, W.L. Chick, P.M. Galletti, A.A. Like, C.K. Colton, M.J. Lysaght, and P.D. Richardson, Trans. Amer. Soc. Artif. Intern. Organs 23:336 (1977).
6. R.A. Knazek, P.M. Guillino, P.O. Kohler, and R.L. Dedrick, Science 178:65 (1972).
7. P.E. Lacy, M.M. Walker, and C.J. Fink, Diabetes 21:987 (1972).

LIVER TUMOR CELLS GROWN ON HOLLOW FIBER CAPILLARIES:

A PROTOTYPE LIVER ASSIST DEVICE

Carl F.W. Wolf

Associate Investigator
The New York Blood Center
New York, N.Y. 10021

INTRODUCTION

Membranes in the form of artificial capillaries or hollow fibers were first conceived as a convenient membrane element for use in differential permeability separators (1) where they have been applied in water desalination and various other industrial separation processes. They were first applied in a biological role as an artificial kidney or dialyzer (2) and as an oxygenator (3). Several years later, their potential as a supporting capillary matrix for the growth of cells in-vitro was realized and apparatus for the growth of cells on artificial capillaries by circumfusion culture was developed. A variety of cell types have been cultured to produce insulin (4), growth hormone (5), or chorionic gonado-trophin (6) and we have grown cells of liver origin able to take up from the medium and detoxify the bile pigment bilirubin (7) (8). With cells grown on the opposite side of the membrane from circu-lating nutrient fluid, it was possible to select membrane permeability so that cells and high molecular weight molecules could not traverse the membrane. There was thus opened the possibility of growing an artificial organ through which blood might be circulated which could exchange nutrients, toxins, hormones and other molecules with the blood but which would be protected from attack by the blood's antibody molecules or circu-lating immunocompetent cells seeking to reject the foreign tissue of the organ. The artificial organ principle has now been demon-strated with insulin-producing cells (9) (10) (11) in which blood glucose levels of diabetic host animals were regulated by the cells, and by our use of cells from rat liver to take up bilirubin bile pigment bound to albumin carrier protein from the blood, con-jugate it with the solubilizing moiety glucuronic acid and make it

557

excretable by the host animal (12). Unlike a device intended only
to produce traces of a necessary hormone such as insulin from a
few pancreatic islet cells, a device to provide liver assistance,
a liver assist device or LAD, based on the artificial capillary
principle, would need a substantial bulk of cells if adequate
amounts of protein were to be synthesized, toxic molecules de-
toxified, and the energy economy of the body regulated as by the
normal liver. The extraordinary demands which must be met by an
LAD have recently been reviewed (13) (14).

EXPERIMENTAL

As shown schematically in Figure 1, nutrient culture medium
was circulated through the interior of the capillaries for days or
weeks, and a suspension of cells seeded onto the capillaries was
maintained on their exterior surfaces. By circulating the nutrient
medium through silicone rubber connecting tubing, sufficient gas
exchange occurred through the tube wall so that adequate oxygen
supply and carbon dioxide removal was achieved. The tubing served
as a lung for the culture system. With certain types of cells, the
cells proliferated to form a solid mass of "tissue" growing between
capillaries, around and into the porous walls of the capillaries.

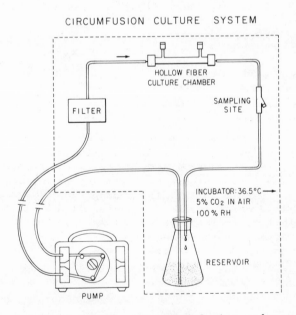

Fig. 1 Schematic diagram of circumfusion culture system.

The cells eventually became visible to the eye as a soft gray white mass on the capillary bundle. Examined microscopically (Figure 2), the cells filled intercapillary spaces and formed nests of cells within the walls. Due to the increasing mass of cells during culture, the consumption of nutrient glucose from the medium also increased proportionately and served as an indirect measure of the viable cell mass present. The technique for tissue culture on capillaries has recently been reviewed (15). The cells we used were the Reuber hepatoma H4-II-EC3 cells from a rat liver tumor or hepatoma and they proliferated readily and by repeated cell divisions formed a tissue-like cell mass.

The capillary membranes studied were mainly modacrylic copolymer of acrylonitrile and vinyl chloride, and a few devices were used with polysulfone capillaries, both types as supplied by Amicon Corporation, Lexington, MA, in their VITAFIBER units. Each unit contained 150 capillary fibers of 200μ I.D. and 60 cm^2 total luminal surface area. Both 3x50 fibers with permeability molecular weight cut-off 50,000 and 3x100 fibers with 100,000 cut-off were used. In most cases, the capillaries had the preferred structure in which a continuous thin inner membranous skin and a porous supporting wall structure with open exterior surfaces allowed cells to grow into the porous wall. In some cases, however, capillaries

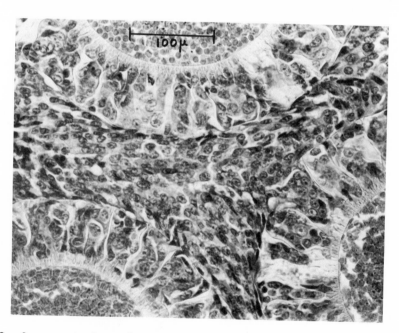

Fig. 2   Cross-section of capillary bundle showing viable cells
between capillaries after 6 hour hemoperfusion. Lumens
contain red cells.

were encountered with a sealed exterior surface which prevented cell in-growth and increased the distance for diffusion of molecules between lumen and cells on the exterior surface (8).

After a period of cell growth and formation of the tissue, the fully grown LAD was removed from circumfusion culture and placed into a blood shunt between the artery and vein of a rat as shown in Figure 3. Any conjugated bilirubin produced by the LAD and returned through the membrane to the circulating blood is promptly concentrated and excreted by the rat's natural liver into the bile in such a system. We chose for our studies an animal with a congenital deficiency of the necessary conjugating enzyme glucuronyl transferase, the Gunn rat, so that glucuronide conjugation performed by the cultured cell/artificial capillary LAD would be detectable.

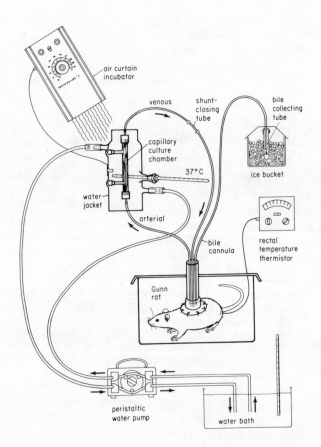

Fig. 3 Schematic diagram of rat/capillary culture LAD system.

RESULTS & DISCUSSION

We analyzed the bile for conjugated bilirubin glucuronide by forming the azo dipyrrole derivatives of all conjugated species with diazotized ethyl anthranilate, separating them with thin layer chromatography and quantitating the glucuronylated derivative by elution and spectrophotometry. A chromatographic separation on a silica gel 60 F-254 plate (E. Merck Co.) is shown in Figure 4. Development was in chloroform-methanol (17:3, v/v) for 2 cm followed by air drying and development in chloroform-methanol-water (65:25:3, v/v/v) for 12 cm. Counting from left to right in the Figure, tracks 1 and 6 are azo derivatives from human bile, track 2 from normal rat bile, and track 3 from control Gunn rat bile. Note the absence of the lowest R_f or "δ" spot (arrow) in track 3 corresponding to absence of the δ glucuronic acid derivative. Culture medium from cultured H4-II-EC3 cells and a similar cell line (MH_1C_1) is shown in tracks 5 and 4, respectively, and δ derivative is clearly visible. As summarized in Table 1, detection of increased azo pigment above background in the δ region in Gunn rat bile occurred when the LAD was used for several hours as diagrammed in Figure 3.

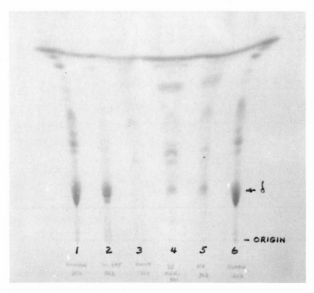

Fig. 4   Thin layer chromatogram of diazotized conjugated bile pigments from: normal human bile, tracks 1 and 6; normal rat bile, track 2; control Gunn rat bile, track 3; culture medium from MH_1C_1 cells, track 4; and culture medium from H4-II-EC3 cells, track 5.

The glucuronide accounted for 20 to 30% of total conjugates produced by the cultured cells. The amount of bilirubin corresponding
to azo derivatives was calculated using the molar extinction coefficient 22.2×10^3 ℓ/mole cm and assuming that all glucuronide
conjugates produced by the LAD were monoglucuronides. In the in-
vivo system, 7.2 to 16.3 µg of bilirubin glucuronide was detected
in the rat's bile after 24 hours of perfusion or 10 to 35% of that
expected based on in-vitro tests of LAD capacity.

TABLE 1

INCREASING PROPORTION OF GUNN RAT BILE BILIRUBIN EXCRETED AS GLUCURONIDE DURING LAD PERFUSION BY GUNN RAT BLOOD

Hours of Perfusion	Percent of azo dipyrrole eluted from chromatogram as glucuronide (δ)		
	24 hr Run*	6 hr Run	17 hr Control Run
0	5 ; 3	2	4
1.7			4
3.5	8 ; 5		
3.8			5
5.9			3
6.0		20	
7.5	7 ; 9		
8.9			3
11.0	13 ; 12		
16.0	23 ; 22		
16.8			10
20.0	18 ; 29		
24.0	18 ; 23		

*Values from duplicate chromatographic separations.

When tested earlier in-vitro without the rat, the culture LAD's had been able to conjugate and return to circulating medium 230 μg of bilirubin in 24 hours, of which 46 to 69 μg (20 to 30%) was probably glucuronide. The discrepancy may be largely due to a 3 to 5 fold decline in conjugating capacity of the cells subsequently found to have occurred during the course of the studies (12). However, given the conjugating capacity measured in-vivo (7 to 16 μg/24 hrs), and assuming that metabolic capacity would scale up proportionately with membrane surface area available for mass transfer and cell growth, we can estimate that a total of 2 to 5 adult size hollow fiber devices of the type now used as artificial kidneys (membrane area 1.8 m^2) would be needed to achieve the amount of liver function (bilirubin conjugation) comparable to that needed by a human newborn infant. It appears, therefore, that an order of magnitude increase in metabolic capacity is needed for a practical device.

Since in the system operated as described transmembrane mass transfer is dependent on diffusion, transport of bilirubin attached to its carrier molecule albumin (MW 66,000 to 69,000 daltons) would be expected to be slow. Theoretical studies and model experiments suggest that ultrafiltration preferentially enhances transport of higher molecular weight molecules and the use of convective transport in these systems may achieve an increase in transmembrane exchange rates (16) (17) (18).

By selecting the optimum membrane hydraulic resistance, convective flow rates and pressures compatible with tissue viability, growth and function may be found. In combination with cells of high metabolic capacity and culture chambers with larger total capacity for cells, useful levels of functional LAD capacities may then be attainable. During the tissue culture phase of LAD preparation, the system is vulnerable to mechanical failures and to infection from the environment because of necessary manipulations such as circulating pump repairs and repeated feeding. It has been suggested that those hazards be reduced by surgically attaching the device for culture and cell growth to the circulatory system of a large animal, thus feeding the cultures continuously in a system protected from infection by the host animal's defenses (19).

Successful further development of a LAD of this type would afford a unique combination of biosynthetic and detoxification abilities, invulnerability to attack by host immune defenses, and membrane geometry with unusual compactness, mechanical stability and susceptibility to analysis, modeling and scale-up.

ACKNOWLEDGEMENTS

Figures 2 and 3 and Table I are reproduced with the permission of the International Journal of Artificial Organs.

REFERENCES

1. Mahon, H.I. U.S. Patent 3,228,876; January 11, 1966.
2. Stewart, R.D., J.C. Cerny and H.I. Mahon Univ. of Mich. Med. Center J. 30:116 (1964).
3. Skiens, W.E., B.J. Lipps, M.E. Clark and E.A. McLain J. Biomed. Material Res. Symp. 1:135 (1971).
4. Chick, W.L., A.A. Like, V. Lauris, P.M. Galletti, P.D. Richardson, G. Panol, T.W. Mix and C.K. Colton Trans. Am. Soc. Artif. Intern. Organs 21:8 (1975).
5. Skyler, J.S., A.D. Rogol, W. Lovenberg and R.A. Knazek Endocrinology 100:283 (1977).
6. Knazek, R.A., P.O. Kohler and P.M. Gullino Exp. Cell Res. 84:251 (1974).
7. Wolf, C.F.W. and B.E. Munkelt Trans. Am. Soc. Artif. Intern. Organs 21:16 (1975).
8. Wolf, C.F.W., C.R. Minick and C.H. McCoy Int. J. Artif. Organs 1:45 (1978).
9. Chick, W.L., J.J. Perna, V. Lauris, D. Low, P.M. Galletti, G. Panol, A.D. Whittemore, A.A. Like, C.K. Colton and M.J. Lysaght Science 197:780 (1977).
10. Sun, A.M., W. Parisius, G.M. Healy, I. Vacek and H.G. Macmorine Diabetes 26:1136 (1979).
11. Tze, W.J., F.C. Wong and L.M. Chen Diabetologia 16 (1979).
12. Wolf, C.F.W., H. Gans, V.A. Subramanian and C.H. McCoy Int. J. Artif. Organs 2:97 (1979).
13. Fischer, J.E. Int. J. Artif. Organs 1:187 (1978).
14. Conn, H.O. and M.M. Lieberthal "The Hepatic Coma Syndromes and Lactulose", The Williams & Wilkins Co., Baltimore (1979).
15. Gullino, P.M. and R.A. Knazek Methods Enzymol. 58:178 (1979).
16. Russ, M.B. M.Ch.E. Thesis, University of Delaware, Dept. of Chemical Engineering, June, 1976.
17. Knazek, R.A., P.M. Gullino and D.S. Frankel U.S. Patent Application 850810, Submitted: Nov. 11, 1977; Issued: approved and pending.
18. Frankel, D.S. Ph.D. Thesis, University of Delaware, Dept. of Chemical Engineering, 1978; Diss. Abstr. Int. 39:1395B (1978).
19. White, D.C., T. Kolobow and R.L. Bowman Int. J. Artif. Organs 1:280 (1978).

APPLICATION OF ULTRAFILTRATION TECHNIQUES TO THE PRODUCTION OF HUMAN PLASMA PROTEIN SOLUTIONS FOR CLINICAL USE

H. Friedli and P. Kistler

Swiss Red Cross Blood Transfusion Service
Central Laboratory
Berne, Switzerland

Ultrafiltration is almost as old as protein chemistry; primitive set-ups have been used for a long time for exchanging salts and concentrating small samples on a laboratory scale. The low efficiency of the dialysis bags, however, prevented applications on a technical scale; alternative methods, e.g., lyophilisation and precipitation, had to be used, therefore, in industrial processes.

In the last few years a new generation of compact, efficient ultrafilters became available. We therefore had to ask ourselves if they could successfully be employed in the preparation of human blood fractions. There are three possible applications for ultrafilters in this field:

- fractionation of plasma proteins according to size
- exchange of solvents
- concentration

a) Fractionation according to size: human plasma is a very complex mixture which contains over 100 different proteins; some of these, although present only in traces, block even large-pore filters. Ultrafiltration is therefore clearly not a suitable basis for a new fractionation scheme. It is also doubtful whether ultrafiltration, which has a relatively low capacity for the separation of substances with similar molecular weights, could be used as a subsystem in a conventional fractionation process.

b) Solvent exchange: the classical application of technical

ultrafiltration in plasma fractionation is the removal of the
precipitant ammonium sulfate. The problems encountered originally with ultrafiltration are at least in part responsible for
the preference of ethanol over other precipitants, since the
former, thanks to its volatility, can easily be removed by
lyophilization. Now that modern efficient ultrafiltration
systems are available, non-volatile precipitants may regain
importance, since removal or exchange of salts or other low
molecular weight-components can be done on a production scale
with diafiltration. Ultrafiltration techniques compete, in
this field, with other modern separation procedures, namely
gel filtration and gel permeation chromatography.

c) Concentration: owing to their high molecular weight, proteins
may be concentrated rapidly by ultrafiltration with only small
losses and more economically than by lyophilization or by thin
layer evaporation. As shall be explained later, certain properties of proteins limit the application range of the method.

It follows that ultrafiltration may be used in two steps in
technical plasma fractionation: as diafiltration for the removal and
exchange of low molecular weight solutes (salts, ethanol), and as a
concentration step. Ultrafiltration may not, however, be used as a
protein separation technique. Let us consider the practical applications, demonstrated by ethanol process.

The fractionation per se is still based on ethanol precipitation.
Fig. 1 shows a simplified scheme of the method of Kistler and Nitschmann (1), which itself is a common modification of Cohn's method (2).

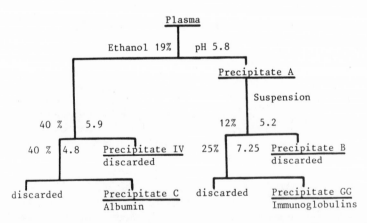

Fig. 1. Ethanol fractionation of human plasma according to Kistler
and Nitschmann (1),(simplified).

The resulting products are essentially pure precipitates of the main plasma proteins albumin and immunoglobulin. Before the albumin can be used clinically, alcohol and part of the salts contained in the mother liquor have to be removed. In the classical procedure the salts were removed by reprecipitation at low ionic strength (Fig. 2a), while alcohol was removed by lyophilization, which is an expensive and work intensive process. Thin layer evaporation (3) greatly simplifies matters, but this procedure removes alcohol only, not the salts (Fig. 2b). Gel filtration (4) removes equally well alcohol and salts, but the resulting solution contains at the most 5 % protein. Prepation of a concentrated solution, containing 20 or 25 % albumin requires an additional concentration step: this can be done advantageously by ultrafiltration (5) (Fig. 2c). It is also possible, as mentioned before, to carry out a diafiltration, which removes salts and alcohol; desalting and concentration may then be carried out without changing equipment (Fig. 2d). A substantial rationalisation can therefore be expected from the introduction of ultrafiltration to the manufacturing of albumin solutions.

What are the requirements an ultrafilter has to meet in this field of application?

1) Protein loss should be kept to a minimum, since the material treated is particularly valuable. Not only should the pore size be well defined, but the dead volume of the equipment should also be as small as possible. If the permeability for albumin

a)	b)	c)	d)
DISSOLUTION	DISSOLUTION	DISSOLUTION	DISSOLUTION
FILTRATION	FILTRATION	FILTRATION	FILTRATION
PRECIPITATION	PRECIPITATION	GELFILTRATION	DIAFILTRATION
CENTRIFUGATION	CENTRIFUGATION		
DISSOLUTION	DISSOLUTION		
FILTRATION	FILTRATION		+
LYOPHILIZATION	THIN LAYER	ULTRA-FILTRATION	ULTRA-FILTRATION
DISSOLUTION	EVAPORATION		
FILTRATION	FILTRATION	FILTRATION	FILTRATION

Fig. 2. Processing of ethanolic albumin precipitate (Cohn fraction IV, precipitate c) to clinical 25 % solution by: a) classical method; b) thin layer evaporation; c) gel- and ultrafiltration; d) dia- and ultrafiltration

is only 1 %, the equivalent of over 50 donations is lost when
preparing one batch of 200 ℓ of 25 % albumin solution. One liter
dead volume corresponds to a loss of approximately 25 (in our
case voluntary) donations. In our hands, losses due to permeation
of albumin are less than 0.5 % of the protein mass, when 5 %
solutions are concentrated to 25 %. Material remaining in the
equipment after the end of the process can be rinsed out and re-
cyled.

2) The possibility of very efficient cleaning and disinfection of
the whole system is an absolute requirement. Since bacteriostatic
additives are not permitted in albumin solutions, ultrafilters
have to be run under sterile or at least virtually sterile conditions. The equipment has to be sterilized either in an autoclave, or chemically; we have successfully used chemical sterilization with formaldehyde.
Certain proteins have the tendency to stick to the membranes and
to clog them; the membranes have therefore to be regenerated
thoroughly after each use. Enzymatic degradation does not help;
besides, products of enzymatic degradation, which cannot be detected easily, may contaminate the final product. A warm solution
of sodium hypochlorite works well, provided the membranes and
other equipment in contact with the solution are made from
resistant material.

3) The proteins treated should not be altered. Next to bacterial
degradation, shearing forces represent the greatest hazard.
Shearing forces occur in the tangential flow alongside the membrane, in the connecting pipes, and mostly in the pump. The risk
of bacterial contamination increases with the duration of the
process; on the other hand, denaturation by shearing forces
increases with pressure and tangential flow. These parameters
have therefore to be balanced carefully and I should like to
devote some time to the problem of their optimization.

The main problem consists in finding the most gentle conditions
that will still effectively prevent concentration polarization. These
conditions depend on the nature of the protein and the construction
of the ultrafilter; they have to be worked out for every system anew.
The practical determination is simple: the feed pressure is kept constant and the flow of ultrafiltrate is measured as a function of the
counter pressure. The completely different behavior of albumin and
immunoglobulin solutions of approximately equal concentration is documented in Fig. 3. In the case of albumin the isobars have a distinct
maximun; with a feed pressure of, e.g., 3 bars a maximum flow of
ultrafiltrate is achieved with a counterpressure of about 1.5 bars.
If the counterpressure is increased, tangential flow diminishes, and

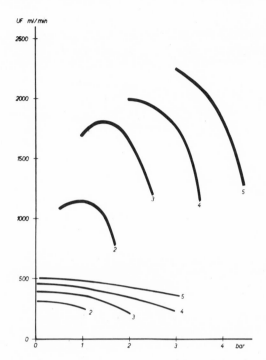

Fig. 3. Ultrafiltration rates as a function of backpressure for 6
percent albumin and immunoglobulin solutions at different
feed pressures. ▬▬ : Albumin;   ▬▬ : Immunoglobulin

concentration polarization prevents further increase of ultrafiltrate
flow. With counterpressures below 1.5 bars the feed pressure results
in unnecessarily high tangential flows and correspondingly high shearing forces. Immunoglobulin solutions generate a different set of isobars; the highest ultrafiltration rates are achieved without counterpressure, apparently reflecting the fact that the gel layer is built
up more rapidly in the case of immunoglobulins. Taking into account
that the membranes used for γ-globulins have an exclusion limit ten
times higher than those used for albumin, a comparison of the ultrafiltrate flows suggest that concentration polarization cannot be
completely prevented in globulin solutions, not even under maximum
tangential flow without counter pressure.

The build-up of a gel layer also depends on the construction of
the ultrafilter. Fig. 4 shows the behavior of a dilute γ-globulin
solution in a cassette and a hollow fiber system of approximately
equal surface; the rate of ultrafiltration is plotted as a function
of the average pressure at selected tangential flow rates. Initially
higher flow rates are obtained with increasing average pressure, as

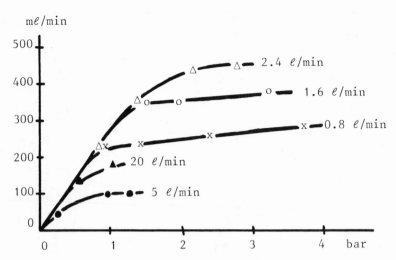

Fig. 4. Gel formation of 2 percent immunoglobulin solution as a
function of ultrafilter construction: o, x, Δ: cassette
module; •, ▲: hollow fiber module

would be expected. Due to concentration polarization a point is
reached, beyond which further pressure increase is practically ineffective. This point obviously depends on the tangential flow: if the
latter is increased, formation of a gel can be avoided under higher
average pressure. The differences between the two systems now become
apparent: in spite of equally good water filtration rates gel formation occurs in the hollow fiber system tested at rates of tangential
flow that still safely prevent gel formation in the cassette system
tested, using the same γ-globulin solution. This example also docu-
ments the strong tendency of even a dilute γ-globulin solution to
form a gel layer. It is therefore doubtful whether ultrafiltration
is an economically feasible alternative for the preparation of
clinical immunoglobulin solutions containing 160 g of protein per·
liter. Immunoglobulins are also quite sensitive to the mechanical
stress imposed by the pump.

Albumin solutions, on the other hand, can easily be concentrated by ultrafiltration. Concentration polarization is suppressed by
rather low tangential flows, and the molecule withstands mechanical
stress much better. As a consequence, ultrafiltration has already
been adopted by many manufacturers of albumin solutions. The use
of ultrafiltration has been described in connection with the ammonium sulfate and heat-ethanol method (6); I should like to discuss its
application to the ethanol fractionation process.

As already mentinoed we have to distinguish between removal of salts and ethanol by diafiltration, and concentration of a protein solution. Normally, the precipitated albumin is redissolved to yield a solution containing approximately 80 to 100 g of protein per liter and around 10% ethanol; after clarification by filtration the pH is adjusted to approximately 7, the value of the final product. This is necessary, since the efficiency of the ultrafilter is greatly reduced when operated around pH 4.8, the isoelectric point of albumin. Optimization of the solvent flow during diafiltration has been analyzed theoretically and practically, using synthetic polymers (7). The alcohol content of albumin solutions causes, however, additional problems. The efficiency of the ultrafilter is lowered by the alcohol content itself; additionally, ethanolic albumin solutions have to be processed at low temperatures to avoid denaturation, thus further reducing the capacity. The protein concentration selected for the diafiltration step is therefore an important choice. If diafiltration is done in the beginning (Fig. 5a), large solvent volumes have to be exchanged, which takes a long time and consumes a lot of water.

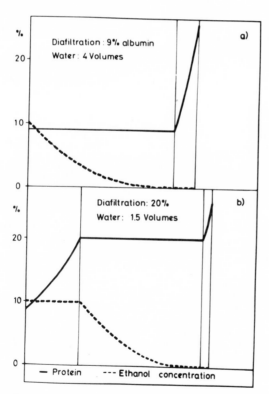

Fig. 5. Combination of concentration and diafiltration at different protein concentrations: a) 9 percent; b) 20 percent.

Concentration, on the other hand, is rapid in the absence of alcohol.
Alternatively (Fig. 5b), concentration can be carried out in the
beginning; it will be slow, due to high ethanol content and low
temperature. Subsequent diafiltration can then be carried out on a
small volume. Optimum concentration has to be worked out for each
system; it usually amounts to approximately 150 g of albumin per
liter. It has been shown in several laboratories that around 5 volumes
of water are required per volume of albumin solution to completely
remove ethanol by diafiltration under constant volume. Since the
final product contains some of the solvent, "water for injection"
has to be used throughout. Diafiltration and gel filtration therefore become competitive processes. Gel filtration has the advantage
of being even more gentle, since no shearing forces occur; additionally, the gel does not have to be regenerated frequently and has a
practically unlimited life span. The investment in a gel filtration
equipment is, however, only justified if some of the albumin eluates
can be processed directly to final products, without further concentration. This is the situation prevailing in the Central Laboratory
of the Swiss Red Cross Blood Transfusion Service, where a large
fraction of the albumin prepared is processed to solutions containing
4 or 5 % protein. The combination of gel filtration and ultrafiltration generates interesting new problems of optimization, which shall
not be further discussed.

Fig. 6 shows a very efficient combination of gel filtration and
ultrafiltration. Within 23 hours, 660 liters of albumin solution

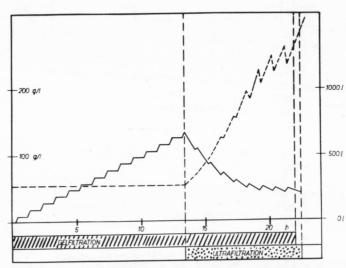

Fig. 6. Combination of gel filtration and ultrafiltration in the
preparation of 25 percent albumin solution. ———:Volume;
————:Protein concentration.

containing ethanol can be processed to yield 200 ℓ of 25% protein solution. The starting solution is first freed of salts and alcohol in a gel filter of 40 cm diameter and 100 cm length; the protein containing eluates are then concentrated in a Pellicon ultrafiltration unit (Millipore), which is charged with 10 cassettes; this corresponds to a membrane surface of approximately 5 m². The solution obtained can easily be sterilized by filtration. The classical approach would have been a reprecipitation of the albumin (which always causes a loss), followed by lyophilization of over 500 kg of solution.

Concentration of albumin solutions does not cause any particular problems. As shown in Fig. 7, the semi logarithmic relation between rate of ultrafiltration and protein concentration is still maintained at a final concentration of 250 g albumin per liter. The risk of bacterial contamination of the solution is low, as long as suitable equipment is used and if the process is carried out properly.

This example demonstrates the substantial rationalization possible in plasma fractionation with the introduction of ultrafiltration techniques.

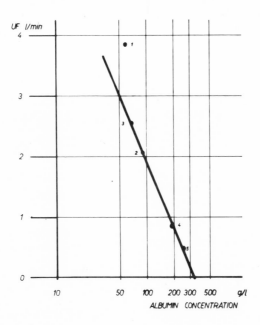

Fig. 7. Relation between albumin concentration and rate of ultrafiltrate in a cassette system (10,000 nmwl, 5 m²) at 5 bar feed pressure and 2.5 bar back pressure.

References

1. P. Kistler and Hs. Nitschmann: Vox Sang. $\underline{7}$, 414-424 (1962).

2. E.J. Cohn, L.E. Strong, W.L. Hughes, D.J. Mulford, N.J. Ashworth, M. Melin and H.L. Taylor, J. Amer. Chem. Soc., 68, 459 (1946).

3. J.K. Smith, J.G. Watt, C.N. Watson and G.G. Mastenbroek, Vox Sang., 22, 120-130 (1972).

4. H. Friedli and P. Kistler, Chimia, 26, 25-27 (1972).

5. H. Friedli, E. Fournier, T. Volk and P. Kistler, Vox Sang., 33, 93-96 (1977).

6. W. Schneider, D. Wolter and L.J. McCarty, Blut, 33, 275-280 (1976).

7. A.R. Cooper and R.G. Booth, Sep. Sci. Techn., 13, 735-744 (1978).

ULTRAFILTRATION AS AN ALTERNATIVE TO REPRECIPITATION AND

LYOPHILIZATION IN COHN FRACTIONATION

S. L. Holst, L. Sarno, M. Martinez

Hyland Laboratories, Inc.
4501 Colorado Blvd., Los Angeles, CA 90039

Almost all fractionators use some modification of Dr. Cohn's Method 6[1] to separate normal serum albumin from plasma. These processes involve selective solubility using a series of precipitation steps with cold ethanol to produce the albumin fraction – Fraction V. In the Cohn process this Fraction V is suspended under specific conditions, filtered and reprecipitated to form the albumin-containing fraction – Fraction V rework (Fig. 1). The teaching in our house had been that the "rework" step was a purification step. This concept paradoxically is true and is not true. There are contaminating salts from components used in the process. On the one hand the rework step reduces the ratio of salts to protein 10 fold (Table 1).

Table 1. Sodium/Protein Ratios in Fraction V and V Rework

	Fraction V	Fraction V rwk
Na^+ conc.	0.33 meq/g	0.03 meq/g

The rework step, on the other hand, does not yield a product which demonstrates appreciable purification over the Fraction V step when viewed with such routine tools as cellulose acetate electrophoresis (Table 2). This is perhaps due to the additional trauma that the proteins are subjected to.

During its early years few options were available to fractionation for concentrating or desalting. Cohn fractionation, with its use of a highly volatile solvent, eliminated the need for the extensive dialysis required in the older ammonium sulphate fractionation schemes. By utilizing alcohol the lengthy and non-sterile

575

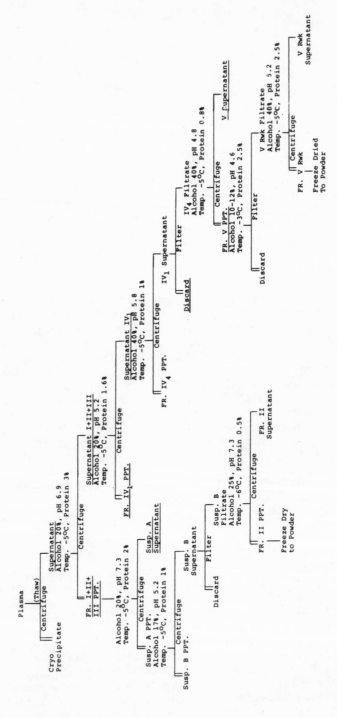

Figure 1. Cohn Method 6

Table 2. Cellulose Acetate Analysis of Fraction V and V rework

	Fraction V	Fraction V rework	
Alb	96 - 98%	97 - 98%	
α	2 - 4%	2 - 3%	Ca
β	< 1%	< 1%	content
γ	< 1%	< 1%	

dialysis step(s) could be replaced by freeze-drying. This is a simple one-step process whereby the precipitating agent (EtOH) could be removed and the proteins concentrated into a relatively stable powder. Rather than depend on dialysis to effect the necessary salt reduction, reprecipitation effectively reduces salts since they remain in solution while albumin is precipitated. The advantage of reprecipitation over dialysis is quickly grasped if one imagines dialyzing thousands of liters of plasma fractions in dialysis bags.

In 1959 Porath and Flodin[2,3] developed techniques for desalting macromolecules using gel permeation chromatography or gel filtration. The gel consisted of cross-linked hydrophilic chains which were later commercially sold under the trade name Sephadex. The good reception this method received prompted the development of other gel materials for the same purpose. In 1972, Friedli and Kistler[4] used this technology for both desalting and alcohol removal. While it is feasible to prepare 5% albumin with this technique, it is not feasible to prepare more concentrated solutions of albumin due to chromatographic anomalies brought on by differential viscosity. For the 5% albumin, an 8% albumin charge solution was sufficient to overcome the dilution effect. To remove alcohol, thin layer vacuum distillation with stringent process control seems to work as a substitute for lyophilization[5].

In 1975, Erikson and von Bockelmann[6] developed a large scale method for concentrating animal serum proteins by ultrafiltration. A year later Friedli et al[7] combined gel filtration and hollow fiber ultrafiltration to prepare 20% albumin solution. Ultrafiltration solved the viscosity constraints. The dilute albumin feed solution was first freed of alcohol and salts by gel filtration then concentrated to 20% by ultrafiltration. A 1977 publication from the same group[8] stated advantages of the Pellicon Cassette System, from Millipore, over the hollow fiber system.

For years the use of semipermeable membranes for ultrafiltration had been merely a laboratory curiosity. The realization of ultrafiltration as an industrial process came in 1960 with the introduction of anisotropic membranes for water desalination. These membranes were developed in the late 1950's by Loeb and Sourirajan of UCLA[9]. This was a major breakthrough in polymer manipulation. Sophisticated film casting procedures produce membranes with a thin skin on a

highly porous support layer. Such skins permit reasonable fluxes.
When examined in cross-section by electron microscopy, these membranes consist of an extremely thin, dense skin layer of 0.1 to 0.5
μm supported by a much thicker (2-5 mils) layer of open celled,
microporous sponge. Because of the extreme thinness and submicro-
scopic porosity of the skin layer, these membranes exhibit the
following characteristics:

1. They have high hydraulic permeability.

2. They have the ability to block passage of molecules
   above a specific size.

In addition despite their thinness:

1. They remain virtually intact and defect-free, ensuring
   maintenance of solute retention throughout the process.

2. They are not plugged or fouled by the solutes they
   retain. Any particle small enough to enter the skin
   layer has no difficulty finding its way out of the
   opposite side, since the support layer contains pores
   several magnitudes larger than those in the skin layer.

Typical UF membranes are made by a "phase inversion" process.
A casting dope, like cellulose acetate, is cast as a thin film and
as the volatile components of the solvent evaporate a thin, dense
polymer layer forms on the surface. Sometimes the film is immersed
in a non-solvent medium, like water, which first gels the surface
into a skin and then gels the entire film. The original cellulose
acetate anisotropic membranes had thermal, chemical, and physical
limitations. Many of these limitations have been eliminated in
other synthetic polymers recently developed. Demand for greater
flexibility in polymer selection led to the development of "composite membranes" in which a dense thin film of 400-800 Å is cast on
another microporous substrate. These membranes are structurally
similar to the anisotropic membranes, but provide the flexibility
in the tailoring of membrane properties. The design of ultrafiltration membranes bears some influence on their concentrating capabilities as well as cleanability, especially on depyrogenation aspects.

The systems we evaluated were: the Dorr-Oliver Iopor System,
type 6 C L. in which the UF membranes are in leaf configuration,
the Amicon DC-30 of hollow fiber UF membranes, the Abcor tubular
ultrafiltration unit and the Millipore Pellicon Cassette System,
which is a modified plate and frame press. Each of these has been
extensively described by the manufacturers.

At Hyland, an ultrafiltration technique has been adapted to
process 25% albumin. The process is similar to that described by

Schmitt-Hauesler[10] which uses ultrafiltration for salt removal and protein concentration. We use UF membranes of 10,000 MW cut-off. The protein loss is 2%, at most. This loss is due primarily to equipment hold up. Our process involves, simply, dialysis as an alternative to reprecipitation for salt content reduction. A simple rapid "forced flow dialysis" is possible utilizing current ultrafiltration technology. By employing this technology, Fraction V source material can be converted into a product that cannot be distinguished from a Fraction V rework source by conventional techniques required for product release. The process is reduced from the 5 steps currently employed to 4 steps, the reprecipitation step being replaced by ultrafiltration. A 5% solution of albumin is prepared from Fraction V powder and concentrated to 25%. The initial phase of the concentration proceeds rapidly with permeation slowing as higher concentrations of albumin are reached (Fig. 2).

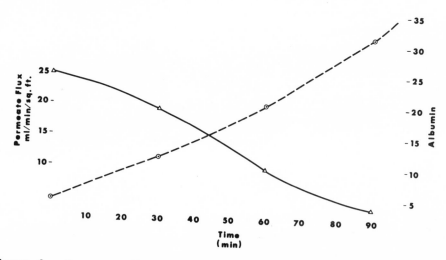

Figure 2.   Permeate flux and albumin concentration as a function of time during ultrafiltration of albumin.

Given these considerations, dialysis should proceed more rapidly with dilute solutions. The case with albumin is less complex. Salts may be adjusted to any molar concentration at a low albumin concentration. Since salts pass through the 10,000 molecular weight filter with the water, the salt concentration remains the same while the protein concentration increases. This, in effect, decreases the ratio of salt to total protein (Fig. 3).

The parameters of flow, polarization, etc., have been extensively covered by others and further discussion will not add to the body of information already presented. The manipulations involved are straightforward and require little imagination to grasp. These too, have been described elsewhere (Metra and Lundblad, 1978 and

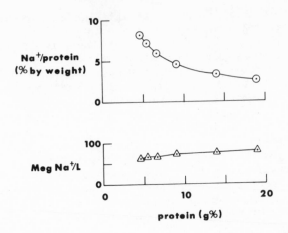

Figure 3. Sodium-Protein ratios during the ultrafiltrative concentration of Fraction V.

Friedli, et al 1977)[8,11]. The proof of the contention that the process is, indeed, an alternative to the Fraction V rework step lies
in comparison of lots produced by the two methods.

We have chosen parameters which offer new dimensions for such
comparisons. The following four areas will be explored in detail:

1. Vasodepressor Activity
2. Molecular Size Distribution
3. Protein Composition
4. Stability

Recently there has been investigation regarding the influences
of process changes on vasodepressor activity. All of the possible
causes of vasodepressor activity in plasma fractions are not fully
understood. Therefore, albumins, derived from modified processes,
must be examined even though they have not previously been associated
with vasodepressor activity. To probe for possible differences,
ultrafiltered lots and Fraction V rework source lots, from the same
parent material, were tested by physiological and radioimmunoassay
techniques. Random lots of each process were also tested. Using
isolated muscle techniques it was not possible to demonstrate any
differential effect of albumins prepared by either process. By
specifically testing ultrafiltered lots for Bradykinin and Prekallikrein activator (PKA) the following results were noted (Table 3).
The Kinin tests were performed by the RIA technique described by
Odya and Goodfriend[12] utilizing reagents prepared in our lab. The
PKA was performed by an esterolytic technique employing a radioactive

Table 3. Bradykinin and Prekallikrein Activator (PKA) in Ultrafiltered Fraction V

Lot No.	Bradykinin (ng/ml)	PKA (% BOB Standard)
A	0.5	None detected
B	0.3	1.0
C	0.2	1.0
D	0.2	None detected

methyl ester and a standard provided by the Bureau of Biologics using the technique of Dr. Pisano at NIH[13]. Comparing ultrafiltered lots with Fraction V rework lots yielded the results summarized in Table 4. They represent the maximum values in each case and similar results are obtained with any combinations of lots tested. From these data, and limited experience with infusion in the intact animal model[14], it would appear that the substitution of ultrafiltration for the Cohn rework process introduces no greater risk in vasodepressor activity.

Table 4. Kinin and PKA in V rework and U.F.-V from the same parent material

	Average Kinin Content in ng/ml	Average PKA % of Standard
V-rework	1 ng/ml	0 - 1%
U.F.-V	1 ng/ml	0 - 1%

Since ultrafiltration is a physical process whose parameters of shear, temperature and pressure may cause protein aggregation or denaturation, we have studied the effect of the process on polymerization. Of those systems tested the turbulent flow systems appear to create the greatest change in polymer content. This data was obtained using hollow fiber and thin channel systems. To evaluate the polymer content elution patterns on Sephadex G-150 were compared for lots from the same parent material[15,16]. To establish that the observed peaks were consistent with the known values of albumin monomer, dimer and polymer, the columns were calibrated with the following protein markers:

Ribonuclease A,	MW	13,700
Aldolase,	MW	158,000
Thyroglobulin,	MW	669,000
Catalase,	MW	232,000
Chymotrypsinogen,	MW	25,000
Ovalbumin,	MW	43,000

The average molecular weights were calculated utilizing the K_{AV}

expression

$$K_{AV} = \frac{V_E - V_O}{V_T - V_O}$$

Where:

K_{AV} = Partition Coefficient

V_E = Elution Volume

V_T = Total Volume of Column

V_O = Void Volume

The peaks identified with the "monomer" were calculated to have a molecular weight of about 61,000, while the "dimer" had a molecular weight of about 155,000. This is reasonably consistent with the known molecular weights of the investigated species[17]. It should be noted that while we identify the various components as monomer, dimer and polymer this is not strictly true. The albumin obtained by the Cohn process is not 100% pure, having up to 4% contaminating alpha globulins. These alpha globulins can be assumed to augment the dimer and polymer peaks. The rework process produces albumin with a G-150 profile typified by Fig. 4a. Note the relative size in the dimer area. Albumin produced by ultrafiltration has a remarkably similar G-150 profile, however, note that the dimer peak is somewhat smaller (Fig. 4b).

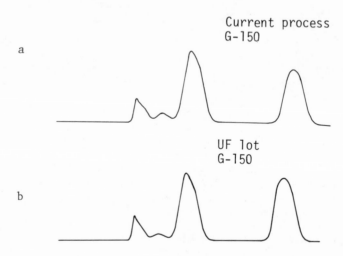

Figure 4.   Gel filtration on Sephadex G-150 of (a) V rework and
            (b) U.F.-V

Quantitation of the V rework peaks reveal that conventionally produced albumin has a composition of approximately 85% monomer, 5% dimer and 10% polymer (Table 5). The columns refer to two different

Table 5. Quantitation of Peaks in Fig. 4a

Lot No.	Current Process					
	% Monomer		% Dimer		% Other	
	(1)	(2)	(1)	(2)	(1)	(2)
1	84.9	85.3	5.0	5.2	10.1	9.5
2	86.0	86.6	5.1	5.0	8.8	8.5
3	86.1	86.2	5.6	5.6	8.3	8.2
4	86.4	86.9	5.1	5.0	8.5	8.2

manners of calculating composition. One uses the chromatogram as the data base, while two uses the total protein of the collected fraction. As can be seen they are not substantially different. The ultrafiltered counterparts (same order) reveal approximately the same composition with slightly reduced dimer content (Table 6).

Table 6. Quantitation of Peaks in Fig. 4b

Lot No.	Ultrafiltration Process					
	% Monomer		% Dimer		% Other	
	(1)	(2)	(1)	(2)	(1)	(2)
A	84.9	85.5	4.4	4.4	10.6	10.1
B	85.7	85.9	3.8	4.0	10.5	10.1
C	85.4	85.9	3.7	3.5	10.9	10.6
D	84.3	85.3	4.6	4.7	10.7	10.1

Attempts to show statistically significant differences were not possible as the number of evaluations increased. Utilization of a different gel permeation medium, Sephacryl S-200, demonstrated the same parameters (Table 7). Again note that albumin derived from the rework process is similar to that obtained by ultrafiltration (Table 8). The dimer area again appears to be decreased when ultrafiltration is employed.

Table 7.   Quantitation of V rework Peaks from Sephacryl S-200

Current Process			
Lot No.	% Monomer	% Dimer	% Other
1	83.2	4.9	11.9
2	83.8	5.6	12.6
3	83.4	7.8	8.9
4	82.5	7.5	9.9

Table 8.   Quantitation of U.F.-4 Peaks from Sephacryl S-200

Ultrafiltration Process			
Lot No.	% Monomer	% Dimer	% Other
A	83.2	4.7	12.1
B	86.6	3.6	9.7
C	83.8	4.8	11.4
D	84.2	4.4	11.4

This time the polymer peak is augmented. S.D.S. Polyacrylamide gel electrophoresis again shows a smaller dimer zone in the ultrafiltered product (Fig. 5).

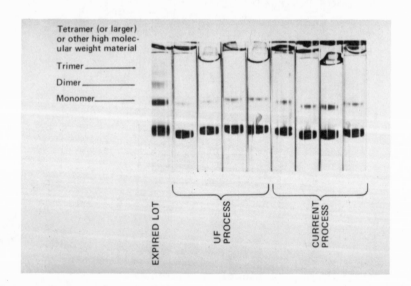

Figure 5.   SDS-Polyacrylamide electrophoresis of V rework and U.F.-V lots.

There appears to be a tendency for the shear forces, or other adverse

physical conditions developed through the process, to convert dimeric
albumin into polymeric albumin, although it has not been possible to
show this statistically. We note that the initial alteration of
albumin from monomeric to dimeric form is brought about by the insult
of freeze drying. Albumin which has not been freeze dried demonstrates a nearly native distribution of molecular size. This observation is supported in a previous publication by Dr. Friedli. His
operation does not employ freeze drying[8].

Changes in procedure, where steps are omitted or replaced,
should produce measurable differences. Conventional techniques,
such as cellulose acetate electrophoresis are inadequate to the task
(Fig. 6). Densitometer scans reveal that one process is indistin-
quishable from another. Clearly more sophisticated techniques are
required to demonstrate difference. Basic polyacrylamide gel electrophoresis failed to show any significant difference.

Figure 6.   Cellulose Acetate Electrophoresis of V rework (Current
Process) and U.F.-V (UF lot).

A knowledge of the distribution of plasma proteins after fractionation with the Cohn process suggested that concentrations of one or
more of the following proteins might vary:

Albumin
α_1-Acid Glycoprotein
α_1-Antitrypsin
Haptoglobin
α_1-Lipoprotein
α_2-Macroglobulin
Ceruloplasmin
Transferrin
α_1-Antichymotrypsin

By using double gel diffusion against specific antisera,[18] albumin, α_1-acid glycoprotein, α_1-antitrypsin, haptoglobin, and transferrin were found to be present in both preparations in sufficent quantity to form a precipitin line. Ceruloplasmin and α_1-lipoprotein were generally found in the ultrafiltered preparation and not in the reworked preparation. α_2-macroglobulin and α_1-antichymotrypsin were not found in either preparation. In order to get some idea of relative differences, cross immunoelectrophoresis was employed[19]. In the current, or rework, process we can see about 9 different contaminants, the major of which appear to be prealbumin.

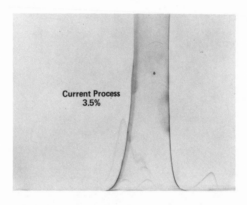

Figure 7. Cross Immunoelectrophoresis of V rework (Current Process)

The ultrafiltered lots present a similar picture, with prealbumin, haptoglobin and transferrin (Fig. 8) as well as an additional peak which also appears as a very minor peak in the rework process material. By utilizing a specific antiserum in conjunction with anti-albumin, this peak can be shown to be α_1-acid glycoprotein (Fig. 9). From the peak heights we would estimate that α_1-acid glycoprotein is present in about 10X the concentration that one finds in the re-worked material. By using a forward light scattering technique test (HYLAND LAS-R^R)[20] the α_1-acid glycoprotein content of the ultrafil-tered material was found to be on the average 374 µg/ml (0.1%). The reworked material was found to have an average of 55 µg/ml (0.02%). These and other differences in the two types of product are not consequences of ultrafiltration, but rather the result of the point in the fractionation process where ultrafiltration is em-ployed. Utilization of a technique that passes molecules of <than 10,000 molecular weight, rather than a precipitate wash technique concentrates macromolecules which are present in solution in the mother liquor. That is, they are trapped in the original

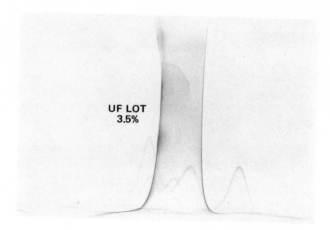

Figure 8.   Cross Immunoelectrophoresis of U.F.-V.

precipitate. Indeed, analysis of albumin solutions prepared by techniques which do not utilize a rework process exhibit similar composition.

   Alterations in protein composition may lead to long and short
term stability differences. Under normal storage conditions ultrafiltered lots have not shown any significant changes in chemistry

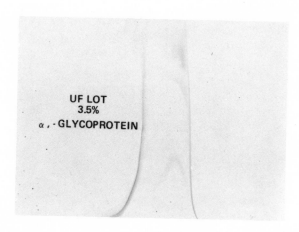

Figure 9. Identification of the minor peak by cross immunoelectrophoresis with antisera to albumin and $_1$-acid glycoprotein only.

or physical appearance for over two years. When subjected to accelerated stability conditions at 60°C the ultrafiltered lots show a greater resistance to increased turbidity and gelation than the reworked material. In the case of turbidity it requires 40% longer time for the ultrafiltered lots to reach the same levels as the reworked lots. And while the reworked process lots gel the ultrafiltered lots did not. This improved stability is positively related to the residual alcohol content of the albumin solutions. Freeze dried powders produce 25% albumin solutions which have about 0.3% residual ethanol. Employing ultrafiltration the alcohol content is reduced to approximately 0.03%.

From these data and from that presented by others we conclude that ultrafiltration can be considered a viable alternative means of salt removal and protein concentration in Cohn fractionation. We have shown that ultrafiltration produces small changes in the composition of 25% albumin. These changes are primarily related to the point of application, rather than the process of ultrafiltration.

References

1. Cohn, E. J., Strong, L. E., Hughes, W. L. Jr., Mulford, D. J., Ashworth, J. N., Melin, M., and Taylor, H. L., : "Preparation and Properties of Serum and Plasma Proteins. IV. A System for the Peparation into Fractions of the Protein and Lipoprotein Components of Biological Tissues and Fluids." J. Am. Chem. Soc., vol. 68 (1946) p. 465.
2. Porath, J. and Flodin, P. "Gel Filtration : A Method for Desalting and Group Separation." Nature, vol. 183 (June 13, 1959) no. 4616, pp. 1657-1659.
3. Flodin, P. "Methodological Aspects of Gel Filtration with Specific Reference to Desalting Operations." Journal of Chromatography, 5 (1961) pp. 103-11 .
4. Friedli, H. and Kistler, P. "Removal of Ethanol from Albumin by Gel Filtration in the Manufacturing of Human Serum Albumin Solutions for Clinical Use." Chimia, 26 : (1972) pp. 25-27.
5. Smith, J. K., Watt, J. G., Watson, C. N. Mastenbrook, G. A. : "Alternatives to Freeze-Drying for the Removal of Ethanol from Plasma Proteins.: I. Vacuum Distillation of Human Albumin." Vox Sang., 22 : 120 1972.
6. Eriksson, G., von Bockelmann, I. : "Ultrafiltration of Animal Blood Serum : Technology and Microbiology." Process Biochemistry September 1975 pp. 11-14.
7. Friedli, H., Fournier, E., Volk, T., Kistler, P., : "Studies on New Process Procedures in Plasma Fractionation on an Industrial Scale. II. Experiments in Concentrating Dilute Albumin Solutions Using Hollow Fiber Ultrafiltration." Vox Sang., 31 : (1976) pp. 283-288.

8. Friedli, H., Fournier, E., Volk, T., Kistler, P., : "Studies On New Process Procedures in Plasma Fractionation on an Industrial Scale. IV. Comparison of Two Semipermeable Membrane Systems For The Concentration of Albumin Solutions." Vox Sang., 33 (1977) pp. 93-106.

9. Loeb, S. and Sourirajan, "Sea Water Demineralization By Means of a Semipermeable Membrane," U.C.L.A. Department of Engineering Report 60-60, (1960).

10. Schmitt-Haeusler, R., "Molecular Filtration in Human Plasma Fractionation," Process Biochemistry (October 1977) pp. 13-16.

11. Lundblad, P. J., Mitra, G. "Optimization of Solute Separation by Diafiltration," Separation Science II (5) pp. 419-502 (1976).

12. Goodfriend, T. L., Odya, C. E., 1974 "Bradykinin" in : "Methods of Hormone Radioimmunoassay" Academic Press, Inc., New York.

13. Imanari, T., Kazu, T., Yoshioa, H., Yates, K., Pierce, J., and Pisano, J. "Radiochemical Assay for Human Urinary, Salivary and Plasma Kallikreins" in Chemistry and Biology of the Kallikrein-Kinin System in Health and Disease. Edited by J. J. Pisano and K. F. Austen (DHEW Publication no. [NIH] 76-794 Washington, D.C., Government Printing Office 1976 pp. 205-213.

14. Department of Pharmacology, University of Edinburgh 1970 Pharmacological Experiments on Isolated Preparations. E. and S. Livingstone, Edinburgh and London.

15. Gelotte, B. and Porath, J., "Gel Filtration in Chromatography" 2nd edition, Reinhold New York 1967.

16. Andrews, P. "Eastman of Molecular Weights of Proteins by Sephadex Gel-Filtration," Biochem. J., 91, 222-223, 1964.

17. Putnam, F. W., The Plasma Proteins Academic Press New York, 1975.

18. Mancini, G., Carbonara, A. O., and Hereman, J. F., "Immunochemical Quantitation of Antigens by Single Radial Immunodiffusion," Immunochemistry, 2, 235-254, 1965.

19. Clarke, H. G., and Freeman, T., : A Quantitative Immunoelectrophoresis Method (Laurell Electrophoresis) in Protides of the Biological Fluids. Ed. H. Peeters. Pergamon Press, Oxford (1971).

20. Ritchie, R. F., : "A Simple, Direct and Sensitive Technique for Measurement of Specific Protein in Dilute Solution." J. Lab. Clin. Med. 70 : 512, 1967.

PRODUCTION OF PROTEIN HYDROLYZATES IN ULTRAFILTRATION-ENZYME

REACTORS

Munir Cheryan and W. David Deeslie

Department of Food Science
University of Illinois
Urbana, Illinois 61801

INTRODUCTION

Chemical or enzymatic modification of proteins has been pract-
ised in one form or another for many centuries, and is the basis for
many traditional fermentation processes, such as cheese making.
From a historical perspective, however, the development of soy sauce
manufacture in the 13th century [1] may be the earliest and most
successful batch-enzymatic hydrolysis process known. Recently,
there has been a great interest in functionally-modified proteins
known collectively as protein hydrolyzates for specific end uses.
The shortage of egg albumin in World War II resulted in the first
development of an enzymatically hydrolyzed protein food ingredient
for purposes of replacing an existing ingredient [2]; its use as a
whipping aid is still a major market today. Protein hydrolyzates,
whether chemically or enzymatically modified, are also used as
flavor enhancers [3].

Another application of enzymatically hydrolyzed proteins is in
"defined formula" diets or "medical foods" for consumption by those
unable to properly digest or absorb whole protein. In clinical cases
of severe pancreatic enzyme insufficiency or malabsorption, it has
been postulated that amino acids are better absorbed from hydrolyzed
protein than from the intact protein [4]. The primary source of
such pre-digested protein today is casein, which has some drawbacks
such as poor palatability and high cost.

Considerable research has also been directed towards producing
an acid soluble hydrolyzate for incorporation into acidic beverages
for nutritional fortification. However, preparing a completely
acid soluble and clear protein requires excessive hydrolysis which

591

is generally accompanied by formation of bitter flavor in the product.

Chemical Hydrolysis: Two types of chemical hydrolysis are in general use today: acid or alkali hydrolysis. In acid hydrolysis, strong acids at high temperatures are used to break the chemical bonds of the protein. While this is an uncomplicated process, and high conversions are usually obtained, the relatively harsh treatment could result in undesirable side-reactions, off-flavors and nutrient losses through degradation of essential amino acids. Alkaline hydrolysis also requires fairly extreme reaction conditions and there is always the danger of the formation of lysinoalanine, a toxic byproduct. In addition, both acid and alkaline hydrolysis results in a fairly high (48-60%) residual ash content in the final product, which limits its use in most food products and clinical applications.

Enzyme hydrolysis: Enzyme hydrolysis is an attractive alternative to modification of proteins because of the mild process

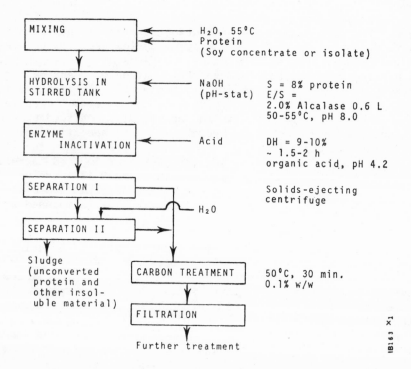

Fig. 1.   A typical batch enzymatic hydrolysis flow sheet for the hydrolysis of soy protein by Alcalase [Adler-Nissen (5)]. Reprinted with permission of Wheatland Journals, Ltd.

conditions employed (Fig. 1). It is also a rapid process where the inherent specificity of various proteolytic enzymes should theoretically afford control of the nature and extent of hydrolysis and thus the functional properties of the product. The major problem with the traditional batch enzymatic process is that it is inefficient and makes poor use of the enzyme. At the end of the process, the enzyme must be inactivited by pH or temperature adjustment, which not only adds to the cost, but implies the enzyme can only be used once in the overall process. In addition, the extent of the reaction must be carefully controlled to avoid development of bitter flavors. The bitter flavor has been attributed to the size of the peptide fractions and their amino acid compositions [5,6] and to the specificity of the enzyme itself [7]. In traditional batch processes, the reaction rarely goes to completion due to product inhibition, resulting in less-than-optimum yields, and, as seen in Fig. 1, extensive post-hydrolysis treatments must be given prior to dehydration to make the product useful.

An alternative approach to enzyme modification of proteins which should overcome many of the problems with batch enzymatic processes is the use of ultrafiltration-enzyme reactors. The concept is illustrated in Fig. 2. In one set-up, a dead-end type of UF cell is used. The cell is charged with the appropriate enzyme solution and substrate is continuously fed in under pressure while permeate is continuously withdrawn through the membrane (i.e., the UF cell is operated in the continuous diafiltration mode). Agitation, necessary to minimize concentration polarization effects, is generally provided with a built-in magnetic stirrer. A slightly different

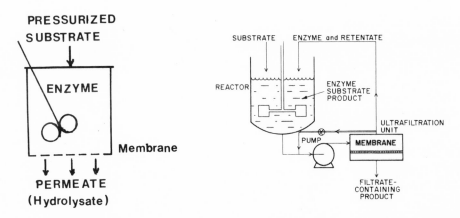

Fig. 2. Schematic representations of ultrafiltration-enzyme reactors. Left: Dead-end cell. Right: Flow-through membrane module with provision for enzyme recycle [From Porter and Michaels (8)]. Copyright © 1972 by American Chemical Society, reprinted from Chem. Tech. Jan. 1972, with permission of American Chemical Society.

configuration was proposed by Porter and Michaels [8] as shown in
the right-hand side of Fig. 2. A continuous stirred tank reactor
(CSTR) is coupled in a semi-closed loop configuration to the membrane
module. The substrate is continuously fed into the CSTR which is
charged with the enzyme. The reaction mixture, consisting of enzyme
unreacted and partially hydrolyzed protein, and hydrolyzed protein
is pumped through the membrane module under pressure. Hydrolyzate
containing peptides with a molecular size smaller than the pore size
of the membrane passes through and is collected as the permeate.
The rest of the reaction mixture, consisting of partially reacted
protein too large to go through the membrane, is returned to the
CSTR for further reaction. By careful design and selection of operating parameters, a steady-state operation can be achieved within
30-60 min of start-up.

     The major advantage of UF-enzyme reactors is that the enzyme is
theoretically retained within the system and can be re-used a number
of times; also, some control is possible over the molecular size of
the product by choosing the appropriate membrane pore size, an important feature since molecular size of the hydrolyzate is presumed
to be a key factor governing the functional properties of the product. However, in practice there is a great deal of difference in
the practical operation of the two types of UF reactors depicted in
Fig. 2. The primary consideration in developing continuous enzyme
reactors should be obtaining high conversion and maintaining steady-
state operation. The former is controlled essentially by the reaction kinetics of the system, i.e., type, specificity and activity of
enzyme, enzyme:substrate ratio, stability of enzyme under long-term
operation. The latter is controlled essentially by the operational
characteristics of the ultrafiltration system, such as the means and
efficacy of controlling concentration polarization, membrane area-to
volume ratio of the membrane module and the fouling characteristics
of the unit. Achieving and maintaining a high conversion is little
problem with enzyme membrane reactors utilizing low molecular weight
substrates and products, where both are membrane permeable, which
affords more flexibility regarding the entrapment of the enzyme.
In protein hydrolysis, however, a large substrate is to be converted
to a smaller product, and both substrate and enzyme are impermeable,
while the product alone is permeable.

     Probably the earliest mention of this concept was by Blatt and
coworkers [9] who demonstrated the feasibility of rapid and continuous ultrafiltration of products of proteolysis as they were being
formed by using a modified dead-end UF cell. Whey from milk was
hydrolyzed with α-chymotrypsin for 12 hours using a UM-10 membrane
to remove the lower molecular size product. Using gel chromatography,
they demonstrated a progessive change towards lower molecular weight
components with increased time of hydrolysis and that intermediate
proteolytic products of desired size range could be isolated by
continuous diafiltration of the reaction mixture. Concurrently,

Table 1. Protein/peptide distribution (as % of initial feed) after
6 hours of hydrolysis of Promine-D [Roozen and Pilnik
(12)]. Reprinted with permission of Wheatland Journals,
Ltd.

Enzyme	pH	Temp (°C)	Permeate (%)	Retentate(%) Soluble	Retentate(%) Insoluble
Bakt proteinase N	7.0	50	32	42	26
FAAN type A	4.0	45	37	33	30
HT proteolytic 200	7.0	45	33	41	26
Pancreas proteinase A	8.0	45	34	36	30
Rhozyme P-53	6.5	50	40	41	19
Saure protease	4.0	45	28	26	46

Wang et al. [10] demonstrated that UF could be used for continuous
production of enzymes in which isolation and purification of microbial enzymes were accomplished simultaneously during fermentation.
Butterworth et al.[11] developed one of the first continuous membrane
reactors using liquid level control for the partial hydrolysis of
starch by α-amyalse. They concluded that molecular weight distribution in hydrolyzed products was dependent upon pore size of the
membrane, transmembrane pressure, enzyme-to-substrate ratio, temperature and residence time of substrate in the reactor. They obtained
a quasi steady-state operation in which the total carbohydrate of
the permeate equaled that of the feed, after about 70 hours of
operation. However, flux steadily declined during operation. This
was attributed to build-up of solids in the reactor and "compaction
of solids on the membrane surface". Despite this, the productivity
(weight of product per unit volume of reactor per unit weight of
enzyme per unit time) was far superior to that of a batch operation.

Roozen and Pilnik [12] used a dead-end cell with a DDS 800 membrane (6000 m.w. cut-off) for the hydrolysis of Promine-D. Various
enzymes as listed in Table 1 were used. An increase in peptide
concentration in the permeate was observed during hydrolysis, but
since overall conversions were low (as indicated by the "permeate"
column) there was a rapid build-up of solids in the retentate, which
essentially stopped permeate flow after about 6 hours of operation.
The fact that a large percentage of the retentate was soluble at the
end of the run (Table 1) indicated that problems were probably due
more to poor control of concentration polarization than loss of
enzyme activity per se. Also, no provision was made for adding
alkali during the reaction, an essential step since protein hydrolysis results in a drop in pH.

Table 2.   Hydrolysis of alfalfa leaf protein concentrate with tryp-
           sin in a UF reactor [14]. (Reprinted in part from J. Food
           Sci. Vol 43(2), 1978. Copyright © by Institute of Food
           Technologists.)

Reaction time period (~ 40 minutes each)	Permeate vol.(ml)	Product	
		Amount(mg)	Conc(mg/ml)
1. 500 mg substrate added			
2. Permeable inhibitors washed out	300	7.4	0.025
3. 0.2% trypsin added			
4.	124	78.2	0.63
6.	84	54.0	0.64
7. 500 mg substrate added			
8.	88	42.0	0.47
10.	86	35.7	0.41
12.	78	31.1	0.39
13. 500 mg substrate added			
14.	76	27.9	0.36
16.	69	27.0	0.39
18.	72	24.3	0.33
19. 500 mg substrate added			
20.	70	21.2	0.30

     More recently, two studies were conducted by researchers at the
University of Wisconsin [13,14] on the hydrolysis of fish protein
concentrate and leaf protein concentrate by trypsin in a dead-end
cell using flat sheet membranes. Their experiments, shown partially in Table 2, illustrate the difficulties associated with the
hydrolysis of very insoluble substrates in dead-end UF cells. An
overall 45% conversion of leaf protein concentrate feed was obtained
but the system required periodic additions of substrate and alternate periods of pressurization and depressurization to depolarize
the membrane of deposited matter. Table 2 indicates a drastic drop
in flux and conversion rates in the initial stages of the process,
but it was also observed that eventually the product output and flux
was almost independent of the build-up of solids in the reactor.
Similar results were obtained for the hydrolysis of fish protein
concentrate [13].

     Cheftel and coworkers used the dead-end cell for continuous
hydrolysis of fish protein concentrate by Pronase [15] and a thin-
channel UF cell for hydrolysis of casein by Alcalase [16]. For the
latter system, the extent of hydrolysis was found to be constant and
equal to 90% and no loss in flow rate was observed for periods up to

150 hours. However, determination of residual enzyme activity in
the UF cell indicated a continuous loss of activity within the UF
reactor itself, as shown in Fig. 3. They also observed that lowering the temperature from the recommended 50°C (curve 2, Fig. 3)
to 40°C (curve 3) improved enzyme stability and that stability was
improved in the presence of the substrate itself (curve 1 vs. 2).
Treatment of Alcalase with glutaraldehyde, a bifunctional cross-
linking reagent, further improved the stability (curves 3 vs. 4)
although this treatment reduced the specific activity slightly.

Cheftel and coworkers' experiments have pointed out three important factors to keep in mind when conducting experiments with
UF reactors that have also been observed in our own work. One, preliminary batch hydrolysis experiments with the native, soluble enzyme
are necessary to determine kinetic parameters such as optimum enzyme-
to-substrate ratio, pH, etc., but are a poor indication of long-term
stability, an essential feature for the UF-enzyme reactor concept
to work. Recommendations of enzyme manufacturers' and published
data on effects of temperature, activators, substrate inhibition
effects, etc. are based on short-term batch hydrolysis experiments,
frequently with low molecular weight synthetic substrates which may
not be directly applicable in real systems. For example, it is
generally advisable to operate at lower temperatures if long-term
stability is important. Two, despite the loss in activity of em-
zyme in the UF reactor, an apparent high conversion (as measured by

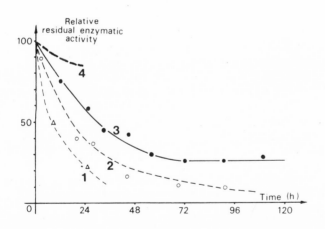

Fig. 3. Residual enzyme activity in the UF reactor during hydrolysis
of casein by Alcalase. Curve 1: Alcalase in phosphate buffer
50°C, no casein. 2: Alcalase-casein, 50°C. 3: Alcalase-casein, 40°C. 4: Glutaraldehyde-treated Alcalase-casein, 40°C.
Boudrant and Cheftel [16], Copyright© 1976 by John Wiley &
Sons. Reprinted with permission of John Wiley & Sons, Inc.

nitrogen in the permeate divided by nitrogen in the feed at the same
flow rate) can be maintained for long periods of time by overloading
the reactor initially with the enzyme. Of course, the practicality
of doing this will depend on the conversion-space time relationship
and the economics of the overall process. Third, the drop in flux,
despite an apparent buildup of solids in the retentate, can be
minimized if the proper means of controlling concentration polarization is used.

Cunningham et al.[17] compared two enzyme reactor systems for
hydrolysis of cottonseed protein using either pepsin or molsin. A
dead-end cell with a 10,000 m.w. cut-off membrane resulted in a
rapidly clogged membrane and reduced flux. Use of a hollow fiber
membrane module (Bio-Fiber 80) improved permeation rates, although
the physical configuration of this particular type of hollow fibers
was not suitable for large scale continuous hydrolysis. It should
be noted that in none of the systems discussed so far was there any
provision for enzyme recycling, nor, with the exception of Cheftel's
work, was there any means for addition of alkali to keep the pH
constant during the reaction.

Our review of the literature to date on a variety of substrates
and enzymes leads us to conclude that the key factors for successful
operation of a continuous ultrafiltration enzyme reactor are (i)
proper design of the UF module and (ii) careful selection of the
enzyme. The UF module should have as high a membrane area-to-volume
ratio as possible and the overall reactor system should incorporate
a mechanism for recycle of enzyme to a separate reaction vessel.
Better control of the reaction is possible if it occurs in a CSTR
physically separate from the UF module, as suggested by Porter and
Michaels in 1972 [8] and shown in the right hand side of Fig. 2.
This will facilitate processing of insoluble substrates, easier control of operating parameters such as temperature, pH, addition of
activators and stabilizers, mixing, etc., and allow higher yields of
products to be obtained with substrate inhibited kinetics, such as
is obtained with the Promine-D--Pronase system [18,19]. The UF
module can then be optimized and controlled independent of the CSTR,
provided the hold-up in the module is small compared to the volume
of the CSTR.

The second key factor is that the enzyme selected should have
as high an activity and, more important, as broad a specificity as
possible to prevent build-up of unhydrolyzable material in the system. This may mean having to use a mixture of exo- and endo-pep-
tidases for the hydrolysis of proteins.

These criteria were applied with success in our work and in the
work of Iacobucci and coworkers published in 1976 [20,21]. The
latter group reported on the pilot-scale continuous production of
soybean protein hydrolyzate with a thermostable fungal acid protease

Table 3.   Comparison of productivity and yields of a UF-enzyme process with the traditional batch process [21]. Reprinted
with permission of Instituto de Agroquimica Y Technologia
de Alimentos.

Parameter	Process	
	Continuous	Batch
pH	3.7	3.7
Temperature (°C)	60	60
Substrate concentration (%w/v)	3	5
Substrate/Enzyme ratio	26	825
Protein yield (Output/input)	94%	72%
Enzymation (contact or reaction) time, hr	1	21
Productivity, hr^{-1}	1144	28

using an experimental laminar flow membrane module with 10 ft^2 of
membrane area, with provision for enzyme recycling. Data from their
studies are summarized in Table 3. The UF-enzyme process was much
more productive than a batch process (productivity was defined as
mass of product produced per unit mass of enzyme replaced per unit
time). However, although conversion remained fairly constant at
90-94% for up to 40 hours of operation time, there was a gradual
build-up of an "unhydrolyzable core" in the reactor, which the
protease they used could not apparently break down. (Independent
investigation suggested that the enzyme itself remained active
throughout the process.) This forced them to bleed their reactor
at various intervals to remove the unhydrolyzable material and to
add enzyme to compensate for losses through bleeding as well as
through leakage through the membrane. Despite this, the productivity of the UF reactor was up to 40 times greater than an optimum
batch process (Table 3).

For our continuous UF-enzyme reactor for the hydrolysis of soy
protein, we chose a CSTR coupled in a semi-closed loop configuration
to a hollow fiber unit. The hollow fibers were chosen over other
modules because of their large area-to-volume ratio and good flux
at relatively low pressures [18]. It was also found advantageous
to pre-heat and pre-filter the soy protein suspension prior to
introduction into the CSTR [18]. By careful selection of operating
parameters, a steady-state operation was achieved within 30-45 minutes of start-up. Fig. 4 shows some of the data obtained.   The
UF-enzyme reactor has been operated for up to 40 hours with no indication of decline in reactor output, although this is controlled
to a large extent by the space time of the system, i.e., the enzyme
concentration, reactor volume, substrate/product flow rate and sub-

strate concentration. Using a 1% soy protein isolate and Pronase
at pH 8.0 and 50°C, a 90-92% yield was obtained after 30 minutes of
start up, and this level of product output was maintained throughout
the run. However, since 100% conversion was not achieved, the total
nitrogen concentration in the reaction vessel slowly increased with
time. However, the build-up of solids in the CSTR had no apparent
influence on the level of reactor output. With much longer processing times, however, substrate inhibition may be significant [18]
and it may be necessary to either shut down the system (if the enzyme costs are small) or to bleed the reactor, centrifuge the reaction mixture to remove excess solids, and return the soluble portion
containing the enzyme back to the CSTR and continue the operation.

Assay of the product indicated an average and consistent protein
content (N x 6.25) of 92% (dry basis) as compared to the raw
material which had 94% protein. Ash content of the hydrolyzate was
slightly higher (7.9%) as compared to the raw material (4.2%), but
this is still much lower than the 16% ash found in commercially
available enzyme hydrolyzed protein produced by the batch method,
and the 48-60% ash content of acid hydrolyzed protein.

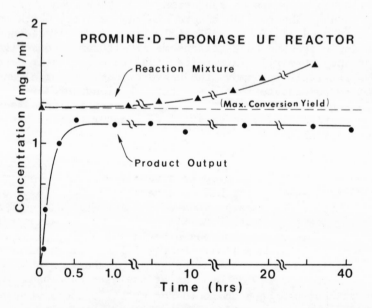

Fig. 4. Product output and reaction mixture concentration build-up
in UF-enzyme reactor

There is considerable interest in using hydrolyzed protein in fortifying acidic beverages and for specific functional end uses. The product from the Pronase-Promine-D UF reactor system is bland and cream colored [18] and promises to have interesting functional properties.

ACKNOWLEDGMENT

This research was partially funded by the Illinois Soybean Program Operating Board.

REFERENCES

1. Y. Sakaguchi, J. Am. Oil Chem. Soc. 56:356 (1979)
2. R. C. Gunther, J. Am. Oil Chem. Soc. 56:345 (1979)
3. K. Prendergast, Food Trade Review 44:14 (1974)
4. M. E. Shils, A. S. Block and R. Chernoff, Clinical Bulletin 6(4):153 (1976)
5. J. Adler-Nissen, Process Biochemistry 12(6):18 (1977)
6. K. H. Ney, Presented at the Symposium on the Contribution of Chemistry to Food Supplies, Hamberg FDR, (1973), Butterworths, London, 1974
7. T. K. Murray and B. E. Baker, J. Sci. Food Agr. 3:470 (1952)
8. M. C. Porter and A. S. Michaels, Chem.Tech.1(1):59 (1972)
9. W. F. Blatt, B. G. Hudson, S. M. Robinson and E. M. Zipilivan, Anal. Biochem. 22:161 (1968)
10. D. I. C. Wang, A. J. Sinskey and T. A. Butterworth, in Membrane Science and Technology, J. E. Flinn, ed., Plenum Press, New York (1970)
11. T. A. Butterworth, D. I. C. Wang and A. J. Sinskey, Biotechnol. Bioeng. 12:615 (1970)
12. J. P. Roozen and W. Pilnik, Process Biochemistry 8(7):24 (1973)
13. S. Bhumiratana, C. G. Hill and C. H. Amundson, J. Food Sci. 42:1016 (1977)
14. R. E. Payne, C. G. Hill and C. H. Amundson, J. Food Sci. 43:385 (1978)
15. C. Cheftel, Annales de Technologic Agricole 21:423 (1972)
16. J. Boudrant and C. Cheftel, Biotechnol. Bioeng., 18:1735 (1976)
17. S. D. Cunningham, C. M. Cater and K. F. Mattil, J. Food Sci. 43:1477 (1978)
18. W. D. Deeslie and M. Cheryan, Presented at the 39th Annual IFT meeting, St. Louis, Mo. (1979), Paper No. 51
19. W. D. Deeslie and M. Cheryan, Illinois Research 21(3):10 (1979)
20. G. A. Iacobucci, M. J. Myers, S. Emi and D. V. Myers, Proc. IV Int. Congr. Food Sci. Technol. 5:83 (1976)
21. D. V. Myers, E. Ricks, M. J. Myers, M. Wilkinson, and G. A. Iacobucci, Proc. IV Int. Congr. Food Sci. Technol. 5:96 (1976)

ULTRAFILTRATION --- THE MEMBRANES, THE PROCESS

AND ITS APPLICATION TO ORGANIC MOLECULE FRACTIONATION

D. Dean Spatz and R. Friedlander

Osmonics, Inc.
15404 Industrial Road
Minneapolis, MN 55343

The use of ultrafiltration membranes for various processes in the chemical industry is now coming of age. This paper discusses ultrafiltration membrane materials, the mechanism of ultrafiltration and gives some application examples. Cellulose acetate and polysulfone membranes are discussed in detail. Their chemical resistance and pore sizes are reviewed, as well as some of the new membrane materials offering an interesting potential for the future. The mechanism of ultrafiltration is reviewed, including the basic criteria used to operate an ultrafiltration membrane. The concept of concentration polarization and the advantages of hydrophilic versus hydrophobic membranes are discussed. The application of ultrafiltration to the food and chemical industry is already becoming widespread. Among the most popular and exciting applications are protein fractionation, latex desalting and oil/water separation. Examples of these applications are presented.

USE OF NEGATIVELY-CHARGED ULTRAFILTRATION MEMBRANES

D. Bhattacharyya, M. G. Balko, C. Cheng,
and S. E. Gentry

Department of Chemical Engineering
University of Kentucky
Lexington, Kentucky 40506

INTRODUCTION

Ultrafiltration with negatively-charged, noncellulosic membranes is a promising technique for the separation and concentration of inorganic salts present in aqueous solution, particularly for industrial waste treatment and in systems designed for water reuse. These membranes provide high water flux at low pressures (5×10^5 N/m^2 to 7×10^5 N/m^2) and good-to-excellent rejections of various metal salts, depending on the process application (1,2,3). The overall rejection characteristics are a function of ion fluxes due to convection, diffusion, and Donnan potential. Because charged membranes reject ionic solutes via repulsion of coions by the fixed charged groups on the membrane skin, the membrane rejection is expected to depend on solute type and coion charge.

An experimental investigation is conducted with a continuous-flow ultrafiltration unit to determine the relative rejection behavior of several metal salts as a function of metal ion (present as chloride or sulfate salts or as complexed ion) characteristics. The application of the process is also experimentally evaluated with acid mine water (at > 95% water recovery) to establish ultrafiltrate reuse.

RESULTS AND DISCUSSIONS

Membrane ultrafiltration was conducted with Millipore PSAL negatively-charged membranes (approximate pore radii 12×10^{-8} cm) containing fixed sulfonic acid groups. The charge capacity of the membranes was determined to be 800 millimoles/liter. The membrane

605

Table 1.   Rejection Behavior of Metal Ions

Pressure = 5.6×10^5 N/m^2
PH = 3.5 - 4.0

Salt	% Rejection	Salt	% Rejection
$BeCl_2$	90	--	--
$MnCl_2$	85	$MnSO_4$	93
$ZnCl_2$	82	$ZnSO_4$	92
$CdCl_2$	81	$CdSO_4$	92
$PbCl_2$	73	--	--

water flux (at 5.6×10^5 N/m^2) ranged between 8×10^{-4} to 12×10^{-4} cm/sec.

The rejection behavior (at 3 mM metal concentration) of simple chloride and sulfate salts of several metals is shown in Table 1. The sulfate salt rejections were independent of cation type because of strong repulsion of SO_4^{2-} ions by the membrane charge groups. With chloride salts the highly hydrated Be^{2+} (Stokes' radius = 4.1×10^{-8} cm) was rejected considerably better than the poorly hydrated Pb^{2+} ion (Stokes' radius = 2.8×10^{-8} cm). The membrane-cation interaction is expected to be a function of metal ion hydration.

For complexed metal ions (such as, metal cyanide complexes), the metal rejections were found to be strongly dependent on the molar feed ratio of the complexing agent to metal ion. Because negatively-charged membranes reject divalent anions better than monovalent anions, the rejection of $M(CN)_4^- > M(CN)_3^-$ would be expected. The complexed metal ion charge is a function of cyanide concentration. Table 2 shows the rejection behavior of some metal cyanide complexes. The high rejections of metal cyanide complexes indicate possible application of this process to metal plating rinse water treatment.

Table 2.   Rejection Behavior of Complexed Metal Ions

Metal Ion	pH	% Rejection
$Zn(CN)_3^-$	10.2	81
$Cd(CN)_4^{2-}$	10.2	94
$Cu(CN)_3^{2-}$	10.2	97

Bhattacharyya, et al. (2,3) have studied the applications of low-pressure charged membrane ultrafiltration to waste treatment involving industrial reuse applications. An application involving reuse of acid mine water for coal conversion processes has been reported (3). High ultrafiltrate recovery with good water flux and ultrafiltrate low in $CaSO_4$ and iron are essential for the purpose of water reuse. With a single-stage ultrafiltration process with no intermediate settling operation only 90% water recovery could be achieved (3). In an effort to increase the water recovery to 97% a new process involving an inter-stage settling operation was studied. The process consisted of ultrafiltration to 80% water recovery, concentrate settling operation and further ultrafiltration of overflow from settler to obtain overall 97% water recovery. This operation produced ultrafiltrate quality considerably better (ultrafiltrate $CaSO_4$ concentration only 2% of saturation concentration) than the single-stage process. The results are shown in Table 3. The overall water flux was also 15% higher than the single-stage operation.

Charged membrane ultrafiltration has shown promising results in several areas:

• Selective recovery of solutes.

• Industrial water reuse operation where complete demineralization is not necessary.

  • metal plating rinse waters.
  • reuse of scrubber and mine waters.
  •. hardness reduction.

• Simultaneous separation of fine solids, ionic solutes, and organic molecules.

• Processing of solutions containing high suspended solids.

Table 3. Ultrafiltration of Acid Mine Water
with Interstage Settling

Overall water recovery = 97%
Membrane operating pressure = 5.6×10^5 N/m^2

Component	% Removal
Al	90
Ca	85
Fe	98
Mn	70
Suspended Solids	100

ACKNOWLEDGEMENTS

This study was supported in part by the Office of Water Research and Technology, Department of the Interior, under the provisions of Public Law 88-379, as Project Numbers B-050-KY and B-059-K. The authors also acknowledge the technical contribution of S. Shelto

REFERENCES

1. D. Bhattacharyya, M. Moffit, and R. B. Grieves, "Charged Membrane Ultrafiltration of Toxic Metal Oxyanions and Cations", Sep. Sci. & Tech., 13: 193 (1979).

2. D. Bhattacharyya, A. B. Jumawan, and R. B. Grieves, "Charged Membrane Ultrafiltration of Heavy Metals from Nonferrous Metal Industries", J. Water Poll. Cont. Fed., 51:176 (1979

3. D. Bhattacharyya, S. Shelton, and R. B. Grieves, "Charged Membrane Ultrafiltration of Multisalt Systems: Application to Acid Mine Waters", Sep. Sci. & Tech., 14:193 (1979).

ELECTRODIALYSIS AND ULTRAFILTRATION

AS A COMBINED PROCESS

RICHARD M. AHLGREN

AQUA-CHEM, INC.
MILWAUKEE, WISCONSIN

The two membrane processes, ultrafiltration and electrodialysis are receiving wide-spread use and publicity as chemical separation tools. The fundamental separations accomplished by the processes are not only technically intriguing but commercially viable in many instances.

Ultrafiltration is a process which is essentially molecular or sub-micron scale filtration. Particles or even specific molecules are separated based on differences in their molecular size or apparent molecular diameter. The fundamental mechanism accomplishing this separation is a microporous film having discreet holes or pores of an intermediate size so that some materials can pass through while others are retained on one side of the membrane. Figure 1 is a simplistic sketch of the type of separation made in an ultrafiltration operation.

A feed stream is directed at a relatively high linear velocity parallel to the membrane surface. A second dimensional flow at a relatively low rate proceeds as an angular vector through the membrane. This flow of solvent through the membrane carries with it molecules or materials of an apparent size smaller than the membrane pores. Particles are higher molecular weight materials are maintained in the original feed chamber and are removed as a partially de-watered concentrate of some of the feed components. This concentrate stream may be either in the form of a solution or suspension of the higher molecular weight or larger particle sizes.

There are several salient features which should be kept in mind regarding ultrafiltration. Usually membranes are microporous polymer films of either a highly jelled or relatively easily water wetted

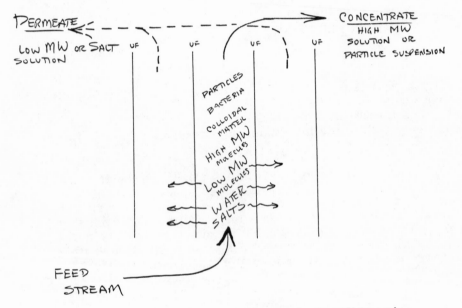

Figure 1. Schematic representation of ultrafiltration

nature. The pores in these films typically range from about .01 micron (or slightly less) up through a few tenths of a micron. This pore size range roughly corresponds to molecular weights between a few thousands and a few million. The physical driving force carrying the solvent and certain feed components through the membrane is a modest hydraulic pressure. All products from an ultrafiltration system must be in the form of solutions or very mobile slurries; in other words, no "filter cake" can be formed and viably removed.

Electrodialysis in contrast to ultrafiltration is a membrane ion exchange process in which ionized particles undergo trans-membrane flow. Ions move in response to a direct electric current imposed across the stack and the flow of electricity is essentially the useful flow of ionized particles through the membranes. Classical membranes are in effect a fixed matrix of either cation or anion exchange sites in a physically and chemically stable film form. Figure 2 is a simplistic diagram of the classic two-stream electrodialysis process. A feed stream containing a mixture of positive and negative ions is directed into a chamber interposed between one cation and one anion selective membrane. Positive ions (sodium) flow toward the cathode through the cation permeable membrane but not through the anion permeable membrane. Conversely, negative ions flow toward the anode and pass through the anion permeable but are repelled by the cation permeable membrane. The salt which has been split and removed from the feed stream is remade as a concentrate stream in the alternate compartments. Like ultrafiltration, it is essential that major product streams from an electrodialysis stack have all components in true solution or a highly mobile slurry form.

Several of the salient features of electrodialysis make it unique from classic fixed bed or resinous ion exchange. Because electric current flow is essentially the "regeneration" mechanism, electrodialysis can be a truly continuous operation. Since salts are not split and exchanged for other ions on discreet ion exchange sites, no regenerant chemicals are required. Related to the fact of the continuous use of ion exchange sites is the fact then that high concentration feed streams can be used in a practical manner with electrodialysis. The typical cost of electric current for stack operation is substantially less than the comparable cost for acid or base chemicals for the regeneration of standard bead resins. With the continuity of the process, there are less product loss or product dilution disadvantages in electrodialysis than are observed for classic bed type ion exchange.

One of the negative features of electrodialysis compared to classic ion exchange is that membrane techniques cannot completely remove ionized constituents from a stream. The influence of the electric current on components in some streams may have a detrimental affect on materials that are preferably unaltered by the process.

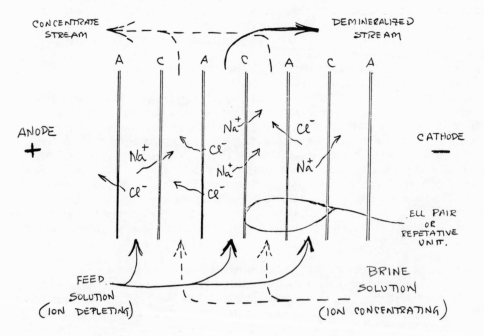

Figure 2. Simplistic diagram of a classic two-stream electrodialysis process.

The combined electrodialysis/ultrafiltration process is an overlay of these two membrane types into a single operating stack. The process is shown in a diagrammatic Figure 3. The feed stream is directed into a chamber having ultrafiltration membranes on one or both sides of that cell. This feed stream is under a modest hydraulic pressure comparable to normal ultrafiltration operation. Cation and anion electrodialysis membranes are interposed outside of the ultrafiltration membranes. A direct electric current is passed between electrodes through all membranes of the cell and the typical ion flow mechanisms are operative. A portion of the solvent and low molecular weight ionized or un-ionized materials flow through the ultrafiltration membranes. The driving force of the organic or un-ionized materials would be the simple hydraulic pressure which is normally operative in an ultrafiltration mode. Two mechanisms are, in fact, operating on the ionized particles from the feed stream. Classic electrodialysis forces are attracting the ions toward their respective electrodes while normal permeation is also removing some of these salts from the feed stream. The salts removed by the ultra-filtration mechanism are, of course, removed as a cation – anion pair through one membrane or the other. Thus, it is necessary that a certain amount of ion transport be, in fact, flow of the improper ion from the permeate stream back through the ultrafiltration membrane reentering the feed stream and consequently out of that stream in the proper direction of flow.

A fundamental separation between the low molecular weight on ionized materials and the ionized salts is made by virtue of the electrodialysis membranes. While low molecular weight materials and negative ions proceed through one ultrafiltration membrane, only the negative ions continue on through the anion exchange membrane toward the anode and into the salt concentrated compartment. The net result of this operation is that the permeate stream has had a substantial amount of the ionized materials removed from it. The salts which have been split from the feed stream are remade in a third or salt collecting compartment.

Some of the salient features of the combined process are the con-current demineralization of the permeate together with additional mineral removal from the feed stream. Because ultrafiltration membranes act as a barrier between the feed stream and the ion exchange membranes, fouling of these ion exchange membranes is reduced. With two mechanisms of salt removal operative on the feed stream a greater portion of the mineral content can be removed. Independent control of the pressure and current flow can be used to obtain a desired balance of removal or fractionation of low molecular weight from high molecular weight materials, together with subse- ‘ quent removal of salts from both streams. In some operations more favorable reaction kinetics can be maintained through the continuous imposition of these two methods of component fractionation.

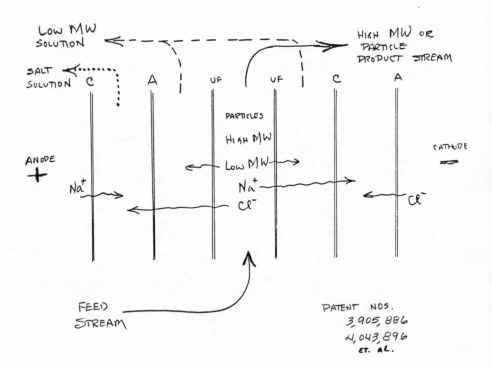

Figure 3. Combined electrodialysis/ultrafiltration

Quantitatively, several variations between the two processes operated independently occur. The electric power consumption of a combined ED/UF system has been observed to be anywhere from 20% to as much as 1000% of the comparable electric power consumption for electrodialysis alone. This says that the protection given by the ultrafiltration membranes to the ion exchange membranes can have a beneficial affect reducing the fouling of these ion selective materials. The flow of liquid from the feed stream into the permeate streams must be monitored to insure proper hydraulics in the permeate desalting chambers. In some cases independent recirculation pumping of this permeate stream is used to maintain a proper liquid velocity across membranes within the cell.

One example of the practical use of combined electrodialysis/ ultrafiltration process would be cheese whey fractionation. Whey is a classic solution of high molecular weight protein material, low molecular weight organics such as lactose and amino acid, together with salts, predominantly potassium chloride. A brief description of this process is shown in Figure 4. The feed whey is directed into a compartment having the ultrafiltration membranes on both sides. High molecular weight protein materials are retained between the UF membranes in a classic manner. Lactose, non-protein nitrogen, and salts pass through the ultrafiltration membranes and are carried into what would be classic permeate compartments of a standard ultrafiltration system. Because of the imposition of the direct electric current and the placement of ion selective membranes outside of these permeate compartments, continuing flow of potassium and chloride ions occurs. This salt is then removed from the permeate stream and remade as a separate salt stream in a third liquid product from the system. The net results of this operation are that the high protein product with a lower content of ash can be produced. Another important benefit of such an operation is that the permeate generated has been partially demineralized without the use of a separate ion exchange step. This can be quite important in the production of high purity lactose from membrane fractionation operations. It should be pointed out that the production from this process is slightly different than what would be observed from independent ultrafiltration and electrodialysis stack operations. The primary logic for the use of this approach would be the desire for a lower ash content specialty high protein product together with the need for a lactose stream having a lower potassium chloride concentration.

The treatment of secondary sewage effluent as a mechanism for tertiary treatment can also be considered by the process. Tests have been performed on secondary effluent with the result that suspended solids, bacteria, inorganics contributing to the chemical oxygen demand, are removed in a feed stream. The permeate through the ultrafiltration membranes which has a reduced organic bacteria content is concurrently partially demineralized through the electrodialysis mechanism. The result is that a clarified and

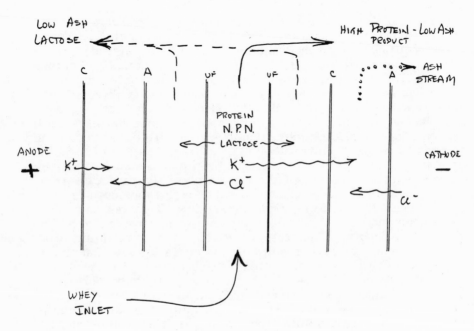

Figure 4. Cheese whey fractionation

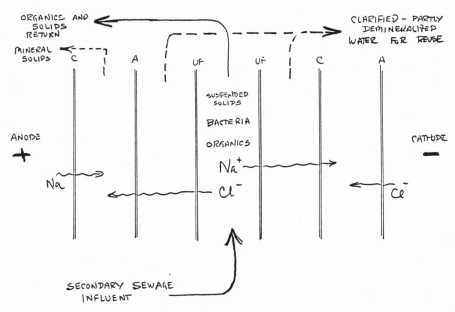

Figure 5. Tertiary sewage treatment

partially· demineralized water can be produced from the permeate
compartments of such a stack operation. It has been observed
that water of this quality looks quite favorable for reuse and recycle
purposes in many industrial operations. The advantage of the
process is that through the control of the current applied and
selection of the membrane porosity, varying qualities of permeate
water can be delivered. The process is shown in Figure 5.

The application of electrodialysis/ultrafiltration is still in the
early development stages. Initial observations of power consumption,
water quality and membrane performance are encouraging. The
overall operating parameters compare favorably with processes of
similar end function such as reverse osmosis or separate use of
electrodialysis and ultrafiltration. The simultaneous transport of
components makes it uniquely capable of performing some tasks
which cannot be accomplished by other membrane or independent
processes.

A STUDY OF THE FOULING PHENOMENON DURING

ULTRAFILTRATION OF COTTAGE CHEESE WHEY

Munir Cheryan and Uzi Merin

Department of Food Science
University of Illinois
Urbana, Illinois 61801

INTRODUCTION

A good example of the successful application of membrane technology is the processing of cheese whey. Table 1 gives some background to the problem. The production of cheese whey is growing at
4% per year and it is estimated that only about 50% of the whey is
utilized in any form either for animal feed or in human foods. The
proximate composition shown in Table 1 emphasizes some of the problems in treating whey. It has a low solids content, and a very un-
favorable protein to lactose ratio of 11:89, which makes it difficult to utilize as animal feed or in human food products. This
explains why reverse osmosis (RO) and ultrafiltration (UF) have
attracted the attention of cheese producers, since it now affords
the possibility of simultaneously purifying, fractionating and concentrating (at low energy cost) the whey and thus enhancing its
utilization and reducing its pollution problems.

The main problem associated with membrane processing of whey
is the fouling phenomenon which results in a drop in output during
processing, i.e., a gradual decline in flux while operating at
steady state conditions. Muller and Harper [1] have recently reviewed the problem and some of the studies in this area. The general concensus appears to be that fouling is due to the deposition
of protein on the membrane surface, and that the microenvironment,
especially pH and type and concentration of salts, plays a very
significant role [1,2,3,4,5]. Recently we reported results of an
investigation into the mechanism of fouling of ultrafiltration
membranes during processing of cottage cheese whey, which appears
to be a particularly vexing problem in the industry. The effects
of salts, individual proteins in cheese whey and combinations of

619

Table 1.   Background to the problem of treating cheese whey

---

10 lb milk → 1-2 lb cheese + 8-9 lb whey

Annual Production: > 75 billion lb (World)
                     > 35 billion lb (USA)

Pollution Load: 30,000 - 50,000 ppm

Composition: Total Solids   = 6.0 - 6.5%
               Lactose   = 4.5 - 5.0%
               Salts   = 0.5 - 0.9%
               Fat   = 0.1 - 0.5%
               Protein   = 0.6 - 0.9%
                 α-lactalbumin   = 0.16%
                 β-lactoglobulin   = 0.4%
                 Bovine serum albumin = 0.04%

---

these on the rate of flux decline in a model UF system were determined and expressed in terms of a mathematical model [2].   We
report here on some further studies into this problem, specifically
the determination of the nature of the fouling deposits on the
membrane with the electron microscope.   The hope was to correlate
the flux decline phenomenon with the physicochemical properties of
the feed [2] and with the manner in which the proteins deposited
on the membrane.

EXPERIMENTAL

     Details of the model ultrafiltration system used to determine
the rate of flux decline are given elsewhere [2,6].   An Amicon 202
diafiltration cell using 64 mm diameter flat sheet membranes was
used.   To evaluate the role of the individual proteins in the overall fouling process, each of the major protein fractions in cheese
whey were made up in their native concentrations as shown in Table
1.   To evaluate the role of salts, two different aqueous environments
were studied; a whey dialystate (WD) buffer, which duplicated the
native whey buffer system, and a salt free (SF) system.   The preparation of the feed solutions is described elsewhere [2,6].   The
membrane used was a Millipore PTGC flat sheet membrane, with a
nominal molecular weight cut-off of 10,000.   Operating conditions
were 50°C, 40 psig pressure and pH 4.6.   Permeate was collected in
tared test tubes at 1 min intervals.   Both cumulative volume and
average (instantaneous) flux could be calculated using the mathematical model shown below.   Statistical treatments of the data are
described elsewhere [2,6].

Preparation of specimens for the electron microscope. For the Transmission Electron Microscope (TEM), the membrane was removed from the UF cell after each run. It was cut into 4 pieces of 1x1 cm and fixated by immersing in 6% glutaraldehyde (Ladd Research Inc., Burlington, VT) in a covered petri dish for 1 hr at room temperature. The pieces were washed 3 times for 10 min each with phosphate buffer and post-fixated in 1% osmium tetraoxide (Stevens Metallurgical Corp., NY) for 1 hr at room temperature. The pieces were then trans-ferred directly to 20% ethanol and dehydrated in a series of 20,40, 60,80,95 and 100% ethanol for 10 min each and 100% ethanol for 1 hr. Embedding was done in low viscosity epoxy resin according to Spurr (7). The blocks were cured in a 70°C oven overnight. The cured blocks were sectioned using glass knives on the LKB Ultratome III (LKB Produkter, Bromma, Sweden). The sections were silver to gold -- 500 - 1100 Å thick (according to Sorvall Microtomes contin-uous interference color for thickness scale for thin sections, Dupont E-21685), and were mounted on 400 mesh copper grids for viewing on the JEOL JEM 100C TEM (JEOL Ltd., Tokyo, Japan), at 40-100 kV accelerating voltage.

For scanning electron microscope studies (SEM), the same pro-cedure as above was followed, except for embedding. The samples were removed before embedding and stored in absolute ethanol until viewed. They were scanned the day of preparation. Membrane pieces of 3x3 mm were mounted on aluminum stubs with Krazy Glue (Krazy Glue Inc., Chicago, IL). The specimens were sputtered (coated) with gold using gold sputter equipment (designed in the Center for Electron Microscopy, University of Illinois, Urbana) and scanned on the JEOL JSM U3 SEM (JEOL Ltd., Tokyo, Japan) at 25-35 kV accelerating vol-tage. Further details on this aspect of the work has been described in detail [6].

RESULTS AND DISCUSSION

A total of eight different feed solutions were studied. Whey, α-lactalbumin, β-lactoglobulin and bovine serum albumin (BSA) were each dissolved in WD and SF environments and processed in the model UF system. Typical results obtained for two of the solutions is shown in Fig. 1. The plotted points are the experimental data and the lines are the fitted curve according to equation 2 (see below) as shown in the figure. That equation was developed from a model which is based on the assumption that the amount of flux decline is a function of the cumulative volume processed [2], as shown below:

$$J = J_o V^{-b} \tag{1}$$

where J is the instantaneous flux (dV/dt) at any time t, V is the cumulative volume permeated and J_o is the initial flux at t → 0. Equation 1 can be integrated and written in terms of V and t as:

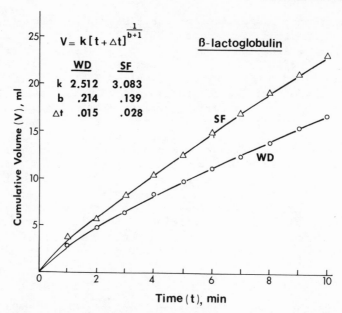

Fig. 1.   Ultrafiltration of β-lactoglobulin in whey dialysate (WD) and salt free (SF) environments.

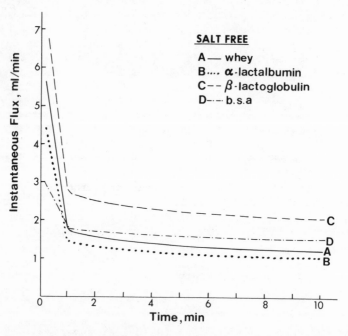

Fig. 2.   Rate of flux decline as a function of time with different proteins in salt free buffer.

$$V = k(t + \Delta t)^{\frac{1}{b+1}} \qquad (2)$$

where $k = [(b+1) e^{\ln J_o}]^{\frac{1}{b+1}}$

$J_o = \dfrac{k^{b+1}}{b+1}$

and Δt was introduced into the model to allow for errors in initial timing during the experiments [2,6].

Data for each protein system was fitted to the model shown in equation 2 using a computer. The mean values of the constants (from replicated runs) k and b were obtained from the regression equation. Table 2 is a listing of the values of J_o and b that completely describe the ultrafiltration behavior of the solutions investigated. (Δt was not treated in the analyses since it does not contribute to the shape of the curve, but merely results in a displacement of the V vs. t curve). From the constants and the differentiated form of the model, the instantaneous flux (dV/dt) at any time t could be calculated and plotted as shown in Fig. 2 for the salt free system. (A similar set of curves were obtained for the whey dialysate system [2].)

The data indicates that fouling occurs in two stages, an initial rapid drop in flux followed by a more gradual decline in flux. We had earlier [2] interpreted the significance of the J_o values in Table 2 as indicative of the first stage flux decline, perhaps due to the resistance of the membrane itself and the resistance of the concentration polarization-gel layer. Lower values of J_o indicate greater tendency for formation of the gel layer. J_o values will then be controlled by the physical properties and operating conditions of the system. The b values, on the other hand, are essentially the slopes of plots of log flux vs. log volume, according to equation 1, and is indicative of the long-term fouling effects, i.e., higher values of b indicate greater fouling. This stage will be dependent on the nature and extent of membrane-solute interactions. When b=0, there is no fouling and flux $J = J_o$ at any time.

In general the values of J_o and b listed in Table 2 suggest that the presence of salts appear to have a detrimental effect on the rate of flux decline. Among individual proteins, α-lactalbumin has the strongest gel-forming tendencies (lowest J_o values) while β-lactoglobulin has the worst long-term fouling effects (highest b values), the latter in essential agreement with other researchers [4,5,8]. The native whey, of course, shows the worst overall fouling pattern, a combined effect of the individual proteins and

Table 2.   Computed values of constants of mathematical model
           (obtained from the regression equation for each
           protein system [6]).

Protein	System	J_o	b
Whey	− Native	1.369	0.297
	− SF	1.905	0.163
α-lactalbumin	− WD	1.722	0.123
	− SF	1.549	0.160
β-lactoglobulin	− WD	2.694	0.214
	− SF	3.163	0.136
Bovine serum albumin	− WD	2.777	0.147
	− SF	1.853	0.069

salts.

The nature of the deposits were then examined under the scanning and transmission electron microscopes. Fig. 3 shows SEM
micrographs of the fouling deposits in WD and SF systems after
ultrafiltering the individual proteins and whey solutions for 10
min.   In the WD pictures (a), two types of deposits can be distinguished: one appears as white clusters on top of another lower
layer of dense material. This is probably not due to differences
in types of deposits, but more likely due to charging under the
electron beam, resulting in "edge effects" causing bright and dark
areas in the micrographs. It also appears from the bottom row of
micrographs (b) in Fig. 3 that removal of salts from the feed solutions results in a different type of deposition. In some cases
(α-lactalbumin and whey), no definite structure was visible on the
membrane surface and hence a break or crack in the membrane was
photographed to show more clearly the SF deposits. Hence, in these
particular micrographs, the voids and void walls of the membrane
ultrastructure are also visible.

The unusual character of SF β-lactoglobulin deposits could be
related to the microenvironment of the feed solution. Apparently,
crystalline β-lactoglobulin aggregates and precipitates out from
solution in deionized water [11] and the granular deposits seen
could be due to these aggregated molecules.

Figs. 4-7 show TEM micrographs of the different feed solution
deposits. Again a clear difference is noticed between the types of

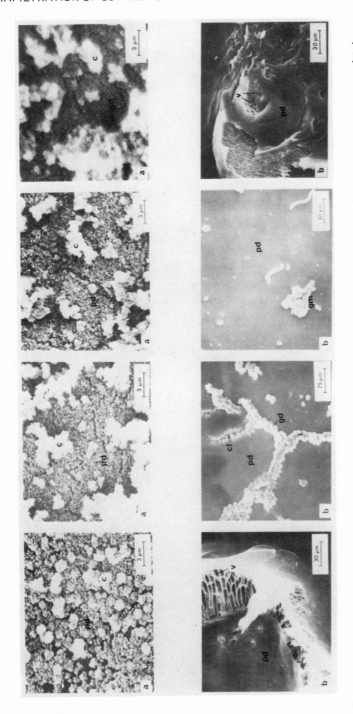

Fig. 3. Scanning electron micrographs of fouling deposits. Top row (a): Whey Dialysates (WD). Bottom row (b): Salt free (SF) systems. Left to right: α-lactalbumin, β-lactoglobulin, bovine serum albumin (BSA), whey. [pd=protein deposit, c=clusters of protein, v=voids, gm/gd=granular material or deposits, cl=canal].

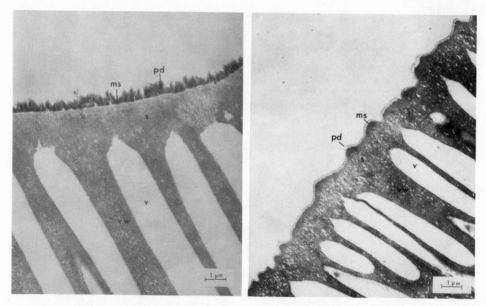

Fig. 4.   TEM micrographs of fouling deposits of α-lactalbumin.
          Left WD, right SF.   [pd=protein deposit, ms=membrane sur-
          face, v=voids, vw=void wall, s="sponge" of membrane].

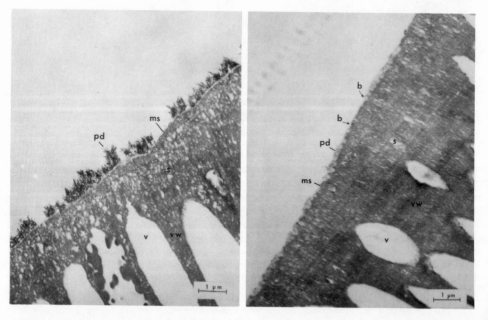

Fig. 5.   TEM micrographs of fouling deposits of β-lactoglobulin. Left
          WD, right SF [Same symbols as Fig. 4., b=break in deposit]

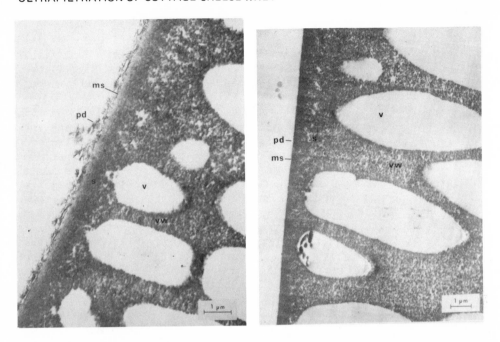

Fig. 6. TEM micrographs of fouling deposits of Bovine Serum Albumin
solutions. Left WD, right SF. [Same symbols as Fig. 4].

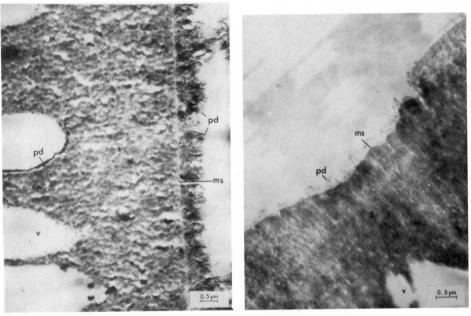

Fig. 7. TEM micrographs of fouling deposits of whey. Left Native
whey, right SF. [Same symbols as in Fig. 4].

deposits occurring in WD and in SF environments. The clusters of
protein deposits visible in the SEM micrographs in Fig. 3 are also
visible in the TEM WD pictures. Some protein was also found within
the voids of the membrane when processing native whey (Fig. 7).
This is not unexpected since some peptides in whey are small enough
to permeate the PTGC membrane.

There is little resemblance between these micrographs and those
published recently by Merson and coworkers [4,5] for a similar
cheese whey system. This could be due to differences in experimental
conditions and different methods of preparation of specimens for the
electron microscope. In all the photographs shown in this work, no
differences could be detected in the nature of the deposits between
individual proteins and the whey. That is, one cannot, on the basis
of these micrographs, pinpoint a single protein as being a major
contributor to flux decline by comparison of its micrograph to that
of whey. The micrographs did indicate some difference in the nature
of the deposits in the presence of salts, which correlates well with
the mathematical model described earlier.

The SEM and TEM micrographs also clearly reveal the anisotropic
nature of the PTGC membrane used in this work. The voids, void walls
and the membrane skin are clearly distinguishable. More detailed
micrographs of the membrane itself are available elsewhere [6].

CONCLUSIONS

From the mathematical model and the electron micrographs presented in this paper, it appears that salts have a profound influence on the rate of fouling. The exact role of salts is not too
clear, but it has been postulated [1,2] that it could change the
conformation of the proteins making it more susceptible to precipitation on the membrane, or it could act as a salt bridge between
membrane and protein, thus leading to greater fouling effects. Removal of salts prior to ultrafiltration has been shown to be beneficial [1,9,10]. However, the cost-benefit ratio may not be favorable and will have to be considered by each processor.

ACKNOWLEDGEMENT

This project was partially supported by the Agricultural Experiment Station, University of Illinois, Urbana. The Center for
Electron Microscopy at the University is thanked for making its
facilities available to us.

REFERENCES

1. L. L. Muller and W. J. Harper, J. Ag. Fd. Chem. 27:662 (1979)
2. U. Merin and M. Cheryan, J. Fd. Processing and Preserv. (1979) In Press
3. T. H. Lim, W. L. Dunkley and R. L. Merson, J. Dairy Sci. 54:306 (1974)
4. D. N. Lee and R. L. Merson, J. Dairy Sci. 58:1423 (1975)
5. D. N. Lee, M. G. Miranda and R. L. Merson, J. Fd. Technol. 10:139 (1975)
6. U. Merin, Ph.D. thesis, Univ. of Illinois, Urbana (1979)
7. A. R. Spurr, J. Ultrastructure Res. 26:31 (1969)
8. P. C. Patel and R. L. Merson, J. Fd. Sci. Technol. 15:56 (1978)
9. J. F. Hayes, J. A. Dunkerley and L. L. Muller, Aus. J. Dairy Technol. 29:132 (1974)
10. L. L. Muller, J. F. Hayes and A. T. Griffin, Aus. J. Dairy Technol. 28:70 (1973)
11. R. McL. Whitney, in "Food Colloids", H. D. Graham, ed. AVI Pub. Co., N. Y. (1977)

ULTRAFILTRATION/ACTIVATED SLUDGE SYSTEM - DEVELOPMENT OF A PREDICTIVE MODEL

A.G. Fane, C.J.D. Fell and M.T. Nor

School of Chemical Engineering
University of New South Wales
Australia, 2033

SUMMARY

A computer model of a combined ultrafiltration and activated sludge wastewater treatment system has been developed following a study of the factors affecting ultrafiltration flux. The results of experiments show that the major resistance to flux is provided by the suspended solids, and that ageing of the gel is important. Correlations are developed for the variation of flux with solids concentration and time.

The combined computer model successfully predicts the changes in flux for conditions of increasing or steady state solids content, but underestimates the flux recovery obtained at low feed conditions.

INTRODUCTION

The treatment of sewage effluents by reverse osmosis or ultrafiltration has been the subject of several studies. Feed streams have varied from raw sewage (1,2), primary or settled sewage (2,3,4), secondary or biologically treated sewage (2,3), to tertiary effluent (2,5). These streams range from difficult (fouling) feeds requiring frequent membrane cleaning and high pumping requirements, to 'clean' feeds containing only residual amounts of solids. On the other hand the 'clean' feeds are obtained at the economic penalty of secondary and/or tertiary pretreatment.

Combination of ultrafiltration with biological treatment by
activated sludge has been pioneered commercially by Dorr Oliver
Inc. (6). Several advantages accrue from this coupling of the two
waste treatment methods. Firstly, the activated sludge process
is freed from the limitations imposed by the 'settling' stage used
conventionally. This means that much higher concentrations of
biological solids may be used and that variations in the input flow
rate are less of a problem. Secondly, the requirements on the
membrane are eased as a result of the biological oxidation of the
sewage components. Finally, the effluent quality of the combined
UF/AS system exceeds that readily obtained with the conventional
AS system (6,7) or readily obtained using membranes alone (8).
The potential applications of the UF/AS system include recreation
areas where human visitation is intermittent (6,9), shipboard
waste treatment (10), and 'in situ' reclamation in individual
buildings or groups of buildings (11).

The system being investigated is depicted in Figure 1 and
comprises a well-mixed activated sludge aeration tank connected
to an ultrafiltration loop, with concentrate from the loop being
returned to the tank. Whilst the separate responses of the
activated sludge aeration stage and the ultrafiltration loop may be
anticipated it is not intuitively obvious how the coupled system
will behave. Therefore the development of a computer model of
the UF/AS system is an important part of this investigation.

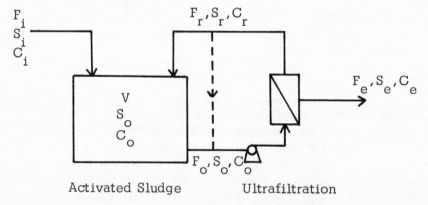

Figure 1.   Schematic Diagram of Coupled AS-UF System

A preliminary model has been presented elsewhere (9), based on a description of the AS by modified Monod kinetics (12,13) and using a conventional gel-polarised representation of the UF (14). Sensitivity analysis of this model indicates that uncertainties in the UF sub-model have a significant impact on the overall system behaviour. This paper describes an evaluation of the factors controlling the UF flux, and shows how these factors may be incorporated into a model of the coupled system.

THE ULTRAFILTRATION OF ACTIVATED SLUDGE MIXED LIQUOR

Determination of the Flux-limiting Species

Realistic modeling of the ultrafiltration of activated sludge mixed liquor is complicated because the liquor is a complex mixture of dissolved and suspended solids. The dissolved substances include high molecular weight fatty acids, proteins, polysaccharides and other bacterial degradation products as well as unreacted substrate and inorganic salts. The suspended solids are composed of bacterial cells and insoluble sludges, and possibly fibrous material and grit particles. (It should be noted that the latter two components were not present in the artificial sewage used in this study). In the activated sludge mixed liquor the suspended solids are deformable and typically 0.5 to 1 μm size. These solids can be expected to behave rather differently from rigid solids, the presence of which have been shown (15,16) to enhance flux for macromolecular solutions.

In order to determine the relative significance of the dissolved and suspended solids in the mixed liquor experiments were made in a stirred cell ultrafilter using Amicon PM30 membranes with a stirring speed of 500 rpm, an applied pressure of 200 kPa and at 30°C. Solutions containing dissolved solids only and suspended solids only were prepared from mixed liquor removed from the pilot plant aeration tank. Supernatant was filtered through a Whatman 41 filter paper to provide a clear solution. Suspended solids were collected by similar filtration and washed repeatedly with distilled water until the filtrate was free from dissolved solutes (checked by evaporation to dryness).

Figure 2 shows how the flux varies with concentration for solutions of dissolved solids and suspensions of suspended solids, plotted according to the semi-log relationship of the conventional

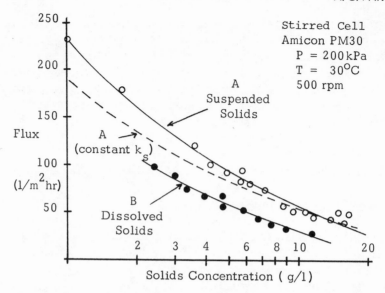

Figure 2.   Ultrafiltration of Activated Sludge Solids

gel-polarised model (14),

$$J = k_S \ln (C_G/C_B) \tag{1}$$

Neither plot is linear over the complete range of concentrations,
with the suspended solids showing the greater deviation from the
semi-log relationship.   This can be partially explained by noting
that the data are for a constant stirring speed rather than a constant
mass transfer coefficient, k_S.   The mass transfer coefficient at
constant stirrer speed will decline as solids concentration
increases.   The broken line on Figure 2 represents the original
line of best fit for the suspended solids data corrected to constant
k_S using the stirred cell mass transfer correlation of Colton (in
Ref. 14), and the relationships between suspended solids concentration and diffusivity and viscosity presented in our earlier paper
(9).   In spite of the correction the plot still deviates from linearity
particularly at high solids concentrations.

      The data presented in Figure 2 may also be expressed in the
form of resistances, defined by the relationship

$$J = \frac{\Delta P}{(R_m + R_{ds} + R_{ss})} = \frac{\Delta P}{R_T} \tag{2}$$

where R_m, R_{ds}, R_{ss} are the resistances offered by the membrane, the dissolved solids and the suspended solids respectively; R_T is the total resistance. Figure 3 is a plot of R_{ds} and R_{ss} as a function of concentration; for comparison the membrane resistance R_m (obtained from distilled water fluxes) is in the range 0.25 to 0.32 $(kPa.m^2.hr/l)$.

The results show that for a given concentration the dissolved solids exert a greater resistance to flow than do the suspended solids. However analyses of the mixed liquors used during the experimental programme show that the dissolved solids content is relatively constant at about 1350 mg/l even though the suspended solids content rises to 10,000 mg/l (Figure 4A). Combining the information in Figures 3 and 4A shows (Figure 4B) that R_{ss} starts to become dominant when the mixed liquor suspended solids (MLSS) exceeds about 3000 mg/l.

To check the additivity of resistances implied by equation (2) ultrafiltrations were made in the stirred cell with liquors containing a fixed concentration of dissolved solids, but with different loadings of suspended solids. The results are shown in Figure 5A, from which it can be seen that at very low loadings of suspended solids (< 1000 mg/l) the effect is negligible, but for higher loadings the flux decreases steadily as suspended solids are increased.

Using values of R_{ds} and R_{ss} from Figure 3, and known values of R_m, the flux has been predicted, using equation (2), for the conditions of the experiment. Figure 5A shows that the resistances-in-series assumption approximates the observed behaviour, with a tendency to underestimate the flux.

The discrepancy between the resistances-in-series model and the actual data can be expressed in terms of ΔR_T, which is the amount by which the model over-estimates the total resistance. Figure 5B shows that ΔR_T has a small (intercept) value at low suspended solids and then increases linearly with loading. The results may be explained by assuming that R_{ss} is unaffected by being in a mixture with dissolved solids, but that R_{ds} is reduced by the presence of suspended solids. Figure 6 depicts the possible situation adjacent to the membrane.

A relatively small amount of suspended solids modifies the surface of the membrane by 'protecting' some of the pores which would otherwise become more effectively obstructed by macro-

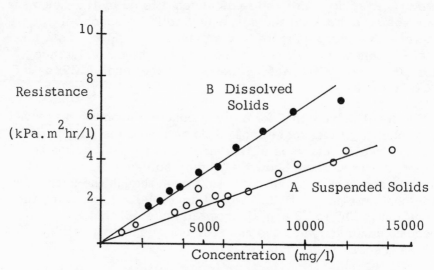

Figure 3.   Effect of Concentration on Resistance
            (A: Suspended Solids in Distilled Water
             B: Dissolved Solids)

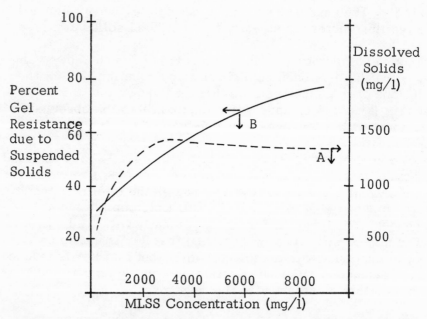

Figure 4A.   Dissolved solids vs. MLSS for Aeration Tank Liquors
Figure 4B.   Percent Gel Resistance due to Suspended Solids

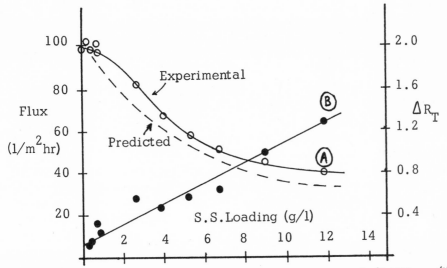

Figure 5A. Flux vs. Added S.S. at Constant D.S. (2400 mg/l)
5B. Error in Total Resistance Predicted vs. Suspended
Solids (ΔR_T = Predicted R_T - Experimental R_T)

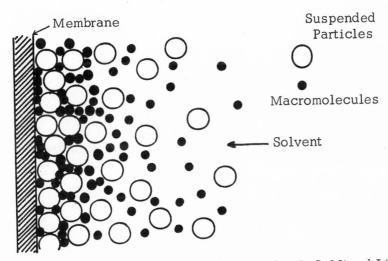

Figure 6. Representation of Gel Layer for A.S. Mixed Liquor

molecules. This accounts for the intercept in Figure 5B. Further increase of suspended solids could have two effects. Firstly, the adsorptive properties of the biological solids could reduce the effective concentration of the free macrosolutes in the gel layer. Secondly, the suspended solids could augment the back-diffusion of macromolecules due to scouring, turbulence promotion and radial migration (17,18). Both effects (adsorption and enhanced back diffusion) would increase with increase in suspended solids.

Whatever the physical picture these experiments indicate that both dissolved and suspended solids can be important in determining the flux. However, since the suspended solids become increasingly dominant, particularly at the concentration levels typical of UF/AS plant operation (> 3000 mg/l), it seems reasonable to attempt to correlate flux against MLSS rather than total solids. Indeed, this is the most convenient approach since MLSS are readily measured and are predictable from AS kinetic models.

The Flux-Concentration Relationship

The variation of flux with mixed liquor concentration was obtained in a small pilot plant comprising an aeration tank of 400 1 capacity coupled to an ultrafiltration loop containing $0.19m^2$ membrane area (membrane type was a Dorr-Oliver XP24, with a nominal molecular weight cut-off of 24,000). The loop had facilities for control and measurement of pressure, recirculation velocity and temperature. Details of the equipment and the synthetic sewage feed (based on glucose) may be found elsewhere (19).

The experimental procedure used was to fix the concentration of mixed liquor suspended solids (MLSS) and the recirculation velocity and measure flux values for increasing transmembrane pressure until the flux became invariant. The procedure which took less than 1 hour, was repeated over a range of concentrations and velocities, with the membrane being cleaned between each test. The reported data are therefore limit (or gel-polarised) fluxes and are also 'initial values' not yet influenced by long term flux decline (see below).

The flux values are correlated in terms of the MLSS concentration only, for the reasons given previously. Figure 7 shows the data plotted according to the semi-log relationship of the

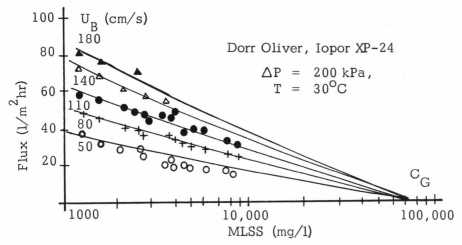

Figure 7. Flux Plotted According to Gel Polarised Model

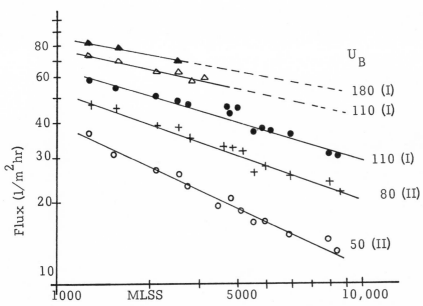

Figure 8. Flux Plotted According to Scour Model.
(I) correlation for $U_B > 90$; (II) correlation for $U_B < 90$.

conventional gel-polarised model (equation (1)), for the range of recirculation velocities used. The correlation curves have been adjusted to constant velocity, rather than constant mass transfer coefficient, using the Dittus-Boelter equation recommended by others (17); details may be found elsewhere (9). Figure 7 is not satisfactory over the whole range of conditions, although the most serious deviations occur at the lowest recirculation velocity which is the least significant from a practical point of view. However these data, and those plotted in Figure 2, question the validity of the semi-log plot for the ultrafiltration of activated sludge mixed liquor.

Figure 8 shows the data plotted according to a scour model (9), which is believed to be appropriate to the ultrafiltration of suspensions. This model predicts a log-log relationship of the form,

$$J \propto (C_B)^{-m} \cdot f(U_B) \tag{3}$$

The data are correlated satisfactorily by the following relationship,

$$J = A \cdot C_B^{-m} \cdot U_B^{a} \tag{4}$$

The factors A and m are also dependent on recirculation velocity for $U_B > 90$ cm/s,

$$A = 6.85 \times 10^4 \text{ and } m = 14 \, U_B^{-0.8} \quad \text{(correlation I)} \tag{4a}$$
$$a = -1.0$$

for $U_B < 90$ cm/s

$$A = 8.0 \times 10^4 \text{ and } m = 3.8 \, U_B^{-0.8} \quad \text{(correlation II)} \tag{4b}$$
$$a = -1.0$$

Whilst it is recognised that equations (4,4a,4b) are empirical and probably specific to a particular geometry, the form of equation (3) provides a useful basis for other systems handling suspensions, such as activated sludge solids.

Long Term Flux Decline

Prediction of the initial flux for given conditions in the ultra-

filtration of activated sludge is only a first step in the modeling.
In order to develop a model which is capable of predicting the
response of the UF/AS system to varying inputs over a period of
time, it is necessary to allow for the decline of flux with time.

The problem of flux decline with various sewage effluents
(raw, primary, secondary etc.) has been the subject of several
investigations.   Sammon and Stringer (8) have reviewed all the
published findings up to 1975 most of which involve reverse
osmosis, rather than ultrafiltration.   They conclude that both
dissolved and suspended organic matter contribute to flux decline,
and that the problem is eased by increasing the cross flow velocity.
In a more recent study Harris et al (20) have shown that for the
ultrafiltration of raw sewage and mixed wastes the flux decline is
related to the nature of the feed.

Typical flux decline data obtained in this study, using the
small pilot plant with the Dorr Oliver membranes, are shown in
Figure 9. It should be noted that the sludge used in these series
of tests was active in the sense that it was in contact with
substrate, and was being aerated in the tank. However by the very
nature of the tests the microbial cells present may have been
stressed and not necessarily exhibiting normal growth characteristics.   The data of Li (21) obtained in similar equipment are also
included.   Decline is rapid initially, and then becomes more
gradual as the flux approaches a lower asymptote.   In common
with earlier studies, flux decline appears to be a function of both
solids concentration and recirculation velocity.

The possible causes of flux decline are,

(i)   membrane compaction or deterioration (hydrolysis etc.),
(ii)   membrane fouling or plugging by macromolecules,
(iii) changes in the gel layer which increases its resistance to flow.

Membrane compaction or deterioration are considered unlikely
causes for flux decline, due to the low transmembrane pressures
used ($\leq$ 200 kPa) and the stability of the membrane.

Plugging of the membrane pores is possible. This phenomenon
is known to be significant in solutions containing a spectrum of
macromolecules (22), as is the case for the activated sludge
mixed liquor. However there is evidence that this effect is not
dominating the long term flux decline. For example Table 1

summarises the results from a 30 day run using the UF/AS pilot
plant. Membrane resistance parameters have been determined as
indicated. The membrane resistance, R_m, is seen to change to a
small extent during the run.

In contrast, the gel resistance increased dramatically during
the 30 days, and this is the most obvious cause of flux decline.
As shown by equation (2) the gel resistance comprises the resistance due to dissolved solids, R_{ds}, and that due to the suspended
solids, R_{ss}. The suspended solids will be deposited as a porous
matrix, for which the permeability P_{ss} may be expressed (23)

$$\bar{P}_{ss} = d_p^2 \, \varepsilon^3 / (180 \, (1 - \varepsilon)^2) \tag{5}$$

This relationship has previously been used (14) to account for flux
decline in ultrafiltration. Thus, noting that $R_{ss} \propto 1/\bar{P}_{ss}$, a decrease
of voidage ε from 0.4 to 0.3, 0.2, 0.1 etc. would increase R_{ss} by
factors of 3, 14, and 140 respectively.

TABLE 1

Change in membrane resistance (R_m) and gel resistance
($R_{ss} + R_{ds}$) over 30 day run using Dorr Oliver XP-24 membrane

	R_m $(kPa.m^2.hr/l)$	$R_{ss} + R_{ds}$ $(kPa.m^2.hr/l)$
Membrane A, t=0	0.34	1.6
t = 30 days	0.50	66.2
Membrane B, t=0	0.9	1.5
t = 30 days	1.0	49.0

Notes (i)   Membrane A was previously unused, B was used
     (ii) $\Delta P = 200$ kPa, recirculation flow 90 l/min.
     (iii) R_m obtained from pure water flux; $R_{ss} + R_{ds}$ from eqn(2)
     (iv)   MLSS change during test from 800 to 2000 mg/l.

The voidage of the 'cake' of suspended solids will probably
be reduced by the combined effects of compression (the cells are
deformable) and biological change. Cells may continue to grow
within the cake, and whilst they maintain their activity they will
continue to be surrounded by a capsule or slime layer (24),

which will help to consolidate the 'cake'.

Correlating Flux Decline

The complex phenomena associated with flux decline have discouraged any attempt to model it mechanistically. Several workers (3,4,20) have used the exponential relationship between flux and time,

$$J_t = J_o \left(\frac{t}{t_o}\right)^{-n} \tag{6}$$

where the exponent n has been related to either recirculation velocity (4) or nature of the feed (3,20). Implicit in this form of correlation is that the flux is asymptotic to zero. Observations in this study (Figure 9) show that flux decline is related to both recirculation velocity and solids concentration. Moreover the flux appears to be asymptotically approaching a minimum, which presumably corresponds to the maximum attainable 'cake' resistance due to compaction and growth.

The correlating method used is based on the resistance concept depicted in Figure 10. This assumes that in the absence of the solids (suspended and dissolved) the flux of solvent is governed simply by the membrane resistance, R_m. As soon as solids are present and the initial gel layer is formed (in a matter of minutes) a resistance due to concentration change, ΔR_c, is added to the flow barrier. The total resistance at this instant is the value calculated using equation (2) and the initial flux, an estimate of which can be obtained from experimental data correlated by either equation (1) or (4).

At time $t = t$, the 'cake' has undergone an ageing process and the total resistance has increased by an additional ΔR_A, to $R_{T,t}$. The 'resistances-in-series' equation (2) may be modified to allow for this,

$$J_t = \frac{\Delta P}{R_{T,t}} = \frac{\Delta P}{R_m + (\Delta R_C + \Delta R_A)} \tag{7}$$

The bracketted term in the denominator combines the effects of dissolved and suspended solids content and of time.

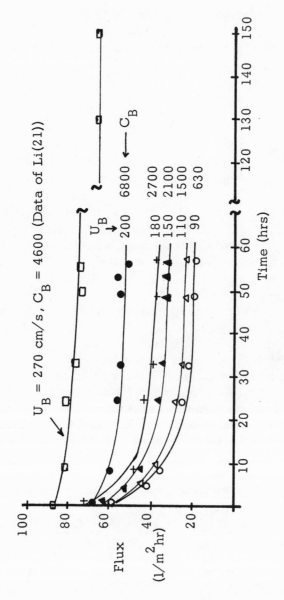

Figure 9.   Flux Decline at Various Recirculation Velocities (U_B) and
MLSS Concentrations (C_B)

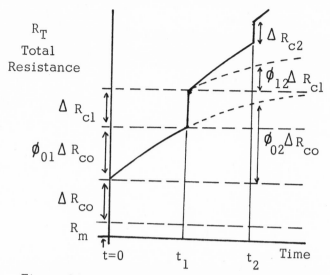

Figure 10. Representation of changes in Resistance
with Time and Concentration

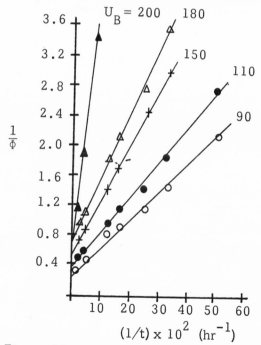

Figure 11. Flux Decline Data Plotted as Eqn. (9)

Assuming that the resistance due to aging, ΔR_A rises from zero at $t = 0$, and approaches a maximum asymptotic value, which is proportional to the initial resistance due to concentration change ΔR_C, leads to the following hyperbolic function,

$$\phi = \frac{\Delta R_A}{\Delta R_C} = \frac{\phi_{max} \cdot t}{K_T + t} \tag{8}$$

or, rearranging

$$\frac{1}{\phi} = \frac{K_T}{\phi_{max}}\left(\frac{1}{t}\right) + \frac{1}{\phi_{max}} \tag{9}$$

Equation (8) shows that at very small values of time ΔR_A is negligible and that at large values of time ΔR_A approaches ϕ_{max} times ΔR_C.

Figure 11 shows the data of Figure 9, reduced by equation (7) and plotted according to equation (9). From Figure 11 the parameters ϕ_{max} and K_T may be correlated as a function of the recirculation velocity,

$$\phi_{max} = 10.4 - 0.04 \, U_B \; ; \; \phi_{max} \not< 1.0 \tag{10}$$

$$K_T = \phi_{max}(0.05 \, U_B - 1.2) \tag{11}$$

A limitation of equation (11) is that it underestimates the K_T value obtained from Li's data (21), which may be because the sludge in his work was less stressed as it was well acclimatised to the conditions imposed.

MODEL OF COMBINED UF/AS SYSTEM

The model is based on the following assumptions;

(i)   the activated sludge kinetics and constituent mass balances are described by the set of equations (12 to 20) in Table 2 (refer to Figure 1). Details may be found elsewhere (9,19).

(ii) increases in MLSS initially reduce the flux according to the flux-concentration relationships (equations (1) or (4)),

Table 2

Summary of Equations Describing the AS Kinetics and System Mass Balances

Kinetics of AS:

$$dC/dt = (\mu - k_d)C \tag{12}$$

$$\mu = (\mu_m \cdot S)/(K_s + S) \tag{13}$$

$$dC/dt = -Y\,dS/dt \tag{14}$$

Mass balances for AS:

Organisms, $d(VC_o)/dt = F_i C_i + F_r C_r + [(\mu_m S)/(K_s + S) - k_d]\,VC_o - F_o C_o$ (15)

Substrate, $d(VS_o)/dt = F_i S_i + F_r S_r - [(\mu_m S)/(K_s + S)]\,(VC_o/Y) - F_o S_o$ (16)

Water $d(V\rho)/dt = \rho F_i + \rho F_r - \rho F_o$ (17)

Mass balances for UF:

Organisms $F_o C_o - F_e C_e - F_r C_r = 0$ (18)

Substrate $F_o S_o - F_e S_e - F_r S_r = 0$ (19)

Water $F_o \rho_w - F_e \rho_w - F_r \rho_w = 0$ (20)

Kinetic Constants for AS: $Y = 0.55$, $K_s = 120$ mg/1 (ref.26); $k_d = 0.005$ hr^{-1} (ref.21)

$$\mu_{max} = 0.3\ \text{hr}^{-1}\text{(batch expt.)}$$

(iii) ageing of the gel layer occurs according to the empirical kinetic relationships (8,10,11),

(iv) decreases in MLSS decrease the flux in accordance with the flux-concentration relationships.

The major computational effort arises from application of the ageing model embodied in assumptions (ii) and (iii). Figure 10 shows how the model pictures the history of changes in the total resistance in response to ageing.

At time zero the total resistance is,

$$R_{TO} = R_m + \Delta R_{co} \tag{21}$$

where ΔR_{co} is the gel resistance $(R_{ds} + R_{ss})$ resulting from the initial application of mixed liquor of a given concentration. The value of ΔR_{co} may be predicted from equation (2) and (1) or (4) assuming a value of R_m.

In the absence of concentration changes the total resistance will increase in accordance with equation (8), so at time t,

$$R_{Tt} = R_m + \Delta R_{co} + \Delta R_{co} \, \phi_{ot} \tag{22}$$

$$= R_m + \Delta R_{co} (1 + \phi_{ot}) \tag{23}$$

The ageing function, ϕ_{ot}, defines the fractional increase in the resistance of the gel originally deposited at t = 0, after a period of time t = t.

Assume now that at t = t, the concentration increases sufficiently to provide an additional resistance, ΔR_{c1}. This increment of resistance is obtained via equation (2) using the difference in flux values $J_{co} - J_{c1}$; these values being calculated from equations (1) or (4). The situation at t = t, is depicted in Figure 10, and likewise at t = t_2 when a further increment of resistance, ΔR_{c2}, is added. The total resistance at t = t_2 is then,

$$R_{T2} = R_m + \Delta R_{co} (1 + \phi_{o2}) + \Delta R_{c1} (1 + \phi_{12}) + \Delta R_{c2} \tag{24}$$

In general form the total resistance is, for t = t_N

$$R_{TN} = R_m + \sum_{i=o}^{N-1} \Delta R_{ci}(1 + \phi_{i,N}) + \Delta R_{cN} \tag{25}$$

This equation can be used to calculate the increase in gel resistance due to both steady increases in concentration and to step
changes. Indeed the model simulates the steady increase
situation as a series of small step changes, in which the ΔR_{ci}
values are arbitrary chosen at a small value. This value of
ΔR_{ci} will determine the time increments which will not necessarily be of equal duration. Equation (25) also caters for decreases
in resistance due to decreases in MLSS concentration. In this
case new ΔR_c terms are negative.

Testing the Model

The model has been tested against short term ($\not> 36$ days)
runs carried out in the authors' laboratory (9,19), with perturbed
feed, and against a longer term (120 day) run at pseudo state-
state reported by Arika et al (11).

Figure 12 shows the experimental and model predicted
values for flux and MLSS for the first 12 days of a typical short
term run. In this situation the sludge solids were acclimatised
to the substrate (glucose plus nutrients (19)), but had not been
subject to the rigors of ultrafiltration for a period of about 14 days.
The decline in flux is predicted reasonably using the values of
ϕ_{max} and K_T from the correlations (equations (10) and (11)). An
improved fit is obtained using the values obtained from Li's data
(21). The model tends to overestimate the MLSS, suggesting
inadequacies in the simple AS kinetics assumed. Commencing
the simulation at day 4 (Figure 13) the model comes closer to the
experimental data.

The conditions examined in Figure 14 are those corresponding
to days 20 to 36 of a short term run, where the system was
initially at steady state with a substrate input concentration of
800 mg/1. From day 23 the substrate was dropped to essentially
zero (20 mg/1) for 6 days, although an aqueous solution of nutrients
was still fed to the system. On day 29 the substrate level was
returned to its original value and the run was terminated on day 36.
This degree of perturbation provides a severe test of the model,
and indeed it can be seen that the rather dramatic improvement in
flux measured experimentally when the substrate is ceased is not

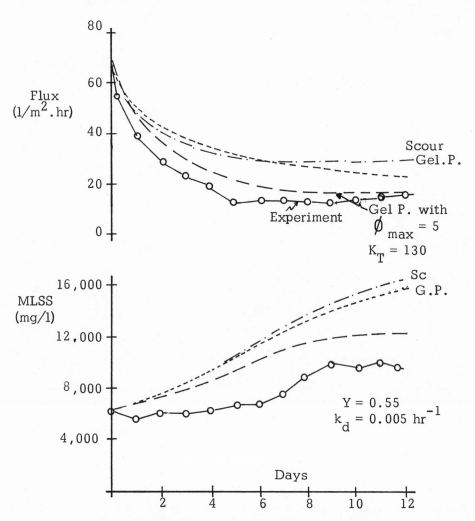

Figure 12.   Comparison of Model with Experiment
Initial Phase (Days 0 → 12)

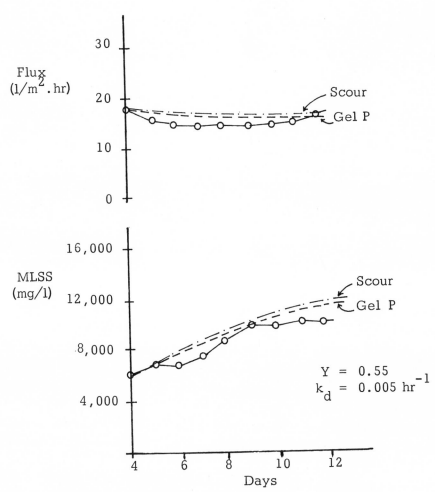

Figure 13. Comparison of Model with Experiment
∿ Days 4 to 12.

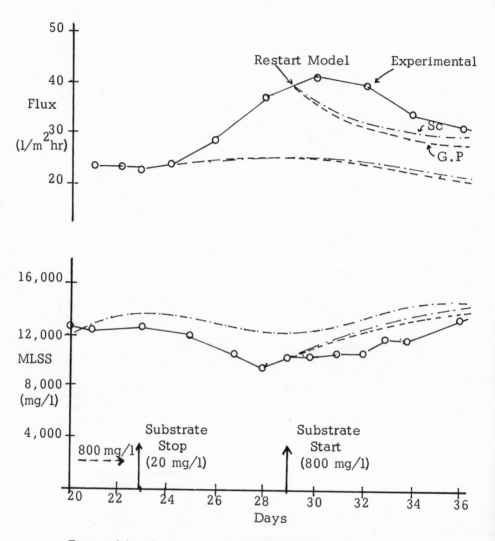

Figure 14. Comparison of Model with Experiment
∿ Perturbed Feed.

predicted by the model. However, by allowing the model to commence afresh from day 29, when substrate is reintroduced, the flux is predicted reasonably. As before the model tends to overestimate the MLSS.

Arika et al (11) have recently reported on the performance of a combined UF/AS system using Dorr Oliver Iopor membranes. Their system is approximately two orders of magnitude larger than that used in this work, although the mode of operation is similar. Figure 15 shows their reported flux and mixed liquor volatile suspended solids (MLVSS), and the values predicted by the model. A number of assumptions have been made in using the model for simulation of Arika et al's data, and they are summarised on the figure (the assumed BOD_5 values are simply smoothed values from the original paper). Linear temperature histories from $10^\circ C$ to $30^\circ C$, and 20 to $30^\circ C$ are assumed based on the reported inlet temperature range ($6^\circ C$ to $26^\circ C$) for a plant located outdoors. It has been necessary to allow for the effect of temperature on flux, and this has been assumed to be an increase in flux of 2.5% per $^\circ C$ (this is the same as the effect of temperature on the reciprocal of the viscosity of water). The model is seen to give reasonable prediction of the smoothed performance for this pseudo steady state long term run, particularly for the 20 to $30^\circ C$ temperature history.

DISCUSSION

The flux declines found in the continuous runs shown in figures 12 to 15 are reasonably well predicted. However the improvement in figure 12 achieved by increasing ϕ_{max} and K_T to values closer to those obtained in the continuous run of Li suggests that there may be some difference between flux decline of actively feeding cells and those which are more stressed (Figures 9 and 11 data).

The flux recovery found experimentally (Figure 14) following non feeding is obviously not simply a reverse ageing process. However, the flux reached after 6 days without substrate is very close to the initial (or unaged) flux at the measured MLSS. It is possible that during prolonged non feeding the cells within the 'cake' lose their slime layer and cohesive flocculant properties and become more susceptible to scour. Further study and modeling of this flux recovery behaviour is required to improve the overall capability of the UF/AS model.

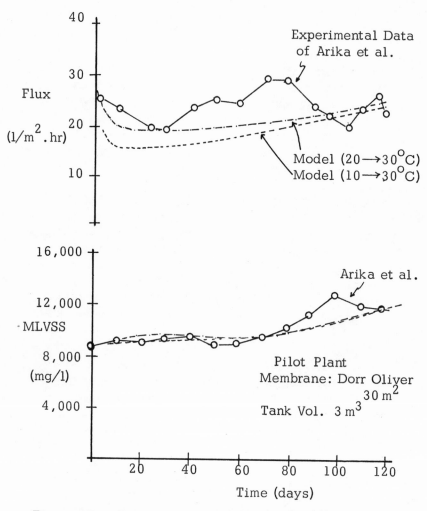

Figure 15.   Comparison of Model with Experiment
            Pseudo Steady State Data of Arika et al (11)

There is little difference between the simulations in figures 12 to 14 using the initial flux data predicted from the gel polarised model (equation (1)) or the scour model (equation (4)). This is presumably because the dominant factor in determination of flux is the ageing rather than the initial value.

The fact that the model predicts the 'steady-state' performance of a larger facility is encouraging. A more detailed simulation using unsmoothed input BOD_5 values appears to be unnecessary in view of the other assumptions involved.

Whilst it would appear that the 'steady state' behaviour of the AS system is reasonably well predicted, it is apparent that agreement between the model and experiment are less satisfactory after a perturbation in the feed concentration has occurred. There is a tendency to over-estimate response and growth (figure 12), and to under-estimate decay (figure 14). These findings are rather similar to those found by Sundstrom et al (25) who imposed sinusoidal variations of substrate concentration onto a biological reactor. In their work the experimentally measured biomass consistently increased less than predicted from their model and exhibited a lag phase. Moreover the apparent value for k_d (the endogenous decay constant) increased up to five times as a result of the perturbations. In this work the value of k_d has been assumed to be $0.005 \, hr^{-1}$, based on the limited experimental data of Li (21). This is approximately double the value typically quoted for activated sludge with domestic waste as substrate (26). However in the UF/AS it is probable that the environmental stress imposed by the UF loop also contributes to an increase in k_d.

From this work, the areas particularly important for further development of the model are seen to be, (i) the influences of UF and perturbed feeding on the AS kinetics, and (ii) the mechanisms and kinetics of flux recovery associated with low substrate feed.

CONCLUSIONS

In the UF/AS process the major flux-limiting species are the suspended solids, with the dissolved solids playing a minor and diminishing role as MLSS increases.

Flux variation with MLSS can be correlated with limited success using the conventional semi-log relationship, or using an alternative (scour model) log-log relationship.

The decline of the initial flux readily occurs and is thought to be caused by changes in the gel layer. It is a function of MLSS concentration, recirculation velocity, and condition of the AS solids. The decline can be modelled using a hyperbolic ageing function to modify the initial gel resistance.

A combined UF/AS model has been developed which success-fully predicts the changes in flux for conditions of increasing or steady state MLSS. It underestimates the flux recovery obtained when low substrate feed causes changes to the gel layer.

Discrepancies between predicted and observed changes in MLSS may be explained by the inability of simple Monod kinetics to account for the effects of substrate perturbation and environmen-tal stresses in the UF loop.

Acknowledgements

The authors wish to acknowledge research funds from the Australian Water Resources Council. One of us (M.T.N) would also like to thank the National University of Malaysia for a scholarship.

Nomenclature

C, C_B, C_G = micro-organism concentration, in the bulk suspension, the gel concentration

d_p = particle size

F = volumetric flow rates (Figure 1)

J, J_t, J_o = water flux, at time t, at zero time

k_d = endogenous decay constant

k_s = solute mass transfer coefficient

K_S = kinetic constant in Monod relationship

K_T = kinetic constant in resistance ageing relationship

ΔP = transmembrane pressure drop

P_{ss} = permeability of a cake of suspended solids

R_{ds}, R_m, R_{ss} = resistances due to dissolved solids, membrane, suspended solids

R_T, R_{Tt} = total resistance, total resistance at time t

S = substrate concentration

t, t_o = time, initial time

U_B = recirculation velocity

V = working volume of aeration tank

Y = biological yield coefficient

ε = voidage of a cake of suspended solids

μ, μ_{max} = specific growth rate, maximum growth rate

ρ_w = density of solution

ϕ, ϕ_{max} = resistance ageing factor, maximum ageing.

REFERENCES

1. Hauck, A.R. and Sourirajan, S., <u>Environ.Sci.Technol.</u>, 12 (3), p 1269, (1969)

2. Feuerstein, D.L. and Bursztynsky, T.A., <u>Chem.Eng.Prog. Symp.Ser.</u> No. 107, Vol.67, p 568, (1971)

3. Cruver, J.E and Nusbaum, I., <u>Journal WPCF</u>, Vol. 46, No. 2, p 301, (1974)

4. Thomas, D.G. and Mixon, W.R., <u>Ind.Eng.Chem.Proc.Des. Dev.</u>, Vol. 11, No. 3, p 339, (1972)

5. Bailey, D.A., Jones, K. and Mitchell, C., <u>Wat.Pollut.Control</u>, 73(4), p 353 (1974)

6. Stavenger, P.L., <u>Chem.Eng.Progr.</u>, No. 67, Vol. 3, p 30 (1971)

7. Gebbie, P., Fane, A.G. and Fell, C.J.D., <u>Water,</u> Vol. 3, No. 2, p 17, (1976)

8. Sammon, D.C. and Stringer, B., <u>Process Biochem.</u>, March 1975, p 4

9. Fane, A.G., Fell, C.J.D. and Nor, M.T., Proceedings APCChE, Jakarta, Nov. 21, (1978), p 117

10. Bemberis, I., Hubbard, P.J. and Leonard, F.B., Paper No. 71 – 878 Presented to American Soc.Agr.Engrs., Chicago, (1971)

11. Arika, M., Kobayashi, H. and Kihara, H., <u>Desalination</u>, 23, p77, (1977)

12. Monod, J., Ann.Rev.Microbiol., 3, 371 (1949)

13. Chui, S.Y., Erickson, L.E., Fan, L.T. and Kao, I.C., <u>Biotech.Bioeng</u>. 14 (2), p 207 (1972)

14. Blatt, W.F., Dravid, A., Michaels, A.S. and Nelsen, L., in "Membrane Science and Technology", (J.E. Flinn, Ed.), Plenum Press, N.Y. (1970) p 47

15. Bixler, H.J. and Rappe, G.C., U.S. Patent 3, 541, 006 (Nov. 17, 1970)

16. Fell, C.J.D. and Fane, A.G., "Ultrafiltration of Wheat Starch Factory Effluents", Report to N.S.W. State Pollution Control Commission, October, 1977

17. Porter, M.C., <u>Ind.Eng.Chem.Prod.Res.Dev.</u>, Vol. 11, No. 3, p 234, (1972)

18. Henry, J.D. Jr. in "Recent Developments in Separation
        Science", Vol. 2, C.R.C. Press, (1972)
19. Nor, M.T., Ph.D. Thesis, University of New South Wales,
        (in preparation)
20. Harris, L.R., Schatzberg, P., Bhattacharyya, D. and
        Jackson, D.F., Water and Sewage Works,
        (Aug. 1978), p 66
21. Li, A.S-k., M.App.Sci. Thesis, Univ. of N.S.W. Australia,
        (1976)
22. Michaels, A.S. in Progress in Separation and Purification,
        Wiley (1968) p 297
23. Carman, P.C., Trans.Inst.Chem.Eng., 15, p 150 (1937)
24. Pelczar, Jr.M.J. and Reid, R.D., "Microbiology", McGraw-
        Hill, N.Y. (1972)
25. Sundstrom, D.W., Klei, H.E. and Brookman, G.T.,
        Biotech. Bioeng. Vol. 18, p 1 - 14 (1976)
26. Metcalf and Eddy, "Wastewater Engineering", McGraw Hill,
        (1972)

HYPERFILTRATION FOR RECYCLE OF 82°C TEXTILE WATER WASH

Craig A. Brandon

CARRE, Inc.
207 North Fairplay Street
Seneca, SC   29678

ABSTRACT

A program to demonstrate high temperature hyperfiltration to renovate hot industrial wastewater for direct recycle is in Phase 2: the installation of a 10 m/hr membrane system to achieve closed cycle operation of a textile dye range. Hydrous-zirconium oxide-polyacrylate (ZOPA) membranes dynamically formed on porous sintered stainless steel tubes arranged in a Single Pass system will be supplied by the Mott-Brandon Corporation. Phase 1 of the program involved on-site pilot testing of several membranes, the procurement of competitive bids for the membrane system, and the complete installation including all interface controls required for the dye operation. The 180°F (82°C) dye wash water contains residual dyes, auxiliary detergent and wetting agents, and guar gum. The total solids in the wash water range from 130 to 7000 mg/ℓ, with the suspended solids ranging from 2 to 90 mg/ℓ. The membrane system will be operated at 96% recovery. The performance of the membranes during the pilot tests, the laboratory evaluation of the reuse, and the complete recovery system are described.

APPLICATION OF ACRYLONITRILE-COPOLYMER

MEMBRANE TO CATIONIC ELECTRO-DEPOSIT COATING

Akio Fujimoto, Kenji Mori, and Kenji Kamide

Textile Research Laboratory
Asahi Chemical Industry Company, Ltd.
Hacchonawate 11-7, Takatsuki, Osaka, Japan

INTRODUCTION

Electrocoating process is widely used for primer-coating of car bodies, parts, appliances, etc. Ultrafiltration is extensively applied in this electrocoating, because there are large merits such as the closed coating system, recovery and reuse of paint, ease of tank liquid composition control (Fig. 1). The process is as follows. By ultrafiltration water, solvent and other low molecular substances are separated from the paint-tank-liquid as filtrate and are used for rinsing the coated parts and then returned to the paint tank; on the other hand, the paint liquid concentrated in the ultrafilter is sent back to the paint tank.

Recently, a cationic electrodeposition paint with higher anti-corrosiveness has been developed, resulting in a rapid conversion from the existing anionic electrodeposition (A-ED) paint to the new cationic electrodeposition (C-ED) paint. Yet, when the conventional membrane for A-ED paint is applied, we have encountered some difficulties such as a decline of filtration performance with the passage of time, a lowering of the performance due to changes in paint conditions, and an electric power failure.

In order to resolve these problems, three approaches can be considered: (1) improvement of paint composition, (2) operation control technology of paint liquid, as well as ultrafiltration system, and (3) improvement of ultrafiltration (UF) membrane. Of these, our attention has been focused on the improvement of the UF membrane. In this report, some experimental results on a new UF membrane developed by Asahi Chemical Industry Co., Ltd., will be presented.

We have already commercialized a hollow-fiber (HF) membrane (Type A) made from random copolymer with more than 90 wt % acrylonitrile. This conventional type is really suitable for A-ED paint,

but not always useful for C-ED paint. Then assuming that the C-ED
paint lowers the UF performance by contaminating the membrane, we
have tried to seek for new materials which might be less contaminated
when exposed to C-ED paint.

EXPERIMENTAL

As a preliminary we prepared the membranes from solutions of
various acrylonitril-copolymers and immersed them in C-ED paint, and
then evaluated the water permeability before and after the immersion.
As a result, we succeeded to find a new membrane from a type of copolymer which showed very small change in water permeability as compared with Type A. The former will be hereafter referred to as Type
Q (Table 1). Table 2 shows the paint conditions employed in the laboratory scale experiments.
A filtration was performed for 3 hr using both Types A and Q
under the same operating condition. Then by changing the pressure,
flux was measured. After this measurement completed, the filtration
was suspended temporarily for 2 hr. Making back-washing for 1 min
with filtrate. Above cycle of the filtration operation (3 hr-filt-
ration at constant pressure, filtration curve measurement, 2 hr-
suspension followed by 1 min back-washing) was repeated again.

RESULTS AND DISCUSSION

The pressure dependence of permeability is illustrated in Fig.2.
The figures at left and right hand sides indicate the filtration
curves obtained at 1st and 2nd filtrations, respectively. When Type
A is employed, an increase in number of cycle brings about a decrease in the slope of the filtration curve and also a decrease in the
effect of flow rate on the curve. In contrast to this, recycled
usage of Type Q makes insignificant change in the filtration curve.
After the completion of the second cycle, the membranes were washed
with pure water and then examined by infrared spectroscopy and the
results obtained were shown in Fig.3. IR spectrum of Type A shows
absorptions at 1510 cm^{-1} and 1234 cm^{-1}, which attribute to Bis-
Phenol A, main chemical component of the paint. In the spectrum of
Type Q these absorptions are not observed.
The filtration curves and the IR spectrum evidence clearly a
gel formation on the membrane surface of Type A. And also it becomes clear that the gel formation causes a prominent decrease of flux.
In addition, the experiments suggest strongly concerning Type Q that
the intermittent rinsing of the membrane surface during the filtra-
tion-operation will be effective for preventing possible gel-forma-
tion on the membrane surface.
Cross-sectional view by electron microscopy for Type Q is shown
in Fig.4. This figure indicates that active thin layers, which play
as a roll of UF function, exist both external and internal surfaces

Table 1. Module specification

Type	Q	A

Hollow fiber size,(inch)
    inner diameter
    outer diameter
	0.03	0.03

Cut off molecular weight	13,000	13,000
Membrane area, (ft^2)	50	50
Water permeability, (gfd)	120	120
Highest pressure, (psig)	42	42
Upper temperature limit (°C)	122	122
PH range		2 - 10

Table 2. Experimental paint conditions

Type;	Nippon paint PTU-30
Non-volatile %;	18-19
PH;	6.4 - 6.6
Temperature, (°F);	80.5 - 82.4

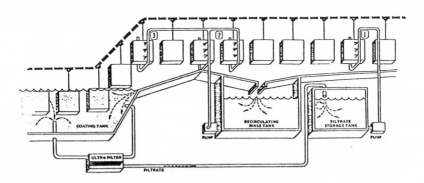

Fig. 1. Typical closed loop ultrafiltration system for
      continuous process electrocoating.
      (Source: technical report of PPG Industries, Inc.,
      J-44139, April, 1972).

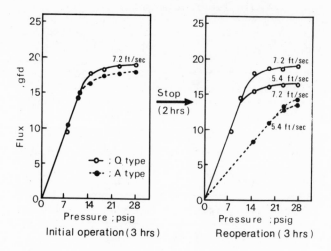

Fig. 2. Effect of operation pressure.

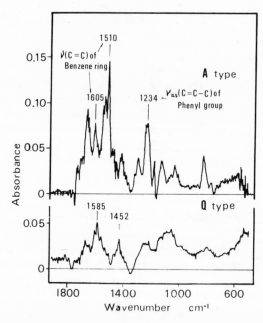

Fig. 3. Analysis of membranes after running by Fourier Transform Infrared Spectroscopy.

Fig. 4. Cross section of hollow fiber.

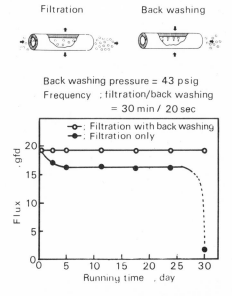

Fig. 5. Effect of periodical back-washing with filtrate.

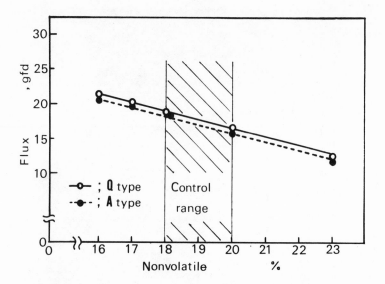

Fig. 6.   Effect of paint concentration on flux.

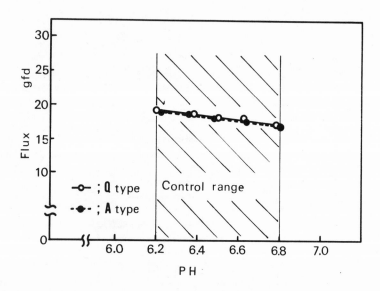

Fig. 7.   Effect of PH on flux.

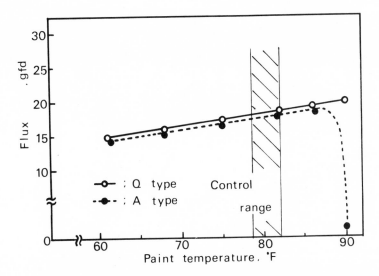

Fig. 8. Effect of paint temperature on flux.

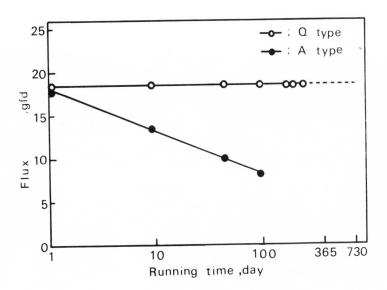

Fig. 9. Flux stability with running time.

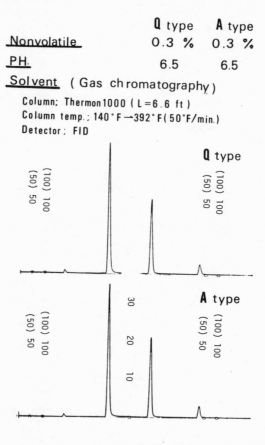

Fig. 10.   Analysis of filtrate.

of hollow fiber. These characteristic features of Type Q enable us to apply successfully back-washing of membrane with filtrate.

Back-washing was carried out for 20 second at interval of every 30 min. under the pressure of 43 psi. The results are shown in Fig.5. With back-washing, a constant flux is obtained, but, on the other hand, without back-washing, the flux lowers to a certain extent at the initial stage and then decreases sharply after about 1 month.

Next, the effects of paint concentration, expressed by wt % of non-volatile components in whole paint liquid, and operating condition including PH and temperature on the flux were examined. The results are shown in Figs. 6,7, and 8. In these figures control range is the range of the operating conditions suggested by a paint manufacturer. All of these effects shows the same paralleled trend for Type A and Q except at higher temperatures. Type Q shows far stabilized flux under the conditions deteriorating the dispersion of the paint, such as 90 °F.

The module of Type Q among Type A modules of the industrial UF line to get the running data on site is installed. Fig.9 shows the permeability stabilities of Type Q and A modules with running time. Type Q has been showing a stable, constant flux for 7 months. The components and quantity contained in the filtrate are measured and the results are shown in Fig.10 as to non-volatile, PH and solvent. Yet no difference is observed between Type A and Type Q.

CONCLUSIONS

In order to resolve the problem with which we have encountered in the ultrafiltration of cationic electrodeposition paint, we prepared a new membrane from solution of a type of acrylonitril-copolymer and examined its applicability to cationic electrodeposition paint, comparing with the conventional product Type A as a control. This new membrane showed a stable flux with lapse of time, and also indicated the stabilized performance to the changes in paint conditions as well as to temporary suspension. Intermittent back-washing with filtrate is effective to prevent from the ppssible gel formation on the membrane surface. The reason why the new membrane has caused such phenomena is a subject to take up hereafter.

THE APPLICATION OF NOVEL ULTRAFILTRATION MEMBRANES

TO THE CONCENTRATION OF PROTEINS AND OIL EMULSIONS

G. B. Tanny and M. Heisler

Gelman Sciences, Inc.
600 South Wagner Road
Ann Arbor, MI   48106

INTRODUCTION:

Amongst the various membrane separation processes, ultrafiltration has enjoyed widespread industrial application, particularly for the concentration of macromolecular species. References to its use are especially found with respect to protein concentration, e.g., in the dairy industry for the concentration of milk[1,2] and cheese whey recovery[3,4] and also in the production of human serum albumin[5]. It has also been successful when applied to industrial waste recovery and/or disposal problems. Spent cooling oil emulsions present heavy industry with a serious waste disposal problem, and their concentration by ultrafiltration[6] can considerably reduce haulage costs. When the process is carried to 25-40% of oil, the emulsion concentrate can be burnt in conventional equipment.

The key element in ultrafiltration equipment is, of course, the membrane performing the separation. In an effort to develop an improved ultrafilter, the following properties were sought:

1. Wet/dry stability without the use of any contaminant humectants or surfactants.

2. Low surface affinity toward adsorption of proteins.

3. Chemical resistance to acids and bases.

4. Mechanical strength and creasability.

5. Steam sterilizability.

Through the use of proprietary technology, these features were substantially achieved in the form of a novel thin film composite membrane structure. It consists of a 0.3 - 0.5 μm thick layer bonded to a 0.2 μm pore size, microporous support. Figure #1 shows a scanning electron micrograph (SEM) of the composite's surface at a point at which the bonded layer has been stripped away to reveal the microporous substructure. One may also note that the membrane surface is not perfectly smooth, as in the case of conventional ultrafilters, but has a somewhat spongy nature. In the sections which follow, we shall examine the implications of this structure for the concentration of proteins and oil emulsions.

EXPERIMENTAL SECTION

Test Apparatus and Solutions:

Stainless steel, pressurizable stirred cells with 10 cm^2 of effective filtration area were used to obtain protein retention and deionized water flow rates. Protein solutions were prepared with Trizma base buffer corrected to a pH 7.5. The con-

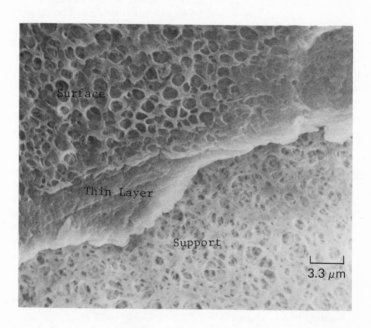

Fig. #1.   SEM of XPTC-10 Ultrafiltration Membrane

centration of the single species protein solutions were: (1) 100 mg/100 cc for cytochrome C and myoglobin; and (2) 170 mg/100 cc for alpha-lactalbumin, ovalbumin and human serum albumin (HSA). Triprotein solutions contained 100 mg of myoglobin, 170 mg. of alpha-lactalbumin and 170 mg of HSA per 100 cc buffer solution.

Analysis of Protein Retention:

Determination of protein retention was carried out by U.V. absorbence measurements at 265 nm for the single species protein solutions. Analysis of the triprotein solution was accomplished using a quantitative electrophoresis technique outlined in Figure #2. A marker protein used as a reference standard (either gamma-globulin or cytochrome C), was added to the retentate and permeate. Thus, to 300 µl of both the retentate and permeate, 100 µl of a 1.0% gamma globulin solution is added. A 10 µl sample is then applied to a mylar supported cellulose acetate strip (Super Sepraphore, Gelman Sciences, Inc.), which has been presoaked in tris-barbital buffer, pH 8.8.

Electrophoresis is carried out at 220 V for 20 minutes in a Sepratek chamber (Gelman Sciences, Inc.), after which it is stained in Ponceau S red dye washed and cleaned according to normal recommended procedure. An ACD-18 scanning densitometer (Gelman Sciences) is used to scan the electrophoresis strip and establish relative peak areas which were then normalized with respect to the size of the gamma-globulin peak. The ratio of the normalized peak of the permeate to that of the retentate yields the percent retention of the species.

Stress Testing the XPTC-10K Membrane:

The membranes were subjected to the following types of stress:

1) Wet/dry stability.
2) Cleaning and sanitizing agents.
3) Sterilization techniques.
4) Chemical compatibility studies.

The conditions for each stress test is explained in Table II (c.f., membrane characterization).

Concentration of HSA:

Two 0.25 ft^2 membranes were placed in a Pellicon cassette plate and frame ultrafiltration system (Millipore Corporation,

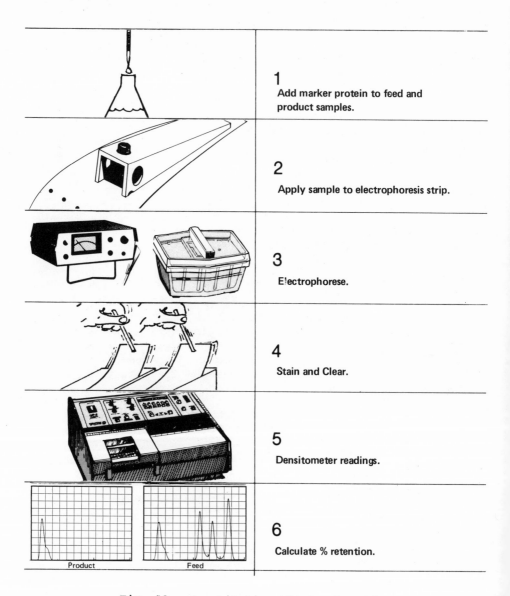

1 Add marker protein to feed and product samples.

2 Apply sample to electrophoresis strip.

3 Electrophorese.

4 Stain and Clear.

5 Densitometer readings.

6 Calculate % retention.

Fig. #2.   Quantitative Electrophoresis

Bedford, MA), using permeate spacer screens for the product channel. A masterflex peristaltic pump (Cole-Palmer Co.) provided the pressure and cross flow velocity which were controlled by a value on the retentate exit flow meter.

HSA feed was diluted with water to 1% from a commercial medical 5% preparation (Hyland Laboratories) and retention measurements were made electrophoretically due to the presence of small stabilizer molecules which absorbed in the U.V.

Measurement of Protein Adsorption:

BSA was radiolabeled with ^{125}I and mixed with a 20 mg/cc solution of nonlabeled BSA to yield a 1.7×10^4 cpm/μl solution. Ten ml of solution was left in contact with 10 cm^2 of XPTC-10 membrane in a cell at 4°C during 18 hours. After washing with two 50 ml aliquots of buffer to remove excess solution, the membrane was removed and counted directly. Irreversible adsorption was 26 μg/cm^2.

Total Organic Carbon (TOC) Determinations:

TOC was determined with a Beckmann TOC analyzer using standard operational procedures. Standards were injected with each set of determinations to check the calibration.

Oil Emulsion Tests:

Oil emulsion ultrafiltration experiments were carried out in a stirred cell described earlier[7]. The cell was modified to accept an internal light pipe and detector and accompanying electronics to allow the stirring speed to be accurately measured and adjusted.

Oil Emulsions:

Two commercially available synthetic cooling oil emulsions were used in the study, a macroemulsion, brand name Cimperial and a microemulsion, brand name SBBP-40. The Cimperial oil emulsion contained 64.6 ppm of organic carbon per 100 ppm of oil emulsion. The SBBP-40 oil emulsion contained 80 ppm of organic carbon per 100 ppm of oil emulsion. Both oil emulsions were of the anionic surfactant type.

RESULTS AND DISCUSSIONS

Membrane Characterization:

In order to define the nominal molecular weight cutoff of the membrane, dilute buffered solutions of test molecules with molecular weights from 1400 (bacitracin) to 67,000 daltons (Human Serum Albumin) were ultrafiltered at 3.4 atm. Typical data from such trials are shown in Table I for three different membranes. Results are shown both for single protein solutions (1 mg/ml) and a "triprotein solution" made up of myoglobin, alpha-lactalbumin and human serum albumin. For the latter, rejection of each species was determined by a quantitative electrophoresis technique (c.f., Fig. #2, experimental).

For the XPTC-10K membrane, relatively little difference was noted between tests with single or multiple protein solutions. However, a material (XPTC-30A) which appears to be a nominal 30K cutoff when tested with triprotein solution, is considerably less selective toward single protein solutions. Examination of a competitive commercial 30K membrane with the same solutions did not show such an effect. The ultrafiltration fluxes with the triprotein solution feed were almost identical for both the competitive 30K membrane and the XPTC-30A, despite the four times larger pure water flow for the competitive membrane.

Table I. Percent Retention of Test Proteins

Test Species	Molecular Weight (Daltons)	XPTC-10		XPTC-30B		XPTC-30A	
		Single Protein	Tri Protein	Single Protein	Tri Protein	Single Protein	Tri Protein
Bacitracin	1,400	42 ± 2					
Cytochrome C	12,400	95					
Myoglobin	17,800	95	100	85 ± 5	100	20 ± 10	77 ± 2
Alpha-Lactalbumin (diamer)	32,000	95	100		100	20 ± 10	93 ± 3
Ovalbumin	45,000	95		100		82 ± 10	
Albumin (Human)	67,000	98	100	100	100	75 ± 10	98
Flow Rate of Deionized Water at 3.4 atm (ml/cm^2-min)		0.18 ± 0.02		0.60 ± 0.10		1.82 ± 0.6	

These effects can be attributed to dynamic membrane formation at the solution/membrane interface, a subject which has recently been reviewed[8]. In the present case, the concentrated
protein boundary layer is the governing resistance both to water
and protein passage. Due to the presence of the porous structure
at the surface, a defined or controlled boundary layer of approximately 5-10 μ is created. Since this protein boundary layer
thickness is similar in magnitude to that obtained even in a
well stirred cell, a similar ultrafiltration rate is observed
with a smooth surface membrane. The existence of a spongy surface structure also serves to physically protect the selective
layer, thereby improving membrane resistance to mechanical damage.

In Table II, ultrafiltration data is shown for the XPTC-10
membrane before and after its subjection to various forms of
stress. As previously indicated, the desired chemical resistance, wet/dry stability and sterilizability are substantially
achieved.

Concentration of Human Serum Albumin:

An important recent application of ultrafiltration involves
the concentration of human serum albumin as a part of the traditional Cohn fractionation process[5]. The albumin paste, which
is obtained in the last step in this fractionation process, must
be freed of alcohol and salt prior to its final concentration.
As this application is particularly suitable for membranes which
can be stored dry and steam sterilized, it seemed worthwhile to
evaluate the membranes with HSA solutions.

The membranes chosen for a performance comparison were XPTC-
10, and the XPTC-30B, which exhibits nominal 30K protein retention with dilute solutions (c.f., Table I). Concentration experiments were carried out with a Pellicon cassette* system in a
batch concentration mode. In Figure #3, the permeate flow rate
is shown as a function of time for both membranes. Since protein
retention was approximately 100% in both cases, it is clear that
the XPTC-30B has an advantage in this application. The data
shown in Figure #3 for the XPTC-30B ultrafiltration rate is similar to values obtained in other studies[5] at similar pressures and retentate flow rates. Simple washing with water restores the membrane to its original water flow, which reflects
the very low extent of irreversible protein absorption. Quantitative adsorption measurements with tagged BSA indicated that no
more than 26 μg of BSA is irreversibly adsorbed per cm^2 of
membrane (c.f., experimental Section E).

---

*   Millipore Corp., Bedford, MA

Table II. The Effects of Stress Testing on the
Performance of the XPTC-10 Membrane

TYPE OF STRESS	TEST CONDITIONS	WATER FLOW @ 3.4 ATM (cc/cm^2-min)		MYOGLOBIN RETENTION	
		Prior to Stress	After Stress	Prior to Stress	After Stress
Wet/Dry Stability	1. Membranes were water flow rate tested. 2. Dried for 3 days at 80°F. 3. Retested for water flow and myoglobin retention	0.17	0.17 ± 0.01	95 ± 2%	93 ± 3%
Cleaning and Sanitizing Agents	1. Antibac B (BASF-Wyandotte Corp.) an organic type chlorine germicide.	0.16	0.15	96%	90 ± 8%
	2. Interest (BASF) concentrated liquid chlorinated cleaner.	0.17	0.15	96%	96 ± 1%
	3. Elevate (BASF) concentrated low foaming liquid acid cleaner (Heated to 120°F).	0.20	0.14	96%	93 ± 2%
Sterilization Techniques					
Ethylene Oxide	1. 20 psi for 6 hours	0.16	0.17	93%	92%
Autoclaving (Wet)	1. 15 min. at 121°C	0.17	0.17	92%	91%
Autoclaving (Dry)	1. 15 min. at 121°C	0.17	0.05	96%	100%
5% Formalin	1. 48 hr. exposure	0.17	0.17	93%	93%
Chemical Compatibility	1. 48 hr. exposure				
	a) 0.1 N NaOH	0.20	0.20	95 ± 2%	84 ± 3%
	b) 0.1 N HCl	0.19	0.19	95 ± 2%	85%
	c) 100% EtOH	0.20	0.19	95 ± 2%	87 ± 1%
	d) Toluene	0.20	0.20	95 ± 2%	89%
	e) 3.0 M UREA	0.20	0.20	95 ± 2%	96%
	f) 2.0 M Ammonium Sulfate	0.20	0.20	95 ± 2%	89%
	g) 0.5% Alconox Soap	0.20	0.19	95 ± 2%	88%

Concentration of Oil Emulsions:

Emulsions of two commercial cutting oils were made up at concentrations of 3% and 5% (vol/vol) which are in their range of normal use. These were tested with the nominal 10K cutoff membrane and the more hydrodynamically permeable XPTC-30A material in order to establish the most suitable cutoff for this application.

Data obtained at 25°C with the XPTC-10 and the XPTC-30A membranes is shown in Table III. The rejection of TOC for the XPTC-10 membrane is found to be independent of the oil feed concentrations, and almost identical for both membranes when the

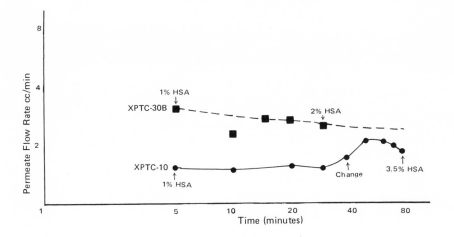

Unit Conditions

	Pressure (atm)		Retentate Flow Rate (cc/min)
	INLET	RETENTATE	OUTLET
Initial XPTC-30B	1.36	0	525 ± 25
XPTC-10	2.04	0	850
After Change	2.04	0.67	600

Fig. #3. Human Serum Albumin Concentration

Table III. Ultrafiltration of Cutting Oil Emulsions
(T = 25°C, Stirred Cells, 700 rpm)

EMULSION	Membrane	Feed Oil Concentration (ppm TOC)	Applied Pressure (atm)	Permeate Flow Rate (cm/hr)	Permeate TOC (ppm)	TOC Rejection (%)
Cimperial	XPTC-10	1.1×10^4	3.4	5.7	640 ± 20	94.2
		2.1×10^4	3.4	3.9	1250 ± 50	94.0
	XPTC-30A	3.35×10^4	1.7	3.1	1450 ± 150	95.7
		3.35×10^4	3.4	3.6	2150	93.6
SBBP-40	XPTC-10	2.4×10^4	3.4	4.7	1720 ± 20	92.8
	XPTC-30A	2.4×10^4	1.7	6.0	1870 ± 50	92.2
	XPTC-30A	2.4×10^4	3.4	6.4	2140 ± 30	91.1

XPTC-30A is operated at 1.7 atmospheres pressure. However, the
XPTC-30A shows evidence of decreased TOC retentivity as the
pressure is increased from 1.7 to 3.4 atmospheres. One also
notes that the retention of the Cimperial macroemulsion is higher than that of the SBBP-40 microemulsion.

As a practical matter, one would prefer to operate at 50°-
60°C both because of increased permeation rates and due to the
considerable temperature rise created by the energy from the
circulation pumps. Figure #4 shows the effects of increasing the
temperature from 25°C to 55°C on the amount of TOC in the permeate from the XPTC-10 and XPTC-30A membranes. The XPTC-10 shows
no oil breakthrough at 55°C. It does show a slight increase in
permeate TOC probably due to a shift in the free-bound surfactant equilibrium. The XPTC-30A shows oil breakthrough with an
increase of TOC in the permeate from 2150 ppm at 25°C to 4700
ppm at 55°C.

To appreciate the results, we must recall that cooling oil
type emulsions (both macro and micro) contain small molecular

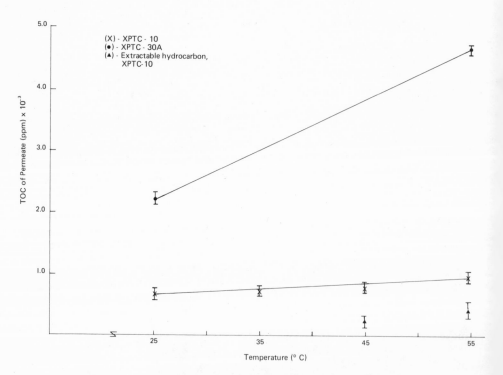

Fig. #4. "Cimperial" Oil Emulsion Ultrafiltration
                (3.4 atm. pressure)

weight surfactant molecules in both a "free" and "bound" state.
The "bound" fraction is associated with the stabilized oil droplets and thereby retained, while the "free" is either molecular-
ly dissolved or in molecular dimensions and able to permeate
through an ultrafiltration membrane. In later experiments, the
conductivity of feed and product was measured for the Cimperial
macroemulsion and approximately 40% of the conducting species
(probably surfactant) permeated the XPTC-10 membrane. Thus, it
is clear that ultrafiltration will never give complete retention
of all organic material from oil emulsions and the permeate may
require additional treatment to meet discharge standards. However, other tests which remove polar molecules have shown that
the permeate from both membranes contains less than 10 ppm of
actual hydrocarbon oil.[9]

Another important point is that the emulsified oil droplets
are liquid, and are retained by a combination of geometry and
surface tension[10]. By analogy to the classical "bubble
point" equation for displacing pore fluid with a gas, the pressure at which oil will be forced through the pores should be of
the form

$$P_B = \frac{4\gamma\cos\Theta}{d} \tag{1}$$

where P_B is the breakthrough pressure, γ is the interfacial
tension between the oil droplet and the aqueous phase, Θ is the
contact angle of the pore fluid with the pore wall, and d is the
pore diameter.

Together with equation (1), the reason why the XPTC-10 had
better oil retention than the XPTC-30A at lower operating temperature and pressure is explained in part by their difference
in pore diameter. When operated at 25°C and 3.4 atmospheres
pressure, the XPTC-30A membrane is apparently in the vicinity of
its oil breakthrough threshold. By increasing the temperature,
γ is decreased further, therefore, P_B is lowered and massive
TOC breakthrough begins. These trends are illustrated in Figure
#5. This schematic demonstrates that the 10K cutoff membrane was
necessary for the desired level of retention and thus further
testing was confined to the XPTC-10 membrane. Figure #6 shows
the ultrafiltration flux as a function of the oil concentration
and the stirring speed at 55°C. This data is typical of systems
which form Class I dynamic membranes[8], in which the ultrafiltration permeate flow rate, J_v, is given by

$$J_v = -k \ln C_B \tag{2}$$

where C_B is the bulk feed oil concentration, and k is the
hydrodynamic coefficient proportional to the inverse of the

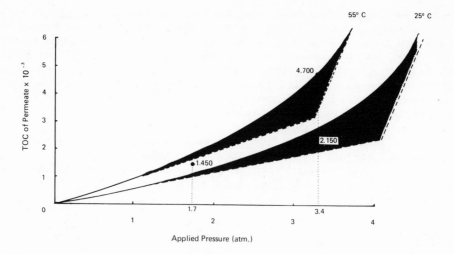

Fig. #5.   Schematic of Oil Breakthrough Process
with XPTC-30A Membrane

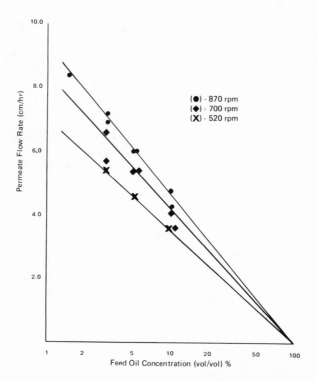

Fig. #6.   Ultrafiltration of "Cimperial" Oil Emulsion
(55°C, 3.4 atm) XPTC-10 Membrane

unstirred layer thickness and directly proportional to the effective diffusivity of the species accumulated at the membrane boundary. For this reason, as the stirring rate is increased, one finds an increase in the steepness of the slope of the plots in Figure #6. The intercept of the oil concentration axis represents the feed concentration at the boundary layer which will cause the water flux to fall to zero, and its value is dependent on the nature of the particular oil.

Due to differences in both the oils used and hydrodynamic conditions, direct comparison of the present data with fluxes and retentions obtained in early work[6] on other membranes is not possible. Nonetheless, present data do appear similar in flux and retention and suggest that the new composite membrane may find wide application.

REFERENCES

1. Settig, D., and Percy, C., Milchwissenschaft, 31, 466 (1976).
2. Bungaard, A.G., Olsen, O.J. and Madsen, R.F., Dairy Ind., 37, 539 (1972).
3. deWit, J.N., and deBoer, R., Neth. Milk Dairy J., 29, 198 (1975).
4. Dejmek, P., Funeteg, B., Hallstrom, B., and Winge, L., J. Food Sci., 39, 1014 (1974).
5. Schmitthausler, R., Process Biochem., Vol. 12, No. 8, Oct. 1977, p. 13.
6. Goldsmith, R.L., Roberts, D.A., and Burre, D.L., Journ. of WPCF, 46, 2183 (1974).
7. Tanny, G., Hoffer, E. and Kedem, O., Experimentia Suppl., 18, 619 (1971).
8. Tanny, G.B., Sep. and Pur. Methods, 7, 183 (1978).
9. Private communication, J. Murkes, Alfa-Laval, Tumba, Sweden.
10. Tanny, G.B., ACS Proceedings, 173, Conf., March 20, p. 56 (1977).

USING INDUSTRIAL MEMBRANE SYSTEMS TO ISOLATE OILSEED

PROTEIN WITHOUT AN EFFLUENT WASTE STREAM

J. T. Lawhon, L. J. Manak and E. W. Lusas

Research Engineer, Research Associate and Director,
respectively
Food Protein R&D Center, Texas A&M University, College
Station, TX 77843

INTRODUCTION

Conventional methods of isolating protein from defatted soy,
glandless cottonseed or peanut flours generate whey-like liquid by-
products, which pose a serious environmental pollution threat unless properly treated before disposal (Goldsmith et al., 1972;
Lin et al., 1974). In addition, the resulting wheys represent a
significant loss of edible protein and other valuable constituents.
Thus, a new approach to the production of protein isolates from
oilseed flours which employs industrial ultrafiltration (UF) and
reverse osmosis (RO) systems was demonstrated by investigators at
Texas A&M University's Food Protein Research and Development Center
(FPRDC) (Lawhon et al., 1977; 1978; 1979). With the FPRDC membrane
isolation process (MIP) protein is extracted from oilseed flours
following conventional procedures. However, the protein is ultrafiltered directly from the liquid extract instead of being removed
by precipitation at its isoelectric point. RO membranes are used
to process UF permeate. A secondary, low-protein product is
obtained by RO and the water effluent (RO permeate) can be reused
as process water.

In addition to avoiding the generation of whey, the MIP offers
these other advantages, (a) shorter processing time, (b) increased
yield of isolate (since whey proteins are recovered along with
protein normally precipitated), (c) isolates with enhanced nitrogen solubility, (d) greatly reduced process water requirements and
(e) products possessing highly desirable functional and nutritional
properties (Manak et al., 1979).

In the work reported here, tubular membrane systems equipped with high performance, noncellulosic UF membranes and cellulose-based RO membranes were used to process protein extracts from soy, glandless cottonseed and peanut flours. An economic evaluation of the MIP applied to soy protein isolation was also performed.

EXPERIMENTAL

Extract Preparation

Protein extracts were prepared in the FPRDC pilot plant by extracting 80 lb. quantities of flour. A single extraction was made with soy and peanut flours as shown in Figure 1. A two-step extraction procedure was used with glandless cottonseed flour, which divided its protein into nonstorage (NSP) and storage protein (SP) fractions (see Figure 2).

Extractions of soy and peanut flour were made with tap water (30:1 solvent-to-flour ratio by weight) adjusted to pH 9 and pH 8, respectively, with either sodium or calcium hydroxide. Extraction continued for 40 minutes at 55°C for soy flour and 60°C for peanut flour. Slurries were subsequently centrifuged to obtain extracts for ultrafiltering.

Cottonseed protein was divided into NSP and SP fractions by extracting the NSP with tap water (18:1 water-to-flour ratio) at 28.5°C for 40 min. NSP extract was separated from insoluble residue by centrifugation, pasteurized and then precipitated at pH 4 using HCl. NSP curd was separated from NSP whey by centrifugation and spray dried.

SP extract was obtained by extracting residue from the initial extraction with water adjusted to pH 9.5 using either potassium or sodium hydroxide. After centrifugation to remove insolubles, SP extract was combined with NSP whey, pasteurized, prefiltered to 100μ and membrane processed (Lawhon et al., in press).

Membrane Systems and Processing Techniques

During developmental work on the FPRDC MIP, basic test units of eight UF systems and five RO systems purchased from leading manufacturers were evaluated for this application. However, data reported here were obtained using the internally-coated tubular UF system of Abcor, Inc., Wilmington, MA and the externally-coated tubular RO system of Western Dynetics, Inc., Newbury Park, CA. The Abcor UF system was equipped with 22 sq. ft. of HFM-180 noncellulosic membrane. The Western Dynetics RO unit was equipped with 5 sq. ft.

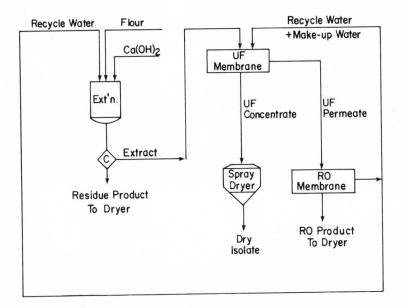

Figure 1. Simplified flow diagram for soybean and peanut protein isolation with UF and RO membranes.

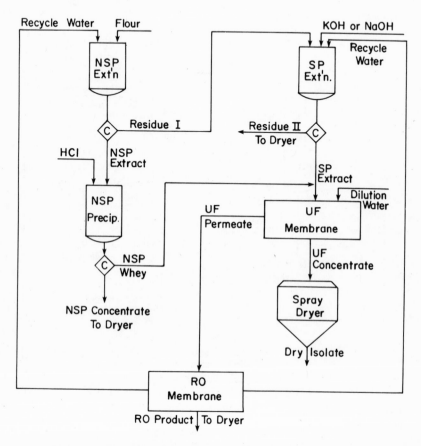

Figure 2.   Simplified flow diagram for cottonseed protein isolation
with UF and RO membranes.

of cellulose acetate membrane having a rejection for 5000 ppm NaCl
of 95% at 500 psi.

After pasteurization and prefiltering, feed solution was pumped
to the UF system at the recommended pressure and flow rate. Extract
was processed in volumes of approximately 250 gallons. A dilution
technique was applied to further purify the retentate after a 4:1
volume reduction in the original feed. Dilution consisted of adding
to the concentrated feed a quantity of filtered water equal to either
3 or 4 times its volume and reconcentrating it. Feed temperature
was maintained at $65^{\circ}C$ throughout the processing cycle to give increased flux and prevent microbial buildup.

RO processing was conducted at 850 psi and $40^{\circ}C$. Feed pH was
maintained below pH 9.

Analytical Measurements and Calculation Procedures

Moisture, total solids, and ash were determined according to
standard AOCS methods (AOCS, 1971). Nitrogen was determined by the
micro-Kjeldahl method. Carbohydrates in terms of glucose were
measured colorimetrically by the phenol-sulfuric acid method of
Dubois et al. (1956). Total phosphorus was determined by the method
according to Summer (1944). Color measurements were made using a
Hunter Digital Color and Color Difference Meter, Model 250. Measurements were first made on products in a dry, powdered form and then
as wet paste prepared by adding water (5:1 water to product ratio
by wt.). Nonprotein nitrogen (NPN) was determined as that nitrogen
soluble in 10% TCA solution. Nitrogen solubility profiles were
prepared on isolates following a modification of a method by Lyman
et al. as previously described by Lawhon et al. (1972). Chemical
oxygen demand (COD) determinations were performed by the standard
method (Am. Pbl. Health Ass'n., 1965). Conductivity measurements
were taken using a Model 31 Conductivity Bridge manufactured by the
Yellow Springs Instrument Co., Inc., Yellow Springs, OH.

Calculations for percentage retention of components by a membrane, mean percent retention and mean permeation rate were performed
as reported by Lawhon et al. (1977). Profitabilities for hypothetical soy plants evaluated were calculated by the discounted-cash-
flow rate of return method (DCFRR), as described by Herbert and
Bisio (1976).

RESULTS AND DISCUSSION

Proximate analyses of extracts ultrafiltered are presented in
Table I. In Table II are data on UF permeates generated. Differences in composition and molecular weights of extract components

Table 1.   Composition of flour extracts processed with UF membranes

Extracts	Total solids %	Ash %	Nitrogen Total %	NPN	Protein (N x 6.25) %	Total sugars %
Soy	2.50	0.28 (11.16)*	0.24 (9.71)*	0.015 (0.61)*	1.52 (60.69)*	0.58 (23.11)*
Cottonseed SP plus NSP whey	2.06	0.18 (8.88)	0.22 (10.79)	0.032 (1.55)	1.39 (67.44)	0.51 (24.76)
Peanut	1.71	0.10 (5.61)	0.22 (12.87)	0.011 (0.64)	1.38 (80.70)	0.34 (19.65)

*Dry wt. basis

caused some variation in flux and permeate solids. Feed and permeate compositions are given on liquid and dry weight bases.

The UF system yielding the data of Table III processed each
feed quite satisfactorily. The system was easily cleaned, returning to its initial flux after a cleaning cycle of about 2 hours.
Analytical data on MIP isolates and a commercial soy isolate are
compared in Table IV. Each MIP isolate is acceptably high in
nitrogen. The excessively high protein content of the peanut isolate results from the 6.25 conversion factor which is less appropriate for peanut protein than for soy and cottonseed. Although
investigators are cognizant of the inexactness of the conversion
it is nevertheless, widely used. The color of the soy UF isolate
compared favorably with that of the commercial isolate. The MIP
soy isolate is shown to be higher in ash than the commercial

Table II.   Data on UF permeates from processing flour extracts

Permeate source	Total solids %	Ash %	Nitrogen Total %	NPN %	Protein (N x 6.25) %	Total sugars %
Soy extract	0.59	0.14 (23.88)*	0.02 (3.22)*	0.015 (2.54)*	0.02 (20.1)*	0.30 (50.85)*
Cottonseed SP plus NSP whey	0.47	0.09 (19.15)	0.02 (4.26)	0.018 (3.83)	0.13 (27.66)	0.20 (42.13)
Peanut extract	0.71	0.16 (22.54)	0.03 (4.23)	0.015 (2.11)	0.20 (28.17)	0.40 (56.33)

* Dry wt. basis

Table III.   Performance of a tubular UF system on
             oilseed flour extracts

Performance characteristics		Extracts Processed		
		Soy	Cottonseed SP plus NSP whey	Peanut
Mean flux, gfd		65.1	52.3	51.6
Mean retention, %	Solids	82.1	84.0	85.3
	N	97.6	92.8	95.5
	Ash	54.6	69.7	60.4
	Sugars	46.3	62.6	22.8

isolate, but as reported previously (Lawhon et al., 1979) a FPRDC
UF isolate may contain more than seven times as much calcium as a
commercial isolate and may contain less than half as much sodium.
This results from using $Ca(OH)_2$ instead of NaOH to extract soy
flour – an option that is assuming added significance as sodium is
increasingly linked to hypertension. $Ca(OH)_2$ may also be used to
prepare MIP peanut isolates.

The color of MIP cottonseed SP is invariably somewhat darker
than soy or peanut isolate. This is true of isoelectric-precipitated
SP as well. Addition of the NSP whey to SP extract as practiced in
the most recent membrane procedure for cottonseed has improved the
color of SP isolate. Nitrogen solubility profiles of Table V reflect
the superior solubility characteristics of soy and peanut MIP isolates as compared to the SP isolate and to the commercial product.

Performance data on a tubular RO system processing soy UF
permeate are given in Table VI. As shown, the RO effluent was lower
in total solids than area tap water employed in the extraction step.
Conductivity measurements on effluent and area tap water are also
given. These data indicate that RO permeate may be recycled for
use as a process water. Material balance calculations show only
about 15% of the process water is lost in the processing cycle.

The secondary products produced by drying RO concentrates
could possibly be marketed as an animal feed ingredient or fertilizer
component. The product from processing soy UF permeate contained

Table IV. Data on protein isolates produced with UF membranes

Product description	Nitrogen			Protein (N x 6.25)	Total P	Total sugars	Color	
	Ash	Total	NPN				Dry	Wet
				% Dry wt. basis				
Soy isolate	6.7	14.82	0.33	92.63	1.33	6.6	82.4	61.6
Cottonseed isolate	5.0	15.05	0.64	94.04	1.42	5.3	65.3	49.5
Peanut isolate	4.0	16.44	0.24	102.74	0.67	3.0	80.4	68.2
Commercial soy isolate	3.8	14.70	0.21	91.84	0.60	4.1	82.5	66.3

46% carbohydrates, 16.7% ash and 8.66% protein on a moisture-free basis.

Figure 3 compares the soy flour solids and nitrogen distributions obtained when processing with the MIP or the conventional isolation method. The MIP increases the yield of soy isolate by about 47%.

Table VII reflects selling prices that were required in Mid-1978 to achieve various rates of return by the DCFRR method. Three sizes of hypothetical plants were designed and priced in considerable detail (Hensley et al., 1979). By comparing these required

Table V. Percentage of soluble nitrogen in UF isolates and a commercial soy isolate

Product description	pH of measurement								
	2	2.5	3	3.5	4	5.5	6	7	9
Soy	94.8	96.1	66.0	9.7	4.8	7.8	83.3	98.4	98.7
Cottonseed SP plus NSP whey	78.2	54.9	54.7	43.5	14.3	16.4	17.2	19.7	49.2
Peanut	87.4	84.5	79.2	41.2	6.0	4.4	9.9	58.0	89.9
Commercial soy isolate	66.2	59.8	47.0	38.3	6.6	26.5	45.1	66.1	69.7

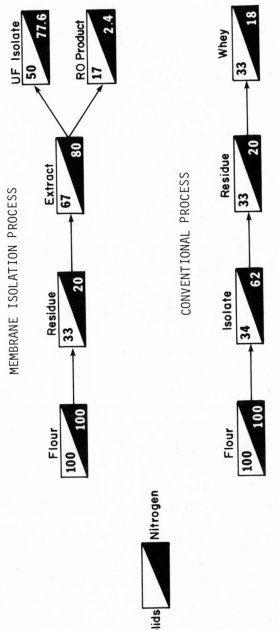

Figure 3. Distribution of soy flour solids and nitrogen for membrane and conventional isolation processes.

Table VI. Performance of a tubular RO system on
soy UF permeate

Performance measurements		
Mean flux, gfd		36
Solids content, %	RO feed	1.24
	RO effluent	0.029
	Area tap water	0.055
Conductivity measurements, μmhos	RO feed	2800 @ 27°C
	RO effluent	209 @ 27°C
	Area tap water	785 @ 27°C
COD measurements, mg/1	RO feed	10,673.
	RO effluent	203.6
% Mean retention %	Solids	98.8
	Ash	98.2

selling prices with the then current prices of commercial soy
isolates (shown at the bottom of Table VII), the MIP is seen to be
economically feasible for use with soy flour. Similar studies are
presently underway for cottonseed and peanut flours'.

CONCLUSIONS

The economic feasibility of the MIP - along with the advantages of improved yields, product functionality, water conservation,
and elimination of a major waste treatment problem by virtue of a
closed-cycle process - makes it a viable and extremely attractive
option for soy protein isolation.

Table VII. Required MIP soy isolate selling
prices for various DCFRR's [1]

DCFRR[2] %	Annual plant isolate production, MM lbs.		
	5	15	20
0	0.67	0.52	0.50
10	0.83	0.63	0.60
20	1.02	0.76	0.73
30	1.24	0.92	0.87

[1] Corresponding selling prices for commercial soy isolates ranged
from 77-92¢/lb. [2] Discounted cash flow rate of return.

This research was funded in part by the National Science Foundation, Research Applied to National Needs (RANN) and in part by the Natural Fibers and Food Protein Commission of Texas.

REFERENCES

Anon. "Standard Methods for Examination of Water and Waste Water", American Public Health Association. 12th ed., N.Y., N.Y. (1965).

AOCS. "Official and Tentative Methods", 3rd ed. Amer. Oil Chemist's Soc., Chicago, IL (1971).

Dubois, M., Gilles, K.A., Hamilton, J.K., Rebers, P.A. and Smith, F. Anal. Chem. 28,350 (1956).

Goldsmith, R.L., Stawiarski, M.M. Wilhelm, E.T. and Keeler, G., "Treatment of Soy Whey by Membrane Processing" in Proceedings of the Third National Symposium on Food Processing Wastes, New Orleans, LA (1972).

Hensley, D.W. and Lawhon, J.T. Food Technology 33,46 (1979).

Herbert, V.D. Jr. and Bisio, A. Chem Tech 6,422 (1976).

Lawhon, J.T., Rooney, L.W., Cater, C.M. and Mattil, K.F. J. Food Sci. 37,778 (1972).

Lawhon, J.T., Mulsow, D., Cater, C.M. and Mattil, K.F. J. Food Sci. 42,389 (1977).

Lawhon, J.T., Hensley, D.W., Mulsow, D. and Mattil, K.F. J. Food Sci. 43,361 (1978).

Lawhon, J.T., Hensley, D.W., Mizukoshi, Masahiko and Mulsow, D. F. Food Sci. 44,213 (1979).

Lawhon, J.T., Manak, L.J. and Lusas, E.W. J. Food Sci. in press

Lin, S.H.C., Lawhon, J.T., Cater, C.M. and Mattil, K.F. J. Food Sci. 39,178 (1974).

Manak, L.J., Lawhon, J.T. and Lusas, E.W. submitted for publication in J. Food Sci.

Summer, J.F. Science 100,413 (1944).

CONCLUDING REMARKS

Anthony R. Cooper

Dynapol
1454 Page Mill Road
Palo Alto, CA 94304

We have experienced six exciting sessions covering many aspects of ultrafiltration membranes and applications. The majority if not all of the presentations will be published in a symposium volume by Plenum Press. This should become an important part of the ultrafiltration literature.

I feel the symposium has been successful in bringing together people from many disciplines with a common interest in the subject, which is already finding a wide diversity of applications. Some of these applications have shown that ultrafiltration has a unique solution to certain problems. In other areas, it is proving to be competitive with established unit operations as factors such as energy costs and the imposition of environmental regulations increase. In some cases, ultrafiltration is able to produce a purer product than that achieved by other processes. Importantly, several completely new products in the medical area based on ultrafiltration are showing promise. There is no doubt that the creativity and imagination of the people gathered at this meeting will combine to improve the quality of life by the successful application of ultrafiltration to solve existing problems and introduce further applications in new areas.

I would like to close this meeting by thanking all the session chairmen, those who presented papers and the audience for enthusiastic attendance.

CONTRIBUTORS

Richard M. Ahlgren, Aqua-Chem, Inc., Milwaukee, Wisconsin

A. R. Akred, Davis Gelatine (Australia) Pty. Ltd., Botany, Australia

C. H. Amundson, University of Wisconsin, Department of Food Science, Madison, Wisconsin

M. G. Balko, University of Kentucky, Department of Chemical Engineering, Lexington, Kentucky

Thomas F. Baltutis, Harza Engineering Company, Chicago, Illinois

N. C. Beaton, Dorr-Oliver, Inc., Stamford, Connecticut

Georges Belfort, Rensselaer Polytechnic Institute, Department of Chemical and Environmental Engineering, Troy, New York

D. Bhattacharyya, University of Kentucky, Department of Chemical Engineering, Lexington, Kentucky

William F. Blatt, Amicon Corporation, Lexington, Massachusetts

Robin G. Booth, Palo Alto, California

J. Bosch, Mount Sinai School of Medicine, New York, New York

Craig A. Brandon, CARRE, Inc., Seneca, South Carolina

Barry R. Breslau, Romicon, Inc., Woburn, Massachusetts

Thomas F. Busby, American Red Cross Blood Services Laboratories, Bethesda, Maryland

Israel Cabasso, Gulf South Research Institute, New Orleans, Louisiana

Franco Castino, American Red Cross Blood Services Laboratories, Bethesda, Maryland

M. Chaudhury, Indian Institute of Technology, Department of Civil Engineering, Kanpur, India

C. Cheng, University of Kentucky, Department of Chemical Engineering, Lexington, Kentucky

699

Munir Cheryan, University of Illinois, Department of Food Science, Urbana, Illinois

William L. Chick, Harvard Medical School, Joslin Research Laboratory, Boston, Massachusetts

Clark K. Colton, Massachusetts Institute of Technology, Department of Chemical Engineering, Cambridge, Massachusetts

Anthony R. Cooper, Dynapol, Palo Alto, California

W. David Deeslie, University of Illinois, Department of Food Science, Urbana, Illinois

P. Dejardin, Centre de Recherches sur les Macromolécules, CNRS, Strasbourg, France

Mahendra R. Doshi, The Institute of Paper Chemistry, Appleton, Wisconsin

Enrico Drioli, University of Naples, Istituto di Principi di Ingegneria Chimica, Naples, Italy

Patricia Duenas, Amicon Corporation, Lexington, Massachusetts

K. Eberle, Gulf South Research Institute, New Orleans, Louisiana

A. G. Fane, University of New South Wales, School of Chemical Engineering, Australia

C. J. D. Fell, University of New South Wales, School of Chemical Engineering, Australia

Kathryn J. Fillmore, Cutter Laboratories, Inc., Berkeley, California

Edward Fitzgerald, Millipore Corporation, Bedford, Massachusetts

Cheryl Ford, Amicon Corporation, Lexington, Massachusetts

R. Friedlander, Osmonics, Inc., Hopkins, Minnesota

H. Friedli, Swiss Red Cross Blood Transfusion Service, Berne, Switzerland

Leonard I. Friedman, American National Red Cross, Bethesda, Maryland

J. P. Friend, Davis Gelatine (Australia) Pty. Ltd., Botany, Australia

Y. Fujii, Toray Industries, Inc., Kanagawa, Japan

Akio Fujimoto, Asahi Chemical Industry Company, Ltd., Osaka, Japan

Pierre M. Galletti, Brown University, Division of Biological and Medical Sciences, Providence, Rhode Island

S. E. Gentry, University of Kentucky, Department of Chemical Engineering, Lexington, Kentucky

R. Geronemus, Mount Sinai School of Medicine, New York, New York

Gary R. Gildert, National Research Council of Canada, Division
of Chemistry, Ottawa, Canada

S. Glabman, Mount Sinai School of Medicine, New York, New York

W. J. Harper, The Ohio State University, Department of Food Science
and Nutrition

M. Heisler, Gelman Sciences, Inc., Ann Arbor, Michigan

C. G. Hill, Jr., University of Wisconsin, Department of Chemical
Engineering, Madison, Wisconsin

F. F. Holland, Gulf South Research Institute, New Orleans, Louisiana

S. L. Holst, Hyland Laboratories, Inc., Los Angeles, California

D. B. Hosaka, University of California, Davis, Department of Food
Science and Technology, Davis, California

John A. Howell, University College of Swansea, Department of
Chemical Engineering, Swansea, United Kingdom

Fu-Hung Hsieh, National Research Council of Canada, Division of
Chemistry, Ottawa, Canada

Kenneth C. Ingham, American Red Cross Blood Services Laboratories,
Bethesda, Maryland

Kenneth Isaacson, Amicon Corporation, Lexington, Massachusetts

David S. Johnson, Millipore Corporation, Bedford, Massachusetts

T. Kahn, Mount Sinai School of Medicine, New York, New York

Kenji Kamide, Asahi Chemical Industry Company, Ltd., Osaka, Japan

P. Kistler, Swiss Red Cross Blood Transfusion Service, Berne,
Switzerland

E. Klein, Gulf South Research Institute, New Orleans, Louisiana

J. T. Lawhon, Texas A&M University, Food Protein R&D Center,
College Station, Texas

M. S. Lefebvre, University of New South Wales, School of Chemical
Engineering, Australia

Bernard LePage, Millipore Corporation, Bedford, Massachusetts

Pak S. Leung, Union Carbide Corporation, Tarrytown, New York

Karl Lin, Millipore Corporation, Bedford, Massachusetts

Miguel Lòpez-Leiva, University of Lund, Division of Food Engineering,
Alnarp, Sweden

E. W. Lusas, Texas A&M University, Food Protein R&D Center, College
Station, Texas

Donald J. Lyman, University of Utah, Department of Materials Science and Engineering, Salt Lake City, Utah

Michael Lysaght, Amicon Corporation, Lexington, Massachusetts

R. F. Madsen, A/S De Danske Sukkerfabrikker, Nakskov, Denmark

Sei-ichi Manabe, Asahi Chemical Industry Company, Ltd., Osaka, Japan

L. J. Manak, Texas A&M University, Food Protein R&D Center, College Station, Texas

Duane M. Martin, Biotek Research, Inc., Shawnee, Kansas

M. Martinez, Hyland Laboratories, Inc., Los Angeles, California

Takeshi Matsuura, National Research Council of Canada, Division of Chemistry, Ottawa, Canada

Jean-Louis Maubois, Dairy Research Laboratory, I.N.R.A., Rennes, France

Gunars Medjanis, Romicon, Inc., Woburn, Massachusetts

Uzi Merin, University of Illinois, Department of Food Science, Urbana, Illinois

R. L. Merson, University of California, Davis, Department of Food Science and Technology, Davis, California

Alan S. Michaels, Stanford University, Department of Chemical Engineering, Stanford, California

Bradford A. Milnes, Romicon, Inc., Woburn, Massachusetts

Gautam Mitra, Cutter Laboratories, Inc., Berkeley, California

Kenji Mori, Asahi Chemical Industry Company, Ltd., Osaka, Japan

Stephen P. Naber, Harvard Medical School, Joslin Research Laboratory, Boston, Massachusetts

L. L. Nelson, Millipore Corporation, Bedford, Massachusetts

W. Kofod Nielsen, A/S De Danske Sukkerfabrikker, Nakskov, Denmark

M. T. Nor, University of New South Wales, School of Chemical Engineering, Australia

G. Paredes, University of California, Davis, Department of Food Science and Technology, Davis, California

E. Pefferkorn, Centre de Recherches sur les Macromolécules, CNRS, Strasbourg, France

W. Pusch, Max-Planck-Institut für Biophysik, Frankfurt am Main, West Germany

John M. Radovich, University of Oklahoma, School of Chemical Engineering and Materials Science, Norman, Oklahoma

A. R. Reti, Millipore Corporation, Bedford, Massachusetts

Peter D. Richardson, Brown University, Divison of Biological and Medical Sciences, Providence, Rhode Island

Ylva Sahlestrom, American Red Cross Blood Services Laboratories, Bethesda, Maryland

Y. Sakai, Toray Industries, Inc., Kanagawa, Japan

L. Sarno, Hyland Laboratories, Inc., Los Angeles, California

Dien-Feng Shieh, University of Utah, Salt Lake City, Utah

K. K. Sirkar, Stevens Institute of Technology, Department of Chemistry and Chemical Engineering, Hoboken, New Jersey

C. A. Smolders, Twente University of Technology, Department of Chemical Technology, Enschede, The Netherlands

Barry A. Solomon, Amicon Corporation, Lexington, Massachusetts

S. Sourirajan, National Research Council of Canada, Ottawa, Canada

Robert E. Sparks, Washington University, Department of Chemical Engineering, St. Louis, Missouri

D. Dean Spatz, Osmonics, Inc., Hopkins, Minnesota

A. Swaminathan, NEERI, Industrial Wastes Division, Nagpur, India

Chieko Takahashi, Harvard Medical School, Joslin Research Laboratory, Boston, Massachusetts

G. B. Tanny, Gelman Sciences, Inc., Ann Arbor, Michigan

H. Tanzawa, Toray Industries, Inc., Kanagawa, Japan

Anthony J. Testa, Romicon, Inc., Woburn, Massachusetts

Yatin B. Thakore, University of Utah, Department of Materials Science and Engineering, Salt Lake City, Utah

C. Toledo, University Complutense of Madrid, Department of Physics, Madrid, Spain

D. R. Trettin, The Institute of Paper Chemistry, Appleton, Wisconsin

H. Tsukamoto, Toray Industries, Inc., Kanagawa, Japan

R. Varoqui, Centre de Recherches sur les Macromolécules, CNRS, Strasbourg, France

Ö. Velicangil, University College of Swansea, Department of Chemical Engineering, Swansea, United Kingdom

B. von Albertini, Mount Sinai School of Medicine, New York, New York

A. G. Waters, University of New South Wales, School of Chemical Engineering, Australia

R. P. Wendt, Gulf South Research Institute, New Orleans, Louisiana

Tom B. Wiltbank, Piedmont Carolinas Red Cross, Charlotte, North Carolina

Carl F. W. Wolf, The New York Blood Center, New York, New York

S. H. Yan, Texas Instruments, Inc., Dallas, Texas

INDEX